MATHEMATICS STANDARD LEVEL

FOR THE

INTERNATIONAL BACCALAUREATE

THIRD EDITION

Alan Wicks

Copyright © 2012 by Alan Wicks

ISBN 978-0-7414-2141-8

Printed in the United States of America

Published July 2012

INFINITY PUBLISHING
1094 New DeHaven Street, Suite 100
West Conshohocken, PA 19428-2713
Toll-free (877) BUY BOOK
Local Phone (610) 941-9999
Fax (610) 941-9959
Info@buybooksontheweb.com
www.buybooksontheweb.com

Preface

This textbook is intended to serve the needs of students studying the International Baccalaureate Mathematics Standard Level (SL) course which begins in September 2012 with first examinations in May 2014.

This third edition contains some significant changes and additions:
- In the new syllabus, matrices have been removed, correlation and regression are introduced, and integration by substitution has been added.
- Complete solutions to all exercises are included at the back of the book. In most instances detailed solutions are given rather than answers only. In some cases, alternative solutions are offered.
- The six topics comprising the syllabus of the Mathematics SL course are arranged into nine units. The calculus is covered in two sections: differentiation in Unit 5 and integration in Unit 7. Statistics and Probability are covered in three sections: statistics in Unit 6, probability in Unit 8, and probability distributions in Unit 9.
- Review exercises are included at the end of Units 1 to 4. A calculus review exercise is included at the end of Unit 7, and a statistics/probability review exercise is included at the end of Unit 9.
- Although the units may be taught in whatever sequence is appropriate bearing in mind the previous knowledge of the students, the author has given some consideration to the order in which the topics are presented.

The mathematical aspects of Theory of Knowledge are not covered in the text; neither is the Internal Assessment component of the course.

The notation used in this textbook adheres to that used by the International Baccalaureate. This notation is given in the IB Group 5 Subject Guide.

It is expected in the Mathematical SL course that students use a graphing calculator. A list of acceptable graphing calculators appears in the Online Curriculum Centre on the IB website. This textbook has used one such acceptable calculator, the TI-84 Plus, in order to illustrate graphing calculator solutions to a wide range of problems.

I am indebted to Elizabeth, my wife, not only for her practical assistance in formatting and proof-reading but also in her advice and encouragement.

AW

i

Contents

UNIT 1: ALGEBRA 1
  1.1 Sequences and Series 1
  1.2 Arithmetic Sequences and Series 2
    1.2.1 Arithmetic Sequences 2
    1.2.2 The Sum of an Arithmetic Series 4
  1.3 Applications of Arithmetic Sequences and Series 6
    1.3.1 Simple Interest 6
    1.3.2 Constant Acceleration 6
  1.4 Geometric Sequences and Series 8
    1.4.1 Geometric Sequences 8
    1.4.2 The Sum of a Finite Geometric Series 10
    1.4.3 The Sum of an Infinite Geometric Series 10
  1.5 Applications of Geometric Sequences and Series 12
    1.5.1 Compound Interest 12
    1.5.2 Compounding Interest 14
    1.5.3 Population Growth 14
  1.6 Exponents and Logarithms 16
    1.6.1 Exponents 16
    1.6.2 Logarithms 19
    1.6.3 Changing the Base of a Logarithm 21
  1.7 The Binomial Theorem 22
    1.7.1 Expansion of Binomial Expressions 22
    1.7.2 Obtaining Specific Terms of a Binomial Expansion 24
  Unit 1 Review Exercises 27
    Part 1 27
    Part 2 28

UNIT 2: FUNCTIONS AND EQUATIONS 30
  2.1 The Nature of a Function 30
  2.2 Use of the Graphing Calculator 34
    2.2.1 Obtaining the Graph of $y = f(x)$ on the Calculator Screen 34
    2.2.2 The Reciprocal Function and Asymptotes 36
    2.2.3 Solution of $f(x) = 0$ to a Given Accuracy 37
    2.2.4 Drawing a Neat Sketch of the Graph of a Function 38
  2.3 Transformations of Graphs 40
    2.3.1 Translations 40
    2.3.2 Stretches 42
    2.3.3 Reflections 44
  2.4 The Quadratic Function $x \mapsto ax^2 + bx + c,\ a \neq 0$ 49
    2.4.1 The Graph of the Quadratic Function 49
    2.4.2 Finding the Vertex and y-Intercept of the Graph of a Quadratic Function
        Using the Graphing Calculator 50
    2.4.3 Completing the Square 51
    2.4.4 Factoring a Quadratic Function 53
  2.5 The General Quadratic Equation $ax^2 + bx + c = 0,\ a \neq 0$ 54
    2.5.1 Solution of Quadratic Equations 54
    2.5.2 Solution of the General Quadratic Equation 55
    2.5.3 The Significance of the Discriminant 57

2.6 The Algebra of Functions 60
    2.6.1 Composite Functions 60
    2.6.2 The Identity Function 61
    2.6.3 The Inverse Function 62
    2.6.4 The Relation between the Graphs of $y = f(x)$ and $y = f^{-1}(x)$ 63

2.7 The Exponential Function $x \mapsto a^x$, $a > 0$, $x \in \mathbb{Q}$ 66
    2.7.1 The Graph of $f : x \mapsto a^x$, $a > 0$, $x \in \mathbb{Q}$ 66
    2.7.2 The Inverse Function $f^{-1} : x \mapsto \log_a x$, $x > 0$ 66
    2.7.3 Solution of Equations of the Form $a^x = b$ 67

2.8 The Functions $x \mapsto e^x$ and $x \mapsto \ln x$, $x > 0$ 69
    2.8.1 The Significance of the Number 69
    2.8.2 The Graph of $y = e^x$ and its Inverse Graph $y = \ln x$ 70
    2.8.3 The Solution of the Equation $a^x = b$ 71
    2.8.4 The Exponential Function Applied to Population Growth 72
Unit 2 Review Exercises 74
    Part 1 74
    Part 2 77

UNIT 3: CIRCULAR FUNCTIONS AND TRIGONOMETRY 80
3.1 Radian Measure 80
    3.1.1 The Definition of a Radian 80
    3.1.2 Length of Arc 80
    3.1.3 Area of a Sector 80
3.2 The Circular Functions 83
    3.2.1 The Sine and Cosine Functions 83
    3.2.2 Values of Sine, Cosine and Tangent for the Interval $0 \leq \theta < 2\pi$ or,
    in Degrees, $0° \leq \theta < 360°$ 85
    3.2.3 Exact Values of $\cos\theta$ and $\sin\theta$ for $\theta = \dfrac{\pi}{6}, \dfrac{\pi}{4}$ and $\dfrac{\pi}{3}$ 86
3.3 Trigonometric Identities 89
3.4 Composite Trigonometric Functions 91
    3.4.1 Transformation (1): Translation 91
    3.4.2 Transformation (2): Stretch 92
3.5 Solution of Trigonometric Equations 95
    3.5.1 Solution of Equations of the Forms $\sin\theta = k$, $\cos\theta = k$ and $\tan\theta = k$ and
    in the Interval $0 \leq \theta < 2\pi$ 95
    3.5.2 Solution of More Advanced Trigonometric Equations 98
    3.5.3 Solution of Trigonometric Equations Using the Graphing Calculator 101
3.6 The Solution of Triangles 102
    3.6.1 Solving Right Triangles and Isosceles Triangles 103
    3.6.2 Solving Scalene Triangles 104
    3.6.3 The Ambiguous Case 107
    3.6.4 The Area of a Triangle 109
Unit 3 Review Exercises 111
    Part 1 111
    Part 2 114

UNIT 4: VECTORS 117
4.1 Elementary Vectors 117
4.2 Vector Algebra 120
    4.2.1 Vector Addition 120
    4.2.2 The Difference of Two Vectors 121
    4.2.3 The Multiplication of a Vector by a Scalar 122
    4.2.4 The Zero Vector 122
    4.2.5 Base Vectors 124
    4.2.6 The Magnitude of a Vector 125
    4.2.7 Unit Vectors 126
    4.2.8 Position Vectors 127
4.3 The Scalar (or Dot) Product 128
    4.3.1 The Scalar Product of Two Vectors in Two Dimensions 128
    4.3.2 The Scalar Product of Two Vectors in Three Dimensions 130
    4.3.3 The Angle between Two Vectors 131
    4.3.4 Perpendicular Vectors 131
    4.3.5 Parallel Vectors 131
4.4 The Vector Equation of a Line 134
    4.4.1 The Derivation of the Vector Equation of a Line 134
    4.4.2 The Vector Equation of a Line in Two-Dimensional Space 135
    4.4.3 Finding a Vector Parallel to a Given Line in a Plane 135
    4.4.4 The Angle between Two Lines in Two-Dimensional Space 136
    4.4.5 An Application to Constant Velocity Motion in Two-Dimensional Space 137
    4.4.6 The Vector Equation of a Line in Three-Dimensional Space 141
    4.4.7 The Angle between Two Lines in Three-Dimensional Space 141
    4.4.8 An Application to Constant Velocity Motion in Three-Dimensional Space 143
4.5 Parallel, Coincident, Intersecting and Non-Intersecting Lines 146
    4.5.1 Parallel and Coincident Lines 146
    4.5.2 Intersecting and Non-Intersecting Lines 147
Unit 4 Review Exercises 150
    Part 1 150
    Part 2 152

UNIT 5: DIFFERENTIATION 155
5.1 Rates of Change 155
    5.1.1 Mean Velocity and Instantaneous Velocity 155
    5.1.2 Differentiation 158
    5.1.4 Differentiation of Polynomial Functions 159
    5.1.5 Differentiation Using the $\dfrac{dy}{dx}$ Notation 161
5.2 Graphical Behavior of Functions 163
    5.2.1 The Second Derivative 163
    5.2.2 Local Maximum and Minimum Points 164
    5.2.3 Points of Inflection 165
    5.2.4 The Derivative of a Graphically Defined Function 166
5.3 Applications of Differentiation 1 169
    5.3.1 Maximum and Minimum Values of a Function 169
    5.3.2 Optimization Problems 1 172
5.4 Differentiation of Other Functions 174
    5.4.1 The Derivative of the Logarithmic Function $f(x) = \ln x,\ x > 0$ 174
    5.4.2 The Derivative of the Exponential Function $f(x) = e^x$ 174

5.4.3 The Derivative of the Trigonometric Function $f(x) = \sin x$ 175
5.5 Techniques of Differentiation 175
    5.5.1 Differentiation of Composite Functions 175
    5.5.2 Differentiation of Product Functions 178
    5.5.3 Differentiation of Quotient Functions 178
5.6 Applications of Differentiation 2 181
    5.6.1 Increasing and Decreasing Functions 181
    5.6.2 The Equation of a Tangent and a Normal at a Point on a Curve 181
    5.6.3 Optimization Problems 2 184
    5.6.4 Kinematics 187

UNIT 6: STATISTICS 190
6.1 Statistics: Basic Concepts 190
    6.1.1 Populations and Samples 190
    6.1.2 Discrete Data 190
    6.1.3 Continuous Data 191
    6.1.4 Grouped Data 191
6.2 Frequency Histograms 193
6.3 Cumulative Frequency 195
    6.3.1 The Cumulative Frequency Table 195
    6.3.2 The Cumulative Frequency Curve 196
    6.3.3 Percentiles and Quartiles 196
6.4 Measures of Central Tendency 199
    6.4.1 The Arithmetic Mean 199
    6.4.2 The Median 202
    6.4.3 The Mode 204
6.5 Measures of Spread 205
    6.5.1 The Inter-Quartile Range 205
    6.5.2 The Box-and-Whisker Plot 206
    6.5.3 The Standard Deviation 207
6.6 Regression and Correlation 210
    6.6.1 Regression 210
    6.6.2 Finding the Regression Line 211
    6.6.3 Correlation 216

UNIT 7: INTEGRATION 221
7.1 Anti-Differentiation 221
7.2 Indefinite Integration 222
7.3 Integration of Composite Functions 223
7.4 Integration by Substitution 224
7.5 Evaluation of the Constant of Integration 226
7.6 Application of Integration to Kinematics 227
7.7 Area under a Curve 229
7.8 Definite Integration 230
7.9 Volumes of Revolution 233
Units 5 & 7 Review Exercises 237
    Part 1 237
    Part 2 239

UNIT 8: PROBABILITY 242
8.1 Elementary Probability 242
    8.1.1 Non-Exclusive Events 247

8.2 Conditional Probability 250
    8.2.1 Contingency Tables 250
    8.2.2 Probability Tree Diagrams 251
8.3 Independent Events 254

UNIT 9: PROBABILITY DISTRIBUTIONS 258
    9.1 Discrete Probability Distributions 258
        9.1.1 Discrete Random Variables 259
        9.1.2 The Expectation of 260
    9.2 The Binomial Distribution 263
        9.2.1 The Expectation of a Binomial Distribution 269
    9.3 The Normal Distribution 271
        9.3.1 A Continuous Random Variable 271
        9.3.2 The Standardized Normal Distribution Function 273
        9.3.3 The Normal Distribution 275
        9.3.4 The Inverse of the Standardized Normal Distribution Function 278
    Units 6, 8 & 9 Review Exercises 281
        Part 1 281
        Part 2 283
        TI-84Plus Simulation Programs 287

Solutions to Unit 1 Exercises 288
    Exercise 1.1 288
    Exercise 1.2 288
    Exercise 1.3 289
    Exercise 1.4 291
    Exercise 1.5 293
    Exercise 1.6 295
    Exercise 1.7 296
    Exercise 1.8 297
    Exercise 1.9 298
    Solutions to Unit 1 Review Exercises 300
        Part 1 300
        Part 2 302

Solutions to Unit 2 Exercises 303
    Exercise 2.1 303
    Exercise 2.2 304
    Exercise 2.3 307
    Exercise 2.4 309
    Exercise 2.5 311
    Exercise 2.6 311
    Exercise 2.7 312
    Exercise 2.8 313
    Exercise 2.9 314
    Exercise 2.10 315
    Exercise 2.11 317
    Exercise 2.12 318
    Solutions to Unit 2 Review Exercises 319
        Part 1 319
        Part 2 322

Solutions to Unit 3 Exercises 326
    Exercise 3.1 326
    Exercise 3.2 327
    Exercise 3.3 328
    Exercise 3.4 329
    Exercise 3.5 330
    Exercise 3.6 332
    Exercise 3.7 334
    Exercise 3.8 335
    Exercise 3.9 337
    Exercise 3.10 338
    Solutions to Unit 3 Review Exercises 339
        Part 1 339
        Part 2 342

Solutions to Unit 4 Exercises 345
    Exercise 4.1 345
    Exercise 4.2 346
    Exercise 4.3 347
    Exercise 4.4 349
    Exercise 4.5 351
    Exercise 4.6 353
    Exercise 4.7 355
    Solutions to Unit 4 Review Exercises 357
        Part 1 357
        Part 2 359

Solutions to Unit 5 Exercises 363
    Exercise 5.1 363
    Exercise 5.2 363
    Exercise 5.3 364
    Exercise 5.4 364
    Exercise 5.5 366
    Exercise 5.6 367
    Exercise 5.7 369
    Exercise 5.8 370
    Exercise 5.9 371
    Exercise 5.10 372
    Exercise 5.11 373
    Exercise 5.12 376

Solutions to Unit 6 Exercises 377
    Exercise 6.1 377
    Exercise 6.2 378
    Exercise 6.3 380
    Exercise 6.4 383
    Exercise 6.5 383
    Exercise 6.6 385
    Exercise 6.7 388

Solutions to Unit 7 Exercises 391
    Exercise 7.1 391
    Exercise 7.2 391

Exercise 7.3 392
Exercise 7.4 393
Exercise 7.5 393
Exercise 7.6 394
Exercise 7.7 395
Exercise 7.8 397
Solutions to Units 5 & 7 Review Exercises 398
    Part 1 398
    Part 2 402

Solutions to Unit 8 Exercises 404
Exercise 8.1 404
Exercise 8.2 405
Exercise 8.3 407
Exercise 8.4 408

Solutions to Unit 9 Exercises 410
Exercise 9.1 410
Exercise 9.2 411
Exercise 9.3 412
Exercise 9.4 412
Exercise 9.5 413
Exercise 9.6 413
Exercise 9.7 414
Solutions to Unit 6, 8 & 9 Review Exercises 416
    Part 1 416
    Part 2 418

INDEX 421

UNIT 1: ALGEBRA

1.1 Sequences and Series

A *sequence* is a set of numbers in which the elements are distinguished by their position. The elements of the sequence are known as *terms*. A sequence $\{u_n\} = \{u_1, u_2, \ldots, u_n\}$ consists of terms which are usually related to one another in some way. For example, the sequence 1, 4, 9, 16, ... can be generated by $u_n = n^2$ for $n \in \mathbb{Z}^+$ and $u_1 = 1$.

A *series* is obtained by summing successive terms of a sequence so that $S_n = u_1 + u_2 + u_3 + \ldots + u_n$.

A series that has a final term, for example, $1 + 4 + 7 + 10 + 13 + \ldots + 100$, is known as a finite series, but if it continues indefinitely, for example, $1 + 4 + 7 + 10 + 13 + \ldots$, it is known as an *infinite series*.

It is often convenient to write the terms of a series using *sigma notation*, $\sum$. *Sigma* is a Greek 'S' denoting 'sum'. We write $\sum_{r=1}^{n} u_r = u_1 + u_2 + u_3 + \ldots + u_n$ so that $\sum_{r=1}^{n} u_r$ means 'add all the terms of the form u_r from $r = 1$ to $r = n$'.

Example 1.1: Write $\displaystyle\sum_{r=1}^{6} r^2$ as the sum of individual terms.

Solution 1.1: $\displaystyle\sum_{r=1}^{6} r^2 = 1^2 + 2^2 + 3^2 + 4^2 + 5^2 + 6^2$

Example 1.2: Express the first n terms of the series $3 + 5 + 7 + 9 + 11 + \ldots$ in sigma notation.

Solution 1.2: In order to write this series using the sigma notation, we need to be able to write a general term of the series in terms of r.

The pattern for the values of the terms is $2r + 1$ because when $r = 1,\ 2,\ 3,\ 4,\ \ldots,$ $u_1 = 3,\ u_2 = 5,\ u_3 = 7,\ u_4 = 9,\ \ldots.$

Hence the first n terms of the series $3 + 5 + 7 + 9 + 11 + \ldots = \displaystyle\sum_{r=1}^{n} (2r + 1)$.

Exercise 1.1
1.   Write the next three terms in each of the following sequences.

(a) $1,\ \dfrac{1}{2},\ \dfrac{1}{4},\ \dfrac{1}{8},\ \dfrac{1}{16}, \ldots$ (b) $1,\ \dfrac{1}{2},\ \dfrac{1}{3},\ \dfrac{1}{4},\ \dfrac{1}{5}, \ldots$ (c) $-2,\ 1,\ 4,\ 7,\ 10,\ \ldots$

(d) $3,\ -3,\ 3,\ -3,\ 3, \ldots$ (e) $1,\ 2,\ 4,\ 8,\ 16,\ \ldots$ (f) $1,\ 8,\ 27,\ 64,\ 125,\ \ldots$

(g) $1, -2, -5, -8, -11, \ldots$ (h) $\dfrac{4}{9}, \dfrac{2}{3}, 1, \dfrac{3}{2}, \dfrac{9}{4}, \ldots$

(i) $1, \dfrac{3}{4}, \dfrac{5}{7}, \dfrac{7}{10}, \dfrac{9}{13}, \ldots$ (j) $1, 1, 2, 6, 24, \ldots$

2. Write the first four terms of the following sequences. In each case, start with $r=1$, $r \in \mathbb{Z}^{+}$.

(a) $u_r = 3r - 1$ (b) $u_r = 2^r - 5$ (c) $u_r = r^2 + r - 1$ (d) $u_r = \dfrac{1}{r+1}$

(e) $u_r = 1 + (-1)^r$

3. Write the following finite series as the sum of individual terms.

(a) $\displaystyle\sum_{r=0}^{4}(r+1)$ (b) $\displaystyle\sum_{r=1}^{5}\dfrac{r}{r+1}$ (c) $\displaystyle\sum_{r=0}^{4}\dfrac{r}{2^r}$ (d) $\displaystyle\sum_{r=1}^{4}u_{2r-1}$

(e) $\displaystyle\sum_{r=1}^{n}u_r^{\,2}$ (f) $\displaystyle\sum_{r=1}^{5}3^r$ (g) $\displaystyle\sum_{r=1}^{6}1$ (h) $\displaystyle\sum_{r=1}^{4}(-1)^{r+1}r$

4. Write each of the following finite series using the sigma notation. In each case, start with $r=1$.

(a) $1+2+3+4+5$ (b) $1+\dfrac{1}{2}+\dfrac{1}{3}+\dfrac{1}{4}+\dfrac{1}{5}$ (c) $1+\dfrac{1}{2}+\dfrac{1}{4}+\dfrac{1}{8}+\dfrac{1}{16}$

(d) $1^2+2^2+3^2+4^2+5^2$ (e) $1+8+27+64+125$ (f) $u_1+u_2+u_3+\ldots+u_n$

1.2 Arithmetic Sequences and Series

1.2.1 Arithmetic Sequences

A sequence $\{u_n\}$, defined by a relation of the form $u_{n+1}=u_n+d$ where d is a constant, is called an *arithmetic sequence*. d is called the *common difference*. The first term is denoted by u_1. For example, the sequence $7, 13, 19, 25, 31, \ldots$ is an arithmetic sequence because $13 = 7 + 6$, $19 = 13 + 6$, $25 = 19 + 6$, $31 = 25 + 6$ and so on. The difference between each successive term is 6, so, in this case, $d = 6$, and the first term is 7, so $u_1 = 7$.

More generally,
$$u_1 = u_1$$
$$u_2 = u_1 + d$$
$$u_3 = u_2 + d$$
$$u_4 = u_3 + d$$
$$\cdots \cdots \cdots$$
$$\cdots \cdots \cdots$$
$$u_n = u_{n-1} + d$$

and, by addition of all these statements,

$$\sum_{r=1}^{n}u_r = \sum_{r=1}^{n-1}u_r + u_1 + d\sum_{r=1}^{n-1}1$$

or
$$u_1 + u_2 + u_3 + u_4 + \ldots + u_n = \left(u_1 + u_2 + u_3 + u_4 + \ldots + u_{n-1}\right) + u_1 + (n-1)d$$
$$\Rightarrow u_n = u_1 + (n-1)d .$$

Therefore, the nth term, u_n, can be expressed in terms of the initial term, u_1, and the common difference, d.

Example 1.3: Find the 10th term of an arithmetic sequence whose first term, u_1 is 6 and whose common difference, d is -7.

Solution 1.3: $u_{10} = 6 + (10-1)(-7) = 6 + 9 \times (-7) = -57$

This calculation may also be done very conveniently using a graphing calculator. Figures 1.01 and 1.02 show how to do this with a TI-84 Plus. Firstly, on the home screen, store the first term 6 to X, then the instruction $X + (-7) \to X$ has the effect of adding the common difference of -7 to the current value of X, and the sequence is generated by applying the ENTER key successively. If the ENTER key is applied nine times the value, -57 is obtained, indicating that $u_{10} = -57$.

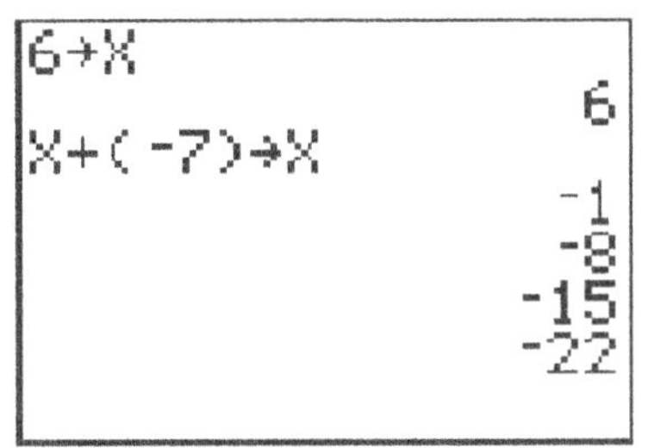

Figure 1.01

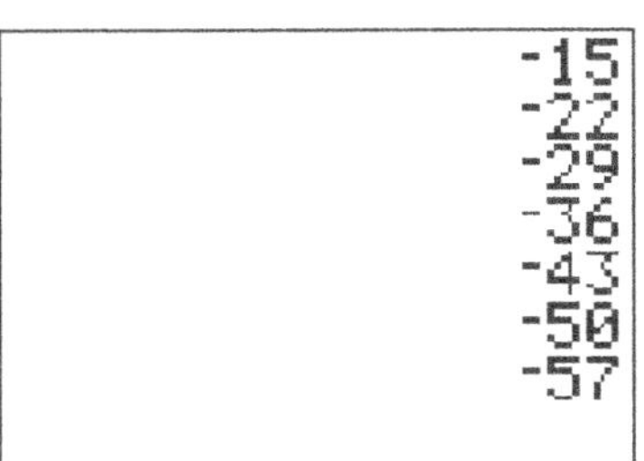

Figure 1.02

Example 1.4: Find the number of terms in the arithmetic sequence $5,\ 5.4,\ 5.8,\ \ldots,\ 28.2$.

Solution 1.4: $u_1 = 5$, $d = 0.4$ and $u_n = 28.2$

So $28.2 = 5 + (n-1)0.4 \Rightarrow 23.2 = 0.4(n-1) \Rightarrow 58 = n-1 \Rightarrow n = 59$

Therefore there are 59 terms in the sequence.

Example 1.5: An arithmetic sequence has first term 5 and common difference 8. Find the smallest term which is greater than 500.

Solution 1.5: $u_n = u_1 + (n-1)d = 5 + 8(n-1) = 8n - 3$

So you need $8n - 3 > 500 \Rightarrow 8n > 503 \Rightarrow n > 62.875$ but $n \in \mathbb{Z}^+$ so that $n = 63$ and the smallest term is $u_{63} = 8 \times 63 - 3 = 501$.

1.2.2 The Sum of an Arithmetic Series

The sum of a sequence of n terms is denoted by S_n. In order to find S_n, we write out the series twice. The second time, the terms are written in reverse order so that u_n appears beneath u_1, and u_{n-1} beneath u_2, and so on.

$$S_n = u_1 + u_2 + u_3 + \ldots + u_{n-1} + u_n$$

$$S_n = u_n + u_{n-1} + u_{n-2} + \ldots + u_2 + u_1$$

By adding columns, we obtain

$$2S_n = \left(u_1 + u_n\right) + \left(u_2 + u_{n-1}\right) + \left(u_3 + u_{n-2}\right) + \ldots + \left(u_{n-1} + u_2\right) + \left(u_n + u_1\right)$$

$$\Rightarrow 2S_n = \left[u_1 + \left(u_1 + (n-1)d\right)\right] + \left[\left(u_1 + d\right) + \left(u_1 + (n-2)d\right)\right] + \ldots + \left[u_1 + (n-1)d + u_1\right]$$

$$= \left[2u_1 + (n-1)d\right] + \left[2u_1 + (n-1)d\right] + \ldots + \left[2u_1 + (n-1)d\right]$$

so that $2S_n$ can be expressed as the sum of a series in which each term is $2u_1 + (n-1)d$. As there are

n terms, $S_n = \dfrac{n}{2}\left(2u_1 + (n-1)d\right)$.

Example 1.6: An arithmetic series has a common difference of 7 and $u_{11} = 53$. Find S_{19}, the sum of the first 19 terms.

Solution 1.6: $u_{11} = u_1 + (n-1)d \Rightarrow 53 = u_1 + (11-1)7 \Rightarrow 53 = u_1 + 70 \Rightarrow u_1 = -17$

$$\Rightarrow S_{19} = \frac{19}{2}\left(2(-17) + 18 \times 7\right) = \frac{19}{2}\left(-34 + 126\right) = 874$$

Example 1.7: The sum S_n of an arithmetic series is given by the formula $S_n = 4n^2 - 9n$. Find
      (a) the sum of the first 15 terms.
      (b) the first term of the series.
      (c) the common difference of the series.

Solution 1.7: (a) $S_{15} = 4 \times 15^2 - 9 \times 15 = 900 - 135 = 765$

        (b) $S_1 = 4 \times 1^2 - 9 \times 1 = 4 - 9 = -5$ but the sum of the first 'one' term is the first term, u_1
           so $u_1 = -5$.

        (c) $S_2 = 4 \times 2^2 - 9 \times 2 = 16 - 18 = -2$ but the sum of the first two terms is
           $u_1 + \left(u_1 + d\right) = 2u_1 + d$ so $2u_1 + d = -2$ and as $u_1 = -5$, $2(-5) + d = -2$ so $d = 8$.

Exercise 1.2

    1.   Write down the 5th, 10th and n^{th} terms of the following arithmetic sequences.
         (a) 1, 3, 5, 7, ...   (b) 5, 12, 19, 26, ...   (c) 12, 10, 8, 6, ...
         (d) 4, -2, -8, -14, ...   (e) $\dfrac{2}{3}, \dfrac{7}{3}, 4, \dfrac{17}{3}, \ldots$

2. Find the number of terms in the following series and find their sum in each case.
   (a) $1+3+5+7+\ldots+99$ (b) $5+12+19+26+\ldots+187$
   (c) $12+10+8+6+\ldots-70$ (d) $4-2-8-14-\ldots-182$
   (e) $\dfrac{2}{3}+\dfrac{7}{3}+4+\dfrac{17}{3}+\ldots+99$

3. An arithmetic sequence has first term $u_1 = 18$. Its 23rd term is 656. Find the common difference.

4. The sum of the first 13 terms of an arithmetic sequence is 572. Its first term is 14. Find the common difference.

5. The 18th term of an arithmetic sequence is 469 and the 33rd term is 784. Find the first term and common difference.

6. If $u_{10} = 25$ and $u_{41} = 304$, find the value of u_1, the first term, and d, the common difference. Hence, find the value of u_{75}.

7. The first term of an arithmetic series is 21, and its common difference is 28. If the nth term of the series is 1197, find   (a) the value of n (b) S_n, the sum of the first n terms.

8. Find the first negative term in the sequence 110, 103, 96, 89, ...

9. Consider the arithmetic sequence 1, 15, 29, 43, ...
   (a) Find the smallest value of n such that, $u_n > 2500$.
   (b) Find the number of terms of the corresponding series needed for the sum to exceed 2500.

10. (a) Show that $\displaystyle\sum_{r=1}^{n} r = \tfrac{1}{2}n(n+1)$.

    (b) If $S_n = 3n^2 + 5n$, write down S_{n-1} in terms of n and hence find u_n. Write down the first term and the common difference of the sequence defined by u_n.

11. Arithmetic sequences are defined as follows: $u_{n+1} = u_n + 7$; $u_1 = 3$ and $v_{n+1} = v_n - 5$; $v_1 = 115$.
    (a) Write down the common difference for each of the sequences.
    (b) Show that $u_n = 7n - 4$ and find v_n in a similar manner.
    (c) Find the smallest value of n such that $u_n > v_n$.

12. An arithmetic sequence is defined by $u_{r+1} = u_r - 4$, $u_1 = 594$. If $u_k = S_k$ where S_k is the sum of the first k terms of the series $\displaystyle\sum_{r=1}^{n} u_r$, find the value of k other than the trivial value, $k = 1$.

1.3 Applications of Arithmetic Sequences and Series

1.3.1 Simple Interest

If A is invested and earns $p\%$ per year, simple interest, then $p\%$ of the initial amount is added to the investment at the end of each year so that the value of the investment at the end of the first year is, in dollars, $A + \dfrac{pA}{100}$.

The value of the investment at the end of the second year is $A + \dfrac{pA}{100} + \dfrac{pA}{100} = A + \dfrac{2pA}{100}$.

The value at the end of the third year is $A + \dfrac{2pA}{100} + \dfrac{pA}{100} = A + \dfrac{3pA}{100}$.

Therefore the value of the investment after n years can be modeled by an arithmetic sequence whose first term is A and whose common difference is $\dfrac{pA}{100}$.

Example 1.8: Brittany invests \$1000 in a bank which offers simple interest of 8% per year. Find how long it takes for her investment to reach \$2200.

Solution 1.8: Considering the value of the investment at the beginning of each year to be an arithmetic sequence whose first term is $u_1 = 1000$ and common difference
$$d = \frac{8 \times 1000}{100} = 80,$$
we need the value of n so that $u_n = 2200$.

Now , $u_n = u_1 + (n-1)d = 1000 + (n-1)80 \Rightarrow 1000 + (n-1)80 = 2200$

$\Rightarrow 80(n-1) = 1200 \Rightarrow n - 1 = 15 \Rightarrow n = 16$.

This means that the 16th term of the sequence is equal to 2200, and as the first term corresponds to the initial investment of \$1000, it will take 15 years for the value of Brittany's investment to reach \$2200.

1.3.2 Constant Acceleration

If air resistance is neglected, then the motion of a particle moving in a straight line under gravity can be modeled by an arithmetic sequence.

Suppose that a stone is thrown vertically upwards at u ms^{-1}. The speed of the stone will decrease by g ms^{-1} during each second of its motion (g is approximately 10 on the surface of the Earth). Therefore, the speed of the stone at the start of each second is given by an arithmetic sequence whose first term is u and whose common difference is $-g$. Then, $u - tg$, the $(t+1)^{\text{th}}$ term of the sequence, will give the speed of the stone at time t seconds.

Example 1.9: A passenger in a car traveling along a straight road at constant speed notes that when her watch reads 14:05 she passes an electric pylon. She further notices that every 14 seconds the car passes other equally spaced pylons.

(a) Find the time when the car passes the 50th pylon.
(b) If the pylons are 200m apart, determine the speed of the car.

Solution 1.9: (a) Let $u_1 = 0$, $d = 14$ and $u_n = u_1 + (n-1)d = 0 + 49 \times 14 = 686$. Therefore, it takes 686 seconds = 11 minutes 26 seconds to pass 50 pylons, and so the time, as she passes the 50th pylon, is 14:16 and 26 seconds.

(b) The distance traveled by the car in 686 seconds is $200 \times 49 = 9800$ m. Therefore, the speed of the car is $\dfrac{9800}{686} = 14.28571... = 14.3$ ms^{-1}.

Exercise 1.3

1. A bank makes the following advertisement: 'We will pay you 12 % interest on a $10 000 investment, over a minimum period of one year.' Considering the value of the investment at the end of each year to be an arithmetic sequence,
   (a) write down the first term and common difference of the sequence
   (b) find the value of the investment after   (i) 5 years   (ii) 10 years   (iii) 25 years.

2. A ball is thrown vertically upwards with an initial velocity of 8 meters per second. The velocity of the ball decreases by 3 meters per second each second. The velocity of the ball at the end of each second is an arithmetic sequence. Find
   (a) the first term and common difference of the sequence.
   (b) the velocity of the ball after (i) 3 seconds (ii) 20 seconds, assuming that the ball does not hit the ground.

3. At a particular instant of time, car A is 800 meters due west of point P and is traveling directly towards P at 10 meters per second. At the same instant, car B is at P and traveling due west at 6 meters per second.
   (a) Write down the distance between car A and car B
          (i) 1 second   (ii) 2 seconds   (iii) 3 seconds   (iv) n seconds   after this instant.
   (b) Find the first term and common difference of the arithmetic sequence you found in (a) and hence find how long, after this instant, they pass each other.
   (c) Can you spot a more direct approach to this problem?

4. A shopping cart has length 0.95m and is designed so that similar carts will fit into it for convenience of storage. When two such carts are 'nested' in this way (figure 1.03), the total length of the two carts is 1.13m.
   (a) Find the length of 12 nested carts.
   (b) The supermarket has a rectangular space for storing carts which is 3m wide and 5m long. Estimate how many carts can be stored in this space if the width of a cart is 0.48m.

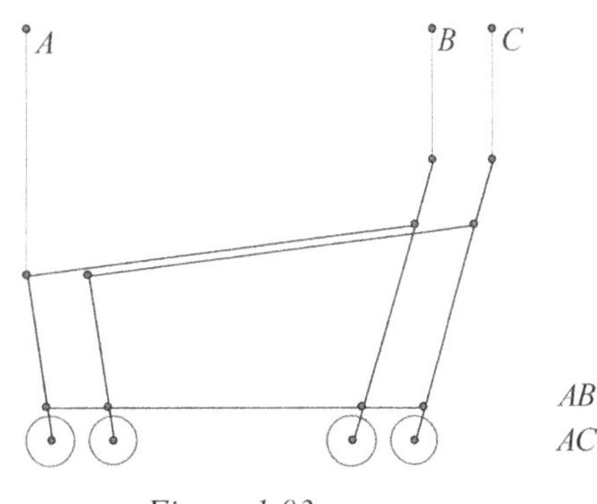

Figure 1.03

7

5.   A city has parking meters which only accept 25¢ coins. One 25¢ coin buys 20 minutes of parking time and each subsequent 25¢ coin buys 2 minutes less parking time than the previous 25¢ coin. Meters do not accept more than ten 25¢ coins at any one time.
(a) The parking time can be modeled by an arithmetic series. Find the first term and common difference of the series measured in minutes.
(b) Show that someone who puts $2 worth of 25¢ coins into a meter will get 1 hour 44 minutes of parking time.
(c) Find the maximum parking time permitted under this system.

6.   Government bonds earn 4% simple interest per year. David buys three $100 bonds in January 2000.
(a) Find the value of his investment in January 2005.
(b) If he buys three bonds each year in January from 2000 to 2004, find the value of his investment in January 2005.

7.   A toy consists of 50 rods of different lengths. The rods have lengths 10cm, 11cm, 12cm, . . . in arithmetic sequence.
(a) Find the length of the longest rod.

A rod may be joined to another rod only if it has a length 1cm longer or shorter than the other rod. Find
(b) the maximum length of 20 joined rods.
(c) the length of 15 joined rods, if the longest rod used is 43cm.
(d) the minimum number of rods needed to make a combined length of 495cm.

8.   An apartment block has 25 floors, access to which is via an elevator. The elevator takes 8 seconds to travel between adjacent floors. The time it takes to reach one of the floors from the first (ground) floor can be represented by a term of an arithmetic sequence.
(a) Write down the first term and common difference of the arithmetic sequence.
(b) Write down the time taken to reach the rth floor.

Twenty-four machines are to be delivered, one to each floor from the second to the twenty-fifth. Weight restrictions demand that only one machine at a time can be taken on the elevator. A machine is loaded into the elevator, taken to the second floor and delivered. Then the elevator returns to the first floor and loads the next machine which it takes to the third floor. This process continues until all twenty-four machines have been delivered.

(c) Given that it takes one minute to load a machine at the first floor and one minute to off-load a machine at the appropriate floor, find how long it takes to deliver all the machines.

1.4 Geometric Sequences and Series

1.4.1 Geometric Sequences

A sequence $\{u_n\}$ is called *geometric* if it is of the form $u_{n+1} = ru_n$ where r is a constant. The first term is denoted by u_1, and r is called the *common ratio*. For example, 6, 9, 13.5, 20.25, 30.375, 45.5625, … is a geometric sequence because $9 = 6 \times 1.5$, $13.5 = 9 \times 1.5$, $20.25 = 13.5 \times 1.5$ and so on. Each term is obtained from the previous term by multiplying it by 1.5. The first term, u_1 is 6, so that $u_1 = 6$ and the common ratio is $r = 1.5$. More generally,

$$u_2 = ru_1 = u_1 r$$
$$u_3 = ru_2 = u_1 r^2$$
$$u_4 = ru_3 = u_1 r^3$$
$$\cdots \cdots \cdots \cdots$$
$$\cdots \cdots \cdots \cdots$$
$$u_n = ru_{n-1} = u_1 r^{n-1}$$

Therefore the nth term of a geometric sequence is $u_n = u_1 r^{n-1}$.

Example 1.10: Find
        (a) the value of the 6th term of the geometric sequence whose 3rd term is 15 and whose 4th is 12.
        (b) the value of the first term.

Solution 1.10: (a) $u_3 = u_1 r^2 = 15; u_4 = u_1 r^3 = 12 \Rightarrow \dfrac{u_1 r^4}{u_1 r^3} = r = \dfrac{12}{15} = \dfrac{4}{5}$.

Therefore, $u_6 = u_1 r^5 = \left(u_1 r^3\right) r^2 = 12\left(\dfrac{4}{5}\right)^2 = \dfrac{192}{25}$.

(b) $u_1 r^2 = 15 \Rightarrow u_1 \left(\dfrac{4}{5}\right)^2 = 15 \Rightarrow u_1 = 15\left(\dfrac{5}{4}\right)^2 = \dfrac{375}{16}$. Therefore, the first term is $\dfrac{375}{16}$.

Example 1.11: A geometric sequence has first term 8 and common ratio 1.3. Find, correct to the nearest integer, the value of u_{11}.

Solution 1.11: Figures 1.04 and 1.05 show how the TI-84 Plus can be used to give successive terms of a geometric sequence. Firstly, on the home screen, store the first term 8 to X. The instruction $1.3X \rightarrow X$ has the effect of multiplying the current value of X by the common ratio of 1.3. The sequence is generated by applying the ENTER key repeatedly. If the ENTER key is applied ten times, the first 11 terms will be generated.

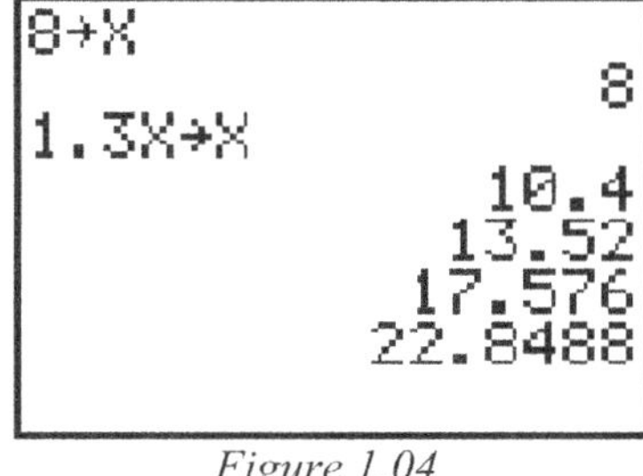

Figure 1.04

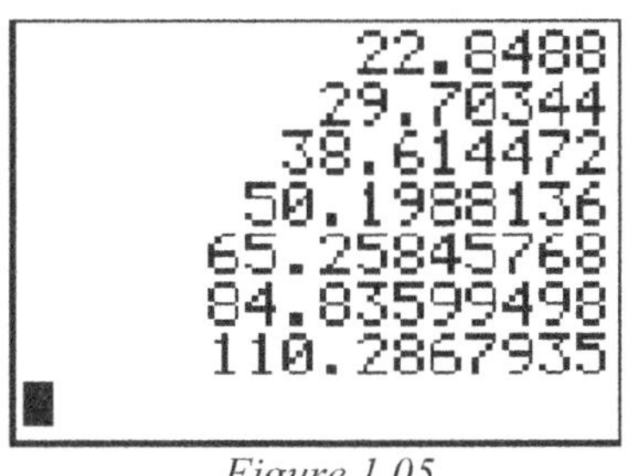

Figure 1.05

Therefore $u_{11} = 110.2867... = 110$

1.4.2 The Sum of a Finite Geometric Series

$$S_n = u_1 + u_2 + u_3 + \ldots + u_n = u_1 + u_1 r + u_1 r^2 + \ldots + u_1 r^{n-1} \ldots \ldots \ldots (1)$$

Multiplying (1) by $r \Rightarrow rS_n = u_1 r + u_1 r^2 + \ldots + u_1 r^{n-1} + u_1 r^n \ldots \ldots \ldots (2)$

Subtracting (2) from (1) gives

$$S_n - rS_n = \left(u_1 + u_1 r + u_1 r^2 + \ldots + u_1 r^{n-1}\right) - \left(u_1 r + u_1 r^2 + \ldots + u_1 r^{n-1} + u_1 r^n\right) = u_1 - u_1 r^n$$

$$\Rightarrow S_n(1-r) = u_1\left(1-r^n\right) \Rightarrow S_n = \frac{u_1\left(1-r^n\right)}{1-r}, \qquad r \neq 1$$

If $r > 1$, both numerator and denominator are negative, and it is more convenient to write the sum as

$$S_n = \frac{u_1\left(r^n - 1\right)}{r - 1}.$$

Of course, if $r = 1$, S_n reduces to the trivial geometric series in which all terms are equal, and it is more useful to think of the series as arithmetic with common difference of 0.

1.4.3 The Sum of an Infinite Geometric Series

When $|r| < 1$, (alternatively written, $-1 < r < 1$) the terms of a geometric series decrease as n increases. In fact, we can make r^n as small as we like, provided $|r| < 1$, by increasing n indefinitely. In this case a series with an infinite number of terms nevertheless has a finite sum, S_∞, and we say that as $n \to \infty$

(as n approaches infinity) the sum *converges*. The sum to infinity, S_∞, is then $S_\infty = \dfrac{u_1}{1-r}$, $|r| < 1$.

On the other hand, if $|r| > 1$ the series *diverges* and it has no sum.

Example 1.12: Find the sum to infinity of the series $6 - 2 + \dfrac{2}{3} - \dfrac{2}{9} + \dfrac{2}{27} - \ldots$.

Solution 1.12: $u_1 = 6$, $r = -\dfrac{1}{3}$ and because $\left|-\dfrac{1}{3}\right| < 1$, the series is convergent and its sum

$$S_\infty = \frac{6}{1-\left(-\frac{1}{3}\right)} = \frac{6}{\frac{4}{3}} = 6 \times \frac{3}{4} = \frac{9}{2}.$$

Example 1.13: A geometric series has first term 3 and common ratio 2. Find $\displaystyle\sum_{r=8}^{15} u_r$.

Solution 1.13: $\displaystyle\sum_{r=8}^{15} u_r = u_8 + u_9 + u_{10} + u_{11} + \ldots + u_{15}$.

$$= \left(u_1 + u_2 + \ldots + u_7 + u_8 + u_9 + \ldots + u_{15}\right) - \left(u_1 + u_2 + \ldots + u_7\right) = S_{15} - S_7$$

$$\text{but } S_{15} = u_1\left(\frac{r^{15}-1}{r-1}\right) = 3\left(\frac{2^{15}-1}{2-1}\right) = 3\times 32767 = 98301 \text{ and}$$

$$S_7 = 3\left(\frac{2^7-1}{2-1}\right) = 3\times 127 = 381. \text{ Therefore } \sum_{r=8}^{15} u_r = 98301 - 381 = 97920.$$

Example 1.14: Write $\dfrac{2}{5} + \dfrac{2}{7} + \dfrac{10}{49} + \ldots$ as a single fraction.

Solution 1.14: $u_1 = \dfrac{2}{5}$, $r = \dfrac{5}{7}$, so the sum to infinity is $S_\infty = u_1\left(\dfrac{1-r^n}{1-r}\right) = \dfrac{2}{5}\left(\dfrac{1}{1-\frac{5}{7}}\right) = \dfrac{2}{5}\times\dfrac{7}{2} = \dfrac{7}{5}.$

Exercise 1.4

1. Write down the first five terms of each of the following geometric sequences.

   (a) $u_1 = 1,\ r = 3$ (b) $u_1 = 32,\ r = \dfrac{1}{4}$ (c) $u_1 = 3,\ r = -\dfrac{2}{3}$ (d) $u_1 = \dfrac{4}{9},\ r = \dfrac{3}{2}$

2. Write down the next three terms of the following geometric sequences and find u_n in each case.

   (a) $1,\ \dfrac{1}{5},\ \dfrac{1}{25},\ \ldots$ (b) $2,\ 6,\ 18,\ \ldots$ (c) $\dfrac{1}{3},\ -\dfrac{2}{9},\ \dfrac{4}{27},\ \ldots$ (d) $1,\ 1.1,\ 1.21,\ \ldots$

   (e) $k,\ 2k,\ 4k,\ \ldots$ (f) $k,\ -\dfrac{1}{2}k^2,\ \dfrac{1}{4}k^3,\ \ldots$

3. In each case, use the method of Example 1.11 to find the minimum value of n such that u_n is greater than the value given in parentheses.

   (a) $1, 2, 4, \ldots$ (100 000)   (b) $3, 11, 40\frac{1}{3}, \ldots$ (1 000 000)   (c) $1, 1.06, 1.1236, \ldots$ (5)

   (d) $\dfrac{1}{2},\ \dfrac{2}{3},\ \dfrac{8}{9},\ \ldots$ (10)   (e) $1, 1.08, 1.1664, \ldots$ (20)

4. In each case, find the sum of the first n terms, where n is given in parentheses. Where an asterisk appears, give the answer exactly; otherwise, give the answer correct to two decimal places.

   (a) $1,\ 2,\ 4, \ldots$ (10)*   (b) $2,\ 1,\ \frac{1}{2}, \ldots$ (5)*   (c) $1,\ 1.06,\ 1.1236, \ldots$ (25)

   (d) $\dfrac{1}{2},\ \dfrac{2}{3},\ \dfrac{8}{9}, \ldots$ (6)   (e) $1,\ \dfrac{1}{2},\ \dfrac{1}{4}, \ldots$ (20)

5. In each case, find the sum to infinity of the series.

   (a) $18 + 12 + 8 + \ldots$ (b) $1 - \dfrac{1}{3} + \dfrac{1}{9} - \ldots$ (c) $5 + 2 + \frac{4}{5} + \ldots$

6.	(a) Show that $1.010101010101\ldots$ can be written in the form $1+\dfrac{1}{100}+\dfrac{1}{10000}+\ldots$ and hence write $1.010101010101\ldots$ as a single fraction.

(b) In a similar manner write (i) $0.444444\ldots$ (ii) $0.91919191\ldots$ (iii) $47.55555\ldots$ as single fractions.

7.	(a) The fifth term of a geometric sequence is 64 and the eighth term is 8. Find the first term and the common ratio of the sequence.

(b) The first term of a geometric sequence is 20 and the 37th term is 30. Find the common ratio correct to four decimal places.

8.	A geometric series is such that $S_n=\frac{1}{2}\left(3^n-1\right)$. Show that the first term is 1 and find u_5.

9.	(a) A geometric sequence has first term 13.5 and common ratio $\frac{1}{3}$. Find the value of n if the nth term of the sequence is $\frac{1}{162}$.

(b) A geometric sequence has first term 10 and common ratio -2. Find the value of n if the nth term is -5120.

10.	A geometric series with common ratio r is such that $S_6=3$ and $S_{10}=18$.

(a) Show that $r^{10}-6r^6+5=0$.

(b) Show that $r=1$ and $r=-1$ (the trivial solutions) satisfy the equation in (a) and find two other, non-trivial values of r which satisfy the equation in (a). (see Unit 2, section 2.2.3)

1.5 Applications of Geometric Sequences and Series

1.5.1 Compound Interest

If $\$A$ is invested and earns $p\%$ per year compound interest, then the investment after one year is $\$\left(\dfrac{pA}{100}\right)$. However, the interest, in dollars, after two years is based not on the initial amount invested, A, but on the value of the investment at the end of the first year, which is $\left(A+\dfrac{pA}{100}\right)=A\left(1+\dfrac{p}{100}\right)$.

Therefore, after two years the interest is $A\left(1+\dfrac{p}{100}\right)\left(\dfrac{p}{100}\right)$ and the value of the investment after two years is $A\left(1+\dfrac{p}{100}\right)+A\left(1+\dfrac{p}{100}\right)\left(\dfrac{p}{100}\right)=A\left(1+\dfrac{p}{100}\right)\left(1+\dfrac{p}{100}\right)=A\left(1+\dfrac{p}{100}\right)^2$. Table 1.1 shows the value of the investment at the end of successive years.

Year	Value of Investment (\$)
0	A
1	$A\left(1+\dfrac{p}{100}\right)$
2	$A\left(1+\dfrac{p}{100}\right)^{2}$
3	$A\left(1+\dfrac{p}{100}\right)^{3}$
n	$A\left(1+\dfrac{p}{100}\right)^{n}$

Table 1.1

Therefore, the value of the investment after n years can be modeled by a geometric sequence whose first term is \$ A and whose common ratio is $\left(1+\frac{p}{100}\right)$.

Example 1.15: A bank offers compound interest of 3.75% per year on deposits. If Ms. Patel invests \$6000,

    (a) find the value of her investment after (i) 1 year (ii) 10 years.

    (b) find the minimum number of years that are needed for the investment to be worth more than \$8000.

Solution 1.15: (a) $u_{1}=6000$, $r=1+\dfrac{3.75}{100}=1.0375$

    (i) As the first term is the initial value of the investment, the value of the investment after one year is $u_{2}=u_{1}r^{2-1}=u_{1}r=6000\times1.0375=6225$, so the value of the investment is \$6225.

    (ii) The value of investment after ten years is $\$6000\times1.0375^{10}=8670.263...$ So the investment is $=\$8670.26$.

    (b) You need to find an n such that $u_{1}r^{n}>8000$. Figures 1.06 and 1.07 show how n can be found.

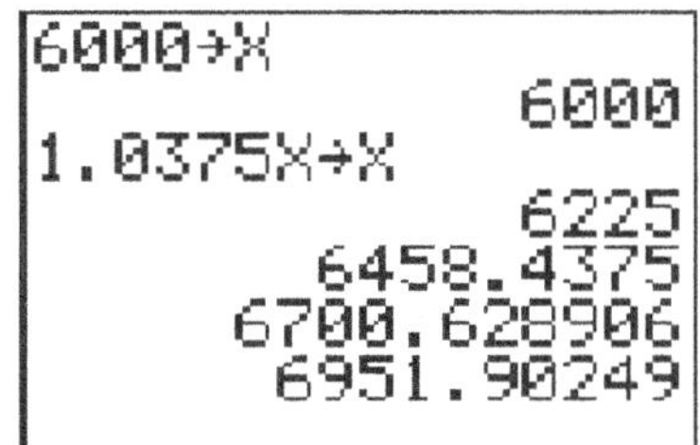

Figure 1.06

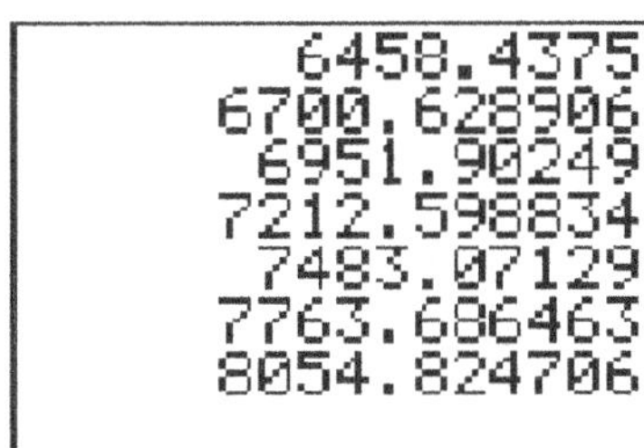

Figure 1.07

Because the number obtained by applying the ENTER key n times after 1. $X \to X$ represents the value of the investment after n years, you need only count how many times you have to apply the ENTER key until you obtain a value in excess of 8000. So 8 years are needed for the value of the investment to exceed $8000.

1.5.2 Compounding Interest

In the previous example, it was assumed that the interest was calculated and added to the amount of the investment at the end of each year, whereas, in reality, this process, known as compounding, is carried out much more frequently. We now investigate the consequences of compounding at more frequent intervals.

Suppose $1000 is invested and earns 12% interest per year compounded yearly. Then after one year the value of the investment is $1000\left(1+\frac{12}{100}\right)=\$1120$.

If, however, $1000 is invested and earns 12% interest per year compounded monthly, then the interest per month is 1%, so after one month, the value of the investment is $\$1000\left(1+\frac{12}{12\times100}\right)=$

$\$1000\left(1+\frac{1}{100}\right)=\$1010$ and therefore after 12 months the value of the investment is

$\$1000\left(1+\frac{1}{100}\right)^{12}=1126.825...=\$1126.83$

Interest added to the initial amount at an early stage has the opportunity to earn more interest later on. This process results in an additional $6.83 being earned during a year.

Example 1.16: The national currency in Swaziland is the lilangeni, whose plural is emalangeni, (E). Sandile invests E1000 in the Bank of Mbabane which offers an interest rate of 3.5% per year, compounded yearly. Makhethe invests E1000 in the National Bank of Swaziland which offers an interest rate of 3.3% per year compounded monthly. Whose investment is worth the most at the end of a year, and by how much?

Solution 1.16: Sandile's investment, in emalangeni, is worth $1000\left(1+\frac{3.5}{100}\right)=1000\times1.035=1035$.

Makhethe's investment, in emalangeni, is worth $1000\left(1+\frac{3.3}{1200}\right)^{12}=1000\left(1+0.00275\right)^{12}$

$=1033.503...=1033.50$. So, Sandile's investment is worth more. The amount by which Sandile's investment is more than Makhethe's is $E1035-E1033.50=E1.50$.

1.5.3 Population Growth

The growth of populations of many living creatures can be modeled by geometric sequences. Suppose a population of n beetles increases by $p\%$ per month. Then after one month, the population of the

beetles has increased by $\frac{np}{100}$ so that the total population after one month is $n+\frac{np}{100}=n\left(1+\frac{p}{100}\right)$.

The population of beetles in subsequent months can be represented by the terms of a geometric

sequence whose first term is n and whose common ratio is $1+\frac{p}{100}$.

Example 1.17: A herd of 360 elephants is known to increase by 3% per year. Find the size of the herd after 5 years.

Solution 1.17: The number of elephants in the herd at the end of each year corresponds to the terms of a geometric sequence in which $u_1 = 360$ and $r = 1 + \dfrac{3}{100} = 1.03$. Therefore, the size of the herd after 5 years is represented by $u_6 = u_1 r^5 = 360 \times 1.03^5 = 417.3386....$ Thus, the size of the elephant herd after 5 years will be 417.

Example 1.18: The population of a city in 1991 was 370 000 and in 2001 it was 428 000. If the population growth rate is constant, find the yearly percentage growth rate during the decade from 1991 and 2001.

Solution 1.18: Let p be the yearly growth rate, expressed as a percentage. The population of the city at the end of each year is represented by a geometric sequence with $u_1 = 370$ (population in thousands) and $r = 1 + \dfrac{p}{100}$ in which $u_1 = 370$, $u_{11} = 428$. Then $370\left(1 + \dfrac{p}{100}\right)^{10} = 428$.

Therefore, $\left(1 + \dfrac{p}{100}\right)^{10} = \dfrac{428}{370} \Rightarrow p = 100\left(\left(\dfrac{428}{370}\right)^{\frac{1}{10}} - 1\right) = 1.466856....$ Therefore the yearly growth rate is 1.47%.

Exercise 1.5

1. The annual growth rate of the population of the United States is thought to be 1.1%. If the current population of the United States is 280 million, estimate its population in
(i) 2 years time   (ii) 10 years time   (iii) 20 years time.

2. A bank announces that it will pay 5.6% interest per year on investments of $40 000 or more and 4.7% interest per year on investments of less than $40 000. Vladimir invests $50 000 for 5 years, Galina invests $15 000 for 20 years. Find the value of the investment at the end of the investment period in each case. Assume that interest is compounded yearly.

3. (a) The number of antelope in a nature reserve is 14 000 and is increasing at a rate of 4.5% per year. How many antelope should be culled (officially killed) each year in order to maintain the herd of antelope at a constant number?
(b) The population of zebras in a large safari park increases at a rate of 4% per year. A survey reveals that the population of zebras is 34 000. Find the population of zebras 6 years later.

4. (i) The population of a city is increasing at 7% per year. How many years will it take for the city's population to double?
(ii) The value of a rare postage stamp appreciates (increases) at a rate of 12% per year. If its value at the beginning of 2001 was $6500, what value can it expect to have at the beginning of 2006?
(iii) A currency depreciates (decreases in value) at a rate of 35% per year relative to the US dollar. In January 2004, there are 10 units of this currency to 1 US dollar. How many units

will there be to 1 US dollar in January 2010?

5. Motor magazines estimate that a Gladiator, a sports utility vehicle, depreciates at 23% per year. Peter buys a Gladiator in June 1997 for $38 000 and sells it in June 2002. Find the value, according to the motor magazines, of Peter's Gladiator when he sells it.

6. Interest of 4.8% per year is paid on an investment.
(a) Find the interest rate (i) per month   (ii) per day.
(b) Find the value of an initial investment of $1000 after 5 years if the interest is compounded
   (i)   monthly   (ii)   daily.

7. It is estimated that the Earth's available quantity of a specific mineral is 1000 billion tonnes. If the current yearly consumption of this mineral is 18 billion tonnes and this consumption is increasing at a yearly rate of 8%, estimate how long it will be before this mineral resource is exhausted.

8. A company, in anticipation of expanding, agrees to buy one photocopying machine at the start of each year for 5 years. The company pays $1000 for each machine. At the start of the sixth year, the company becomes bankrupt and has to sell all five machines. The machines depreciate (lose value) at the rate of 10% per year.
(a) Show that the first machine purchased is worth $590.49 at the time of bankruptcy.
(b) Show that the total value of all five machines at the time of bankruptcy can be modeled by the sum of a geometric series whose first term is $900 and whose common ratio is 0.9.
(c) Find the total value of all five machines at the time of bankruptcy to the nearest $100.

9. An island is populated by two rodent species: rats and mice. The rat population is 160 000 and is increasing at a rate of 1.7% per year. The mouse population is 70 000 and is increasing at a rate of 3.8% per year. If it is assumed that these rates of increase are maintained, how many years will elapse before the population of mice exceeds the population of rats?

10. Colette borrows $1000 and agrees to repay it in two equal payments, the first after 6 months and the second after one year. She is required to pay interest at 12% per year, compounded monthly. If each payment is P show that

(a) the amount owed after 6 months is $\left(1000(1.01)^6 - P\right)$.

(b) $P = 546.60$.

1.6 Exponents and Logarithms

1.6.1 Exponents

If $a \in \mathbb{R}^+$, then $a \times a \times a = a^3$ and $a \times a = a^2$. Therefore, $(a \times a \times a) \times (a \times a) = a \times a \times a \times a \times a$

$\Rightarrow a^3 \times a^2 = a^5 = a^{3+2}$, which suggests that $a^x \times a^y = a^{x+y}$.

Similarly, $\dfrac{a \times a \times a \times a \times a}{a \times a \times a} = a \times a \Rightarrow \dfrac{a^5}{a^3} = a^2 \Rightarrow a^5 \div a^3 = a^2 = a^{5-3}$ which suggests that

$a^x \div a^y = a^{x-y}$.

Furthermore, $(a \times a \times a) \times (a \times a \times a) = (a \times a \times a \times a \times a \times a) \Rightarrow (a^3)^2 = a^6$ which suggests that $(a^x)^y = a^{x \times y}$.

The following three laws determine the manner in which the algebra of exponents operates.

$$a^x \times a^y = a^{x+y} \quad \ldots\ldots\ldots\ldots (1)$$
$$a^x \div a^y = a^{x-y} \quad \ldots\ldots\ldots\ldots (2)$$
$$(a^x)^y = a^{xy} \quad \ldots\ldots\ldots\ldots (3)$$

Only positive rational values of a and only rational values of x, y are considered. That is $a \in \mathbb{Q}^+$, $x, y \in \mathbb{Q}$.

Special Cases: From (1): If $y = 0$, then $a^x \times a^0 = a^{x+0} = a^x \Rightarrow a^0 = 1$.

From (2): If $y = -x$ then $a^x \times a^{-x} = a^{x-x} = a^0 = 1 \Rightarrow a^{-x} = \dfrac{1}{a^x}$.

From (3): If $x = \dfrac{1}{2}$, $y = 2$ then $\left(a^{\frac{1}{2}}\right)^2 = a^{\frac{1}{2} \times 2} = a^1 = a$. By taking the positive square

root of both sides, $a^{\frac{1}{2}} = \sqrt{a}$.

Example 1.19: Simplify $\sqrt{\dfrac{a^3 \times (ab)^2}{a}}$.

Solution 1.19: $\sqrt{\dfrac{a^3 \times (ab)^2}{a}} = \left(a^3 \times (ab)^2 \times a^{-1}\right)^{\frac{1}{2}}$ but $(ab)^2 = a^2 b^2$ so that

$$\left(a^3 \times (ab)^2 \times a^{-1}\right)^{\frac{1}{2}} = \left(a^3 \times a^2 \times b^2 \times a^{-1}\right)^{\frac{1}{2}} = \left(a^{3+2-1} \times b^2\right)^{\frac{1}{2}} = \left(a^4 b^2\right)^{\frac{1}{2}} = \left(a^4\right)^{\frac{1}{2}} \left(b^2\right)^{\frac{1}{2}} = a^2 b.$$

Example 1.20: In each case, write the expression as an integer.

$$\text{(a)} \quad 27^{\frac{2}{3}} \qquad\qquad \text{(b)} \quad \left(\frac{1}{6}\right)^{-2} \qquad\qquad \text{(c)} \quad 2^{\frac{1}{2}} \times 3^{\frac{5}{2}} \times 6^{\frac{3}{2}}$$

Solution 1.20: (a) $27^{\frac{2}{3}} = \left(27^{\frac{1}{3}}\right)^2 = 3^2 = 9$ $\qquad$ (b) $\left(\dfrac{1}{6}\right)^{-2} = \left(6^{-1}\right)^{-2} = 6^{-1 \times -2} = 6^2 = 36$

(c) $2^{\frac{1}{2}} \times 3^{\frac{5}{2}} \times 6^{\frac{3}{2}} = 2^{\frac{1}{2}} \times 3^{\frac{5}{2}} \times (2 \times 3)^{\frac{3}{2}} = 2^{\frac{1}{2}} \times 3^{\frac{5}{2}} \times 2^{\frac{3}{2}} \times 3^{\frac{3}{2}} = 2^{\frac{1}{2}+\frac{3}{2}} \times 3^{\frac{5}{2}+\frac{3}{2}} = 2^2 \times 3^4$

$\qquad = 4 \times 81 = 324$

Example 1.21: Solve the equation $4^x = 2^{1-x}$.

Solution 1.21: $4^x = \left(2^2\right)^x = 2^{2x} = 2^{1-x} \Rightarrow 2x = 1-x \Rightarrow 3x = 1 \Rightarrow x = \dfrac{1}{3}$

Exercise 1.6

1. Show that (i) $\left(abc^3\right) \times \left(ab\right)^2 = \left(abc\right)^3$ (ii) $\sqrt{\dfrac{a^5 b^3 c}{abc^3}} = a^2 bc^{-1}$

2. Simplify the following expressions.

   (i) $\left(ab^2\right) \times \left(2a^2\right)$ (ii) $\dfrac{\left(abc\right)^3}{c^2 b}$ (iii) $a^2 \times \left(2b\right)^{-1} \times \left(4c\right)$ (iv) $\sqrt{\dfrac{a^4 b^2 c}{c^5}}$

   (v) $\dfrac{a^{-3} b}{a^2 b^3}$ (vi) $\left(a^3 b^4 a^{\frac{1}{2}}\right)^{-2}$ (vii) $\dfrac{\left(4a\right)^{-2} \times \left(ab\right)^3}{\left(2b\right)^2}$

3. In each case, write the problem as an integer or simplified fraction, as appropriate.

   (a) $64^{\frac{1}{2}}$ (b) $2^{-3} \times 4^4$ (c) $\left(\dfrac{8}{27}\right)^{\frac{2}{3}}$ (d) $\left(\dfrac{1}{8}\right)^{-\frac{1}{3}}$

   (e) $\left(\dfrac{1}{3}\right)^{-2}$ (f) $\left(\dfrac{4}{25}\right)^{\frac{1}{2}}$ (g) $\left(\dfrac{125}{64}\right)^{\frac{4}{3}}$ (h) $\left(0.25\right)^{-2}$

   (i) $\left(\dfrac{4}{9}\right)^{\frac{3}{2}}$ (j) $\left(\dfrac{1}{2}\right)^{-2} \times \left(\dfrac{3}{4}\right)^3 \times \left(\dfrac{2}{5}\right)^2$ (k) $12^{\frac{3}{2}} \times 15^{-\frac{3}{4}} \times \left(0.6\right)^{\frac{1}{4}}$

4. In each case except (h) solve for x by writing both sides of the equation in terms of the same base and then by equating the exponents as in Example 1.21.

   (a) $3^{-x} = \dfrac{1}{9}$ (b) $2^{3x+1} = 16$ (c) $4^{3-2x} = 8^{x+1}$ (d) $2^{3-2x} = 16^{x+1}$

   (e) $\left(\dfrac{1}{8}\right)^{x-3} = 2^{5x-4}$ (f) $\left(0.5\right)^{x-1} = \left(0.125\right)^{2+5x}$ (g) $5^x \times 25^{1-2x} = 1$

   (h) $2^{x+2} + 3^x = 5^x$ (Hint: Use inspection. In other words, guess values of x and substitute these into the equation to see if they satisfy it.)

1.6.2 Logarithms

In question 4 of Exercise 1.6, it is possible to find the exact value of x because both sides of the equation can be expressed in terms of the same base. In general, however, this is not possible. For example, to solve the equation, $6^x = 49$ another method needs to be used because 49 cannot be written as a rational exponent of 6. In such situations we need to use logarithms.

If $a^x = b$, then we can write $x = \log_a b$ and we say that x is the logarithm to the *base* a of b. Graphing calculators are equipped to provide logarithms to base 10 (and logarithms to base e) so that $10^x = 74 \Rightarrow x = \log_{10} 74 = 1.869231...$. Although this can be very useful, on its own it will not help in solving, for example, the equation $6^x = 49$. However, the three laws of exponents may be restated as laws of logarithms which will enable equations such as $6^x = 49$ to be solved.

From exponent law (1), $a^x \times a^y = a^{x+y}$, let $a^x = A$, $a^y = B$, $a^{x+y} = C \Rightarrow AB = C$. Then $x = \log_a A$; $y = \log_a B$; $x + y = \log_a C = \log_a AB \Rightarrow \log_a A + \log_a B = \log_a AB$.

From exponent law (2), $a^x \div a^y = a^{x-y}$, and exponent law (3), $\left(a^x\right)^y = a^{xy}$, we get, in a similar manner,

$\log_a A - \log_a B = \log_a \dfrac{A}{B}$ and $\log_a A^n = n \log_a A$.

Therefore, the laws of logarithms are

$$\log x + \log y = \log xy \quad \dotfill (1)$$

$$\log x - \log y = \log \dfrac{x}{y} \quad \dotfill (2)$$

$$\log x^n = n \log x \quad \dotfill (3)$$

The base is omitted because the laws are true for all bases a, $a \in \mathbb{Z}^+$.

Provided $a > 0$, $a^x > 0$ for all x. Therefore, if $a^x = b$, then $b > 0$ and, as $x = \log_a b$, it is not possible to take the logarithm of a negative number.

Special Cases: $\qquad a^0 = 1 \Rightarrow \log_a 1 = 0$

$$a^1 = a \Rightarrow \log_a a = 1$$

$$a^{-1} = \frac{1}{a} \Rightarrow \log_a\left(\frac{1}{a}\right) = -1$$

Example 1.22: If $u = \log x$, $v = \log y$. Write $\log\left(\dfrac{x^4}{y^3}\right)$ in terms of u and v.

Solution 1.22: $\log\left(\dfrac{x^4}{y^3}\right) = \log x^4 - \log y^3$ (by law (2) of logarithms).

$$= 4 \log x - 3 \log y \text{ (by law (3) of logarithms).}$$

So $\log\left(\dfrac{x^4}{y^3}\right) = 4u - 3v$.

Example 1.23: Write $\log_3 9$ as an integer.

Solution 1.23: If $x = \log_3 9$ then $3^x = 9 = 3^2 \Rightarrow x = 2$. Alternatively, $\log_3 9 = \log_3 3^2 = 2\log_3 3$ and as $\log_3 3 = 1$, $\log_3 9 = 2$.

Exercise 1.7

1. In each case, write x in terms of logarithms to the appropriate base. For example, $2^x = 5 \Rightarrow x = \log_2 5$.

(a) $2^x = 3$ (b) $3^x = 2$ (c) $6^x = 5$ (d) $10^{x+1} = 7$

(e) $5^{-x} = \dfrac{1}{4}$ (f) $4 \times 3^{2x} = 7$ (g) $2^{x-3} = 9$ (h) $\dfrac{1}{3^x} = \dfrac{2}{5}$

(i) $3^{\frac{1}{x}} = 11$ (j) $4^{2-3x} = 12$ (k) $6^{x^2} = 3$

2. In each case, solve for x. For example, $\log_2 x = 5 \Rightarrow x = 2^5 = 32$

(a) $\log_3 x = 2$ (b) $\log_4 x = 3$ (c) $\log_9 x = \dfrac{3}{2}$ (d) $\log_6 2x = 3$

(e) $\log_2 (x+1) = 5$ (f) $\log_5 \dfrac{1}{x} = 3$ (g) $\log_2 \left(\dfrac{x-1}{2} \right) = 7$ (h) $\log_4 \sqrt{x} = 2$

3. If $u = \log x$, $v = \log y$, write the following expressions in terms of u and v.

(a) $\log(xy)$ (b) $\log\left(\dfrac{x}{y}\right)$ (c) $\log\left(x^2 y^3\right)$ (d) $\log\left(\dfrac{y}{x^2}\right)$ (e) $\left(\log x\right)^2 + \log\left(y^2\right)$

(f) $3\log(xy) - \log y$ (g) $\log xy + \log\left(\dfrac{x^3}{y^2}\right)$ (h) $2\log(3x) - \log(9y)$

4. In each case, write x in terms of base 10 logarithms and then use the logarithm key on your graphing calculator to solve for x.

(a) $10^x = 491$ (b) $10^x = 18$ (c) $10^x = 1.294$ (d) $10^x = 3.1 \times 10^6$

(e) $10^{x-1} = 2097$ (f) $4 - 10^x = 2.916$ (g) $100^x = 78.5$ (h) $0.1^x = 19$

5. Write the following as single logarithms. For example, $\log 3 + \log 8 = \log 24$.

(a) $\log 2 + \log 3$ (b) $\log 15 - \log 3$ (c) $\log 4 + \log 3 - \log 2$

(d) $\log 8 - \log 12$ (e) $\log 66 - \log 11 - \log 2$ (f) $2\log 15 - \log 9$

(g) $\dfrac{1}{2}\left(\log 9 + \log 4\right)$ (h) $3\log 2 + 2\log 3$ (i) $\log 9 + \log 6 - 3\log 3$

(j) $\log \dfrac{1}{2} + \log \dfrac{3}{4}$ (k) $\log 2a + \log 4a$, $a > 0$

6. Write each of the following as a rational number.

(a) $\log_5 5$ (b) $\log_2 16$ (c) $\log_4 2$

(d) $\log_6 1$ (e) $\log_5\left(\dfrac{1}{125}\right)$ (f) $\log_3\left(\dfrac{1}{3}\right)$

(g) $\log_3 81$ (h) $2\log_2\left(\dfrac{1}{16}\right)$ (i) $\log_{a^2} a + \log_a a^2$, $a > 0$

7.   Solve the simultaneous equations $\log_2\left(x^3 y^2\right) = 28$, $\log_2\left(\dfrac{x^4}{y^3}\right) = -8$ by putting $u = \log_2 x$, $v = \log_2 y$ and writing the equations in terms of u and v.

1.6.3 Changing the Base of a Logarithm

Suppose $a^x = b$; then, by taking the logarithm to base c of both sides, you get $\log_c a^x = \log_c b$.

Therefore, $x \log_c a = \log_c b$ by law (3) of logarithms $\Rightarrow x = \dfrac{\log_c b}{\log_c a}$. However, $x = \log_a b$, so

$\log_a b = \dfrac{\log_c b}{\log_c a}$. So, for example, $\log_7 3 = \dfrac{\log_{10} 3}{\log_{10} 7} = 0.5645750\ldots$.

The ability to change the base of a logarithm is clearly very useful when solving equations of the kind

$6^x = 49$. $6^x = 49 \Rightarrow x = \log_6 49 = \dfrac{\log_{10} 49}{\log_{10} 6} \Rightarrow x = 2.172066\ldots$.

The ability to change the base of a logarithm is also valuable where logarithms to different bases occur in the same equation.

Example 1.24: Show that the solution of the equation $\log_3 2x + \log_9 x^2 = 1$ is $x = \sqrt{\dfrac{3}{2}}$.

Solution 1.24: $\log_9 x^2 = \dfrac{\log_3 x^2}{\log_3 9} = \dfrac{\log_3 x^2}{\log_3 3^2} = \dfrac{\log_3 x^2}{2\log_3 3} = \dfrac{1}{2}\log_3 x^2$

$\Rightarrow \log_3 2x + \dfrac{1}{2}\log_3 x^2 = 1$

$\Rightarrow \log_3 2x + \log_3 x = 1$ (by law (3) of logarithms)

$\Rightarrow \log_3 2x^2 = 1 = \log_3 3$ (by law (1) of logarithms)

$\Rightarrow 2x^2 = 3 \Rightarrow x^2 = \dfrac{3}{2} \Rightarrow x = \pm\sqrt{\dfrac{3}{2}}$, but $x > 0$ therefore $x = \sqrt{\dfrac{3}{2}}$.

Example 1.25: Find the range of values of x which satisfy the inequality $1.05^x > 20$.

Solution 1.25: $\log_{10} 1.05^x > \log_{10} 20$ (taking logarithms to base 10)

$$\Rightarrow x \log_{10} 1.05 > \log_{10} 20 \text{ (law (3) of logarithms)}$$

$$\Rightarrow x > \frac{\log_{10} 20}{\log_{10} 1.05} = 61.40033... \text{ (dividing by } \log_{10} 1.05)$$

Therefore $x > 61.4$.

Exercise 1.8

1. In each case, solve the equations by writing them in logarithmic form and then changing the base to 10. For example, $3^x = 11 \Rightarrow x = \log_3 11 = \dfrac{\log_{10} 11}{\log_{10} 3} = 2.182658... = 2.18$.

   (a) $2^x = 5$ (b) $3^x = 88$ (c) $4^{x+1} = 15$ (d) $2^{3-x} = 22$

   (e) $5^{x-2} = 338$ (f) $3^{-x^2} = \dfrac{1}{2}$

2. . If $t = \log_3 x$, write the following in terms of t.

   (a) $\log_3 x^2$ (b) $\log_3 3x$ (c) $\log_3\left(\dfrac{x^3}{3}\right)$ (d) $\log_9 x$ (e) $\log_x 3$ (f) $\log_{3x} x$

3. (a) (i) Show that $\log_2 4 = 2$

   (ii) By changing the base of the right hand side find the solution of $\log_2 x = \log_4 25$

   (b) By changing an appropriate base find both solutions of $\log_3 x = \log_9 (7x - 6)$

   (c) Find the values of x which satisfies the equation $\log_x 81 = \log_3 x$

4. Solve each equation by changing the base of the logarithm to 10.
   (a) $\log_2 x = \log_5 3$ (b) $\log_7 x = \log_2 9$ (c) $\log_2 x + \log_4 x = \log_2 5$
   (d) $\log_3 x = 5 \log_{10} 2$

5. Find the range of values of x which satisfy the given inequalities.
   (a) $2^x > 5000$ (b) $3^x > 697$ (c) $1.01^x > 10$
   (d) $1.15^{x+1} > 33$ (e) $0.99^x < 0.5$ (f) $0.5^x < 0.001$.

1.7 The Binomial Theorem

1.7.1 Expansion of Binomial Expressions

The expansions of $(x+y)^2$ and $(x+y)^3$ can be carried out in the following ways:

$$(x+y)^2 = (x+y)(x+y) = x(x+y) + y(x+y) = x^2 + xy + yx + y^2 = x^2 + 2xy + y^2$$

$$(x+y)^3 = (x+y)(x+y)(x+y)$$
$$= (x+y)(x^2 + 2xy + y^2) = x(x^2 + 2xy + y^2) + y(x^2 + 2xy + y^2)$$
$$= x^3 + 2x^2y + xy^2 + yx^2 + 2xy^2 + y^3$$
$$= x^3 + 3x^2y + 3xy^2 + y^3$$

To go on expanding higher two-term or *binomial expressions* in this way gets increasingly tedious. The binomial theorem provides a convenient way of expanding such two-term expressions. We can develop the binomial theorem in the following way.

Summarizing these expansions and adding the trivial results of expanding $(x+y)^0$ and $(x+y)^1$, we obtain the following:

$$
\begin{aligned}
(x+y)^0 &= & & 1 \\
(x+y)^1 &= & & 1x \;+\; 1y \\
(x+y)^2 &= & & 1x^2 \;+\; 2xy \;+\; 1y^2 \\
(x+y)^3 &= & & 1x^3 \;+\; 3x^2y \;+\; 3xy^2 \;+\; 1y^3
\end{aligned}
$$

It is not difficult to see that the sum of the powers of terms in the expansion must be equal to the power applied to the binomial expression. For example, in the expansion of $(x+y)^3$, the sum of the powers of x and y in each term, x^3, $3x^2y^1$, $3x^1y^2$, y^3 is 3.

What is not so easy to determine are the coefficients (numbers) associated with each term. If we write out the triangle formed above omitting the x's and y's, it is easier to see the pattern, and we see that the sum of each row is 2^n, where n is the power to which $(x+y)$ is raised.

$$
\begin{array}{ccccccc}
 & & & 1 & & & \\
 & & 1 & & 1 & & \\
 & 1 & & 2 & & 1 & \\
1 & & 3 & & 3 & & 1
\end{array}
\qquad
\begin{aligned}
1 &= 2^0 \\
2 &= 2^1 \\
4 &= 2^2 \\
8 &= 2^3
\end{aligned}
$$

With this in mind, and following the observed pattern already established, the next line is

$$
\begin{array}{ccccc}
1 & 4 & 6 & 4 & 1
\end{array}
\qquad 16 = 2^4,
$$

and the next is

$$1 \quad 5 \quad 10 \quad 10 \quad 5 \quad 1 \qquad 32 = 2^5.$$

We see that each coefficient is obtained by summing the two adjacent coefficients in the row above. This triangle of numbers, known as *Pascal's triangle*, is shown below and provides an easy method of finding the coefficients of binomial expansions.

$$
\begin{array}{ccccccccccccccc}
& & & & & & & 1 & & & & & & & \\
& & & & & & 1 & & 1 & & & & & & \\
& & & & & 1 & & 2 & & 1 & & & & & \\
& & & & 1 & & 3 & & 3 & & 1 & & & & \\
& & & 1 & & 4 & & 6 & & 4 & & 1 & & & \\
& & 1 & & 5 & & 10 & & 10 & & 5 & & 1 & & \\
& 1 & & 6 & & 15 & & 20 & & 15 & & 6 & & 1 & \\
1 & & 7 & & 21 & & 35 & & 35 & & 21 & & 7 & & 1 \\
\end{array}
$$

$$1 \quad 8 \quad 28 \quad 56 \quad 70 \quad 56 \quad 28 \quad 8 \quad 1$$

Example 1.26: Find the expansion of $(x+y)^5$.

Solution 1.26: $(x+y)^5 = x^5 + 5x^4 y^1 + 10x^3 y^2 + 10x^2 y^3 + 5x^1 y^4 + y^5$

Example 1.27: Write down and simplify the expansion of $(a-3b)^4$.

Solution 1.27: It is convenient to write $(a-3b)^4 = (a+(-3b))^4$

$$\Rightarrow (a-3b)^4 = 1a^4 + 4a^3(-3b)^1 + 6a^2(-3b)^2 + 4a^1(-3b)^3 + (-3b)^4.$$

Now $(-3b)^2 = (-3)^2 \times (b)^2 = 9b^2$ and, in a similar manner,

$(-3b)^3 = (-3)^3 \times (b)^3 = -27b^3$ and $(-3b)^4 = (-3)^4 \times (b)^4 = 81b^4$ so that $(a-3b)^4$

$$= a^4 - 12a^3 b + 54a^2 b^2 - 108ab^3 + 81b^4.$$

1.7.2 Obtaining Specific Terms of a Binomial Expansion

It is convenient to be able to identify specific *binomial coefficients*. For example, the binomial coefficient of the term $x^2 y^3$ in the expansion of $(x+y)^5$ is denoted by $\binom{5}{3}$. Your graphing calculator is able to find the coefficient although the notation will probably be different. A TI-84 Plus uses the notation nCr instead of $\binom{n}{r}$, where n is the exponent of the binomial expression and r is the exponent of y.

Figures 1.08 and 1.09 show how the TI-84 Plus can be used to find $\binom{5}{3}$.

Firstly, on the home screen, enter the number 5, then go to the MATH menu and select PRB. Finally, enter the number 3 and press ENTER.

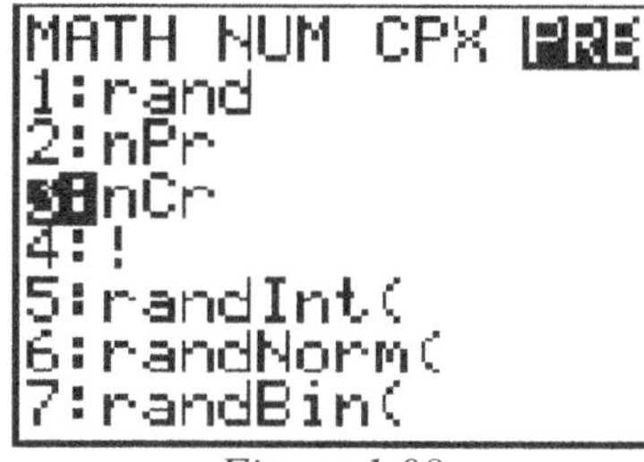

Figure 1.08

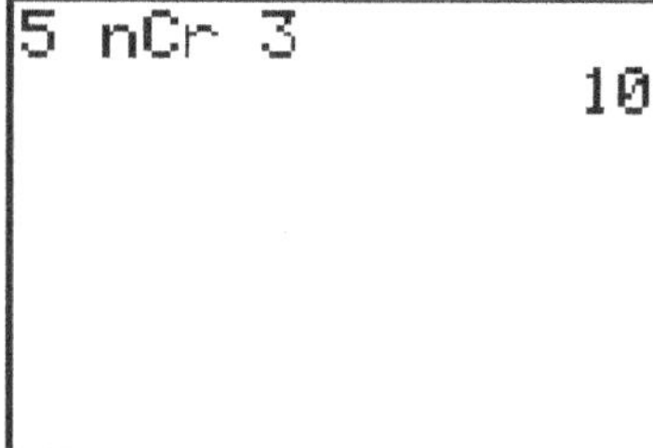

Figure 1.09

Therefore $\binom{5}{3}=10$.

Example 1.28: Find the coefficient of x^7 in the expansion of $(2+x)^{11}$.

Solution 1.28: As you need to find the term involving x^7, the other part of the binomial expression, 2 will need to be raised to the power of $11-7=4$ and the binomial coefficient will be $\binom{11}{7}$. Therefore the term involving x^7 is $\binom{11}{7}(2)^4(x)^7$, and the required coefficient is $\binom{11}{7}(2)^4 = 330 \times 16 = 5280$.

Exercise 1.9

1. In each case, expand the binomial expression.

(a) $(x+y)^4$ (b) $(1+y)^3$ (c) $(1+x)^7$ (d) $(1-x)^6$ (e) $(k-1)^8$

2. In each case, simplify the given expression. For example, $(2x)^3 = 2^3 x^3 = 8x^3$.

(a) $(3x)^2(2x)^4$ (b) $(2a)^3(5b)^2$ (c) $(-2)^5(6)^2$ (d) $\left(\dfrac{x}{2}\right)^3(2y)^2$

(e) $(-1)^8(2x)^3$ (f) $\left(\dfrac{1}{3}a\right)^4\left(-\dfrac{3}{2}b\right)^3$

3. In each case, expand the binomial expression, giving your answer in simplified form.

(a) $(x+2y)^4$ (b) $(1+2x)^5$ (c) $\left(2-\dfrac{k}{2}\right)^3$ (d) $(3+2x)^3$

(e) $\left(1+\dfrac{2}{3}a\right)^4$ (f) $\left(\dfrac{1}{3}x+\dfrac{3}{2}y\right)^3$ (g) $\left(x^2-\dfrac{1}{x}\right)^4$ (h) $\left(\dfrac{x}{2}+\dfrac{2}{x}\right)^6$

4. Evaluate the following binomial coefficients.

(a) $\dbinom{9}{2}$ (b) $\dbinom{14}{3}$ (c) $\dbinom{12}{6}$ (d) $\dbinom{20}{2}$ (e) $\dbinom{15}{8}$

5. Find the coefficient of the specified term in each of the given binomial expressions.

(a) $(x+y)^{10}$; x^7y^3 (b) $(1+x)^{18}$; x^{11} (c) $(1-x)^{14}$; x^9 (d) $(a-b)^{21}$; $a^{15}b^6$

(e) $(2+k)^{15}$; k^{10} (f) $(1+2y)^9$; y^5 (g) $\left(2-\dfrac{1}{2}x\right)^{16}$; x^6 (h) $\left(3+\dfrac{2}{x}\right)^{11}$; $\dfrac{1}{x^4}$

6. Find the specified term in each of the given binomial expressions.

(a) $(1+3x)^9$; term in x^3 (b) $(2-x)^{13}$; term in x (c) $(1+x)^{25}$; term in x^2

(d) $(1-x)^{19}$; term in x^{16} (e) $\left(a+\dfrac{b}{3}\right)^5$; term in a^3b^2 (f) $\left(3+\dfrac{1}{2}x\right)^{11}$; term in x^4

(g) $\left(3x-\dfrac{1}{3}\right)^7$; term in x^5 (h) $\left(2x+\dfrac{y}{2}\right)^6$; term in x^2y^4

7. (a) Expand $(1+x)^4$. By putting $x=0.1$, show that the value of $1.1^4=1.4641$.

(b) Expand $(1+x)^7$. By putting $x=0.01$, find the exact value of 1.01^7.

(c) <u>Write down</u> the exact value of 1.01^8.

8. (a) Find the term in x^5 in the expansion of $\left(x+\dfrac{1}{x}\right)^7$.

(b) Find the term in x^4 in the expansion of $\left(x^2+\dfrac{2}{x}\right)^5$.

(c) Find the term in x^{-2} in the expansion of $\left(2x-\dfrac{3}{x^2}\right)^4$.

9. In each case, for the given binomial expression and the given term, find the value of k.

(a) $\left(\dfrac{1}{k}+kx\right)^5$; the coefficient of x^4 is 625

(b) $(k-x)^7$; the coefficient of x^3 is -560, $(k>0)$

(c) $(1+kx)^{12}$; the coefficient of x^7 is $\dfrac{11264}{243}$

10. (a) Expand $(1+y)^4$.

(b) By letting $y=x+x^2$, show that
$$(1+x+x^2)^4 = 1+4x+10x^2+16x^3+19x^4+16x^5+10x^6+4x^7+x^8.$$

Unit 1 Review Exercises

Part 1

No graphing calculators should be used to answer questions in Part 1.

1. Consider the arithmetic sequence $3,\ 5,\ 7,\ 9,\dots$
   (a) Find u_{51}
   (b) If $u_n = 87$ find n.

2. If a geometric sequence has first term 9 and fourth term $\dfrac{8}{3}$ find

   (a) the common ratio
   (b) S_∞, the sum to infinity.

3. (a) Find the sum of the infinite geometric series
   $$\text{(i)}\ \ 1+\frac{1}{2}+\frac{1}{4}+\frac{1}{8}+\dots \qquad \text{(ii)}\ \ 1-\frac{1}{2}+\frac{1}{4}-\frac{1}{8}+$$
   (b) Find the sum of the infinite geometric series
   $$\text{(i)}\ \ 1+x+x^2+x^3+\dots \qquad \text{(ii)}\ \ 1-x^2+x^4-x^6+x^8-\dots$$
   and, in both cases, state for what values of x the sum is valid.

4. In each case, find the exact value of x.
   (a) $\ 4^{x-1}=64$
   (b) $\ 3^{2-x}=81$
   (c) $\ 4^x=\dfrac{1}{32}$
   (d) $\ a^{3x-2}=\dfrac{1}{a}$

5. In each case, find the exact value(s) of x.
   (a) $\log_3(2x+1)=2$ (b) $\log_2(4x^2)=1$ (c) $\log_2(3x-2)=2\log_2 x$
   (d) $\log_a 5x = 3$

6. If $a=\log_2 x$ and $b=\log_2 y$ write the following in terms of a and b.

   (a) $\log_2(xy)$ (b) $\log_2\left(\dfrac{x^2}{y}\right)$ (c) $\log_2\left(2\sqrt{xy}\right)$

   (d) $\log_4(xy)$ (e) $\log_x(xy)$

7. Show that the expression $(x+y)^4 - (x-y)^4 = 8xy(x^2+y^2)$.

8. Complete the expansion of the binomial expression $(x-ky)^5 = x^5 - 5kx^4y + \ldots - k^5y^5$ by finding the three missing terms.

9. (a) Expand $(x-2)^3$ and simplify your result.

(b) Hence find the x^3 term in the expansion of $(x-2)^3(3x+1)$

Part 2

Graphing calculators will usually be needed to answer questions in Part 2.

10. An arithmetic series given by $\displaystyle\sum_{r=1}^{n}(3r+5)$

(a) Write down the first four terms of this series
(b) Find the sum of the first 15 terms
(c) Find, S_n the sum of the first n terms of the series.

(d) Find the smallest value of n which satisfies $\displaystyle\sum_{r=1}^{n}(3r+5) > 200$

11. If $S_n = 3n^2 + 7n$,

(a) find S_6, the sum of the first six terms.
(b) find S_7, the sum of the first seven terms.
(c) Use your answers from (a) and (b) to find u_7, the value of the seventh term.

12. Let an arithmetic sequence be 8, 13, 18, 23, ...
Find (a) u_{31}, the value of the thirty first term

(b) S_{20}, the sum of the first 20 terms.

13. An arithmetic sequence has a first term of -4 and a common difference of 3.
(a) Show that the eighteenth term, u_{18} is 47.
(b) Find the value of the lowest term which is greater than 300.

14. Consider the geometric series $8 - \dfrac{16}{3} + \dfrac{32}{9} - \dfrac{64}{27} + \ldots$. Find

(a) correct to three decimal places, the value of u_{10} the tenth term.
(b) correct to three decimal places, S_{10}, the sum of the first ten terms.
(c) S_∞, the sum to infinity.

15. Jeffrey invests $8000 in a bank which offers an interest rate of 5.4% per annum (year). Find

(a) the interest rate per month
(b) the value of the investment after 5 years if
      (i) the interest is compounded yearly.
      (ii) the interest is compounded monthly.

16.   An island has a population of 500 rats which increase at 3% every year.
    (a) Find the rat population after 10 years.

    Now suppose 100 rats are killed at the end of each year.
    (b) After how many complete years will there be no more rats?

17.   Consider the expansion of $(2x+y)^{12}$.
    (a) Write down the number of terms in the expansion.

    The first three terms in the expansion are $4096x^{12}+ax^{11}y+bx^{10}y^2+\ldots$.
    (b) Find the values of a and b.

    A term in the expansion is kx^6y^6, where k is a constant.
    (c) Find the value of k.

18.   Consider the expansion of $\left(x+\dfrac{1}{y}\right)^9$.

    (a) Show that one of the terms in the expansion of $\left(x+\dfrac{1}{y}\right)^9$ is $126\dfrac{x^5}{y^4}$.

    (b) If $y=x$ find the term in x^3.

19.   Juan invests 8000 pesos at an annual rate of 6.23 percent, compounded yearly.
    (a) Find, correct to the nearest peso, the value of Juan's investment after 10 years.
    (b) How many complete years are needed for Juan's initial investment to exceed 50 000 pesos?
    (b) Find the interest rate which would make Juan's initial investment worth 40 000 pesos after 10 years.

UNIT 2: FUNCTIONS AND EQUATIONS

2.1 The Nature of a Function

The following six examples are all *functions*.

1) A company manufactures batteries. The cost of producing each battery is $1.20. Then the total cost, C, of producing batteries is a function of the number, n, of batteries produced.

2) A function behaves rather like a machine. Figure 2.01 shows an 'Add 3' machine. The function accepts input of a number x and outputs the number $x+3$.

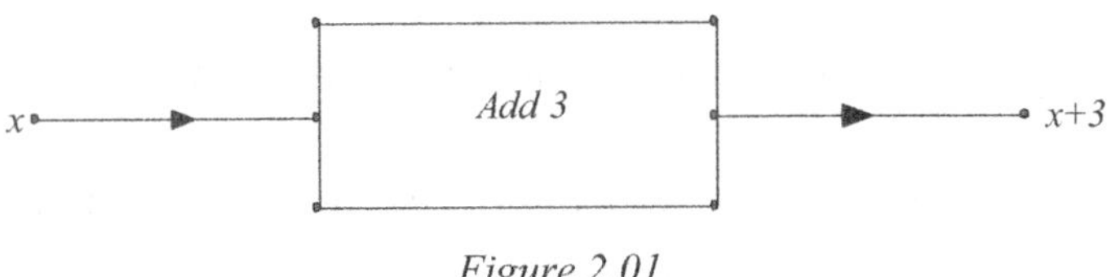

Figure 2.01

3) $y = x^2 + 7$ is a function which assigns to each value of x one and only one value of y, according to the given formula so that when $x = 1$, $y = 8$, and when $x = 4$, $y = 23$ and so on.

4) Figure 2.02 shows a function expressed graphically. Each value of x on the x-axis is assigned to one, and only one, point on the y-axis according to the position of the point P whose coordinates are (x_1, y_1).

5) A function can be illustrated diagrammatically as shown in figure 2.03. Each element in set A is mapped to one and only one element in set B. The arrowed lines show the *mappings*.

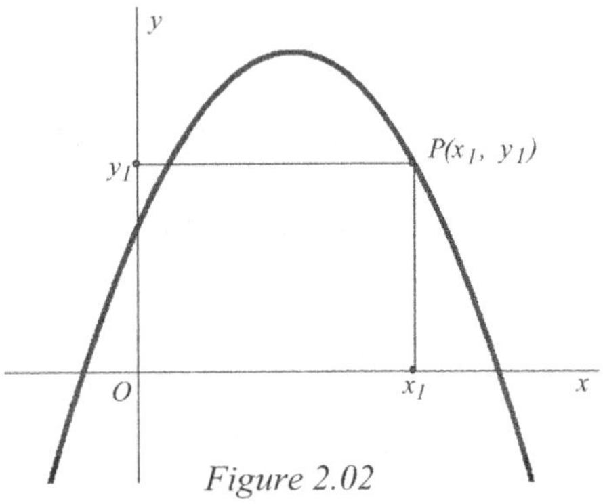

Figure 2.02

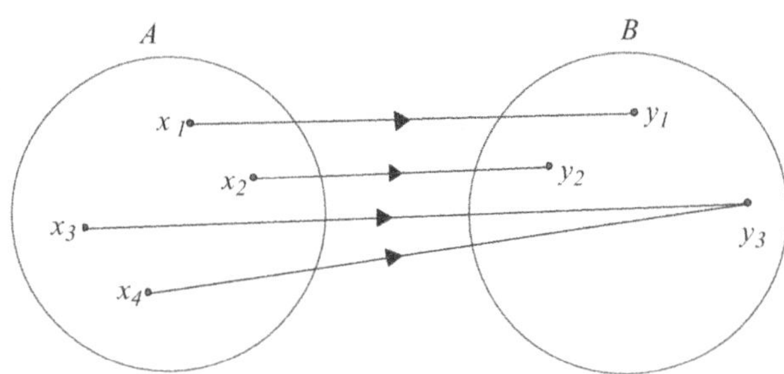

Figure 2.03

6) A game is played in which a regular six-sided die is thrown. If a 'one', 'two' or 'three' is obtained, then the player wins nothing, but if a 'four' or 'five' is obtained, the player wins $1. If a 'six' is obtained, then the player wins $5. This situation can be represented by a function f. Elements from the set $A = \{1, 2, 3, 4, 5, 6\}$ (the possible scores on the die) are mapped by f to elements of set $B = \{0, 1, 5\}$ (the possible winning amounts in dollars). f defines a function between the elements in A and the elements in B such that $f(1) = f(2) = f(3) = 0$

$f(4) = f(5) = 1$ and $f(6) = 5$.

The following four examples are <u>not</u> functions.

7) $y^2 = x$. Each value of x, except $x = 0$, is assigned to two values of y. For example, when $x = 9$, $y = -3$ or $y = 3$; therefore, x is mapped to two values of y.

8) Figure 2.04 shows a graph of $y = \dfrac{1}{x-2}$ in which all values of x, except $x = 2$, are mapped to a value of y. 2 is not mapped to any value of y.

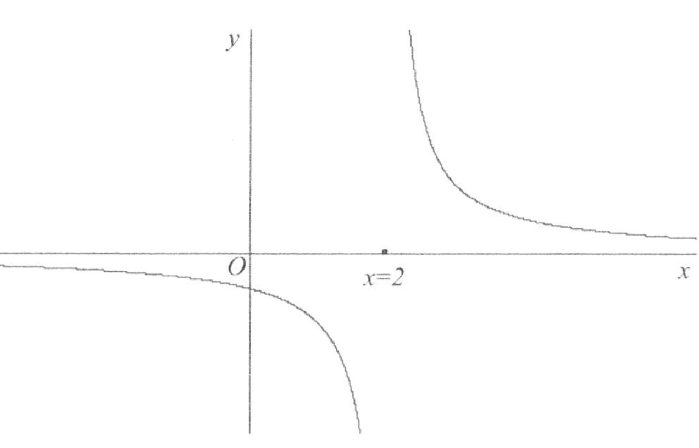

Figure 2.04

9) Figure 2.05 shows a diagram in which elements in set A are mapped to elements in set B. The element x_3 is not mapped to any element in set B, and the element x_4 is mapped to two elements in set B.

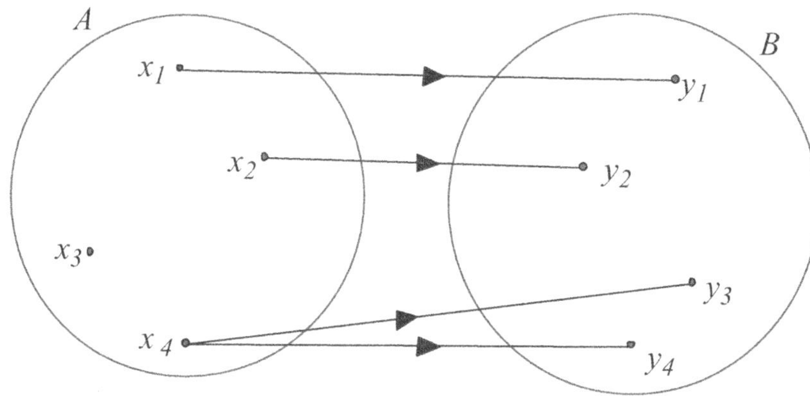

Figure 2.05

10) Figure 2.06 shows a graph in which the x values less than 4 are mapped to two values of y and x values greater than 4 are mapped to no values of y.

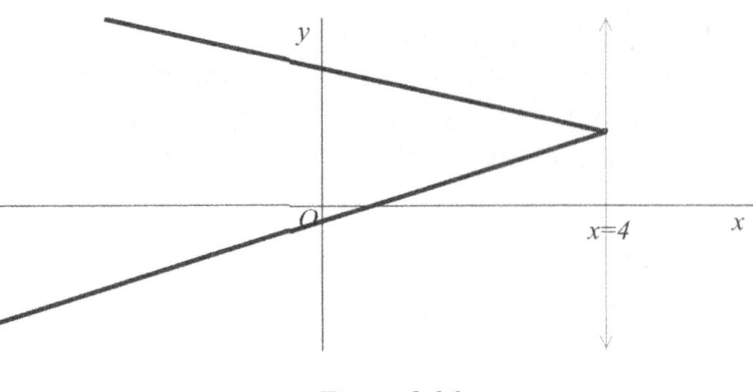

Figure 2.06

What are the common features that make examples 1) to 6) functions and examples 7) to 10) not functions?

All ten examples have a set of x values or equivalent, which we call the *domain*, and a set of y values or equivalent, which we call the *range*. In examples 1) to 6), all the elements of the domain are mapped to one, and only one, element of the range. However, in examples 7) to 10) either an element in the domain maps to two or more elements in the range or an element of the domain is not mapped to any element in the range. Example 8 can be made into a function by removing $x = 2$ from the domain so that now all the x values are mapped to one, and only one, y value.

A function, f, is a mapping from the domain set A to the range set B in which an element, $x \in A$, is mapped to $f(x) \in B$. $f(x)$ is called the *image* of x. The situation with regard to example 8 makes it clear that a function is not fully stated unless its domain is given. For example, the function $y = 3x^2 + 4, \ -2 \le x \le 6$ gives the rule that maps the x values to the y values, and, in addition, it states the domain which is the values of x between -2 and 6. If no statement of the domain is given, it is assumed to be $\mathbb{R}$, all the real numbers.

A function can be written using a variety of notations. For example, the function f that assigns x to $3x^2 - 4$ for the domain $x \in \mathbb{R}$ can be written in three different ways:

$$(1) \quad f : x \mapsto 3x^2 - 4$$
$$(2) \quad f(x) = 3x^2 - 4$$
$$(3) \quad y = 3x^2 - 4$$

Notation (1) is usually used when explaining the theory of functions but is not a suitable format for carrying out algebra. Notations (2) and (3), on the other hand, are suitable formats for doing algebraic manipulation. Notation (3) is particularly useful when describing a function graphically.

Example 2.1: Identify which of the following relations are functions. For those which are functions, state the range. For those which are not, explain why they fail to qualify.

(a) $\quad f : x \mapsto \sin x, \ -\pi \le x \le \pi, \ x \in \mathbb{R}$

(b) $\quad g : x \mapsto \dfrac{1}{x}$

(c) $\quad h : x \mapsto$ the largest integer less than or equal to x

(d) $\qquad y = \log_{10} x, \ x > 0$

(e) $\qquad x^2 + y^2 = 4, \ -2 \le x \le 2$

(f) $\qquad y = 1$

Solution 2.1: Remember that if the domain is unstated then you should assume it is all the real numbers. It is necessary to establish that each element of the domain has a unique image if the relation described is a function.

(a) A function whose range is $(-1, \ 1)$.

(b) Not a function because 0 is an element of the domain but $f(0)$ does not exist because $\dfrac{1}{0}$ is undefined.

(c) A function whose range is the integers, $\mathbb{Z}$.

(d) A function whose range is the real numbers, $\mathbb{R}$.

(e) Not a function because some values of the domain do not have a unique image but have two images. For example, when $x = 0$, $y = 2$ and $y = -2$.

(f) A function whose range is $\{1\}$.

Exercise 2.1

1. Explain why each of the following relations denoted by k fails to be a function.

(a) $k : x \mapsto \dfrac{x+1}{x-1}$

(b) $k : x \mapsto \pm\sqrt{4x}, \ x > 0$

(c) $k : x \mapsto \sqrt{2-3x}$

(d) For $n \in \mathbb{Z}^+$, k assigns n to the largest prime number less than n

(e) For $n \in \{0, 1, 2, \ldots, 10\}$, k assigns to each n the factors of n

(f) For $x \in \mathbb{R}, \ 0 \le x \le 1$ k assigns to each x the integer closest to x

2. If $f : x \mapsto 1 - 3x$, find (a) $f(0)$ (b) $f(1)$ (c) $f(5)$

3. For the function $f : x \mapsto x^2 + 4x - 12$, find the image of (a) 3 (b) 2 (c) $-\dfrac{2}{5}$ (d) $\sqrt{10}$

4. If $f(x) = \dfrac{1+x}{x^2}, \ x \ne 0$ find (a) $f(-3)$ (b) $f(2)$ (c) $f(10)$

5. Find the range of f in each case, given that $x \in \mathbb{R}$.

(a) $f(x) = 2x + 3, \ -5 \le x \le 5$ (b) $f(x) = 5 - x, \ 0 < x < 10$

(c) $f(x) = x^2, \ -10 \le x \le 10$ (d) $f(x) = 1 - 2x^2, \ -1 < x < 1$

6.　　f is the function which assigns to each value of the domain, $\mathbb{R}$, the value of x given correct to three significant figures. For example, $f(2.9876) = 2.99$. Find

(a) $f(0.0036577)$　　　　(b) $f\left(\sqrt{3}\right)$　　　　(c) $f\left(\log_{10} 5\right)$　　　　(d) $f\left(3^{-0.2}\right)$

7.　　f is the function which assigns to each value of the domain x where $x \in \mathbb{R}$, the largest integer, $\mathbb{Z}$, less than or equal to x. For example, $f(5.2) = 5$. Find

(a) 7.6　　　　(b) π　　　　(c) 0.038　　　　(d) 6　　　　(e) -2.25

2.2 Use of the Graphing Calculator

2.2.1 Obtaining the Graph of $y = f(x)$ on the Calculator Screen

It is expected that you will be able to draw graphs of functions like the quadratic function and the exponential function, which are specifically mentioned in the course, but it is also expected that you will be able to draw graphs of many other functions as well. This can be done by first obtaining a suitable graph of the function on the screen of your graphing calculator and then by making a copy of this graph on paper.

The following are three examples to explain how to obtain a screen display of a function. However, it is important to realize that the graphing calculator screen usually does not have good definition and so a certain level of intelligent interpretation is needed in order to be able to make a satisfactory copy of the graph of a function.

1) $y = \sin(1 - \cos x)$,　$0 \le x \le 2\pi$

The domain indicates that x should be in radians rather than degrees, so the mode of your graphing calculator should be changed accordingly. You can set the x values of the window at 0 and 2π, as indicated by the domain.

In addition, as the range of the cosine function is $-1 \le \cos x \le 1$, the range of $\sin(1 - \cos x)$ will be $\sin 0 \le y \le \sin 2$, so that you know that y is always positive and you can set the y values of the window at $0 \le y \le 1.1$. Of course, you might find it easier to use trial and error to find the best range for y, but it can be useful to use your mathematical knowledge to find a suitable window for your graph. Figure 2.07 shows a suitable window, and figure 2.08 shows the graph.

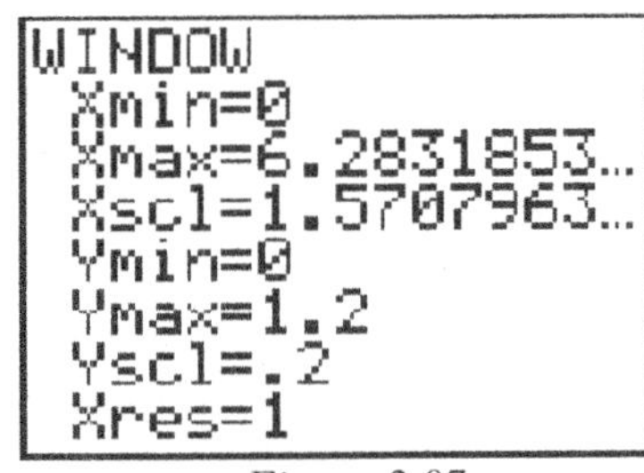

Figure 2.07

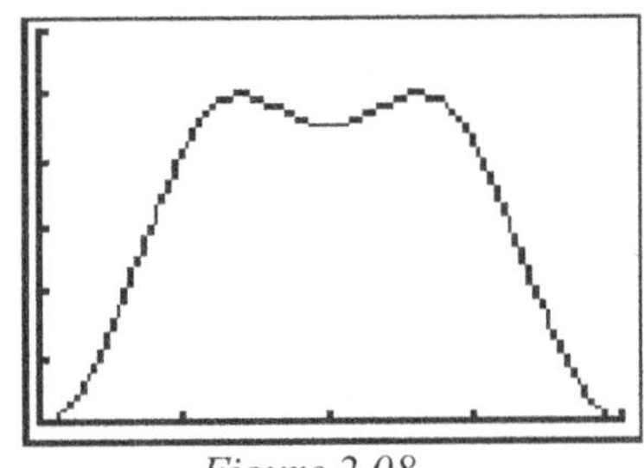

Figure 2.08

2) $y = x^4 - 8x^3 + 140$

In this case the domain is all real numbers $\mathbb{R}$, and, as you cannot include the entire domain in the graph, you need to choose that part of the domain which shows all the interesting features of the graph. You can do this by using a standard domain, $-10 \leq x \leq 10$ (figure 2.09), and a standard range, $-10 \leq x \leq 10$, and then editing the window to produce a curve which shows everything of interest.

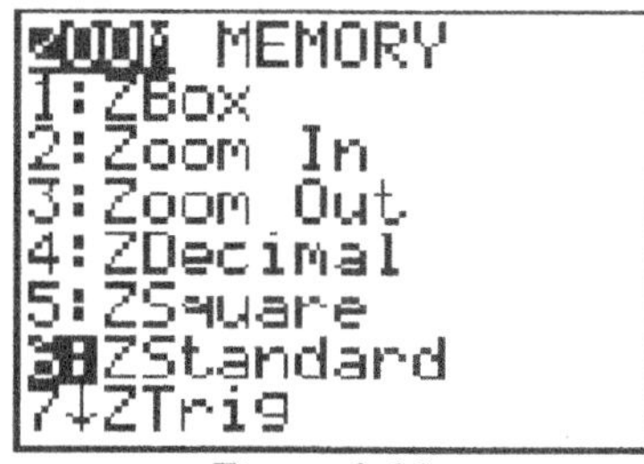

Figure 2.09

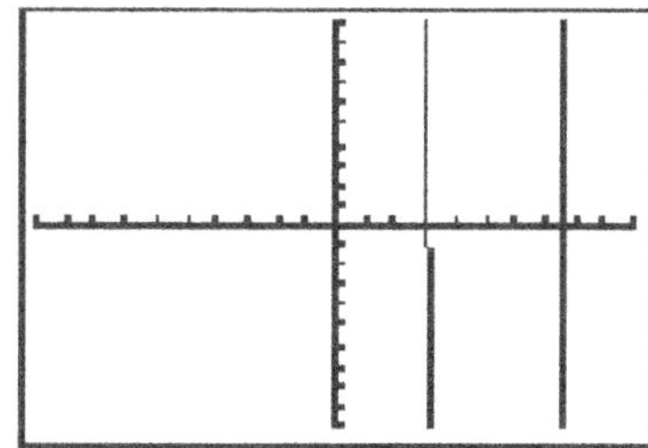

Figure 2.10

The screen (figure 2.10) shows only a portion of the curve, and it is necessary to increase the y interval.

Figures 2.11 and 2.12 show a new window and graph in which the y-interval is increased.

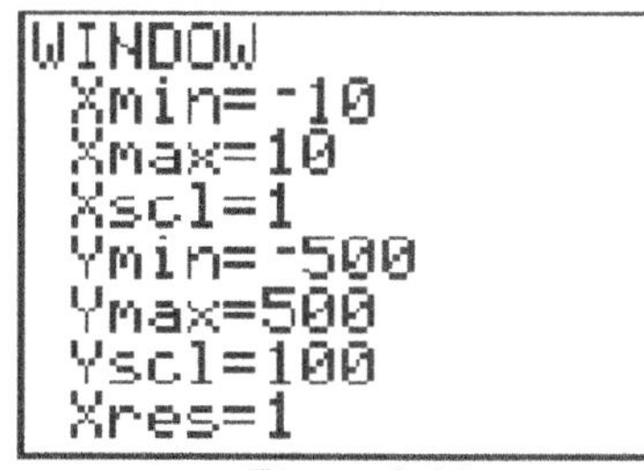

Figure 2.11

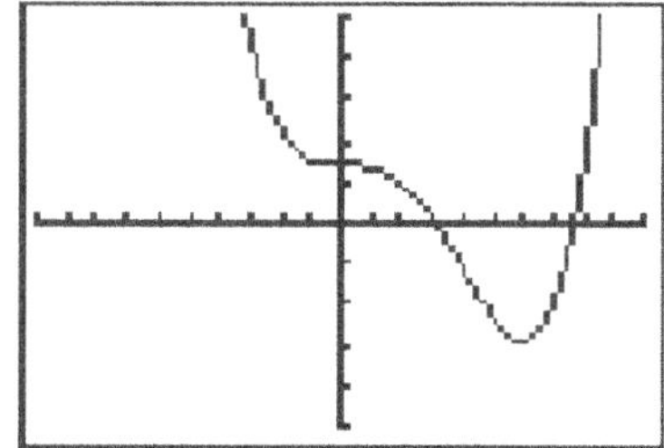

Figure 2.12

The important features of the curve are now displayed and it remains only to make minor adjustments of the window (figures 2.13 and 2.14) to get the best view possible.

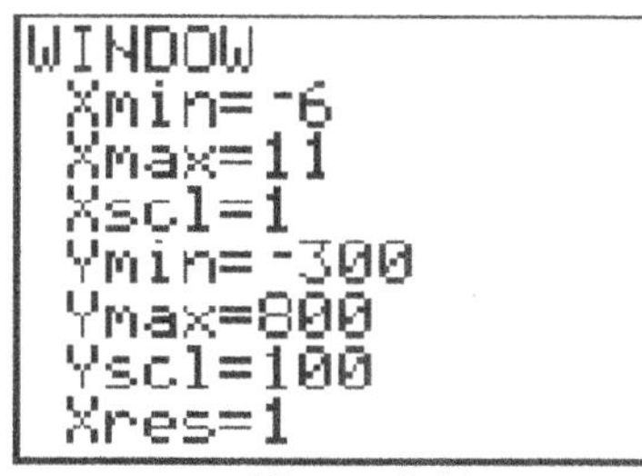

Figure 2.13

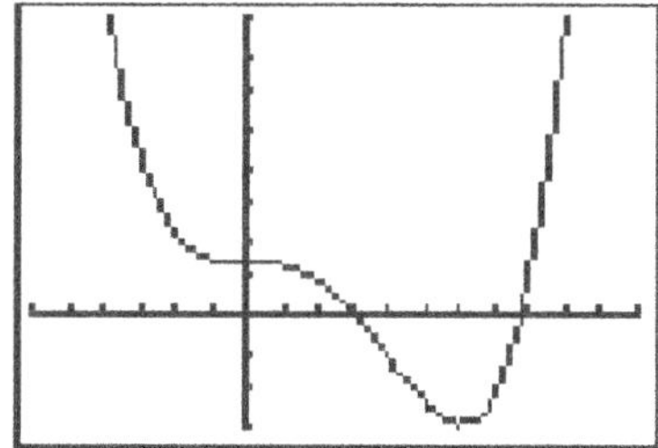

Figure 2.14

Of course, unless you know roughly what to expect, there is no real way of knowing if your window includes all important features of the curve and knowing what to expect only comes with experience.

3) $y = \dfrac{1}{x(1-x)}, \quad x \neq 0, x \neq 1, \ -1 \leq x \leq 2$

It is clear that the x values of the window should be -1 to 2, as indicated by the domain. The range of this function is $\mathbb{R}$, so it is not clear what window for y you should choose. Trial and error indicates that a suitable interval for y is $(-10, 10)$ (figures 2.15 and 2.16).

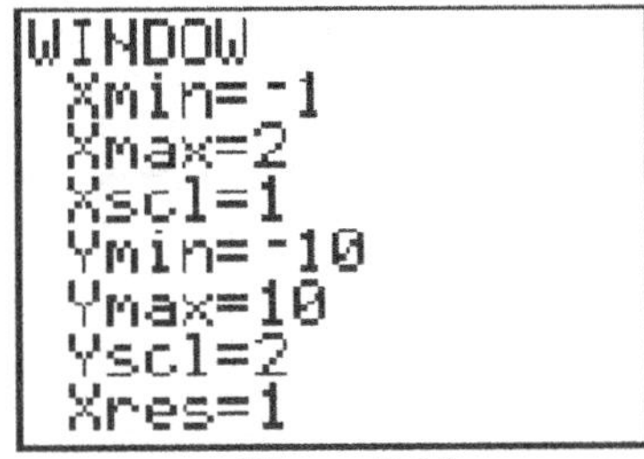

Figure 2.15

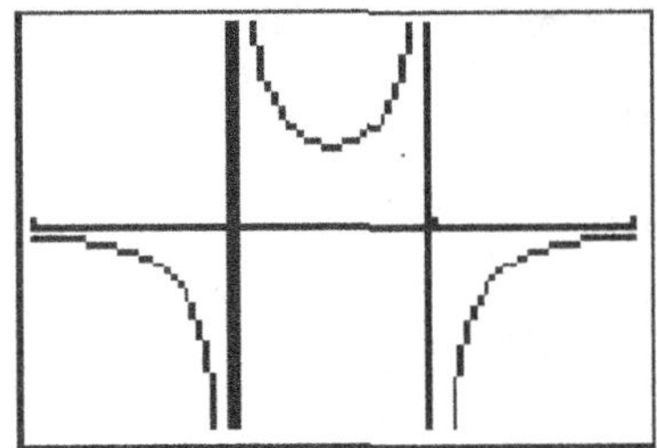

Figure 2.16

However, the graph shows some vertical lines at $x = 0$ and $x = 1$ which are not part of the graph. These lines are called vertical *asymptotes*.

2.2.2 The Reciprocal Function and Asymptotes

The graph of the reciprocal function

$$f(x) = \dfrac{1}{x}, \ x \neq 0 \text{ is shown in figure 2.17.}$$

There is a discontinuity at $x = 0$. As the curve approaches $x = 0$, it is 'pushed' towards large positive or large negative values of y. In fact, we should not be surprised that this happens because as

$x \to 0$, $\dfrac{1}{x}$ becomes very large positive if

x is positive and very large negative if x is negative. When $x = 0$, there is no image

because $\dfrac{1}{0}$ is undefined. We call the

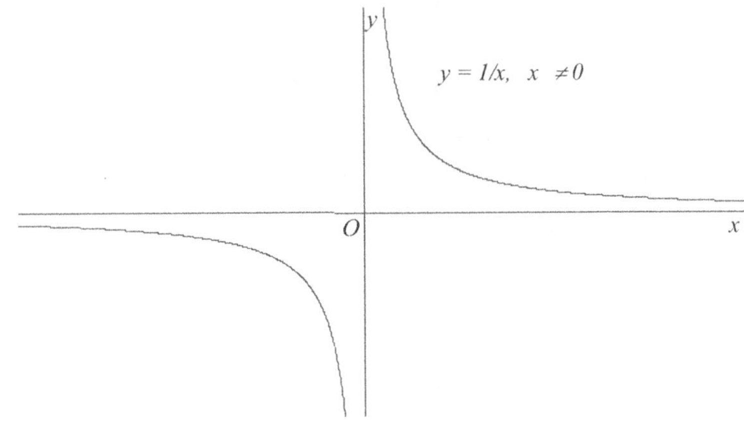

Figure 2.17

vertical line where the function has no image a *vertical asymptote*. A vertical asymptote is a line parallel to the y-axis which occurs at values of x where the denominator of $y = f(x)$ is zero.

For $y = \dfrac{3x+1}{2x-1}$, there is a vertical asymptote at $x = \dfrac{1}{2}$ because when $2x - 1 = 0$, $x = \dfrac{1}{2}$. The vertical asymptote acts rather like a barrier which separates the graph of the function into different sections. The asymptote is not part of the graph of the function, and the only reason that these lines sometimes appear is that if the calculator is in 'connected' mode, wherein it will try to make the curve continuous. Figure 2.18 shows how to put the calculator into 'dot' mode, and figure 2.19 shows the graph without the asymptotes.

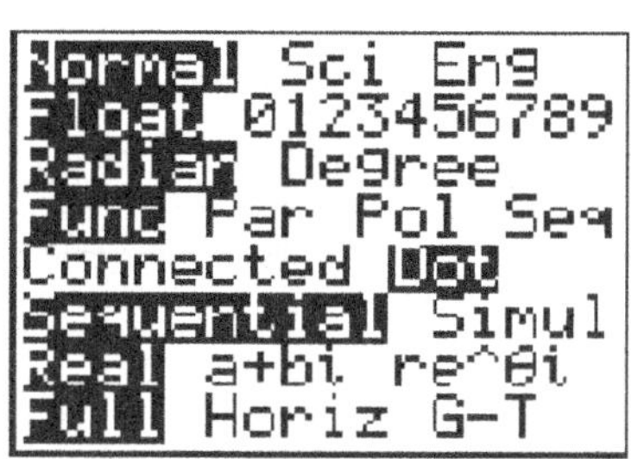

Figure 2.18

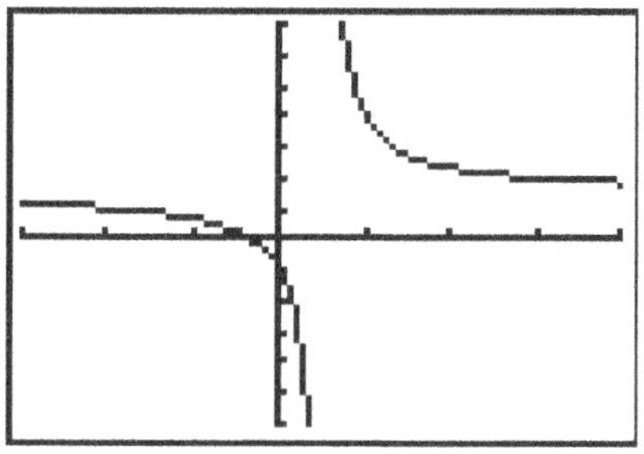

Figure 2.19

A *horizontal asymptote* is a line parallel to the x-axis occurring at that value of y which the curve approaches as x gets large positive or large negative. The graph of $y = \dfrac{3x+1}{2x-1}$ has a horizontal asymptote at $y = \dfrac{3}{2} = 1.5$. The equation of the horizontal asymptote can be found by considering what happens to the value of the function when x gets large. As x gets large, we write $x \to \infty$, and $\dfrac{1}{x}$ gets small and we write $\dfrac{1}{x} \to 0$. By rewriting the function in the form $y = \dfrac{3 + \frac{1}{x}}{2 - \frac{1}{x}}$ then, as $x \to \infty$, $\dfrac{1}{x} \to 0$ so that the horizontal asymptote is $y = \dfrac{3}{2} = 1.5$. Alternatively the equation of the horizontal asymptote can be obtained using the table on your graphing calculator to find what value the function approaches as x gets large.

Figures 2.20 and 2.21 show this method applied to $y = \dfrac{3x+1}{2x-1}$. As you scroll down the table, it is easy to see that the horizontal asymptote is $y = 1.5$.

Figure 2.20

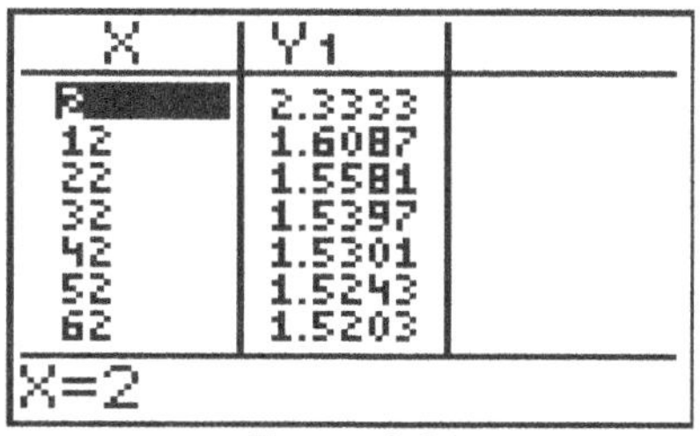

Figure 2.21

2.2.3 Solution of $f(x) = 0$ to a Given Accuracy

The graphing calculator can be conveniently used to find the approximate solution of the equation $f(x) = 0$ to a given level of accuracy. When the graph of $y = f(x)$ cuts the x-axis, $y = 0$. Therefore, at this value of x, $f(x) = 0$. Suppose that $f(x) = x^4 - 8x^3 + 140$; now you can solve the equation $x^4 - 8x^3 + 140 = 0$, for example, by finding where the graph of $y = f(x)$ cuts the x-axis.

Figures 2.22 to 2.26 show a sequence of calculator screen displays which demonstrate how to find the smallest positive solution of the equation $x^4 - 8x^3 + 140 = 0$, correct to four decimal places.

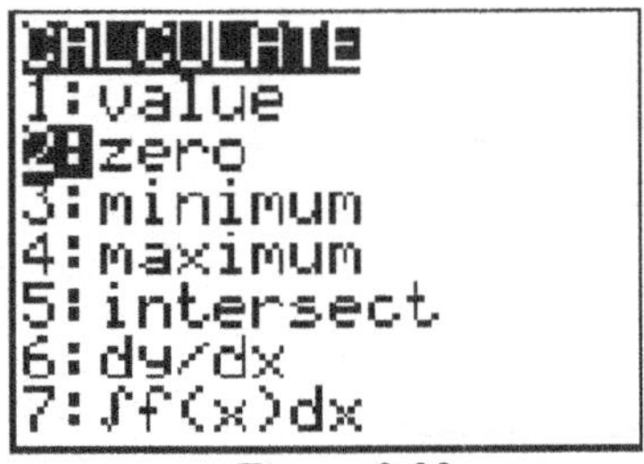

Figure 2.22

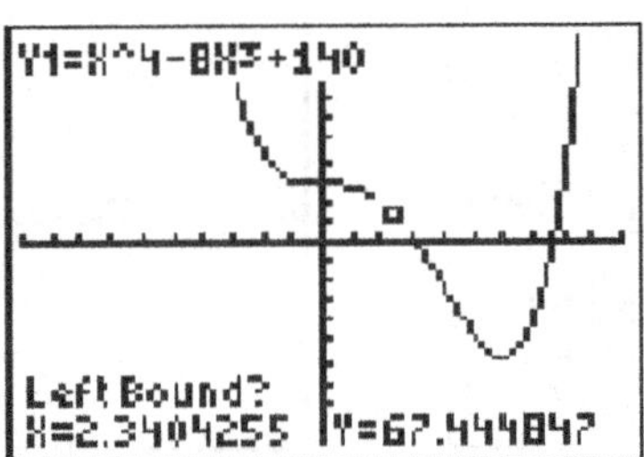

Figure 2.23

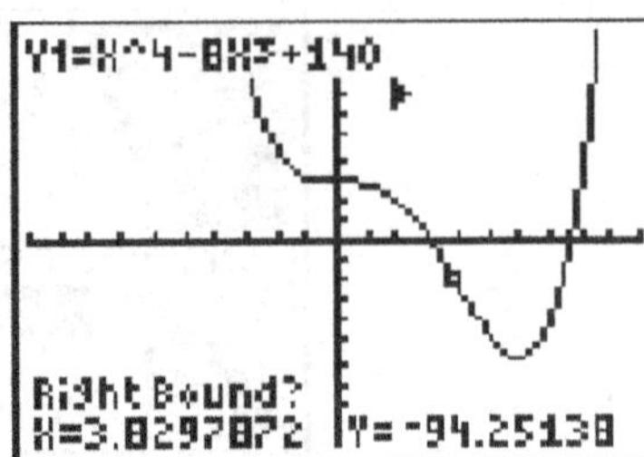

Figure 2.24

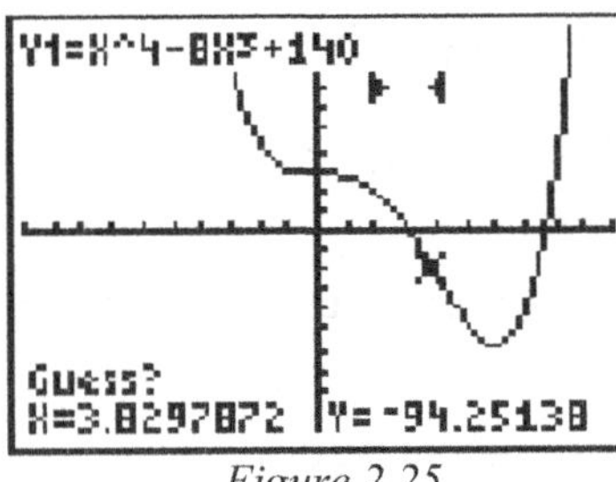

Figure 2.25

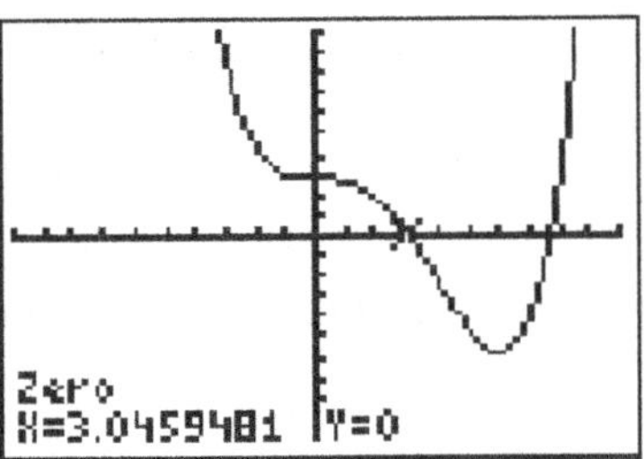

Figure 2.26

Therefore, the required solution is $x = 3.0459$ correct to four decimal places.

2.2.4 Drawing a Neat Sketch of the Graph of a Function

You are expected to be able to sketch a reasonably neat and large graph of a function on paper. It is generally anticipated that you will first obtain a screen display of the function on your graphing calculator before copying it on to paper. The following gives suggestions for sketching a useful graph:

- axes should be drawn with a straight edge, preferably a ruler
- axes should be labeled appropriately
- some indication of scale should be given on each axis
- sketches should be at least ¼ page in size
- the curve should be drawn roughly to scale and show the given information
- sketches of graphs should display any symmetry possessed by the function being graphed
- a pencil rather than a pen should be used to sketch the graph so that parts of the curve with which you are dissatisfied after the first attempt can be erased easily and improved upon

Exercise 2.2

Some of the functions appearing in this exercise may be unfamiliar to you, but understanding their relevance is less important at this stage than using them as a vehicle for gaining experience in obtaining graphs on your calculator screen and sketching them.

1.   Use your graphing calculator to obtain, over an appropriate portion of the domain, the graph of the following functions, and then make a sketch of each graph.

     (a) $y = x^2 - 9$ (b) $y = -11 - 5x^2$ (c) $y = x^3 - x - 15$

     (d) $y = x^3 + 8x^2 - 11x + 15$ (e) $y = 3x^4 - 5x^3 + x - 15$ (f) $y = 10x + 4x^2 - \dfrac{x^3}{2} - \dfrac{x^4}{18}$

2. Obtain the graph of the following functions for the given domain on your graphing calculator and then, in each case, make a sketch of the graph.

(a) $y = 2\cos(2 - \sin x),\ 0 \le x \le 2\pi$

(b) $y = e^{1-\sin x},\ 0 \le x \le 10$

(c) $y = \ln(x^2 + 1),\ -5 \le x \le 5$

(d) $y = 8xe^{1-x},\ 0 \le x \le 5$

(e) $y = \sin(e^{0.7x} - 1),\ 0 \le x \le \pi$

(f) $y = e^{-x}\sin 2x,\ 0 \le x \le \pi$

(g) $y = \dfrac{x^2 - 1}{x^2 + 1},\ -4 \le x \le 4$

(h) $y = x\sin 2x,\ 0 \le x \le 2\pi$

3. In each case find the equations of (i) vertical asymptotes and (ii) horizontal asymptotes. Then obtain the graph of the following functions for the given domain on your graphing calculator. Finally draw, as dotted lines, the asymptotes then make a neat sketch of the graph for the given domain.

(a) $f(x) = \dfrac{2}{1 - x},\ x \ne 1,\ -4 \le x \le 6$

(b) $f(x) = \dfrac{x^2}{1 - x^2},\ x \ne \pm 1,\ -5 \le x \le 5$

(c) $f(x) = \dfrac{2x + 1}{3x - 5},\ x \ne \dfrac{5}{3},\ -2 \le x \le 5$

(d) $f(x) = 4 - \dfrac{x}{x + 2},\ x \ne -2,\ -8 \le x \le 6$

(e) $f(x) = \ln|x|,\ x \ne 0,\ -4 \le x \le 4$

(f) $f(x) = \dfrac{x^2 + x}{x^2 + 1},\ -10 \le x \le 10$

4. Figure 2.27 shows the graph of $f(x) = \dfrac{x + b}{x - a},\ x \ne a,\ a, b \in \mathbb{Z}$

The graph cuts the x-axis at the point $(-2,\ 0)$ and the vertical asymptote has equation $x = 3$. The equation of the horizontal asymptote is $y = k$.
(a) Find the value of k
(b) Find the value of b
(c) Find the coordinates of the y-intercept.

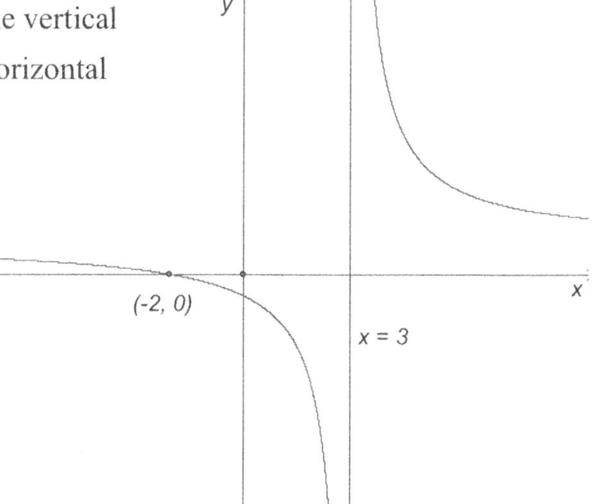

Figure 2.27

5. Let f be the function $f(x) = \dfrac{ax + 2}{x + b},\ x \ne b$
(a) Find, in terms of a and b, the equation of
    (i) the vertical asymptote
    (ii) the horizontal asymptote

Given that $b = -1$ and the graph of f passes through the point $(5,\ 3)$
(b) find the value of a

6. Let $f(x) = \dfrac{2x^2 - 7}{x^2 + k}$, $k \geq 0$

   (a) Explain why the graph of f has no vertical asymptote.

   (b) Find the equation of the horizontal asymptote.

If $k = 4$

   (c) State the range of f.

   (d) Write down the number of solutions of the equation $f(x) = 0$

7. Let $f(x) = \dfrac{x^2 + 2k}{(x-k)^2}$, $x \neq k$

   (a) Show that the equation of the horizontal asymptote is $y = 1$.

Given that the point P with coordinates $(0,\ 1)$ lies on the graph of f

   (b) Find the value of k, given that $k \neq 0$.

   (c) Make a sketch of the graph of f.

2.3 Transformations of Graphs

A *transformation* is a function whose domain and range are sets of points. Translations, stretches and reflections are all examples of transformations. In this section's figures, the graph drawn with a thick line is the image of the graph drawn with the thin line.

2.3.1 Translations

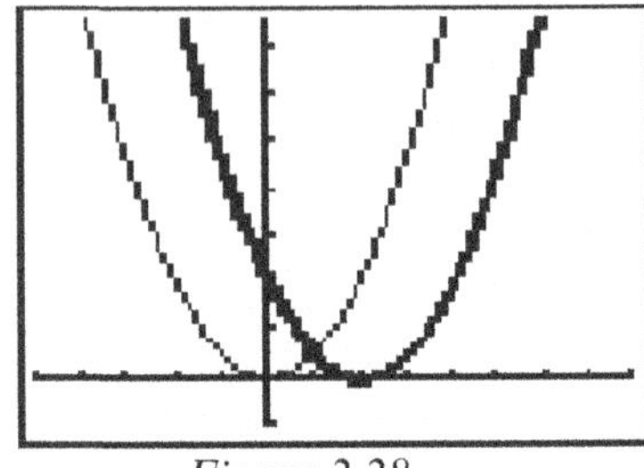 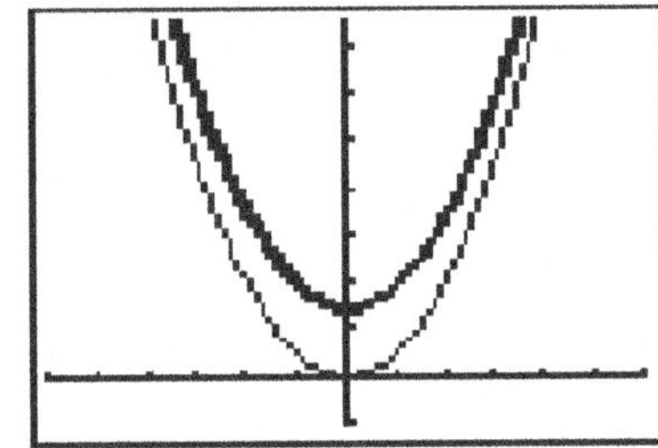 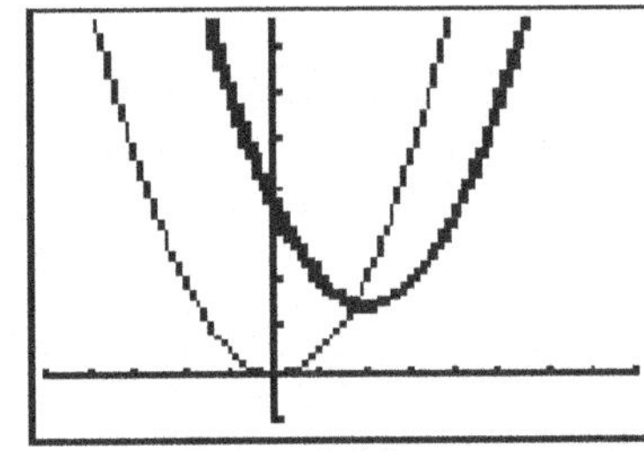

Figure 2.28   *Figure 2.29*   *Figure 2.30*

Figure 2.28 shows the graphs of the functions $y = x^2$ and $y = (x-2)^2$. The graph of $y = x^2$ has been slid, we use the word translated, 2 units in the positive x direction.

Figure 2.29 shows the graphs of the functions $y = x^2$ and $y = x^2 + 3$. The graph $y = x^2$ has been translated 3 units in the positive y direction.

Figure 2.30 shows the graphs of $y = x^2$ and $y = (x-2)^2 + 3$. The graph $y = x^2$ has been translated 2 units in the positive x direction and 3 units in the positive y direction. A more concise notation is to use the *column vector* $\begin{pmatrix} 2 \\ 3 \end{pmatrix}$ to express the translation (see Unit 4).

More generally, a translation of a in the positive x direction and b in the positive y direction transforms the graph of $y = f(x)$ to the graph of $y = f(x-a) + b$.

Example 2.2: In Figure 2.31, what column vector translates the parabola with vertex A to the parabola with vertex B?

Solution 2.2: The translation in the positive x direction is $3 - (-1) = 4$, and the translation in the positive y direction is $1 - 6 = -5$. So, expressed as a column vector, the translation is $\begin{pmatrix} 4 \\ -5 \end{pmatrix}$.

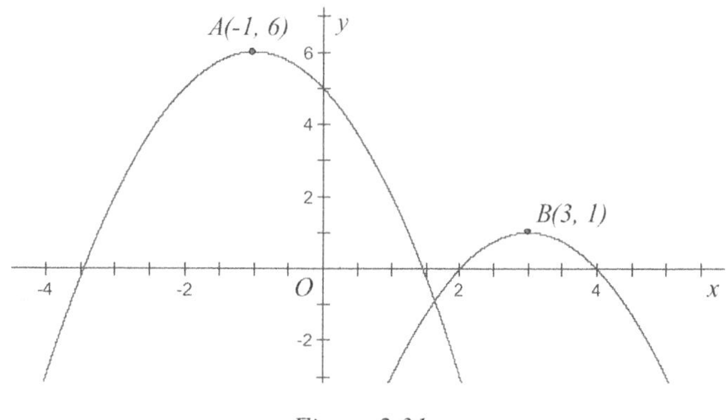

Figure 2.31

Example 2.3: The graph of $y = x^2$ is translated 5 units in the positive x direction and -4 units in the positive y direction. Find the equation of the transformed graph.

Solution 2.3: $y = x^2$ is transformed by a translation $\begin{pmatrix} a \\ b \end{pmatrix}$ to the graph whose function is $y = (x-a)^2 + b$. As $a = 5$ and $b = -4$, the required function is $y = (x-5)^2 - 4$.

2.3.2 Stretches

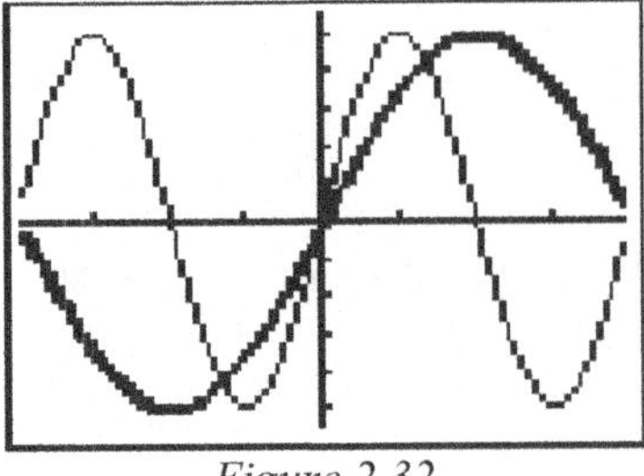

Figure 2.32

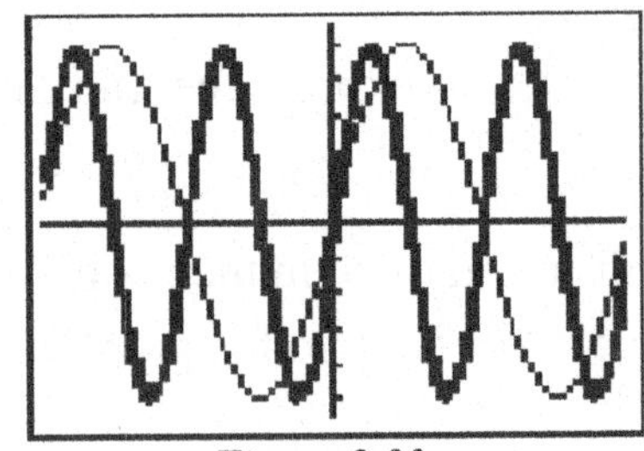

Figure 2.33

Figure 2.32 shows the graphs of the functions $y = \sin x$ and $y = \sin \dfrac{x}{2}$. The graph of $y = \sin x$ has

been stretched parallel to the x-axis so that a point on $y = \sin \dfrac{x}{2}$ is now twice as far from the y-axis

as its corresponding point on $y = \sin x$. This is called a stretch of factor 2 parallel to the x-axis.

Figure 2.33 shows the graphs of the functions $y = \sin x$ and $y = \sin 2x$. The graph of $y = \sin x$ has
been compressed parallel to the x-axis so that a point on $y = \sin 2x$ is now half the distance from the

y-axis as its corresponding point on $y = \sin x$. So $y = \sin x$ has undergone a stretch of factor $\dfrac{1}{2}$

parallel to the x-axis.

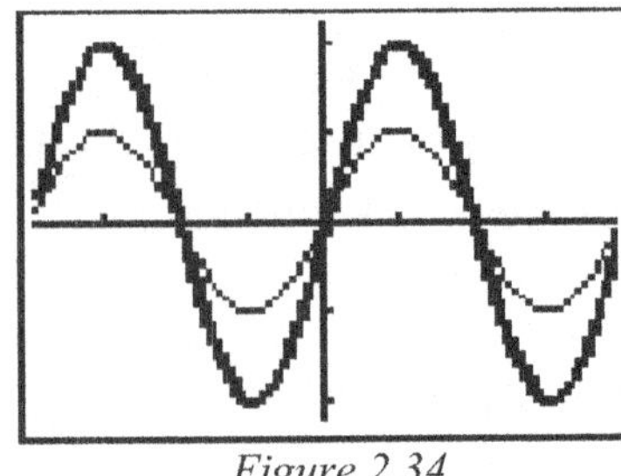

Figure 2.34

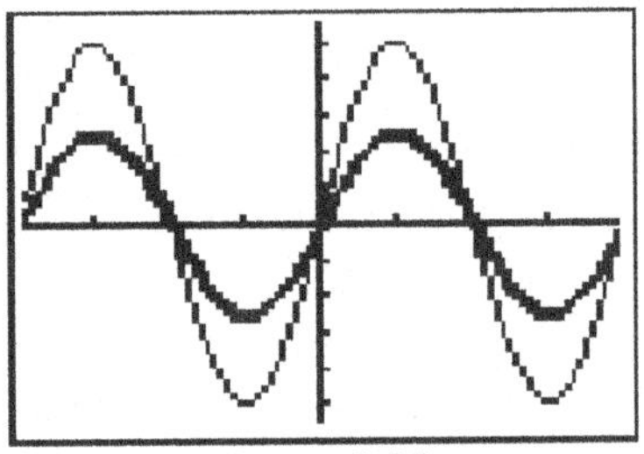

Figure 2.35

Figure 2.34 shows the graphs of the functions $y = \sin x$ and $y = 2\sin x$. The graph of $y = \sin x$ has
been stretched parallel to the y-axis by a factor of 2.

Figure 2.35 shows the graphs of the functions $y = \sin x$ and $y = \dfrac{1}{2}\sin x$. The graph of $y = \sin x$ has

been stretched parallel to the y-axis by a factor of $\dfrac{1}{2}$.

Example 2.4: Find the transformation which takes the point $P(1.6,\ 1)$ on curve A to the point $Q(4.8,\ 1)$ on curve B. Curves A and B are shown in Figure 2.36.

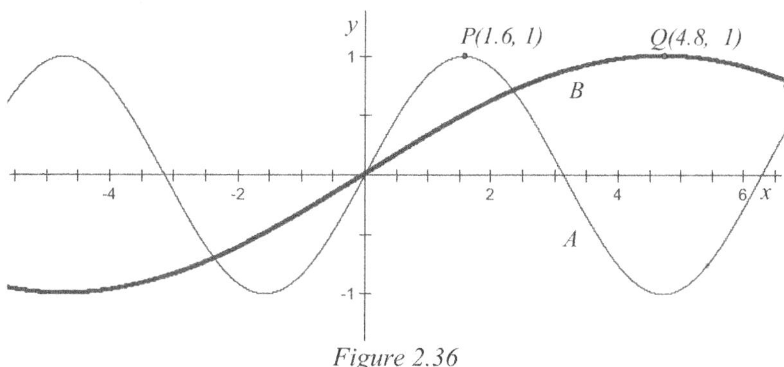

Figure 2.36

Solution 2.4: The transformation is a stretch parallel to the x-axis. The point Q is three times as far from the y-axis as P and therefore the stretch factor is 3.

Example 2.5: Draw the graph of $f(x) = x^2$, and then, on the same axes, draw the graph of the function g, which is obtained from f by a stretch of factor 3 parallel to the y-axis. Then, again on the same axes, draw the graph of the function h, obtained from g by a translation of -2 parallel to the x-axis. Find h.

Solution 2.5: The graphs of $y = f(x)$, $y = g(x)$ and $y = h(x)$ are shown in figure 2.37.

$$f(x) = x^2 \Rightarrow g(x) = 3x^2 \Rightarrow h(x) = 3\left(x - (-2)\right)^2 = 3(x+2)^2$$ so the function h is given by $h : x \mapsto 3(x+2)^2$.

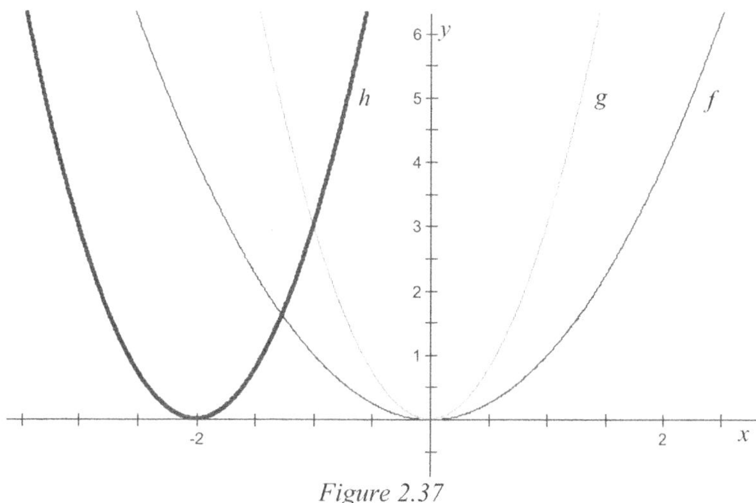

Figure 2.37

In general, a stretch with factor a parallel to the x-axis transforms the graph of $y = \sin x$ to the graph of $y = \sin\left(\dfrac{x}{a}\right)$ and a stretch with factor b parallel to the y-axis transforms the graph of $y = \sin x$ to the graph of $y = b\sin x$.

Example 2.6: Show that the transformation which takes the graph of $y = x^2$ to the graph of $y = \dfrac{x^2}{4}$ can be described by two different stretches.

Solution 2.6: In figure 2.38, the point A on the graph of $y = x^2$ is transformed to the point B which is twice as far from the y-axis as A so the transformation can be described as a stretch parallel to the x-axis with factor 2.

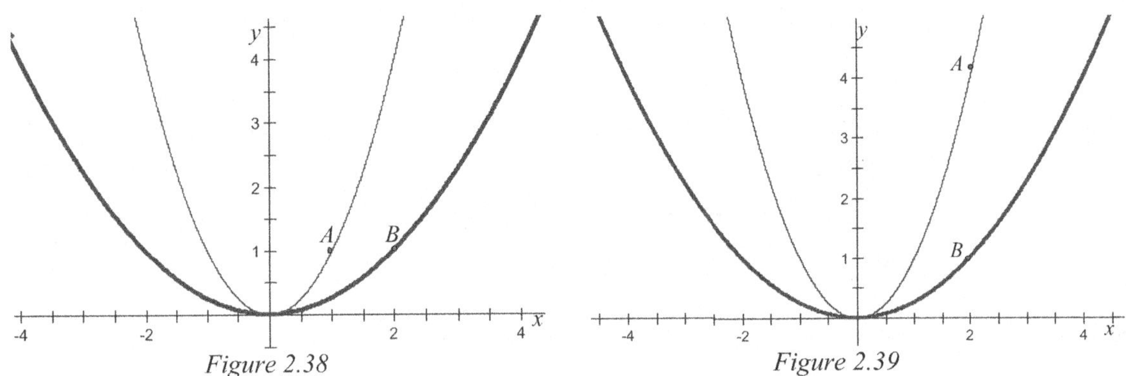

Figure 2.38   Figure 2.39

In figure 2.39, the point A on the graph of $y = x^2$ is transformed to the point B which is one fourth as far from the x-axis as A so the transformation can be describes as a stretch parallel to the y-axis with factor $\dfrac{1}{4}$.

2.3.3 Reflections

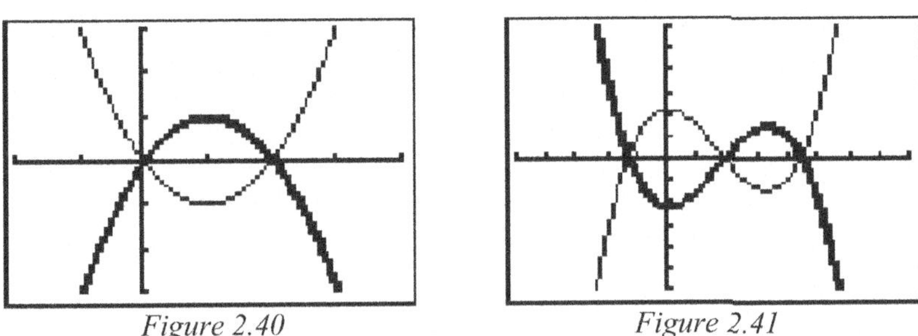

Figure 2.40   Figure 2.41

Figure 2.40 shows the graphs of $y = x^2 - 2x$ and $y = 2x - x^2$. The graph of $y = x^2 - 2x$ has been reflected in the x-axis.

Figure 2.41 shows the graphs of $y = x^3 - 5x^2 + 11$ and $y = -x^3 + 5x^2 - 11$. The graph of $y = x^3 - 5x^2 + 11$ has been reflected in the x-axis. For a reflection in the x-axis the sign of the y ordinate needs to be reversed. Therefore the equation of the reflected curve is $-y = x^3 - 5x^2 + 11$ $\Rightarrow y = -x^3 + 5x^2 - 11$.

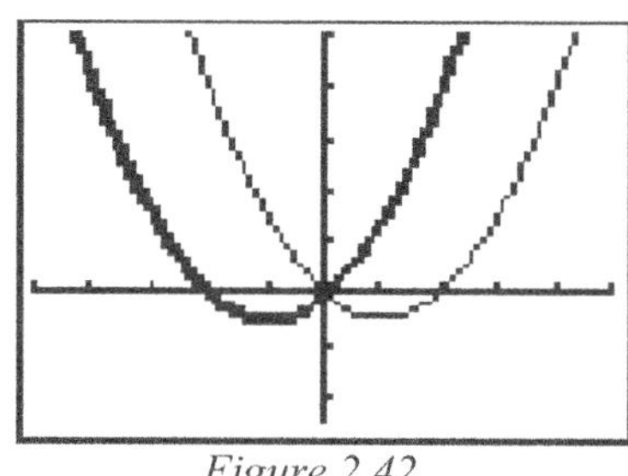

Figure 2.42   Figure 2.43

Figure 2.42 shows the graphs of $y = x^2 - 2x$ and $y = 2x + x^2$. The graph of $y = x^2 - 2x$ has been reflected in the y-axis.

Figure 2.43 shows the graphs of $y = x^3 - 5x^2 + 11$ and $y = -x^3 - 5x^2 + 11$. The graph of $y = x^3 - 5x^2 + 11$ has been reflected in the y-axis. For a reflection in the y-axis, the sign of the x ordinate needs to be reversed. Therefore the equation of the reflected curve is $y = (-x)^3 - 5(-x)^2 + 11$ $\Rightarrow y = -x^3 - 5x^2 + 11$.

In general, a reflection in the x-axis transforms $y = f(x)$ to $y = -f(x)$ and a reflection in the y-axis transforms $y = f(x)$ to $y = f(-x)$.

Example 2.7:   Figure 2.44 shows the graph of $y = x^3 + x + 2$ and its reflection in the y-axis. Write down the function whose graph is this reflection.

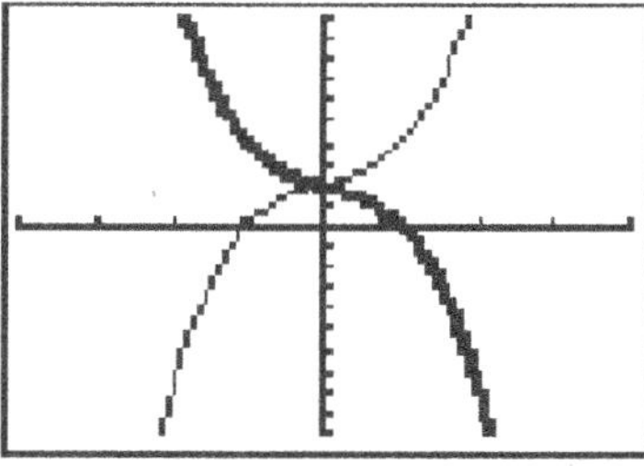

Figure 2.44

Solution 2.7:   A point with coordinates (x, y) is transformed to the point $(-x, y)$ by a reflection in the y-axis. Therefore, the required function is $y = (-x)^3 + (-x) + 2$ $\Rightarrow y = -x^3 - x + 2$.

Exercise 2.3

1.   On squared paper, copy the graph $y = f(x)$ shown in figure 2.45. On the same axes, draw the graph of $y = f(x-2) + 4$. State what transformation changes $y = f(x)$ to $y = f(x-2) + 4$.

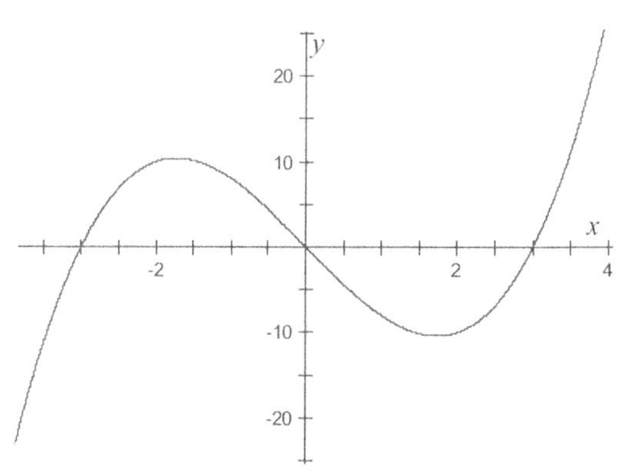

Figure 2.45

45

2.	Figures 2.46 to 2.49 each show a transformation of $y = x^2$. Describe completely the transformation in each case.

(a)

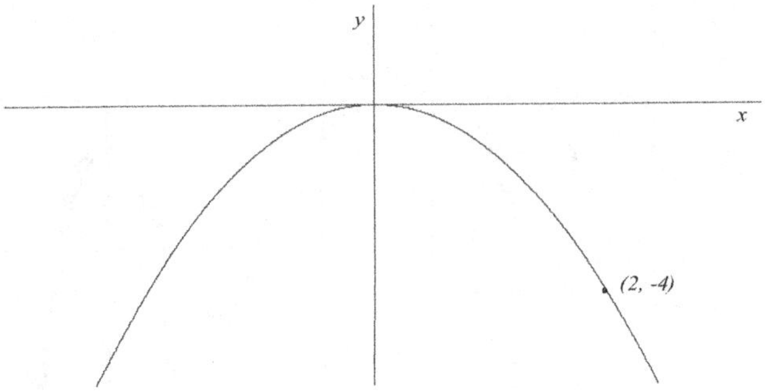

Figure 2.46

(b)

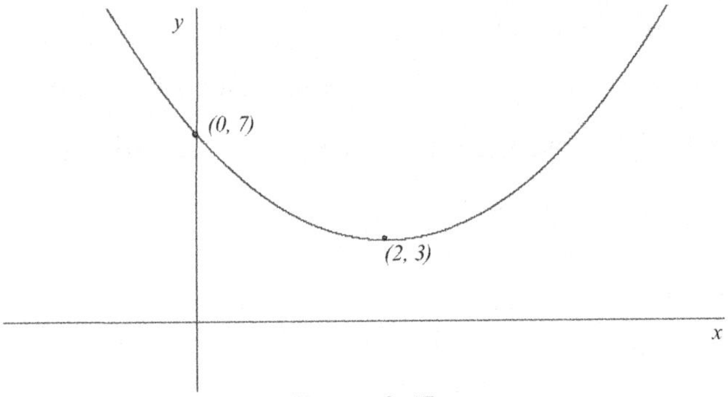

Figure 2.47

(c)

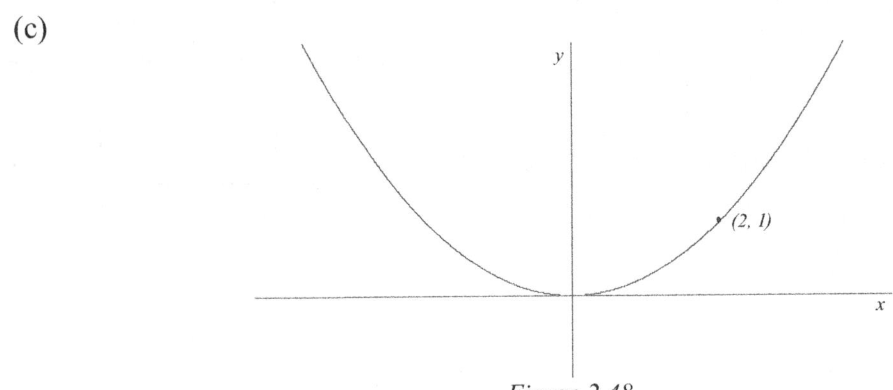

Figure 2.48

(d)

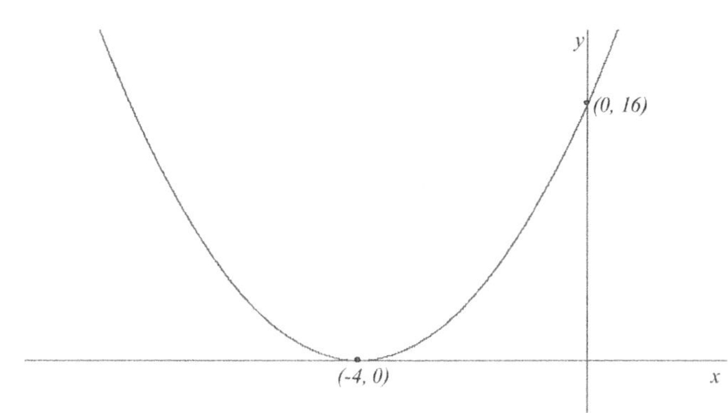

Figure 2.49

3. State which transformation changes the function f in Column A to the function g in Column B.

<u>Column A</u>   <u>Column B</u>

(a) $f(x)=x^2$ $g(x)=(x+3)^2+1$

(b) $f(x)=x^2$ $g(x)=\dfrac{1}{3}x^2$

(c) $f(x)=\sin x$ $g(x)=\sin\left(x-\dfrac{\pi}{6}\right)+1$

(d) $f(x)=x^3$ $g(x)=(x-4)^3+5$

(e) $f(x)=x^2$ $g(x)=4-x^2$

(f) $f(x)=(x-1)^2+5$ $g(x)=x^2$

4. Figures 2.50 to 2.52 each show a transformation. In each case, state the transformation which changes the graph on the left to the graph on the right.

(a)

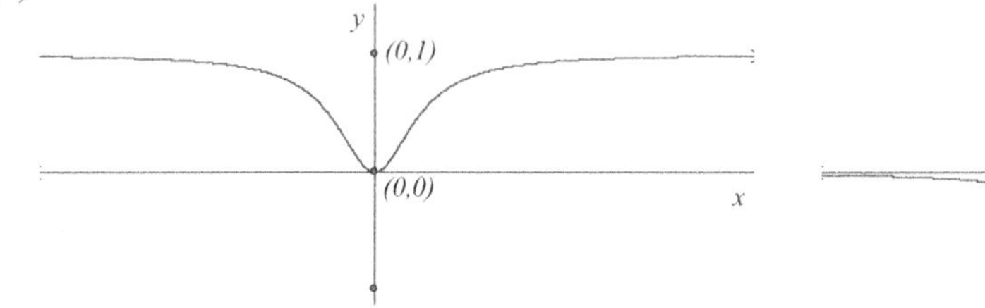

Figure 2.50

47

(b)

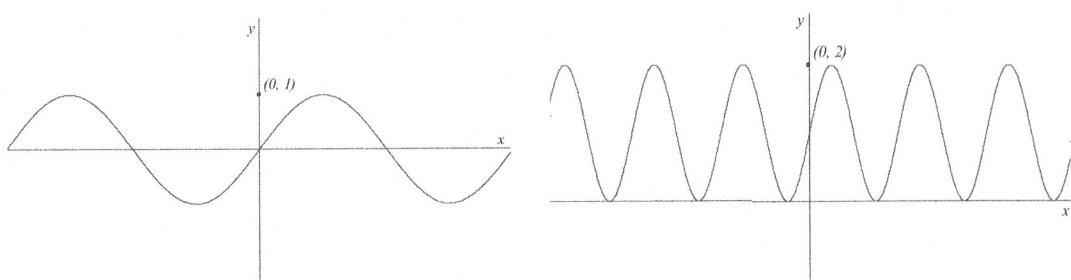

Figure 2.51

(c)

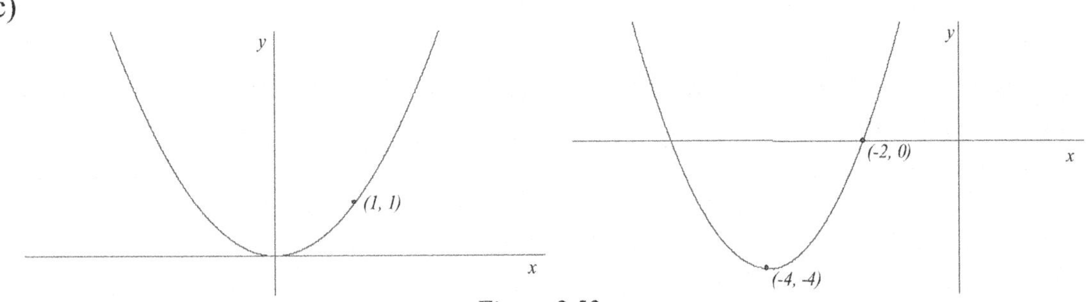

Figure 2.52

5. Figure 2.53 shows the graph of a function $f(x)$ whose domain is $1 \le x \le 4$.

Copy the graph of f on to squared paper and add to it the graphs of the following functions:

(a) $f(x-2)$

(b) $3f(x)$

(c) $f\left(\dfrac{x}{2}\right)+1$.

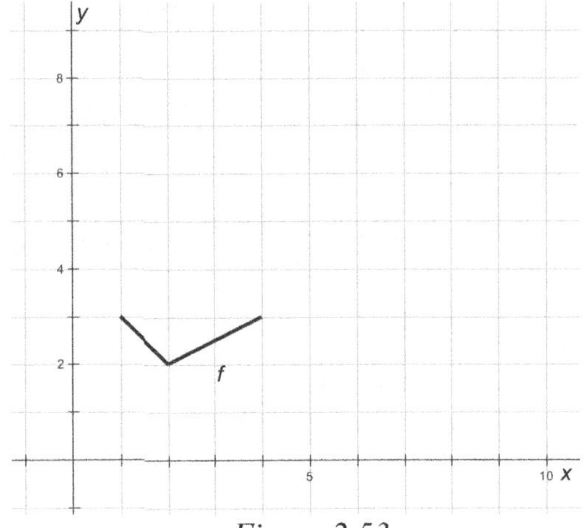

Figure 2.53

6. Figure 2.54 shows the graph of the function $f(x)$, for $2 \le x \le 5$. Copy the graph of f on to squared paper and add to it the graphs of the following functions.

(a) $-f(x)$

(b) $f(x+2)$

(c) $f(-x)-2$

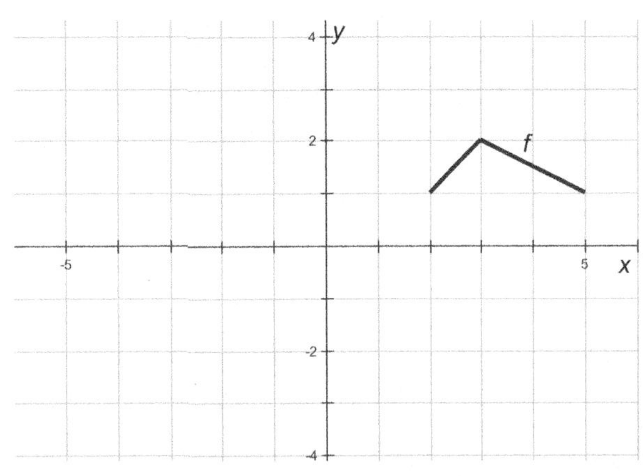

Figure 2.54

48

7. Figure 2.55 shows the functions f and g, with domains $4 \le x \le 7$ and $-1 \le x \le 2$ respectively.

   (a) Given that $g(x) = f(x-a) + b$ find the value of a and b.

   (b) Copy the graph of f on to squared paper and add to it the function h, where

$$h(x) = f(x+3)$$

   (c) Given that $h(x) = g(x-c) + d$ write down the values of c and d.

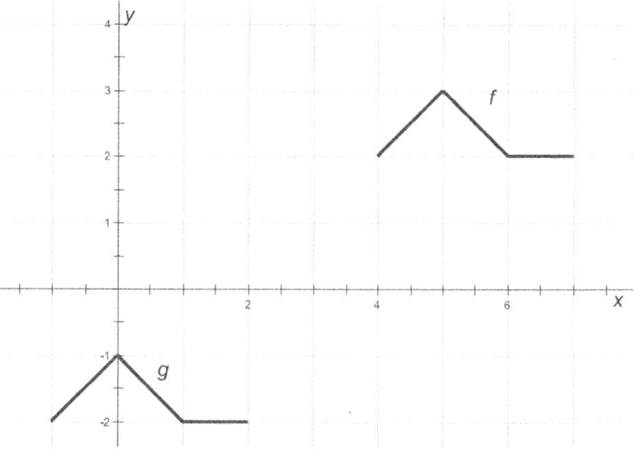

Figure 2.55

2.4 The Quadratic Function $x \mapsto ax^2 + bx + c,\ a \ne 0$

2.4.1 The Graph of the Quadratic Function

Figure 2.56 shows that the graph of $f_1 : x \mapsto x^2$ is a parabola whose vertex is located at the origin $(0,0)$

If f_1 is transformed to f_2 by a translation of h units parallel to the x direction and k units parallel to the y direction, then

$$f_2 : x \mapsto (x-h)^2 + k \ \Rightarrow f_2 : x \mapsto x^2 - 2hx + \left(h^2 + k\right)$$

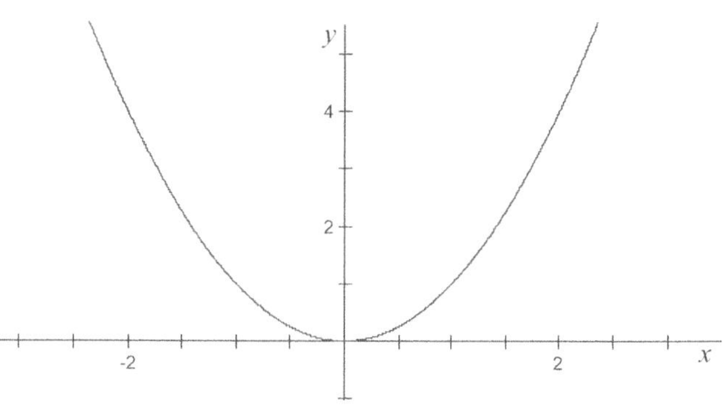

Figure 2.56

If f_2 is further transformed to f_3 by a stretch of factor a parallel to the y direction, then $f_3 : x \mapsto a\left((x-h)^2 + k\right) \Rightarrow f_3 : x \mapsto ax^2 - 2ahx + a\left(h^2 + k\right)$

and, by putting $b = -2ah$ and $c = a\left(h^2 + k\right)$, $f_3 : x \mapsto ax^2 + bx + c$ is obtained. Therefore, the general quadratic function, $y = ax^2 + bx + c$, can be obtained from the function $y = x^2$ by a translation and a stretch.

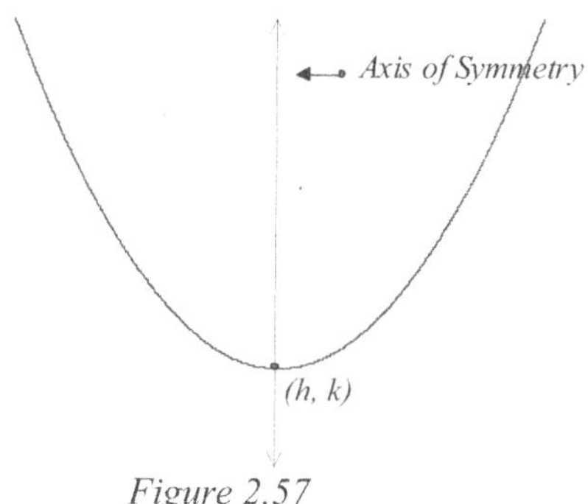

Figure 2.57

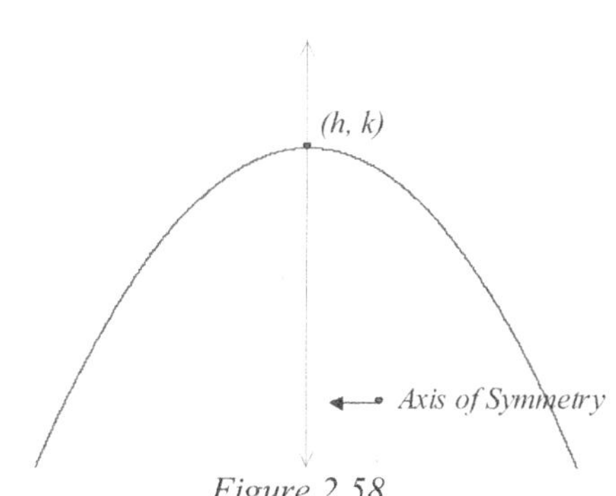

Figure 2.58

If $a>0$, the graph of the quadratic function is a parabola in which its vertex forms a minimum point (figure 2.57).

If $a<0$, the graph of the quadratic function is a parabola in which its vertex forms a maximum point (figure 2.58).

The axis of symmetry of the parabola is parallel to the y-axis and passes through the vertex $(h,\ k)$ so the equation of the axis of symmetry is $x=h$. However, $b=-2ah$, so that the equation of the axis of symmetry is $x=-\dfrac{b}{2a}$.

2.4.2 Finding the Vertex and y-Intercept of the Graph of a Quadratic Function Using the Graphing Calculator

Consider the function $f(x)=x^2-9x+5,\ -2\le x\le 12$ whose graph is a portion of a parabola. You can use the graphing calculator to find the approximate coordinates of the vertex of the parabola. Figures 2.59 to 2.64 show how this is done using a TI-84 Plus.

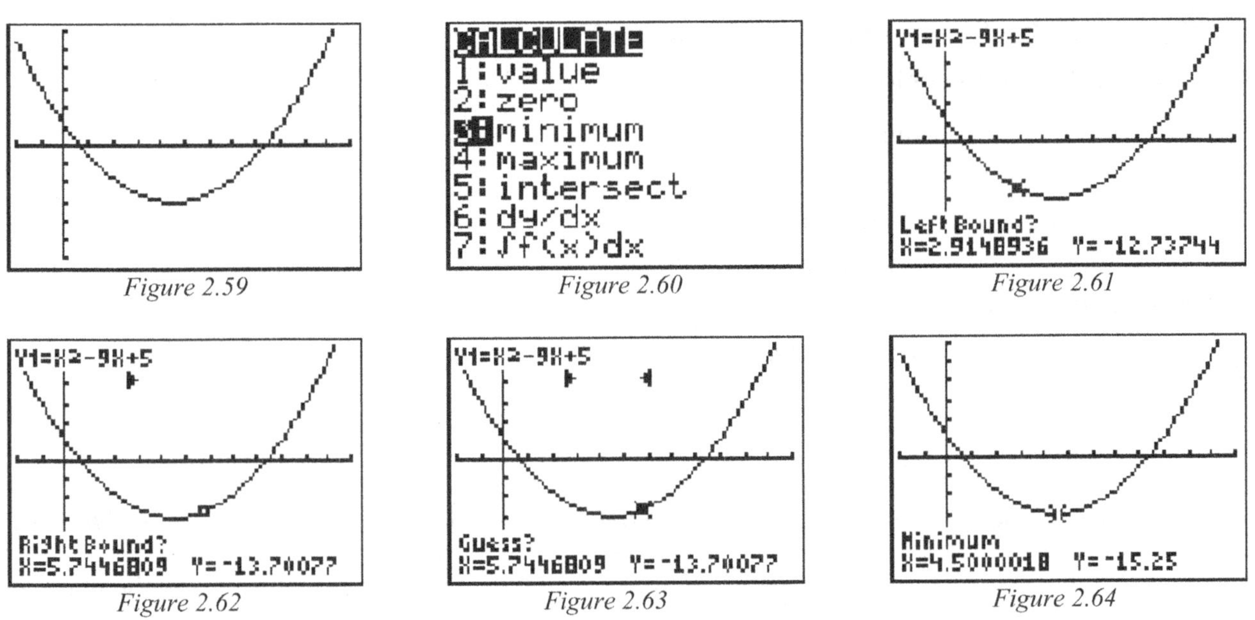

Figure 2.59   Figure 2.60   Figure 2.61

Figure 2.62   Figure 2.63   Figure 2.64

So the coordinates of the vertex are $(4.50,\ -15.25)$, correct to two decimal places. It is important to understand that the minimum value obtained in this way is generally only approximate, although the accuracy is usually correct to six or seven significant figures. You might like to repeat the operation a few times to see that the x-coordinate given by the graphing calculator varies in the sixth and seventh significant figure.

The y-intercept is the point where the parabola cuts the y-axis, and so its coordinates are $(0, f(0))$. You can use the graphing calculator to find $f(0)$, but it is quicker to find $f(0)$ by substituting $x=0$ into the given function. In general, for the function $f(x)=ax^2+bx+c$, $f(0)=c$ so that the y-intercept is $(0,\ c)$.

2.4.3 Completing the Square

The process of writing the general quadratic function $x \mapsto ax^2 + bx + c$ in the form $a(x-h)^2 + k$ is called 'completing the square'. This process is illustrated by the example below:

$$f(x) = x^2 - 8x + 3 = \left(x^2 - 8x\right) + 3$$

The term in parentheses is made into a perfect square by adding to it a suitable constant and subtracting an equal constant outside the parentheses. In order to find the suitable constant, it is necessary to look more closely at a perfect square:

$$(x+b)^2 = (x+b)(x+b) = (x+b)x + (x+b)b = x^2 + bx + xb + b^2 = x^2 + 2bx + b^2$$

Therefore $x^2 + 2bx$ can be made into a perfect square by the addition of b^2 and in the example $f(x) = x^2 - 8x + 3$ as the term in parentheses is $x^2 - 8x$, $2b = -8 \Rightarrow b = -4$. Therefore, it is necessary to add $(-4)^2 = 16$ inside the parentheses and subtract 16 outside the parentheses:

$$f(x) = \left(x^2 - 8x + 16\right) - 16 + 3 = (x-4)^2 - 13$$

You can see from section 2.4.1 that this represents a translation of $y = x^2$ by 4 units parallel to the x direction and -13 units parallel to the y direction. If you obtain the graph of $f(x) = x^2 - 8x + 3$ on your graphing calculator, you will see that the coordinates of the vertex of the parabola are $(4, -13)$, so this enables you to check your algebraic method.

Example 2.8: (a) Write $f(x) = x^2 + 14x - 3$ in the form $f(x) = (x-h)^2 + k$ and hence write down the equation of the axis of symmetry of $y = f(x)$.

(b) Show that $f(x) = 15 + 2x - x^2$ can be written in the form $f(x) = k - (x-h)^2$.

Solution 2.8: (a) $f(x) = \left(x^2 + 14x + 49\right) - 49 - 3 = (x+7)^2 - 52 = (x-(-7))^2 - 52$. Therefore the axis of symmetry has equation $x = -7$.

(b) $f(x) = 15 + 2x - x^2 = 15 - \left(x^2 - 2x\right) = 15 - \left(x^2 - 2x + 1\right) + 1 = 16 - (x-1)^2$. Note that 1 is added inside the parentheses but as the expression inside the parentheses is subtracted from 15 it is necessary to <u>add</u> 1 outside the parentheses in order to compensate.

Example 2.9:   Write $x^2 + 7x + 13$ in the form $(x - h)^2 + k$.

Solution 2.9: $x^2 + 7x + 13 = \left(x^2 + 7x + \left(\dfrac{7}{2}\right)^2 \right) - \dfrac{49}{4} + 13 = \left(x + \dfrac{7}{2} \right)^2 + \dfrac{52 - 49}{4} = \left(x + \dfrac{7}{2} \right)^2 + \dfrac{3}{4}$

Exercise 2.4

1. Use an algebraic method to write the following quadratic expressions in the form $(x + a)^2 + b$. Check each answer by obtaining the graph of the corresponding function on your graphing calculator and by finding the coordinates of the vertex of the parabola.

(a) $x^2 + 4x - 6$ (b) $x^2 + 12x + 15$ (c) $x^2 - 2x + 9$ (d) $x^2 + 6x - 11$

(e) $x^2 - 8x - 1$ (f) $x^2 - 4x + 19$ (g) $x^2 + 30x + 205$

2. Use an algebraic method to write the following quadratic expressions in the form $a - (x + b)^2$. Check each answer with your graphing calculator as you did in question 1.

(a) $2 + 4x - x^2$ (b) $9 + 6x - x^2$ (c) $6 - 18x - x^2$ (d) $5 + 16x - x^2$

(e) $4x - 7 - x^2$ (f) $17 - 8x - x^2$ (g) $-10 - 2x - x^2$

3. Write the following expressions in the form $(x + a)^2 + b$, and check each answer using your graphing calculator.

(a) $x^2 + 3x - 5$ (b) $x^2 + 5x + 2$ (c) $x^2 - x - 11$ (d) $x^2 - 7x + 17$

(e) $x^2 + 11x - 5$ (f) $x^2 - 15x + 32$ (g) $x^2 + 17x + 73$

4. Write the following expressions in the form $a - (x + b)^2$, and check each answer using you graphing calculator.

(a) $7 + 3x - x^2$ (b) $1 - x - x^2$ (c) $5 + 5x - x^2$ (d) $1 - 3x - x^2$

(e) $x - 10 - x^2$ (f) $-8 + 7x - x^2$ (g) $21 - 7x - x^2$

5. Show that the coordinates of the vertex of the parabola given by $y = x^2 - 10x + 12$ are $(5,\ -13)$ and write down the equation of the axis of symmetry of the parabola.

6. The graph of a quadratic function f has axis of symmetry whose equation is $x = \dfrac{5}{2}$ and passes through the points $(0,\ 6)$ and $(3,\ 0)$. Find f.

The examples used so far all have 1 as their x^2 coefficient. For quadratic expressions of the form $ax^2 + bx + c$ where $a \neq 1$, a further step is required in order to complete the square. This process is demonstrated in Example 2.10.

Example 2.10: Write $4x^2 - 24x - 9$ in the form $a(x-h)^2 + k$.

Solution 2.10: $4x^2 - 24x - 9 = 4(x^2 - 6x) - 9 = 4(x^2 - 6x + 9) - 4 \times 9 - 9$

$$= 4(x-3)^2 - 9 - 36 = 4(x-3)^2 - 45$$

In general, if $y = ax^2 + bx + c = a\left(x^2 + \dfrac{b}{a}x\right) + c$

$$= a\left(x^2 + \dfrac{b}{a}x + \dfrac{b^2}{4a^2}\right) + \dfrac{4ac - b^2}{4a} = a\left(x + \dfrac{b}{2a}\right)^2 - \left(\dfrac{b^2 - 4ac}{4a}\right)$$

Exercise 2.5

Write the following expressions in the form $a(x-h)^2 + k$ or $k - a(x-h)^2$ as appropriate.

1. $2x^2 - 4x + 1$
2. $4x^2 + 6x - 9$
3. $11 - 4x - 2x^2$
4. $3x^2 - 10x + 15$
5. $1 - 2x - 4x^2$
6. $-5 + 8x - 2x^2$

2.4.4 Factoring a Quadratic Function

Method 1: Using the graphing calculator

Although, in general, a quadratic function $y = ax^2 + bx + c$, $a,b,c \in \mathbb{Z}$ cannot be written in the form $y = a(x-p)(x-q)$, $p,q \in \mathbb{Z}$, it is useful to be able to factor those quadratic functions for which this form is possible. If $y = a(x-p)(x-q) = 0$, then either $x - p = 0$ or $x - q = 0$. Therefore, $x = p$ or $x = q$, so that the x-intercepts of the graph of $y = a(x-p)(x-q)$ are $(p, 0)$ and $(q, 0)$.

The graphing calculator can be used to find p and q, and, hence, to factor the quadratic function. Figures 2.65 to 2.70 show the procedure using a TI-84 Plus.

Suppose that $y = x^2 + 9x - 22 = 0$. If this function is graphed over a suitable domain, the x-intercepts can be found and the function factored. The diagrams show the procedure for finding one of the intercepts.

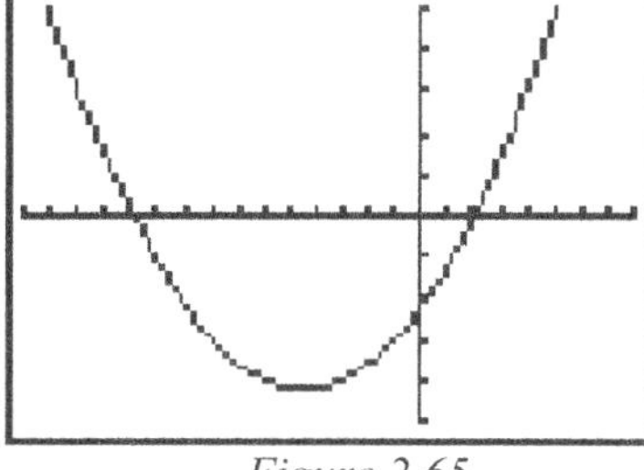

Figure 2.65

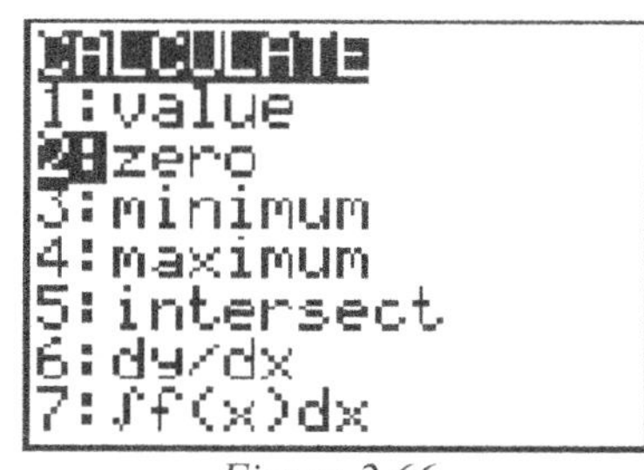

Figure 2.66

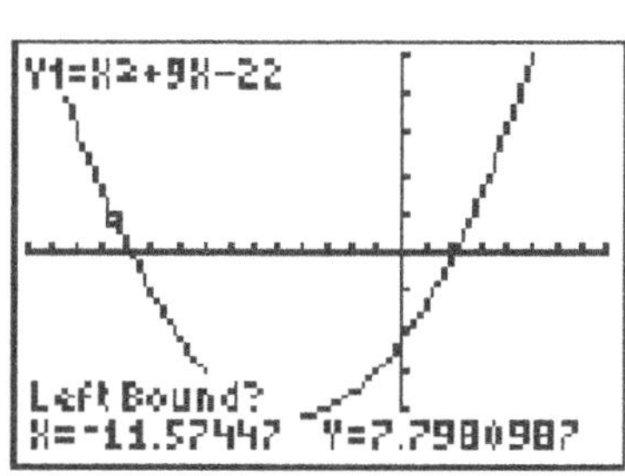

Figure 2.67

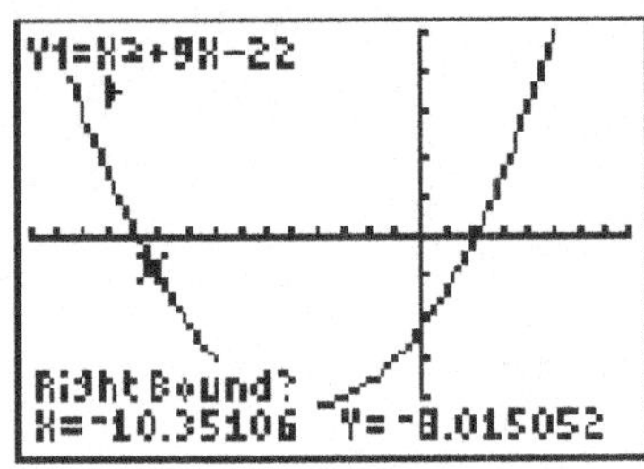
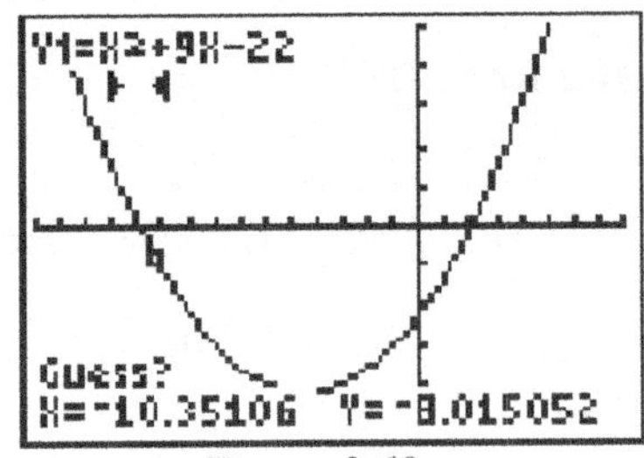
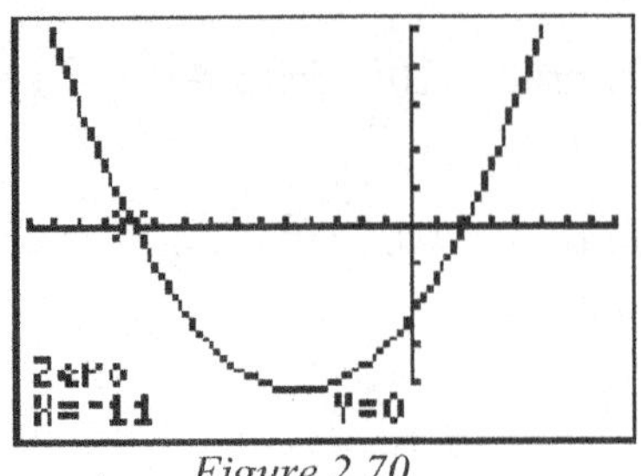

Figure 2.68 *Figure 2.69* *Figure 2.70*

In a similar way, you can find that $q = 2$. Then $y = \left(x - (-11)\right)(x - 2) \Rightarrow y = (x + 11)(x - 2)$.

Method 2: Algebraic – splitting the middle term

In order to factor the quadratic function $y = x^2 + bx + c$, there needs to be integers p and q such that $pq = c$ and $p + q = b$. So, in this case, you need to find two integers whose product is -22 and whose sum is 9. Solution of these simultaneous equations is best done by inspection (guess and check). Clearly, $p = 11$ and $q = -2$, so then $y = x^2 + 11x - 2x - 22 \Rightarrow y = x(x + 11) - 2(x + 11)$

$\Rightarrow y = (x + 11)(x - 2)$.

Exercise 2.6

1. Use your graphing calculator to find the x-intercepts of the graphs of the following quadratic functions and hence factor them.

   (a) $x^2 - 7x + 12$ (b) $x^2 + 5x + 6$ (c) $x^2 + 5x - 36$ (d) $x^2 - x - 90$

   (e) $x^2 - 2x + 1$ (f) $15 - 2x - x^2$ (g) $6 - x - x^2$ (h) $7 + 6x - x^2$

   (i) $x^2 - 3x - 10$ (j) $5 - 4x - x^2$ (k) $x^2 + 8x - 33$ (l) $x^2 + 11x + 28$

2. Use the algebraic method of splitting the middle term to factor the quadratic functions in question 1.

2.5 The General Quadratic Equation $ax^2 + bx + c = 0$, $a \neq 0$

2.5.1 Solution of Quadratic Equations

There are numerous ways of solving the equation $f(x) = 0$ where $f(x)$ is a quadratic function.

1) The graphing calculator may be used as described in section 2.2.3

2) If $f(x)$ can be written in factored form, then $f(x) = a(x - p)(x - q)$ and method 2, described in section 2.4.4, can be used. If $f(x) = 0$ then $a(x - p)(x - q) = 0$

   $\Rightarrow x - p = 0,\ x - q = 0 \Rightarrow x = p,\ x = q$.

54

3) As all quadratic functions can be written in the form $f(x) = a(x-h)^2 + k$, the equation $f(x) = 0$ can be solved by completing the square.

Example 2.11: Solve the equation $x^2 - 6x - 16 = 0$ by using all three methods outlined above.

Solution 2.11: 1) Figures 2.71 to 2.76 show the graphing calculator solution.

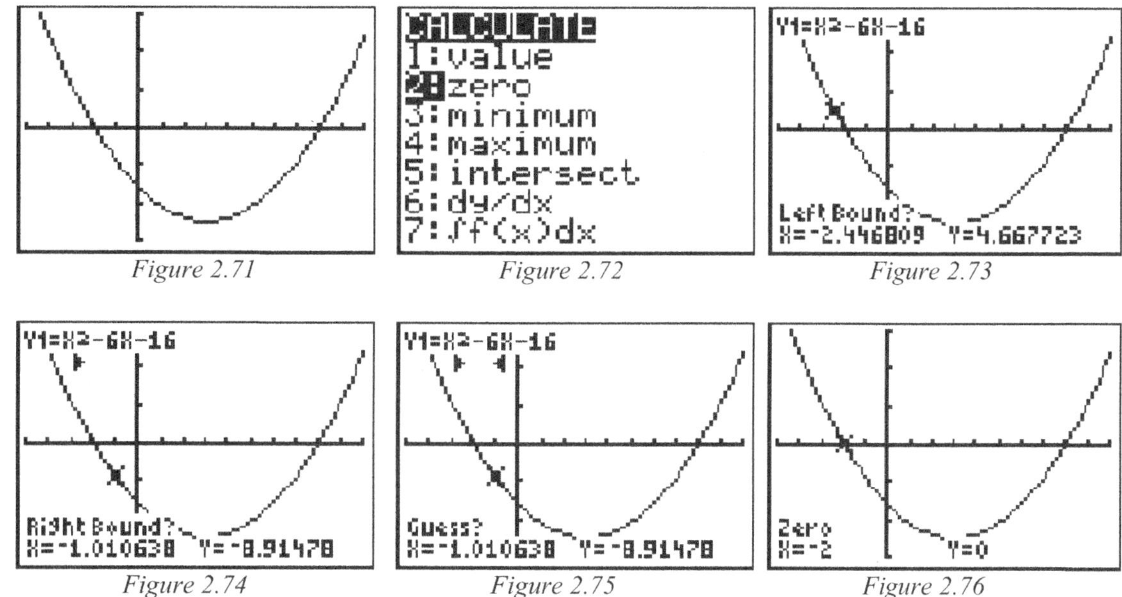

Figure 2.71   Figure 2.72   Figure 2.73

Figure 2.74   Figure 2.75   Figure 2.76

So $x = -2$ is a solution. A similar process will give the positive solution, $x = 8$.

2) You need to find p and q such that $pq = -16$ and $p + q = -6$ so $p = -8$ and $q = 2$. Therefore, $x^2 - 6x - 16 = x^2 - 8x + 2x - 16 = 0 \Rightarrow x(x-8) + 2(x-8) = 0 \Rightarrow$
$(x-8)(x+2) = 0 \Rightarrow x - 8 = 0, \ x + 2 = 0 \Rightarrow x = 8, \ x = -2$.

3) $x^2 - 6x - 16 = \left(x^2 - 6x + 9\right) - 9 - 16 = (x-3)^2 - 25$.

Therefore $x^2 - 6x - 16 = 0 \Rightarrow (x-3)^2 - 25 = 0 \Rightarrow (x-3)^2 = 25 \Rightarrow (x-3) = \pm 5$
$\Rightarrow x = 3 + 5, \ x = 3 - 5 \Rightarrow x = 8, \ x = -2$.

2.5.2 Solution of the General Quadratic Equation $ax^2 + bx + c = 0$, $a \neq 0$

The method of completing the square, when used on the general quadratic equation, provides a formula for solving $ax^2 + bx + c = 0$.

$$ax^2 + bx + c = 0, \ a \neq 0 \Rightarrow a\left(x + \frac{b}{2a}\right)^2 - \left(\frac{b^2 - 4ac}{4a}\right) = 0 \quad \text{(see section 2.4.3)}$$

$$\Rightarrow a\left(x+\frac{b}{2a}\right)^2 = \frac{b^2-4ac}{4a} \Rightarrow \left(x+\frac{b}{2a}\right)^2 = \frac{b^2-4ac}{4a^2} \Rightarrow x+\frac{b}{2a} = \pm\frac{\sqrt{b^2-4ac}}{2a}$$

$$\Rightarrow x = \frac{-b+\sqrt{b^2-4ac}}{2a} \text{ and } x = \frac{-b-\sqrt{b^2-4ac}}{2a}$$

Example 2.12: Use the quadratic formula to solve the equation $x^2-10x+6=0$, giving your answers in the form $p\pm\sqrt{q}$, $p,\,q\in\mathbb{Z}$. Write these solutions correct to three significant figures and check your answers by using your graphing calculator to produce a graph of $y=x^2-10x+6$ and by finding the x-intercepts.

Solution 2.12: $x^2-10x+6=0 \Rightarrow x = \dfrac{10\pm\sqrt{100-24}}{2} \Rightarrow x = \dfrac{10+\sqrt{76}}{2} = \dfrac{10\pm\sqrt{4\times19}}{2} = \dfrac{10\pm\sqrt{4}\sqrt{19}}{2}$

$\Rightarrow x = \dfrac{10\pm2\sqrt{19}}{2} = 5\pm\sqrt{19}$. Therefore, $x=5+\sqrt{19}$ and $x=5-\sqrt{19}$. These solutions can be written, correct to three significant figures, as $x=0.6411010... = 0.641$ and $x=9.358898...=9.36$. Figures 2.77 to 2.82 check these solutions using the graphing calculator.

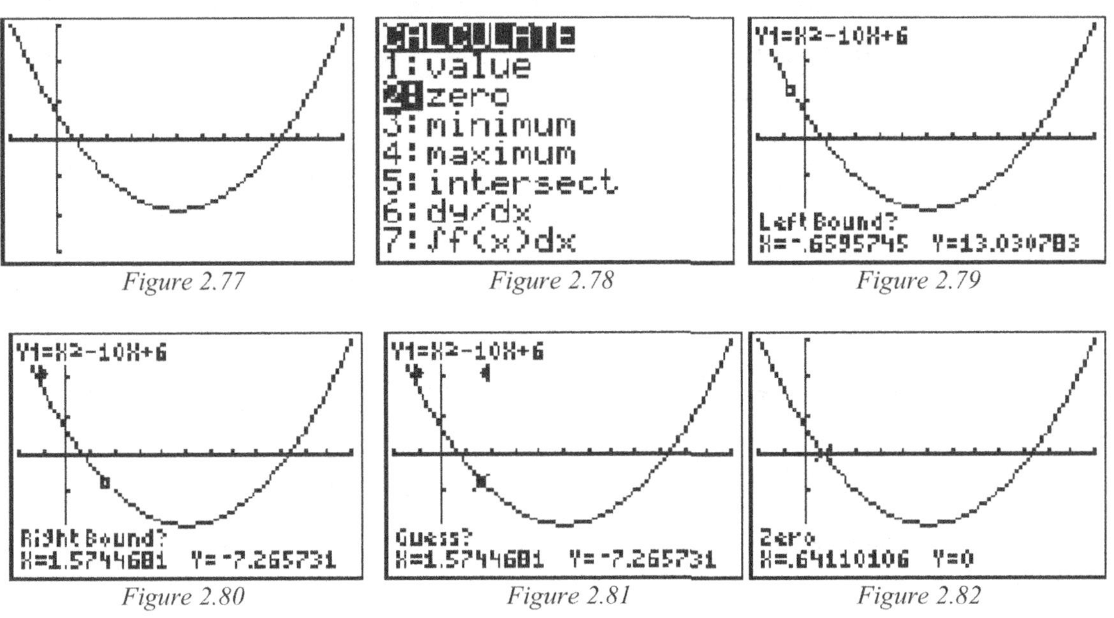

Figure 2.77	Figure 2.78	Figure 2.79
Figure 2.80	Figure 2.81	Figure 2.82

So, $x=0.641$, correct to three significant figures. A similar process can be used to find the other solution, which is $x=9.36$ to a similar accuracy.

Exercise 2.7

1. In each case, solve the equation by rearranging if necessary and factoring.

(a) $x^2 + 8x - 20 = 0$

(b) $x - \dfrac{72}{x} = 1$

(c) $x(3-x) = 9(x+1)$

(d) $(x-4)^2 = 8x - 47$

2. In each case, write the solution of the equation in the form $p + k\sqrt{q}$, where $p, k \in \mathbb{Q}$ and $q \in \mathbb{Z}$.

(a) $3x^2 + 8x - 1 = 0$ (b) $x^2 + 7x + 11 = 0$ (c) $5x^2 - 3x - 1 = 0$ (d) $2x^2 - 17x + 34 = 0$

3. Solve the equations of question 2 using your graphing calculator and give the solutions correct to three significant figures.

4. (a) Show that the equation $x^2 - 4x - 11 = 0$ can be written in the form $(x-a)^2 = b$ where $a, b \in \mathbb{Z}$. Hence show that the solutions of the equation are $x = 2 - \sqrt{15}$ and $x = 2 + \sqrt{15}$.

(b) Show that the equation $x^2 - 10x + 23 = 0$ has solutions $x = 5 - \sqrt{2}$ and $x = 5 + \sqrt{2}$.

2.5.3 The Significance of the Discriminant

As you discovered in section 2.2.3, the solution of $f(x) = 0$ is obtained by finding where the graph of $f(x)$ intersects the x-axis. Figure 2.83 shows the graphs of (a) $y = x^2 - 4x + 3$, (b) $y = x^2 - 4x + 4$ and (c) $y = x^2 - 4x + 5$.

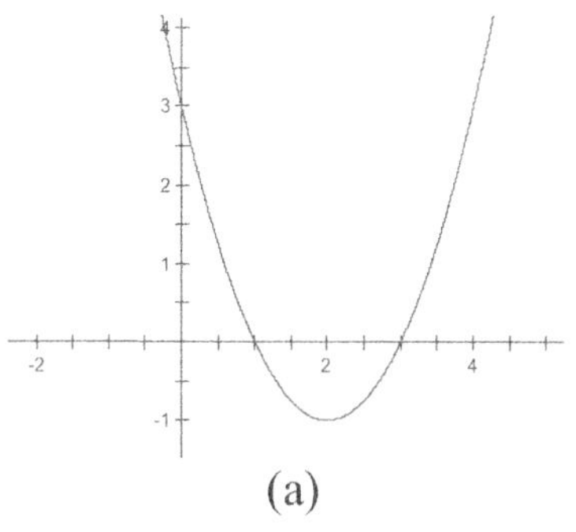

(a)

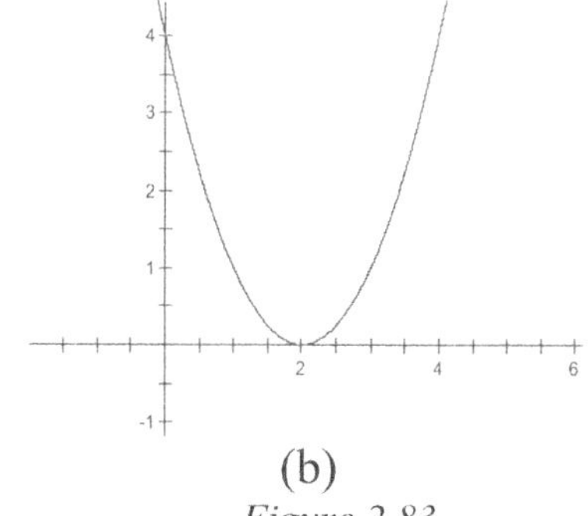

(b)

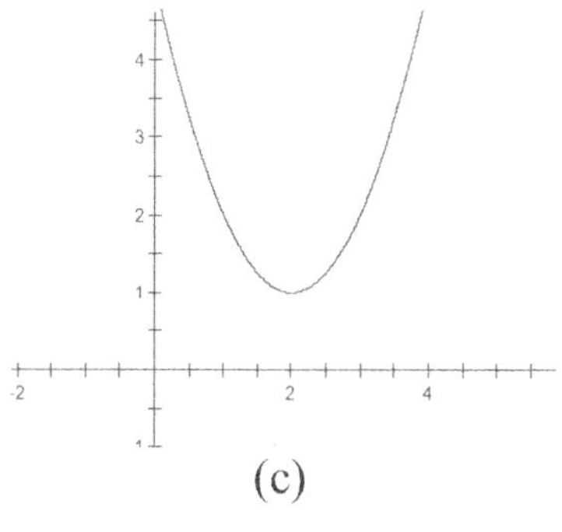

(c)

Figure 2.83

Graph (a) shows two different solutions, graph (b) shows one solution (or two equal solutions), and graph (c) shows no solutions.

We can use $x = \dfrac{-b \pm \sqrt{b^2 - 4ac}}{2a}$, which is the solution of $ax^2 + bx + c = 0$ that we found in section 2.5.2, to solve these three equations:

$$\text{for (a),} \quad x = \frac{4 \pm \sqrt{16-12}}{2} = \frac{4 \pm 2}{2} \Rightarrow x = 1, \ x = 3$$

$$\text{for (b),} \quad x = \frac{4 \pm \sqrt{16-16}}{2} = \frac{4 \pm 0}{2} \Rightarrow x = 2, \ x = 2$$

$$\text{for (c),} \quad x = \frac{4 \pm \sqrt{16-20}}{2} = \frac{4 \pm \sqrt{-4}}{2} \Rightarrow \text{there are no solutions for } x.$$

It is clear that the different situations occur as a result of what is happening under the square root sign. The expression under the square root sign, $b^2 - 4ac$, therefore determines whether the quadratic equation has two different, two equal, or no solutions.

We write $\Delta = b^2 - 4ac$ and call Δ the *discriminant*. We now look more closely at the three situations which Δ discriminates.

Case 1: $\Delta > 0$

The graph of $y = f(x)$ is either a 'valley' which intersects the x-axis in two distinct points or it is a 'hill' which also intersects the x-axis in two distinct points so that, in either event, $f(x) = 0$ has two distinct (different) solutions (figure 2.84).

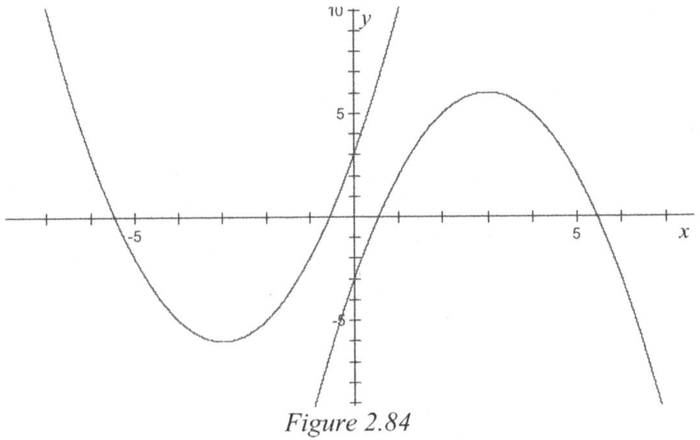

Figure 2.84

For example, the solution of $11x^2 + 3x - 1 = 0$ is

$$x = \frac{-3 \pm \sqrt{3^2 - 4 \times 11 \times (-1)}}{2 \times 11} = \frac{-3 \pm \sqrt{9 + 44}}{22} = \frac{-3 \pm \sqrt{53}}{22}$$

$$\Rightarrow x = 0.1945504... = 0.195 \text{, and } x = -0.4672777... \ x = -0.467$$

Case 2: $\Delta = 0$

The graph of $y = f(x)$ either just touches the x-axis from above or just touches it from below so that, in either event, $f(x) = 0$ has two equal solutions (figure 2.85).

58

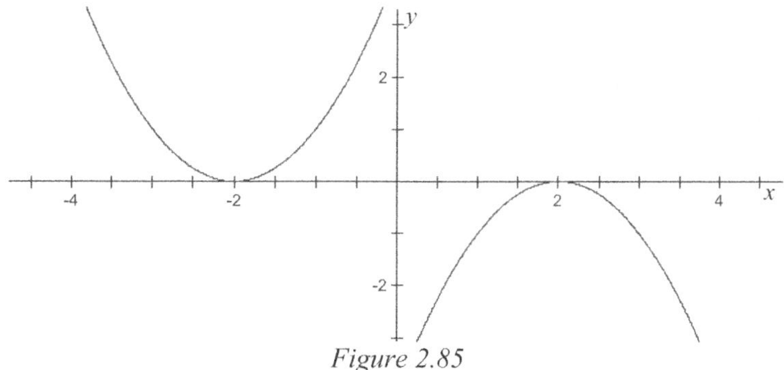

Figure 2.85

For example, the solution of $25x^2 - 30x + 9 = 0$ is $x = \dfrac{30 \pm \sqrt{900 - 4 \times 25 \times 9}}{50} = \dfrac{30 \pm 0}{50} \Rightarrow x = \dfrac{3}{5},\ x = \dfrac{3}{5}$.

Case 3: $\Delta < 0$

The graph of $y = f(x)$ lies either entirely above the x-axis or entirely below the x-axis so that, in either event, the graph has no intercepts on the x-axis and so no real solutions (figure 2.86).

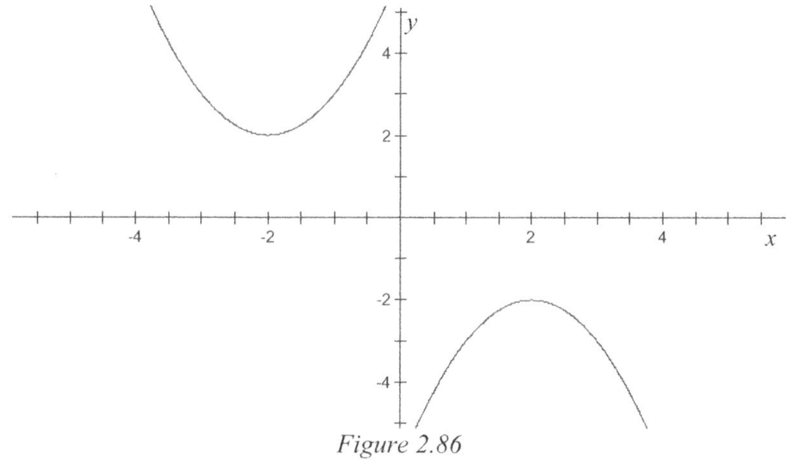

Figure 2.86

For example, if you try to solve $5x^2 - x + 6 = 0$, then $x = \dfrac{1 \pm \sqrt{1 - 120}}{10} = \dfrac{1 \pm \sqrt{-119}}{10}$. However, as $\sqrt{-119}$ is not a real number, there are no real solutions.

Example 2.13: Find all the values of k for which the equation $x^2 - 2kx - k = 0$ has two equal solutions.

Solution 2.13: For equal solutions, $\Delta = 0$; that is, $b^2 - 4ac = 0$. Now $a = 1,\ b = -2k$ and $c = -k$

$$\Rightarrow (-2k)^2 - 4 \times 1 \times (-k) = 0 \Rightarrow 4k^2 + 4k = 0 \Rightarrow 4k(k+1) = 0 \Rightarrow k = 0 \text{ and } k = -1.$$

Exercise 2.8

1. If necessary, write the equation in the form $f(x)=0$, where $f(x)$ is a quadratic function. Then calculate the discriminant of f and hence state which equations have two real solutions, no real solutions, and which have equal solutions.

   (a) $x^2+6x+7=0$ (b) $x^2+6x+9=0$ (c) $x^2+6x+10=0$

   (d) $7x^2+3=0$ (e) $x^2=5x-3$ (f) $5x=2-3x^2$

   (g) $x+3=\dfrac{5}{x}$, $x\neq 0$ (h) $11x^2+13x+3=0$ (i) $49x^2+64=112x$

2. Show that the equation $kx^2=x+1$ has no real solutions if $k<-\dfrac{1}{4}$.

3. Show that if $k=-4$ or $k=1$ then $kx^2+4x+k+3=0$ has two equal solutions.

4. Show that if the line $y=kx-k$ touches the parabola with equation $y=x^2+5x+10$ then $k=-1$ or $k=15$.

5. (a) Show that the graphs of the functions $y=x^2+\dfrac{1}{8}$ and $y=3x-x^2-1$ touch, and find the coordinates of their point of contact.

   (b) Find the values of k for which $y=x^2+kx+k$ touches $y=-2-2x-x^2$. Find also the coordinates of the point where they touch.

2.6 The Algebra of Functions

2.6.1 Composite Functions

More complicated functions can be formed from simpler functions using a process called *functional composition.*

Consider the functions $f:x\mapsto x+3$ and $g:x\mapsto 5x$. As you saw in section 2.1, a function can be likened to a kind of machine or production process. In this context, f is an 'add three' machine and g is a 'multiply by five' machine. This can be shown diagrammatically by figures 2.87 and 2.88.

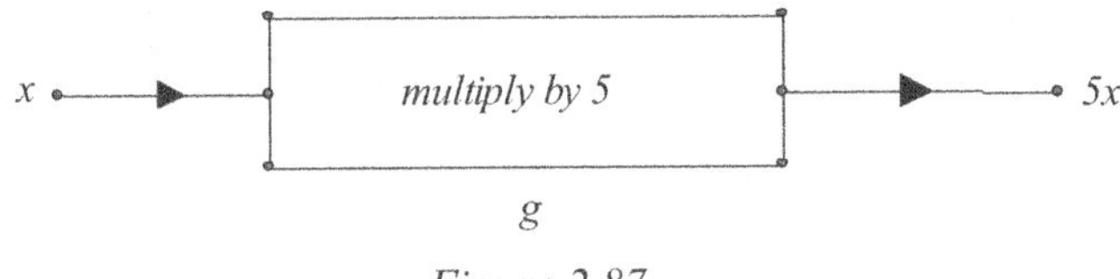

Figure 2.87

60

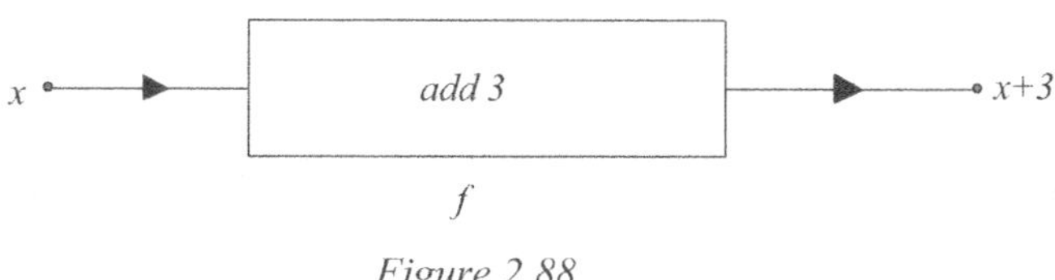

Figure 2.88

If the two machines are linked so that the input of g is attached to the output of f, a composite function, $g(f(x))$, written $g \circ f$, is formed. The output of the composite function $g \circ f$ is shown by the composite machine in figure 2.89.

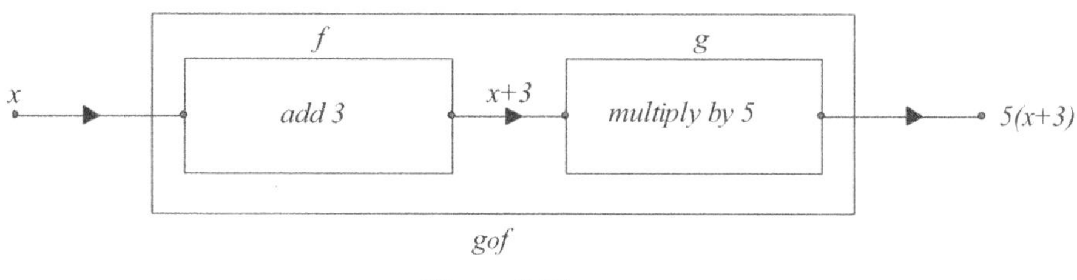

Figure 2.89

The composite function $g \circ f$ is $x \mapsto 5(x+3) = 5x+15$ so that $g \circ f : x \mapsto 5x+15$. Notice that $g \circ f$ couples the functions g and f in the order f first and g second. This is because g operates on the output of f.

Example 2.14: If $f : x \mapsto 3-x$ and $g : x \mapsto 4x$,

    (a) Find (i) $(f \circ g)(1)$ (ii) $(g \circ f)(5)$

    (b) Show that $(g \circ f)(x) = 12 - 4x$

    (c) Find $(f \circ g)(x)$

Solution 2.14: (a) (i) $(f \circ g)(1) = f(g(1)) = f(4) = -1$; (ii) $(g \circ f)(5) = g(f(5)) = g(-2) = -8$

    (b) $(g \circ f)(x) = g(f(x)) = g(3-x) = 4(3-x)$ so $(g \circ f)(x) = 12 - 4x$

    (c) $(f \circ g)(x) = f(g(x)) = f(4x) = 3 - 4x$

2.6.2 The Identity Function

The identity function $i : x \mapsto x$ assigns each element of the domain to itself. For example, $i(7) = 7$, $i(2.6) = 2.6$ and so on. Therefore,

$$(i \circ f)(x) = i(f(x)) = f(x) \text{ and}$$

$$(f \circ i)(x) = f(i(x)) = f(x)$$

$$\Rightarrow f \circ i = i \circ f = f$$

Example 2.15: If $f : x \mapsto 5-2x$ and $g : x \mapsto \dfrac{1}{2}(5-x)$, show that $f \circ g = i$.

Solution 2.15: $(f \circ g)(x) = f(g(x)) = f\left(\dfrac{1}{2}(5-x)\right) = 5 - 2\dfrac{1}{2}(5-x) = 5 - (5-x) = x \Rightarrow (f \circ g)(x) = x$

and, because, by definition, $i(x) = x$, $f \circ g = i$.

2.6.3 The Inverse Function

In section 2.1, a function was likened to a machine. If f is an 'add three machine', then the inverse of f, written f^{-1}, is a 'subtract three machine'. If f is a 'multiply by five machine' then f^{-1} is a 'divide by 5 machine' or a 'multiply by one fifth machine'. The inverse function, f^{-1} , undoes whatever f does. For example, the function $f : x \mapsto 7x+2$ can be described by the machine shown in figure 2.90.

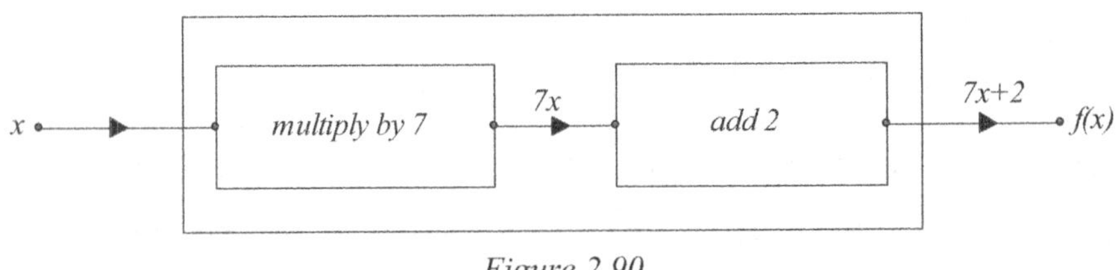

Figure 2.90

The inverse machine f^{-1}, shown in figure 2.91, undoes this process.

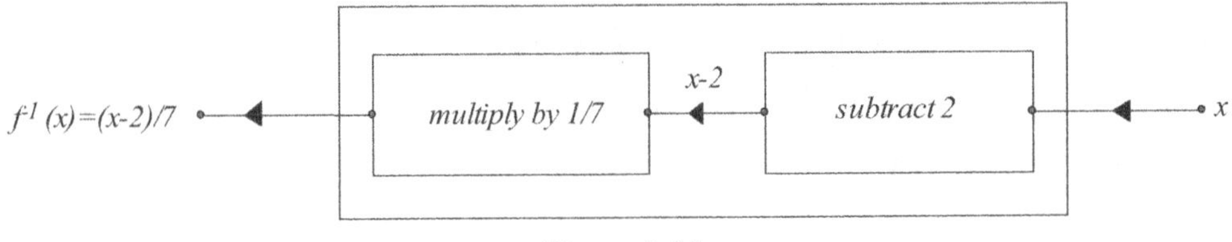

Figure 2.91

So, if $f : x \mapsto 7x+2$, then the inverse function is $f^{-1} : x \mapsto \dfrac{1}{7}(x-2)$.

The inverse function can also be obtained in a more practical, if less illuminating, manner. The inverse f^{-1} is obtained by interchanging x for y and then by making y the subject of the equation:

$$f : x \mapsto 7x+2 \Rightarrow y = 7x+2$$

$$x = 7y+2 \Rightarrow 7y = x-2 \Rightarrow y = \frac{1}{7}(x-2)$$

$$\text{Therefore } f^{-1} : x \mapsto \frac{1}{7}(x-2).$$

Example 2.16: If $f(x) = 4 - 3x$,

      (a) find $f^{-1}(x)$

      (b) show that (i) $f \circ f^{-1} = i$ (ii) $f^{-1} \circ f = i$

Solution 2.16: (a) $y = 4 - 3x$. Interchange x and y and then make y the subject.

$$x = 4 - 3y \Rightarrow 3y = 4 - x \Rightarrow y = \frac{4-x}{3}. \text{ Therefore, } f^{-1}(x) = \frac{4-x}{3}.$$

(b)   (i) $\left(f \circ f^{-1}\right)(x) = f\left(f^{-1}(x)\right) = f\left(\frac{4-x}{3}\right) = 4 - 3\left(\frac{4-x}{3}\right) = 4 - (4-x) = x$ but

      $i(x) = x$ therefore $f \circ f^{-1} = i$.

    (ii) $\left(f^{-1} \circ f\right)(x) = f^{-1}\left(f(x)\right) = f^{-1}(4 - 3x) = \frac{4 - (4 - 3x)}{3} = \frac{3x}{3} = x$. Therefore

      $f^{-1} \circ f = i$.

2.6.4 The Relation between the Graph of $y = f(x)$ and the Graph of $y = f^{-1}(x)$

The following three examples (figures 2.92, 2.93 and 2.94) are intended to show the relation between the graph of a function and its inverse. In each case, the graph of $i(x) = x$ is also shown as a thin line.

(a) $f(x) = 7x + 2$, $f^{-1}(x) = \frac{1}{7}(x - 2)$

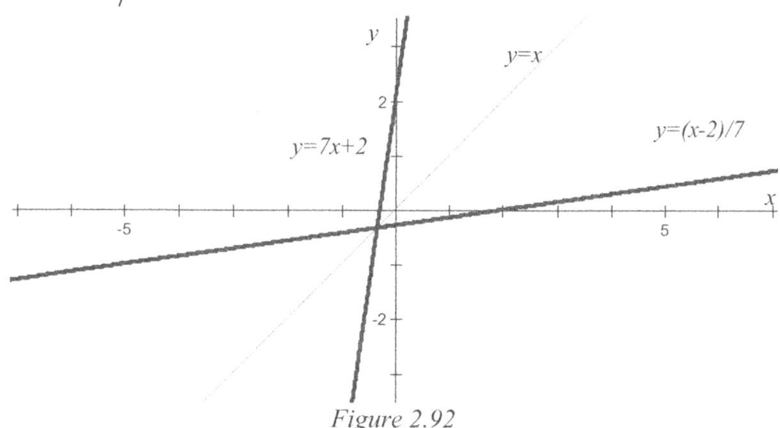

Figure 2.92

(b) $f(x) = x^2 + 3,\ x \geq 0$, $f^{-1}(x) = \sqrt{x-3},\ x \geq 3$

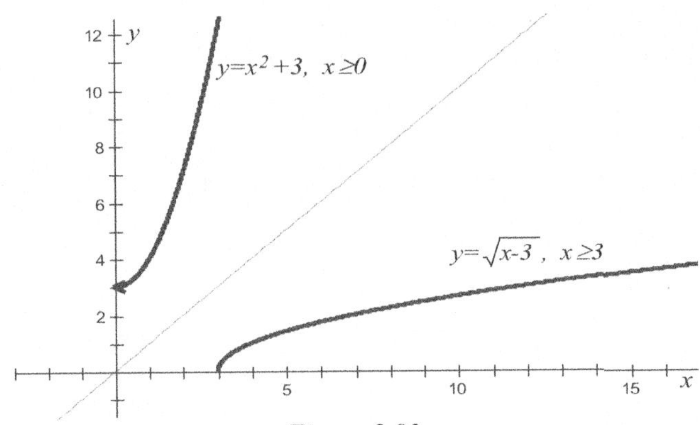

Figure 2.93

(c) $f(x) = x^{\frac{1}{3}} + 2$, $f^{-1}(x) = (x-2)^3$

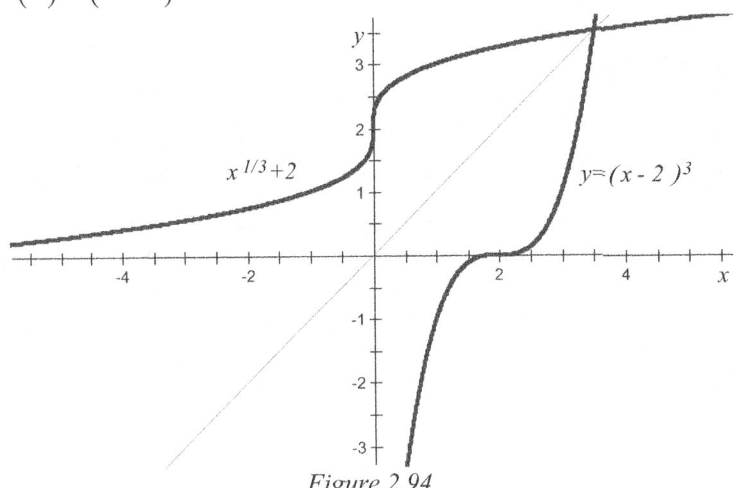

Figure 2.94

It is clear that the graph of a function and the graph of its inverse are reflections of each other in the line $y = x$.

Example 2.17: Find f^{-1}, the inverse of the function $f : x \mapsto x^2,\ x \geq 0$ and draw, on the same axes, a sketch of $y = f(x)$ and $y = f^{-1}(x)$.

Solution 2.17: $f : x \mapsto x^2 \Rightarrow y = x^2$. Exchange x and y so that $y^2 = x \Rightarrow y = \sqrt{x} \Rightarrow f^{-1}(x) = \sqrt{x}$. Therefore, the inverse of f is $f^{-1} : x \mapsto \sqrt{x},\ x \geq 0$, as shown in figure 2.95.

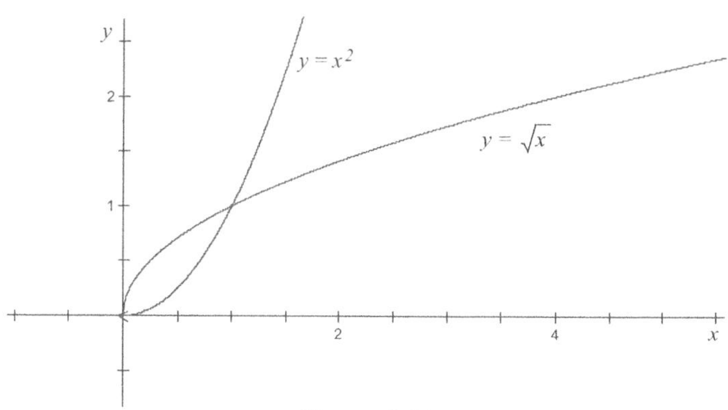

Figure 2.95

Exercise 2.9

1.   For the functions $f : x \mapsto 3x+7$ and $g : x \mapsto 4-x$ evaluate

    (a) $(f \circ g)(1)$ (b) $(g \circ f)(4)$ (c) $(f \circ f)(2)$ (d) $(g \circ g)(-1)$

2.   For the functions $f(x) = x+1$ and $g(x) = 3x$, show that

    (a) $(g \circ f)(x) = 3x+3$ (b) $(f \circ g)(x) \neq f(x) \times g(x)$

    (c) $f^2(x) = x+2$ (Note that $f^2 = f \circ f$.)

3.   If $f(x) = 5x-1$, $g(x) = \dfrac{1}{2}x+4$, $h(x) = 1-2x$ find

    (a) $(g \circ f)(x)$ (b) $(f \circ g)(x)$ (c) $(g \circ h)(x)$ (d) $(f \circ h)(x)$

4.   The functions f, g and h are defined as follows: $f : x \mapsto 2-3x$, $g : x \mapsto x^2-3$,

    $h : x \mapsto \dfrac{1}{x}$, $x \neq 0$. Find

    (a) $f \circ g$ (b) $g \circ h$ (c) $h \circ f$ (d) $h \circ h$ (e) $g \circ g$

5.   For $f(x) = 2x$, $g(x) = x+5$ and $h(x) = 1-x$, show that $f \circ (g \circ h) = (f \circ g) \circ h$.

6.   (a) Show that if $f : x \mapsto 2x-1$ and $g : x \mapsto \dfrac{1}{2}x+\dfrac{1}{2}$, then $g \circ f = i$.

    (b) Show that if $f : x \mapsto 2-x$ then $f^2 = i$.

7.   Find the inverses of the given functions.

    (a) $f(x) = 3x+1$ (b) $g(x) = 3-x$ (c) $h(x) = 1-4x$

    (d) $f(x) = \dfrac{3}{x}$, $x \neq 0$ (e) $g(x) = \dfrac{2x+1}{3}$ (f) $h(x) = \dfrac{x+4}{x-2}$, $x \neq 2$

8.   If $f : x \mapsto x+5$, $g : x \mapsto 4x$, $h : x \mapsto 2-x$ find

    (a) $f^{-1}(4)$ (b) $g^{-1}(-1)$ (c) $h^{-1}(2)$

    (d) $(f \circ g)^{-1}(3)$ (e) $(h^{-1} \circ g^{-1})(-1)$

9.   If $f:x\mapsto 5x$, $g:x\mapsto 1-2x$, show that $\left(f\circ g\right)^{-1}=g^{-1}\circ f^{-1}$.

2.7 The Exponential Function $x\mapsto a^x$, $a>0$, $x\in\mathbb{Q}$

2.7.1 The Graph of $f:x\mapsto a^x$, $a>0$, $x\in\mathbb{Q}$

Some exponents of negative real bases, such as $(-3)^{\frac{1}{2}}$ and $(-5)^{\frac{1}{4}}$, have non-real values. Therefore, a is restricted to positive values. In addition, you should note that the domain is the set of rational numbers, $\mathbb{Q}$.

Figure 2.96 shows the graph of $y=a^x$ for different values of a .

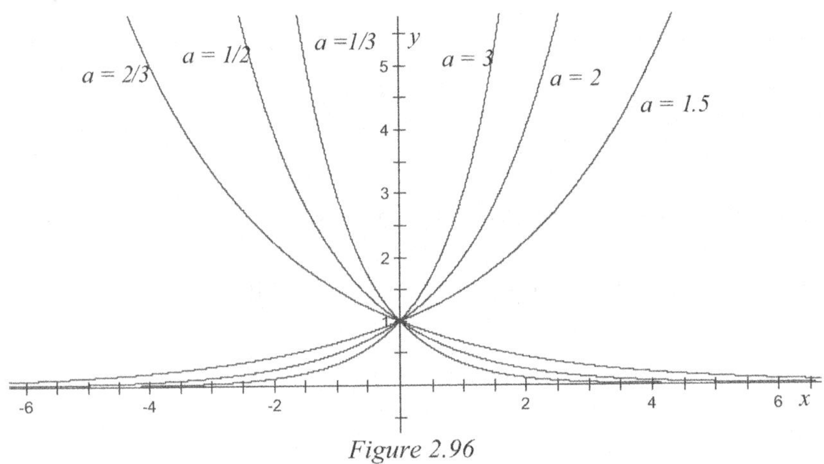

Figure 2.96

From figure 2.96, you can see that the graph of $y=a^x$
- passes through the point $(0,\ 1)$ for all values of $a>0$
- has a horizontal asymptote, $y=0$ for all values of $a>0$, $a\neq 1$
- approaches $y=0$ as x becomes large in the negative direction if $a>0$
- approaches $y=0$ as x becomes large in the positive direction if $0<a<1$
- has a range $y>0$ for all values of $a>0$, $a\neq 1$
- is a straight line $y=1$ if $a=1$

2.7.2 The Inverse Function $f^{-1}:x\mapsto\log_a x$, $x>0$

The laws of exponents and logarithms from Unit 1 indicate that if $f:x\mapsto a^x\Rightarrow y=a^x$
$\Rightarrow x=a^y\Rightarrow y=\log_a x$. So, $f^{-1}:x\mapsto\log_a x$, $x>0$, $(a>0)$, and as you should expect from section
2.6.4, the graph of $y=f^{-1}(x)$ is the reflection of the graph of $y=f(x)$ in the line $y=x$.

Figure 2.97 shows the graphs of $y=a^x$, $y=\log_a x$ and $y=x$, for $a=2$ and you can see that
$y=\log_a x$ is the reflection of $y=a^x$ in the line $y=x$ and also vice versa.

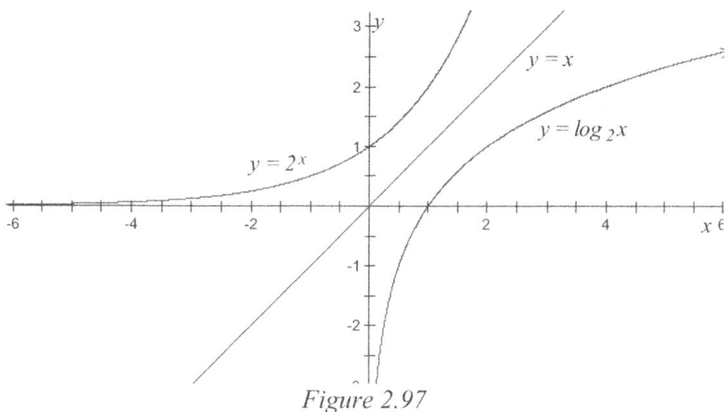

Figure 2.97

Also, from section 2.6.3, $\left(f \circ f^{-1}\right)(x) = i(x) = x$.

If this is applied to the exponential function $f : x \mapsto a^x$, then $\left(f \circ f^{-1}\right)(x) = f\left(f^{-1}(x)\right)$

$= f\left(\log_a x\right) = a^{\log_a x} = x$.

If this is applied to the logarithmic function $f : x \mapsto \log_a x$, then $\left(f \circ f^{-1}\right)(x) = f\left(f^{-1}(x)\right)$

$f\left(a^x\right) = \log_a a^x = x$.

2.7.3 Solution of Equations of the Form $a^x = b$

The logarithmic function can be used to solve equations of the form $a^x = b$. Suppose you want to solve the equation $5^x = 33$. Then let $f(x) = 5^x$ so that $f^{-1}(x) = \log_5 x$ and hence $x = \log_5 33$.

If you need to evaluate x, then it is necessary to change the base of the logarithm from 5 to 10 (or e) (see Unit 1): $x = \log_5 33 = \dfrac{\log_{10} 33}{\log_{10} 5} \Rightarrow x = 2.172502... = 2.17$, correct to three significant figures

Of course, if the right hand side of the equation is expressible as a power of the base of the left-hand side, then the equation can be solved without logarithms by using the laws of exponents (see Example 2.18(a)).

Example 2.18: (a) Solve the equation $3^{2x-1} = \dfrac{1}{9}$.

        (b) Solve the equation $3^{2x-1} = 11$

            (i) by using the logarithmic function

            (ii) by using your graphing calculator to find where the function $y = 3^{2x-1}$ intersects the line $y = 11$

67

Solution 2.18: (a) $3^{2x-1} = \dfrac{1}{9} = 3^{-2} \Rightarrow 2x-1 = -2 \Rightarrow 2x = -1 \Rightarrow x = -\dfrac{1}{2}$

(b) (i) $\log_3\left(3^{2x-1}\right) = 2x-1 = \log_3 11 \Rightarrow x = \dfrac{1}{2}\left(\log_3 11 + 1\right) = \dfrac{1}{2}\left(\dfrac{\log_{10} 11}{\log_{10} 3} + 1\right)$

$x = 1.591329... = 1.59$, correct to three significant figures.

(ii) Figures 2.98 to 2.103 show the procedure.

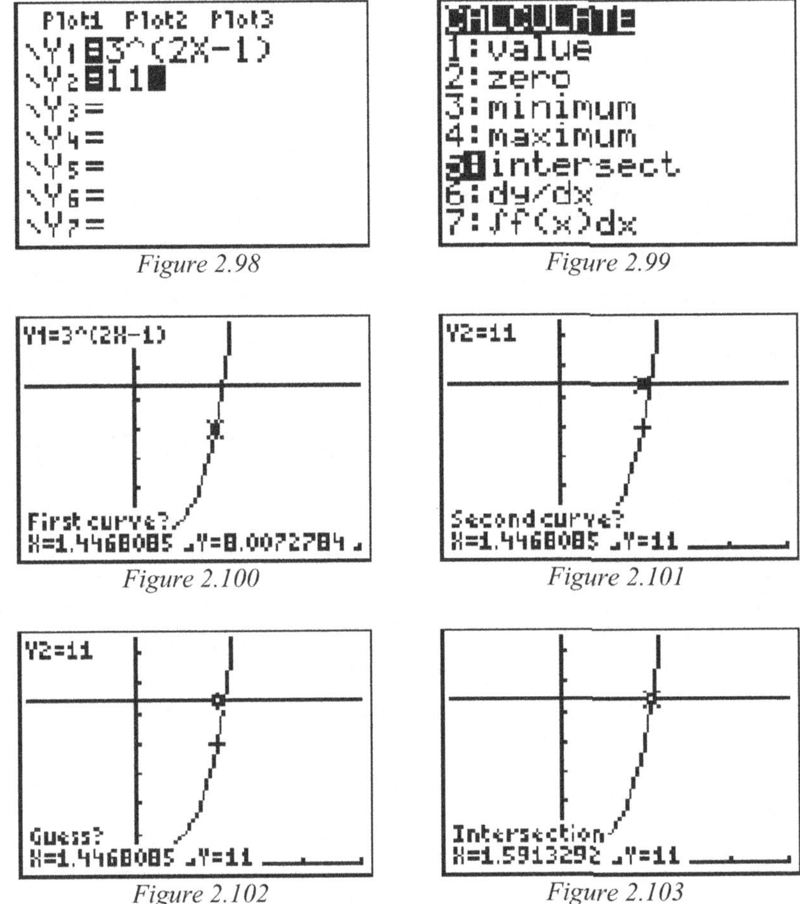

Figure 2.98 *Figure 2.99*

Figure 2.100 *Figure 2.101*

Figure 2.102 *Figure 2.103*

Exercise 2.10

1. Draw the graph of $y = f(x)$ for the given domain.

(a) $f(x) = 2^x$, $-3 \le x \le 3$ (b) $f(x) = 2^{1+x}$, $-2 \le x \le 2$

(c) $f(x) = 3^{x^2}$, $-1.5 \le x \le 1.5$ (d) $f(x) = (5.8)^{-x^2}$, $-2 \le x \le 2$

(e) $f(x) = 2^{-x} + 3^x$, $-3 \le x \le 2$

2. Draw, on the same axes, the graphs of $y = 2^x$ and $y = x$ for the domain $-4.5 \le x \le 4.5$. Use the property that the graph of $y = f^{-1}(x)$ is the reflection of $y = f(x)$ in the line $y = x$ to draw the graph of $y = \log_2 x$, $x > 0$.

3. Draw the graph of $y = 3^{\left(1-\frac{x}{2}\right)}$ for the domain $-3 \leq x \leq 6$ and use it to draw the graph of $y = 2\left(1 - \log_3 x\right)$.

4. Find the value of

(a) $a^{\log_a 5}$ (b) $a^{3\log_a 2}$ (c) $a^{-2\log_a 3}$ (d) $a^{\frac{1}{2}\log_a 16}$ (e) $a^{\log_{a^2} 3}$

5. If $\log_a 3 = 4.931$, evaluate   (a) $\log_3 a$ (b) $\log_a 27$ (c) $\log_{3a} a^2$

6. Solve the following equations by using the laws of exponents, if possible, and giving the answer exactly, or, if not possible, by using the logarithmic function and then by changing the base, giving your answer correct to three significant figures. If you use logarithms, carry out a check of your answer using your graphing calculator.

(a) $2^x = 5$ (b) $5^{2x-3} = 39$ (c) $3^x = 243$ (d) $3^{1-x} = 4$ (e) $2^{3-x} = \dfrac{1}{16}$

2.8 The Functions $x \mapsto e^x$ and $x \mapsto \ln x,\ x > 0$

2.8.1 The Significance of the Number e

Suppose the number of bacteria, n_0, in a dish doubles in unit time (the actual time period is unimportant). If a very simple growth model is adopted, you can think of the number of bacteria as remaining constant during the period $0 \leq t < 1$ and then, at $t = 1$, the number doubles instantaneously.

This, of course, is not very realistic. To make it more realistic, suppose that the number of bacteria remains constant during the periods

$$0 \leq t < 0.1, \quad 0.1 < t < 0.2, \quad 0.2 < t < 0.3, \ldots, 0.9 < t < 1,$$

and the number of bacteria grows by 10% (one tenth) at times

$$t = 0.1, \quad t = 0.2, \quad t = 0.3, \ldots, t = 0.9, \quad t = 1.$$

Then, when $t = 1$, the number of bacteria is $n_0 \left(1 + \dfrac{1}{10}\right)^{10} = 2.593742 n_0$, and you can see that the number has more than doubled. In fact, it has increased by a factor of almost 2.6 because of the compounding effect (see Unit 1).

A better growth model can be obtained by reducing the length of the time intervals and increasing the number of intervals. For example, suppose that the number of bacteria remains constant during the periods

$$0 \leq t < 0.01, \quad 0.01 < t < 0.02, \ldots, \quad 0.99 < t < 1,$$

and the number of bacteria grows by 1% (one hundredth) at times

$$t = 0.01, \quad t = 0.02, \ldots, t = 0.99, \quad t = 1.$$

Then, when $t=1$, the number of bacteria is $n_0\left(1+\dfrac{1}{100}\right)^{100} = 2.704814\,n_0$ and the number has increased by a factor of more than 2.7.

The growth model can be refined further by reducing the length of the time intervals and increasing the number of intervals correspondingly so that a model of continuous growth is achieved. When this is done, you will see that the compounding effect does not cause growth to continue indefinitely but to reach a limit.

Suppose the number of bacteria remains constant during each of the n time intervals and the number of bacteria increases by one n^{th} at the end of each of the n intervals. Then, when $t=1$, the number of bacteria is $n_0\left(1+\dfrac{1}{n}\right)^{n}$. Table 2.1 shows the value of $\left(1+\dfrac{1}{n}\right)^{n}$ for various large values of n.

Table 2.1 shows that as n gets larger so the value of $\left(1+\dfrac{1}{n}\right)^{n}$ gets closer and closer to a fixed number that, correct to eight decimal places, is 2.71828183. This number is called 'e', and it is significant because it is closely related to models of continuous or natural growth.

n	$\left(1+\dfrac{1}{n}\right)^{n}$
10	2.59374246
10^3	2.716923932
10^5	2.718268237
10^7	2.718281693
10^9	2.718281827
e	2.718281828

Table 2.1

2.8.2 The Graph of $y=e^x$ and its Inverse Graph $y=\ln x$

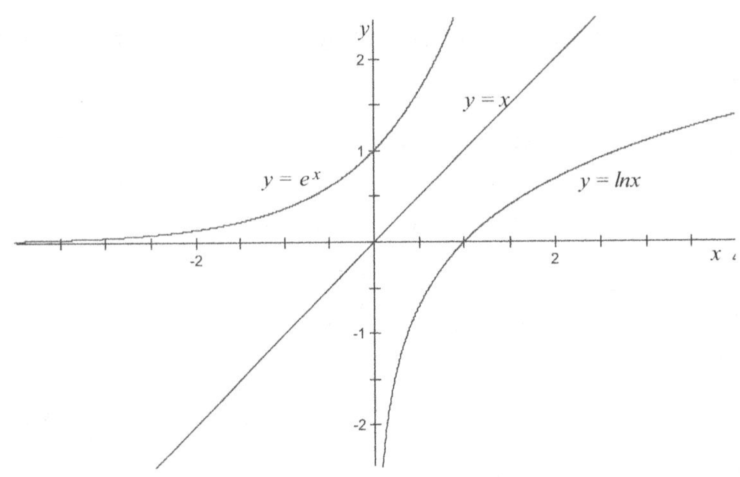

Figure 2.104

Figure 2.104 shows the graphs of $y=e^x$ and $y=\ln x$, $x>0$. "$\ln x$" is just a convenient way of writing $\log_e x$. Because $y=\ln x$, $x>0$ is the inverse of $y=e^x$, its graph is the reflection of the graph of $y=e^x$ in the line $y=x$. Equally, you could regard $y=e^x$ as the inverse of the $y=\ln x$, $x>0$ so that the graph of $y=e^x$ is the reflection of the graph of $y=\ln x$ in the line $y=x$.

Example 2.19: If $f(x) = e^{2x-5}$,

  (a) find $f^{-1}(x)$ and state any restrictions necessary on the domain.

  (b) draw, on the same axes, the graphs of $y = f(x)$ and $y = f^{-1}(x)$.

Solution 2.19: (a) $y = e^{2x-5}$ and, exchanging x and y $\Rightarrow x = e^{2y-5} \Rightarrow \ln x = \ln e^{2y-5} \Rightarrow \ln x = 2y - 5$

$\Rightarrow y = \dfrac{1}{2}(\ln x + 5)$. The domain is restricted to positive real values because it is

possible only to take the logarithm of a positive number. Therefore,

$f^{-1}(x) = \dfrac{1}{2}(\ln x + 5)$, $x > 0$.

  (b) Figure 2.105 shows the graphs of $y = f(x)$ and $y = f^{-1}(x)$.

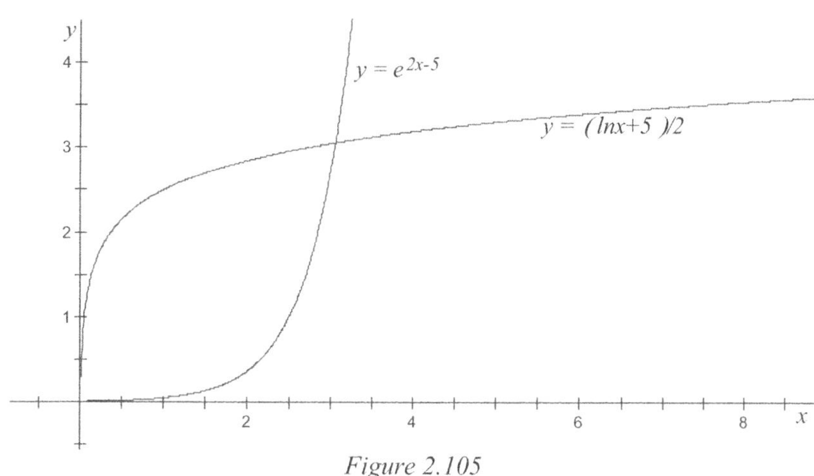

Figure 2.105

2.8.3 The Solution of the Equation $a^x = b$

It is useful to be able to write the function $x \mapsto a^x$ in terms of e. You saw in section 2.7.2 that $u = a^{\log_a u}$ so that, putting $a = e$, $u = e^{\ln u}$ and, putting $u = a^x$, you get $a^x = e^{\ln a^x}$. An alternative solution to the one given in Unit 1 for equations of the form $a^x = b$ is offered here, based on this process.

If $a^x = b$, then, as $a^x = e^{\ln a^x} = e^{x \ln a}$, $e^{x \ln a} = b$, and if natural logarithms are taken, $\ln e^{x \ln a} = \ln b$

$\Rightarrow x \ln a = \ln b \Rightarrow x = \dfrac{\ln b}{\ln a}$.

Example 2.20: Solve the equation $7^{4-3x} = 25$, giving your answer correct to four decimal places.

Solution 2.20: $7^{4-3x} = e^{\ln 7^{(4-3x)}} = e^{(4-3x)\ln 7} \Rightarrow e^{(4-3x)\ln 7} = 25$

$\Rightarrow \ln e^{(4-3x)\ln 7} = \ln 25 \Rightarrow (4 - 3x)\ln 7 = \ln 25$

$$\Rightarrow 4-3x = \frac{\ln 25}{\ln 7} \Rightarrow x = \frac{1}{3}\left(4 - \frac{\ln 25}{\ln 7}\right) = 0.7819417... = 0.782 \text{ , correct to three significant}$$

figures.

Exercise 2.11

1. If $f : x \mapsto e^{3x}$, find f^{-1} and draw, on the same axes, the graphs of $y = f(x)$ and $y = f^{-1}(x)$. State the transformation that maps $y = f(x)$ to $y = f^{-1}(x)$.

2. The function g is such that $g(x) = e^{\frac{1}{x}}$, $x \neq 0$.

(a) Find $g^{-1}(x)$.

(b) Draw, on the same axes, the graphs of $y = g(x)$ and $y = g^{-1}(x)$, stating the transformation that maps $y = g(x)$ to $y = g^{-1}(x)$.

3. Find the inverse function, f^{-1}, given the following functions, f.

(a) $f(x) = 3e^{-x}$ \qquad (b) $f(x) = 2\ln(x+1)$, $x > -1$

(c) $f(x) = e^{-x^2}$, $x > 0$ \qquad (d) $f(x) = \ln(3x-4)$, $x > \dfrac{4}{3}$

4. Write the following functions in the form e^{kx}, where k is written correct to three significant figures. For example, $2^x = e^{\ln 2^x} = e^{x\ln 2} = e^{0.693x}$.

(a) 3^x \qquad (b) 4^{-2x} \qquad (c) $10^{0.1x}$ \qquad (d) $7^{0.5x}$ \qquad (e) $\left(\dfrac{2}{5}\right)^x$

5. Solve the following equations by using natural logarithms and give your answers correct to three significant figures.

(a) $2^{x+1} = 5$ \qquad (b) $3^{2x} = 22$ \qquad (c) $1.3^{-2x} = 0.165$ \qquad (d) $5 \times 4^{2x-3} = 12$

2.8.4 The Exponential Function Applied to Population Growth

In Unit 1, geometric sequences were used to model population growth, and, in section 2.8.1, the number e was obtained from a geometric model by compounding growth over shorter and shorter periods of time in order to obtain a model for continuous or natural growth.

Suppose that the population of the United States increases by 1.1% per year and its population in 2003 is 280 million people. Then the estimated population, in millions, in subsequent years, can be modeled by a geometric sequence whose first term is 280 and whose common ratio is

$$\left(1 + \frac{1.1}{100}\right) = 1 + 0.011 = 1.011$$

For $a = 280$, $r = 1.011$ the geometric sequence is a, ar, ar^2, ar^3, ... so that the population of the United States in the years 2003, 2004, 2005, ... could be represented by the sequence

$$280, \ 280 \times 1.011, \ 280 \times 1.011^2, \ 280 \times 1.011^3, \ \ldots \ = 280, \ 283.08, \ 286.19, \ 289.34, \ldots$$

However, the function $f(t) = 280e^{0.011t}$ offers an alternative and more refined model. Using this model, the estimated population of the United States in the years 2003, 2004, 2005, ... is

$$f(0), \quad f(1), \quad f(2), \quad f(3), \ldots = 280, \ 283.10, 286.23, \ 289.39, \ldots$$

The exponential function can be used to estimate increases or decreases in a wide variety of population models.

Example 2.21: The number of elm trees in a forest is declining. The function $f(t) = n_0 e^{-kt}$ models this decline, where n_o is the number of trees at the beginning of 2003, when $t = 0$, and time, t, is measured in years. Find

(a) the number of trees at the beginning of 2007 if $n_0 = 4500$ and $k = 0.0113$.

(b) approximately when the number of trees is expected to fall below 4000.

Solution 2.21: (a) $n = 4500e^{-0.0113 \times 4} = 4301.128\ldots$, so the number of elm trees at the beginning of 2007 is estimated to be 4300.

(b) $4500e^{-.0113t} < 4000 \Rightarrow e^{-.0113t} < \dfrac{8}{9} \Rightarrow \ln\left(e^{-.0113t}\right) < \ln\dfrac{8}{9}$

$\Rightarrow -0.0113t < -\ln\dfrac{8}{9} \Rightarrow t > \dfrac{\ln(8/9)}{-0.0113} \Rightarrow t > 10.42327\ldots$. Therefore, one would expect the number of elm trees to fall below 4000 after 11 years, that is by 2014.

Exercise 2.12

1. The population of an organism at time t, measured in days, is modeled by the function $f(t) = 100e^{0.5t}$, $t \geq 0$. Find the population of the organism after

(a) 3 days   (b) 1 week   (c) 4 weeks

2. An endangered species of bird is declining in numbers. A survey reveals that its population can be modeled by the formula $n = 250e^{-0.08t}$, $t \geq 0$ where n is the number of birds at time t, measured in years from January 2002. Estimate

(a) the number of birds in January 2002.

(b) the decline in the number of birds during the year 2005.

(c) when the population of birds falls below 100.

3. A bank offers 4.3% interest per year on investments. Jim invests $10 000. Jim uses an exponential model to work out the interest he earns over a five-year period. Find his estimate for the interest earned during a five-year period and compare this with the interest he would actually earn if the interest was compounded (a) annually and (b) monthly (see Unit 1).

4. An island is populated by two species of seal, A and B. The number of A seals is given by the function $n_A = 5000e^{0.02t}$ and the number of B seals is given by the function $n_B = 3500e^{0.025t}$, where t is measured in years. Find

(a) the initial number of A seals and the initial number of B seals.
(b) the number of A seals after 10 years.
(c) how many years (to the nearest year) it will take for the number of B seals to double.
(d) how many years it will take for the population of A and B seals to be equal.

5. A statistical survey reveals that, in an election, candidate A has the support of 28% of the electorate and his support is increasing at the rate of 0.4% per day. Candidate B, according to the survey, has support of 40% of the electorate, but her support is declining at the rate of 1.2% per day.

If the election is due to take place three weeks after the survey, find out, assuming that the survey is accurate, which candidate is likely to win. Use the model $p = p_0 e^{kt}$, where t is time measured in days, p_o is the current percentage support ($t = 0$) and p is the percentage support at time t.

6. Thirty trout are introduced into a large pond. Given that the number of trout in the pond at time $t = 0$ is modeled by the function $f : t \mapsto n_0 \left(1 - 0.9e^{-0.15t}\right)$, $t \geq 0$, where time t is measured in weeks,

(a) show that the value of $n_0 = 300$.

(b) draw a sketch of $n = f(t)$ for the domain $0 \leq t < 50$ where n is the number of trout in the pond at time t.

(c) estimate the number of trout in the pond after 6 weeks.
(d) estimate the number of trout in the pond after a long time.

Unit 2 Review Exercises

Part 1

No graphing calculators should be used to answer questions in Part 1.

1. Consider the function $f(x) = x^2 - 4x + 13$

(a) Given that f can be written in the form $f(x) = (x - a)^2 + b$ find a and b.

(b) The function $g(x) = x^2 + px + q$ is formed from f by a translation of $\begin{pmatrix} 2 \\ -5 \end{pmatrix}$. Find the value of p and of q.

(c) State the range of f.

2. Figure 2.106 shows part of the graph of

the function $y = a + \dfrac{bx}{x^2 - c}$, $x \neq \pm\sqrt{c}$.

The asymptotes, whose equations are
$x = -2$, $x = 2$ and $y = 2$ are show as
dotted lines.

(a) Write down the value of a and c.

(b) Given that the graph passes through

the point P with coordinates $(3,\ 5)$

find the value of b.

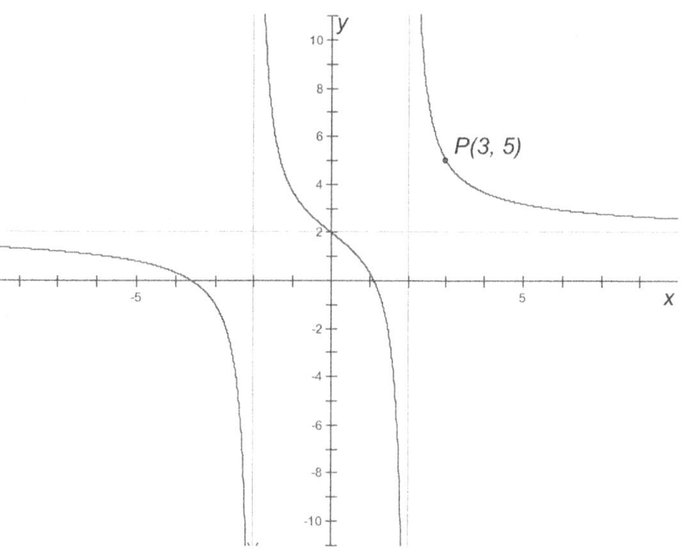

Figure 2.106

3. Let $f(x) = 2 - 3x$, $g(x) = 1 - \dfrac{3}{x}$, $x \neq 0$

(a) Given that $(f \circ g)(x) = \dfrac{k - x}{x}$, $x \neq 0$ find the value of k.

(b) Show that $(g \circ f)(x) = \dfrac{3x + 1}{3x - 2}$

(c) Solve the equation $(f \circ g)(x) = 3$

4. The graph of $y = a(x - b)^2 + c$, $-1 \leq x \leq 7$ is shown in

the figure 2.107. The x-intercepts are $(2,\ 0)$, $(4,\ 0)$ and

the y-intercept is $(0,\ 24)$.

(a) Write down the equation of the line of symmetry of
    the graph.

(b) Write down the value of b.

(c) Find the value of a and of c.

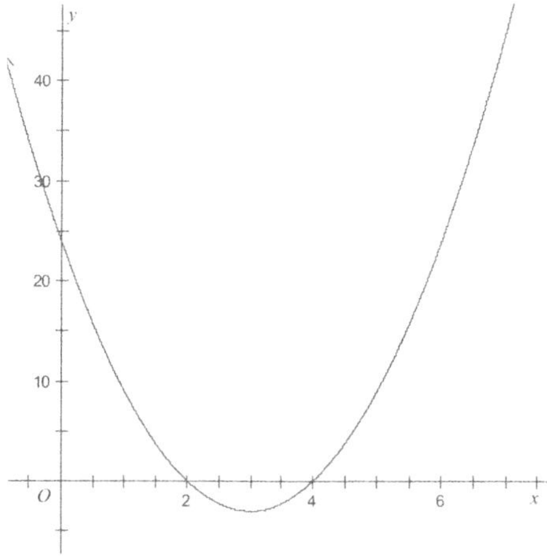

Figure 2.107

5. The graph of $f(x) = -k + kx - x^2$ is shown in figure 2.108

(a) Given that P has coordinates $(0, -4)$ and Q has coordinates $(2, 0)$, show that $k = 4$.

The graph of f is given a stretch of factor 2 parallel to the x-axis to form the graph of function g.
(a) Copy figure 2.108 and add to it a sketch of the graph of g.
(c) Find the equation of the function g.

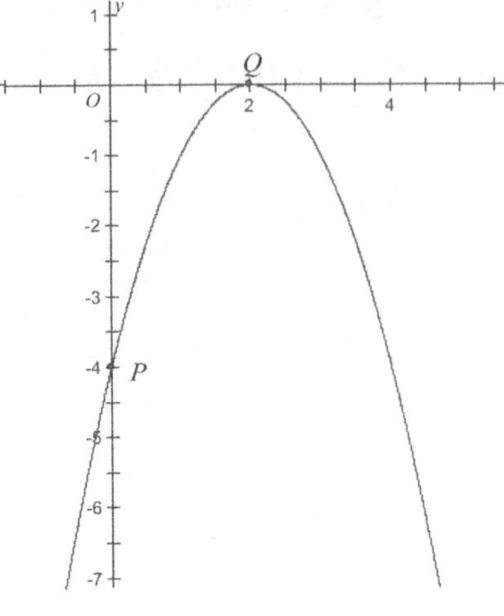

Figure 2.108

6. The function $f(x) = x^2 - 6x + 8$ can be written in the form

$$f(x) = (x-a)(x-b)$$

(a) Find the value of a and b.
(b) Write down the x-intercepts of the graph of f.

The graph of f is translated by $\begin{pmatrix} 0 \\ k \end{pmatrix}$ to the graph of function g

shown in figure 2.109.
(c) Find the value of k.

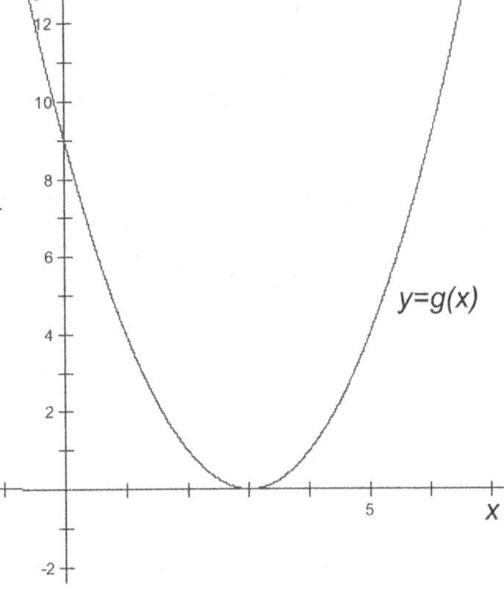

Figure 2.109

7. Consider the functions f and g where $f(x) = x + 4$ and $g(x) = 2x + 1$. Find

(a) f^{-1}, the inverse of function f.
(b)   (i) $(f^{-1} \circ f)(x)$

        (ii) $(g^{-1})^{-1}$
(c) $(g \circ f)(x)$

8. Find the value, or values, which k should take so that the equation $x^2 + kx - k + 3 = 0$ has only one solution.

9. The graph of $f(x) = \dfrac{1}{x}$, $x \neq 0$ is transformed by a translation of $\begin{pmatrix} 2 \\ 1 \end{pmatrix}$ to the graph of

$g(x) = \dfrac{x-a}{x-b}$, $x \neq b$.

(a) Find the value of a and of b.

(b) Sketch graphs of both f and g on the same axes and show all the asymptotes as dotted lines.

10. Let $f(x) = 2x^2 - 4x - 16$

(a) Show that $f(x) = 2(x-1)^2 - 18$

(b) For the graph of f write down the coordinates of
      (i)   the x-intercepts
      (ii)   the y-intercept
      (iii)   the vertex

(c) Sketch the graph of f.

(d) The graph of g passes through the origin and is obtained from the graph of f by a translation of $\begin{pmatrix} c \\ 0 \end{pmatrix}$. Find a value for c.

Part 2

Graphing calculators will usually be needed to answer questions in part 2

11. Consider the function $f(x) = \dfrac{3}{x-2} + \dfrac{1}{3}x^2$, $x \neq 2$.

(a) Sketch the curve of f for $-8 \leq x \leq 6$, including also the asymptote.

(b) Write down the equation of the asymptote.

(c) Find the x-intercept and the y-intercept.

12. Consider the function $f(x) = \dfrac{1}{e^x(2x-5)} + 1$, $x \neq \dfrac{5}{2}$

(a) Sketch the curve of f for $-5 \leq x \leq 7$ including any asymptotes.

(b) Write down the equation of the vertical asymptote.

(c) Find the values, correct to 7 significant figures, of
     (i) $f(3)$ (ii) $f(5)$ (iii) $f(10)$ (You will need to extend the domain for (iii))

(d) Hence, or otherwise, write down the equation of the horizontal asymptote.

13. Let $f(x) = 2 - e^{-x}$ and $g(x) = \ln(x+3)$, $x > -3$.

(a) Find the range of f.

(b) Sketch the graphs of $y = f(x)$ and $y = g(x)$ on the same axes for the domain, $-5 \leq x \leq 5$.

(c) Find all the solutions of the equation $f(x) = g(x)$.

14. Initially an iceberg has mass 30 000 kg. The mass m kg of an iceberg after time t days is modeled by the function $m(t) = 30000e^{-kt}$. The mass of the iceberg after 10 days is 12 000 kg.
(a) Find the value of k.
(b) Find the mass of the iceberg after 7 days.
(c) After how many complete days is the mass of the iceberg less than 1000 kg?

15. Let $f(x) = 4\ln(x+2) - \dfrac{7}{x+2}$, $\quad x > -2$

(a) Sketch the graph of f for the domain $-5 \le x \le 10$
(b) Write down the equations of any asymptotes.
(c) Find the coordinates of the y-intercept.
(d) Solve the equation, $f(x) = 10 - x$.

16. Consider the function $f(x) = ax^2 + bx + c$. The graph of f passes through points $A(4,\ 0)$, $B(0,\ 20)$ and $C(2, -10)$.

(a) Show that $c = 20$
(b) Form a pair of simultaneous linear equations with a and b.
(c) Solve the equations from (b) and hence show that $f(x) = 5x^2 - 25x - 20$.

17. Let $g(x) = \dfrac{x^2 + 3}{x^2 + 1}$.

(a) Sketch the graph of g for the domain, $-5 \le x \le 5$.
(a) Find the equation of the horizontal asymptote.
(c) Write down the range of g.
(d) Solve the equation $g(x) = x^2$.

18. The graph of $f(x) = x^2$ is transformed to the graph of g by a translation of 5 units in the positive x-direction.
(a) Write down the function g.

The graph of g is now transformed to h by a scale factor of 3 in the y-direction.
(b) Write down the function h.

Finally h is transformed to j by a translation of 2 units in the positive y-direction.
(c) Write down the function j.

If the three transformations are carried out on f in the reverse order the function obtained is k.
(d) Find the function k.

19.   100 green frogs and 100 yellow frogs are introduced into a pond. The number n_1, of green frogs after time t weeks, is modeled by the function $n_1 = \dfrac{200e^{0.3t}}{1+e^{0.3t}}$ and the number n_2, of yellow frogs after time t weeks, is modeled by the function $n_2 = \dfrac{200}{1+e^{0.3t}}$.

(a) Sketch, on the same axes, graphs of the two models for a sufficient period of time so that the population trends of both green and yellow frogs are clear.

(b) How many of each type of frog are there after 5 weeks?

(c) Describe what happens to the two frog populations.

(d) Show that the total population of frogs remains constant.

UNIT 3: CIRCULAR FUNCTIONS AND TRIGONOMETRY

3.1 Radian Measure

3.1.1 The Definition of a Radian

Angles have been measured in degrees since Babylonian times about 4000 years ago, but the fact that a right angle contains $90°$ is quite arbitrary, and it is necessary to adopt a more fundamental measure of angle. This measure is known as a *radian*, and it is defined as the angle at the center of a unit circle made by a point which rotates 1 unit around that circle. Figure 3.01 shows a radian obtained in this way.

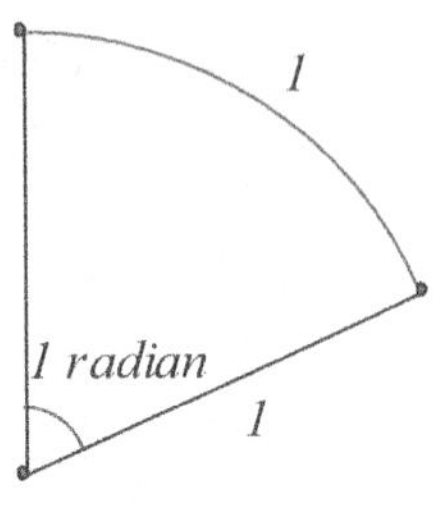

Figure 3.01

3.1.2 Length of Arc

The formula for the length of an arc of a circle arises directly from the definition of a radian. If a radius of length r is rotated through an angle of θ radians, then the length, l, of the arc generated is $r\theta$. Therefore, $l = r\theta$. A circle of radius r has circumference $2\pi r$, and so, according to the formula $l = r\theta$, the radian measure of a complete rotation is 2π; therefore, 2π radians is equivalent to $360°$.

3.1.3 Area of a Sector

The area of a circular disc of radius r is πr^2. Figure 3.02 shows a sector whose angle, in radians, is θ and whose radius is r. The ratio of the angle, θ of a sector is proportional to its area, A. Therefore, because a circle disc can be thought of as a sector whose angle is 2π radians and whose area is πr^2,

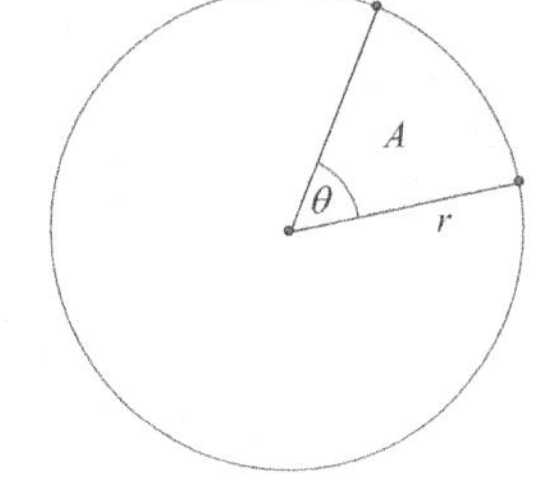

$\dfrac{\theta}{A} = \dfrac{2\pi}{\pi r^2} = \dfrac{2}{r^2}$. Therefore, $2A = r^2\theta \Rightarrow A = \dfrac{1}{2}r^2\theta$.

Figure 3.02

When an angle is given in radians, no units are used; however, when an angle is given in degrees, the symbol for degrees, $°$, must be used. It is convenient to write the radian measure of angles in terms of π. Table 3.1 shows some common angles measured in degrees with their radian equivalents given in terms of π.

Angle in Degrees	Angle in Radians
$180°$	π
$90°$	$\pi/2$
$60°$	$\pi/3$
$45°$	$\pi/4$
$30°$	$\pi/6$

Table 3.1

The use of degrees as a measure of angle is appropriate for many practical situations such as surveying, navigation and the solution of triangles. However, for most of the trigonometry encountered in this course, radian measure is used. In general, unless stated otherwise, use radians to measure angles.

Example 3.1: A circular disc has area 10cm^2. Find the angle of a sector of this disc with area 3cm^2.

Solution 3.1: As the angle of the sector is proportional to the area of the sector, $\dfrac{3}{10} = \dfrac{\theta}{2\pi}$, where θ is

the angle of the sector. Therefore, $\theta = \dfrac{6\pi}{10} = \dfrac{3}{5}\pi$ or $108°$.

Exercise 3.1

1. Write the following angles in degrees.
   (a) $\dfrac{2\pi}{3}$ (b) $\dfrac{2\pi}{5}$ (c) $\dfrac{5\pi}{6}$ (d) $\dfrac{4\pi}{9}$ (e) 3π (f) $\dfrac{3}{10}\pi$

   (g) $\dfrac{\pi}{12}$ (h) $\dfrac{7}{18}\pi$ (i) $\dfrac{19\pi}{24}$

2. Write the following angles in radians, giving your answers as a multiples of π.
   (a) $120°$ (b) $100°$ (c) $18°$ (d) $1°$ (e) $408°$ (f) $35°$
   (g) $21°$ (h) $55.5°$ (i) $210°$

3. Find the length of an arc of a circle of radius 5cm which is generated by rotating a radius through an angle of
   (a) 2 (b) $\dfrac{\pi}{9}$ (c) $40°$ (d) $137°$

4. A sector of a circle of radius 158mm has the given arc lengths, measured in millimeters. Find, in each case, the angle of the sector. Give your answer in both radians and degrees.
   (a) 10 (b) 150 (c) 35.4 (d) 78.9

5. Find the area of a sector of a circle of radius 6cm if its angle is
   (a) $\dfrac{1}{2}\pi$ (b) $\dfrac{1}{3}\pi$ (c) $\dfrac{5}{12}\pi$ (d) 1.29 (e) $18°$ (f) $73°$

6. A circular sector has a radius of 12m. Find the angle, in radians, of the sector given that its area, measured in square meters, is
   (a) 100 (b) 75 (c) 62.7

7. Figure 3.03, which is not drawn to scale, shows the sector of a circle. The center of the arc which defines the sector lies on the surface of the Earth, and the arc passes through the center of the moon. The angle of the sector is $0.5167°$ and its radius is 382100km. Find the diameter of the moon, in kilometers.

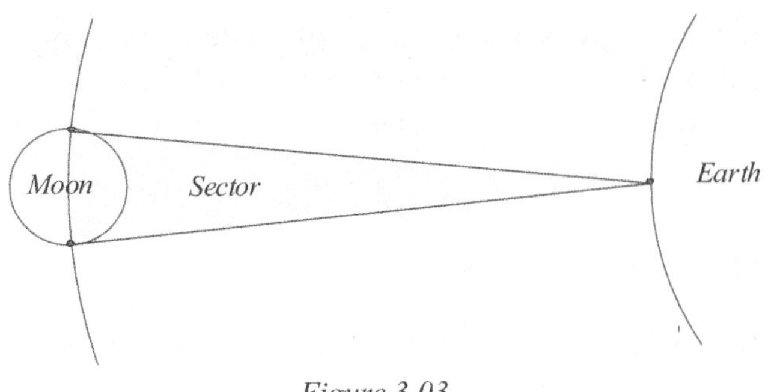

Figure 3.03

8. An equilateral $\triangle ABC$ of side 6cm rests on horizontal ground with AB in contact with the ground. The triangle is rotated in a clockwise direction about B until BC is in contact with the ground. Then it is rotated about C until CA is in contact with the ground. Finally, it is rotated about A until AB is again in contact with the ground. Find the length of the path traveled by A.

9. In the year 240 BC, Eratosthenes observed that in Syene, Egypt, a vertical stick had no shadow at noon on the summer solstice, but at Alexandria, on the same meridian as Syene, the sun's rays were inclined at 1/50 of a complete circle to the vertical. The distance from Syene to Alexandria was 850km. Figure 3.04, which is not drawn to scale, shows (1) a section of the Earth through the meridian containing Syene and Alexandria, (2) the vertical stick and (3) the direction of the sun's rays.

(a) Find the angle, in radians, between the sun's rays and the vertical stick.

(b) Show that the radius of the Earth is approximately 6760km.

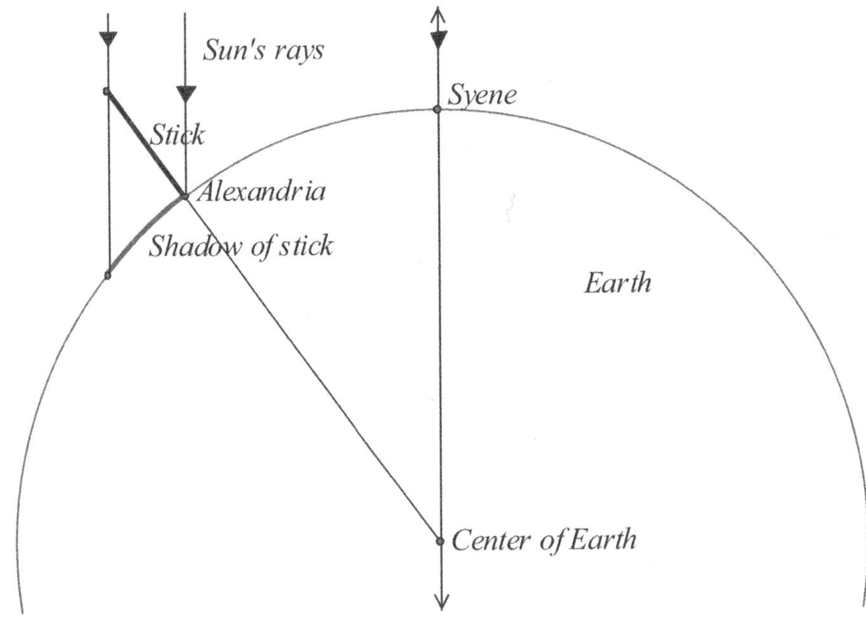

Figure 3.04

3.2 The Circular Functions

3.2.1 The Sine and Cosine Functions

Consider a circle of unit radius, center O, with an arm OP, rotating counter-clockwise. Initially the arm OP lies on the positive x-axis.

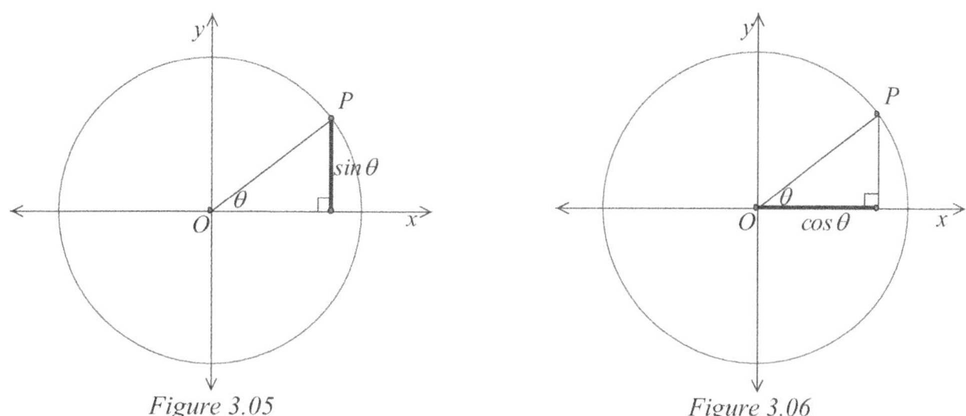

Figure 3.05   Figure 3.06

Figure 3.05 shows the arm as it is rotating. The *sine* function $f : \theta \mapsto \sin\theta$ maps the angle θ, which OP makes with the positive x-axis, to the perpendicular distance of the point P from the x-axis.

Figure 3.06 shows the same unit circle. The *cosine* function $g : \theta \mapsto \cos\theta$ maps θ to the perpendicular distance of the point P from the y-axis.

Figure 3.07 shows three positions of P; P_1, P_2 and P_3 at different points around the unit circle. Q_1 is the point on the graph of the sine function which corresponds to P_1. R_1 is the point on the graph of the cosine function which corresponds to P_1. Similarly, Q_2 and R_2 correspond to P_2. Q_3 and R_3 correspond to P_3.

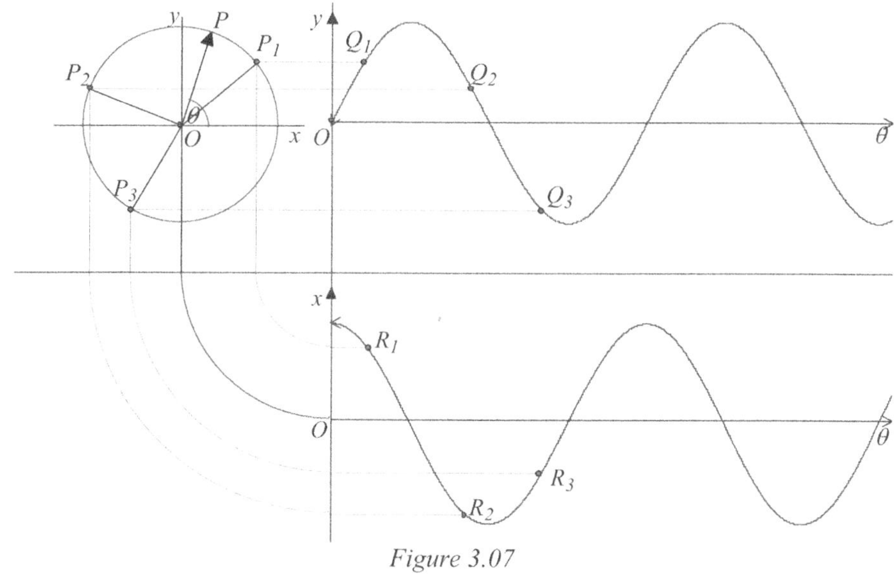

Figure 3.07

Figure 3.07 shows that as OP rotates around the unit circle, the graph of $f(\theta) = \sin\theta$ describes a wave. After one complete rotation of 2π radians of the arm OP this wave has gone through one complete cycle. If the domain is $\mathbb{R}$, then the graph of $y = \sin\theta$ is a continuous wave in which each cycle is repeated at intervals of 2π radians. The interval over which the cycle is repeated is called the *period* of the function so that the period of the sine function is 2π. Functions which exhibit cyclic behavior are called *periodic functions*.

Figure 3.08 shows the graph of $y = \sin\theta$ with domain $-2\pi \le \theta \le 2\pi$.

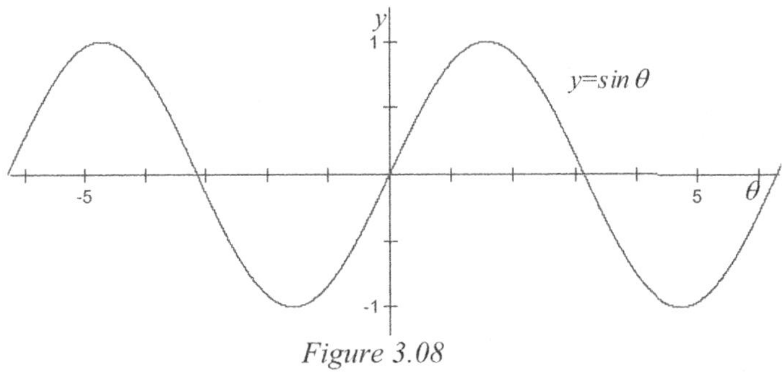

Figure 3.08

In a similar way, as OP rotates around the unit circle, the cosine function, $f(\theta) = \cos\theta$, describes a wave. After one complete rotation of 2π radians, this wave has gone through one complete cycle so that the period of the cosine function is also 2π. Note, however, that the cosine function is a sine wave which has been translated through a distance of $-\dfrac{\pi}{2}$ parallel to the θ direction.

Figure 3.09 shows the graph of $y = \cos\theta$ with domain $-2\pi \le \theta \le 2\pi$.

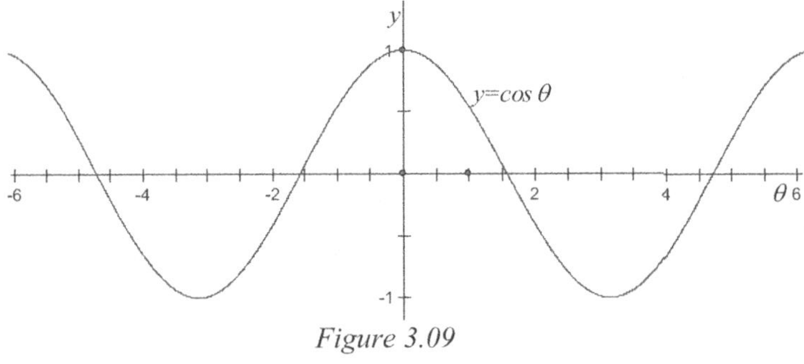

Figure 3.09

The *tangent* function, h, maps the angle θ to $\tan\theta$ so that $h : \theta \mapsto \tan\theta$. By definition, $\tan\theta = \dfrac{\sin\theta}{\cos\theta}$ so that $\tan\theta$ is undefined for values of θ which make $\cos\theta = 0$. This means that the graph of $y = \tan\theta$ has vertical asymptotes wherever $\cos\theta = 0$. In the graph of $y = \cos\theta$ (figure 3.09), $\cos\theta = 0$ when $\theta = \left\{\ldots, -\dfrac{5\pi}{2}, -\dfrac{3\pi}{2}, -\dfrac{\pi}{2}, \dfrac{\pi}{2}, \dfrac{3\pi}{2}, \dfrac{5\pi}{2}, \ldots\right\}$, so there are asymptotes in the graph of $y = \tan\theta$ at these values of the domain.

84

Figure 3.10 shows the graph of $y = \tan\theta$ with domain $-2\pi \le \theta \le 2\pi$. As can be seen, the tangent function is also a periodic function even though it is not a wave. Its period is π.

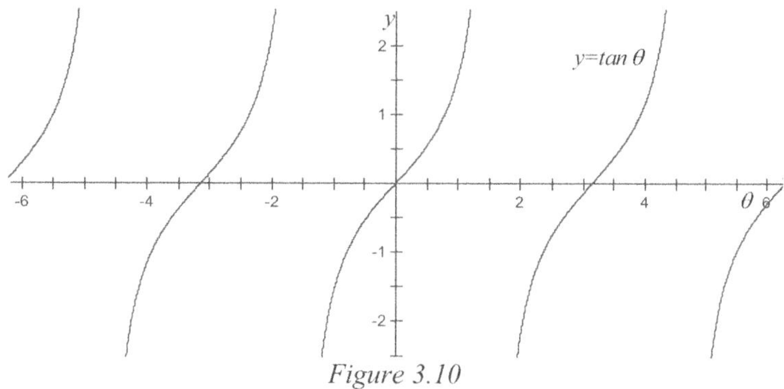

Figure 3.10

3.2.2 Values of Sine, Cosine and Tangent for the Interval $0 \le \theta < 2\pi$ or, in Degrees, $0° \le \theta < 360°$

The four regions numbered in figure 3.11 are known as *quadrants*. An angle between 0 and $\dfrac{\pi}{2}$ lies in the first quadrant, an angle

between $\dfrac{\pi}{2}$ and π lies in the second quadrant, an angle between π

and $\dfrac{3\pi}{2}$ lies in the third quadrant and an angle between $\dfrac{3\pi}{2}$ and 2π

lies in the fourth quadrant. It is useful to be able to express sines, cosines and tangents of angles in the second, third and fourth quadrants in terms of sines, cosines and tangents of angles in the first quadrant.

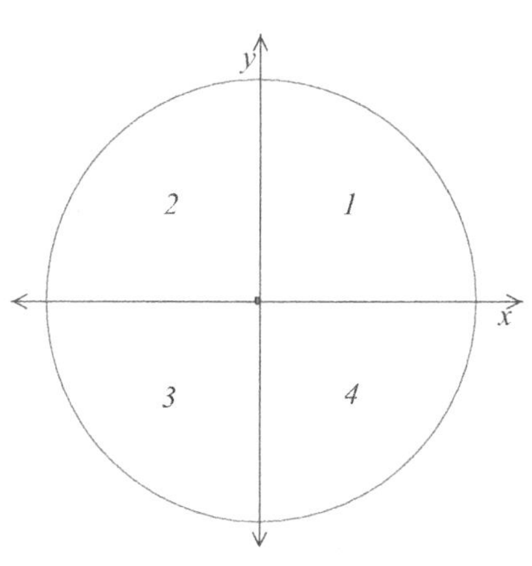

Figure 3.11

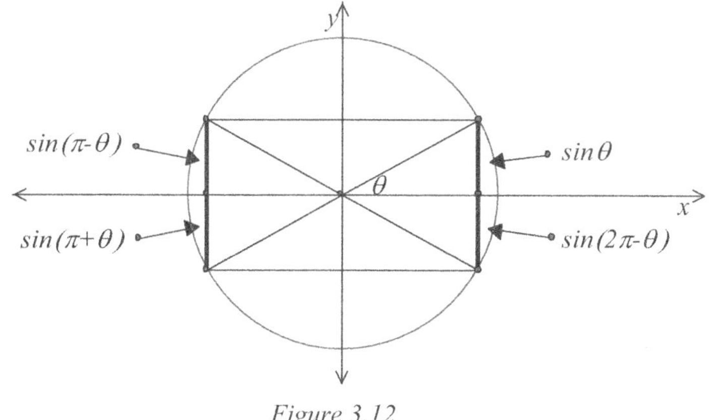

Figure 3.12

For **Sines**: Figure 3.12 shows the unit circle and the magnitudes of $\sin\theta$, $\sin(\pi - \theta)$,

$\sin(\pi + \theta)$ and $\sin(2\pi - \theta)$, where $0 \le \theta < \dfrac{\pi}{2}$.

By symmetry, $\sin(\pi - \theta) = \sin\theta$,

$\sin(\pi + \theta) = -\sin\theta$, $\sin(2\pi - \theta) = -\sin\theta$.

For example, if $\theta = \dfrac{\pi}{6}$, then $\sin\dfrac{7\pi}{6} = \sin\left(\pi + \dfrac{\pi}{6}\right) = -\sin\dfrac{\pi}{6}$, and, in degrees, if $\theta = 30°$, then

$\sin 210° = \sin(180° + 30°) = -\sin 30°$.

For **Cosines**: Figure 3.13 shows the unit circle and the magnitudes of $\cos\theta$, $\cos(\pi-\theta)$, $\cos(\pi+\theta)$ and $\cos(2\pi-\theta)$, where $0 \le \theta < \dfrac{\pi}{2}$. By symmetry,

$$\cos(\pi-\theta) = -\cos\theta, \quad \cos(\pi+\theta) = -\cos\theta,$$
$$\cos(2\pi-\theta) = \cos\theta.$$

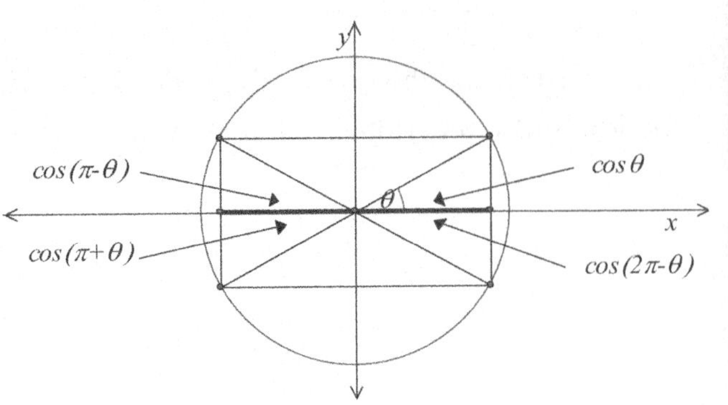

Figure 3.13

For example, if $\theta = \dfrac{\pi}{4}$, then $\cos\left(\dfrac{3\pi}{4}\right) = \cos\left(\pi - \dfrac{\pi}{4}\right) = -\cos\left(\dfrac{\pi}{4}\right)$, and, in degrees, if $\theta = 45°$, then

$$\cos 135° = \cos(180° - 45°) = -\cos 45°.$$

For **Tangents**: As $\tan\theta = \dfrac{\sin\theta}{\cos\theta}$,

$$\tan(\pi-\theta) = \frac{\sin(\pi-\theta)}{\cos(\pi-\theta)} = \frac{\sin\theta}{-\cos\theta} = -\tan\theta$$

$$\tan(\pi+\theta) = \frac{\sin(\pi+\theta)}{\cos(\pi+\theta)} = \frac{-\sin\theta}{-\cos\theta} = \tan\theta$$

$$\tan(2\pi-\theta) = \frac{\sin(2\pi-\theta)}{\cos(2\pi-\theta)} = \frac{-\sin\theta}{\cos\theta} = -\tan\theta$$

For example, if $\theta = \dfrac{\pi}{3}$, then $\tan\left(\dfrac{5\pi}{3}\right) = \tan\left(2\pi - \dfrac{\pi}{3}\right) = -\tan\dfrac{\pi}{3}$, and, in degrees, if $\theta = 60°$, then

$$\tan 300° = \tan(360° - 60°) = -\tan 60°.$$

3.2.3 Exact Values of $\cos\theta$ and $\sin\theta$ for $\theta = \dfrac{\pi}{6}, \dfrac{\pi}{4}$ and $\dfrac{\pi}{3}$

Figure 3.14 shows a right isosceles $\triangle ONP$ in which $OP = 1$, $N\hat{O}P = \dfrac{\pi}{4}$ and $ON = NP$. Therefore, if $ON = NP = u$, by Pythagoras' theorem

$$ON^2 + NP^2 = OP^2 \Rightarrow u^2 + u^2 = 1$$
$$\Rightarrow 2u^2 = 1 \Rightarrow u^2 = \frac{1}{2} \Rightarrow u = \pm\frac{1}{\sqrt{2}}, \quad u > 0$$

so $u = \dfrac{1}{\sqrt{2}}$ and therefore $\cos\dfrac{\pi}{4} = \dfrac{1}{\sqrt{2}}$ and $\sin\dfrac{\pi}{4} = \dfrac{1}{\sqrt{2}}$.

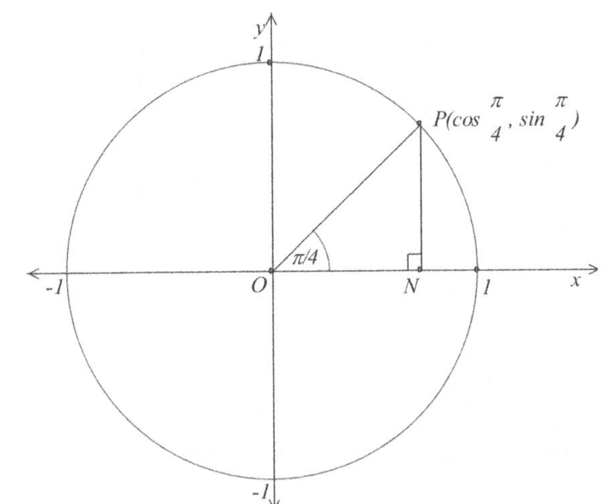

Figure 3.14

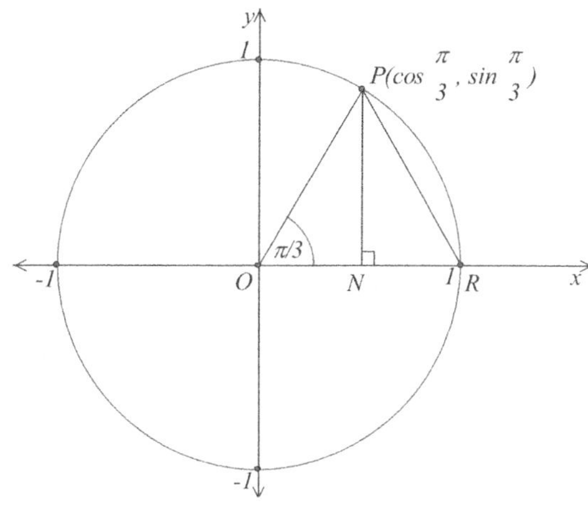

Figure 3.15

Figure 3.15 shows equilateral ΔOPR in which all sides are of unit length. PN bisects OR at N so $ON = \dfrac{1}{2}$ and $N\hat{O}P = \dfrac{\pi}{3}$, $O\hat{P}N = \dfrac{\pi}{6}$. Therefore,

$$\cos\frac{\pi}{3} = \frac{ON}{OP} = \frac{1}{2}.$$ Then, by Pythagoras' theorem,

$$ON^2 + NP^2 = OP^2 \Rightarrow \left(\frac{1}{2}\right)^2 + NP^2 = 1^2$$

$$\Rightarrow \frac{1}{4} + NP^2 = 1 \Rightarrow NP^2 = \frac{3}{4} \Rightarrow NP = \pm\sqrt{\frac{3}{4}} = \pm\frac{\sqrt{3}}{2}$$

but $NP > 0$ so $NP = \dfrac{\sqrt{3}}{2}$ and $\sin\dfrac{\pi}{3} = \dfrac{\sqrt{3}}{2}$

Figure 3.16 shows equilateral ΔOPQ in which all sides are of unit length. ON bisects PQ so that $PN = \dfrac{1}{2}$. Then, by Pythagoras' theorem, $ON^2 + NP^2 = OP^2$

$$\Rightarrow ON^2 + \left(\frac{1}{2}\right)^2 = 1^2 \Rightarrow ON^2 = \frac{3}{4} \Rightarrow ON = \pm\sqrt{\frac{3}{4}} = \pm\frac{\sqrt{3}}{2}$$

but $ON > 0$. $ON = \dfrac{\sqrt{3}}{2}$. So, the coordinates of P are $\left(\dfrac{\sqrt{3}}{2}, \dfrac{1}{2}\right)$ and hence $\cos\dfrac{\pi}{6} = \dfrac{\sqrt{3}}{2}$ and $\sin\dfrac{\pi}{6} = \dfrac{1}{2}$.

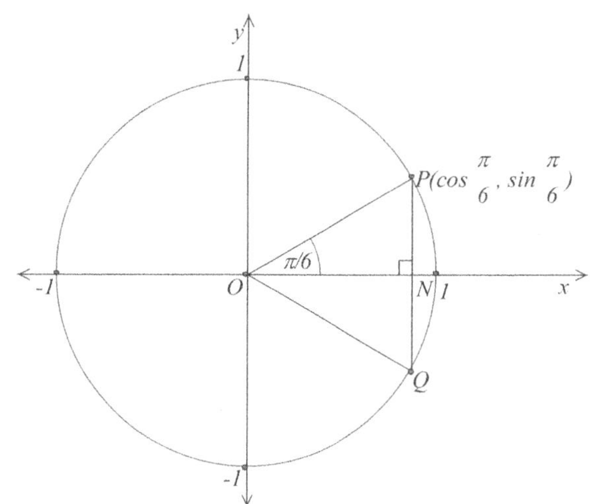

Figure 3.16

Table 3.2 summarizes these results.

θ	$\cos\theta$	$\sin\theta$
$\dfrac{\pi}{6}$	$\dfrac{\sqrt{3}}{2}$	$\dfrac{1}{2}$
$\dfrac{\pi}{4}$	$\dfrac{1}{\sqrt{2}}$	$\dfrac{1}{\sqrt{2}}$
$\dfrac{\pi}{3}$	$\dfrac{1}{2}$	$\dfrac{\sqrt{3}}{2}$

Table 3.2

Exercise 3.2

1. In each case, write the following as sines of angles between 0 and $\dfrac{\pi}{2}$ inclusive or between $0°$ and $90°$, as appropriate. For example, $\sin\dfrac{5\pi}{4} = -\sin\dfrac{\pi}{4}$.

   (a) $\sin\dfrac{3\pi}{4}$ (b) $\sin\dfrac{5\pi}{8}$ (c) $\sin\dfrac{13\pi}{10}$ (d) $\sin\dfrac{7}{4}\pi$

   (e) $\sin 300°$ (f) $\sin 135°$ (g) $\sin 236°$ (h) $\sin 337°$

2. Write the following as cosines of angles between 0 and $\dfrac{\pi}{2}$ inclusive or between $0°$ and $90°$, as appropriate.

   (a) $\cos\dfrac{3\pi}{2}$ (b) $\cos\dfrac{4\pi}{5}$ (c) $\cos\dfrac{7\pi}{6}$ (d) $\cos\dfrac{5}{6}\pi$

   (e) $\cos 110°$ (f) $\cos 210°$ (g) $\cos 310°$ (h) $\cos 133°$

3. Give the following as tangents of angles between 0 and $\dfrac{\pi}{2}$ inclusive or between $0°$ and $90°$, as appropriate.

   (a) $\tan\dfrac{9\pi}{4}$ (b) $\tan\dfrac{4\pi}{3}$ (c) $\tan\dfrac{7}{8}\pi$ (d) $\tan\dfrac{7}{6}\pi$

   (e) $\tan 130°$ (f) $\tan 320°$ (g) $\tan 260°$ (h) $\tan 350°$

4. Write down the exact value of the following.

   (a) $\sin\dfrac{3\pi}{4}$ (b) $\sin\dfrac{5\pi}{6}$ (c) $\sin\dfrac{5\pi}{3}$ (d) $\sin\dfrac{5}{4}\pi$

   (e) $\cos\dfrac{3\pi}{4}$ (f) $\cos\dfrac{4\pi}{3}$ (g) $\cos\dfrac{11\pi}{6}$ (h) $\cos\dfrac{5}{3}\pi$

   (i) $\tan\dfrac{7\pi}{4}$ (j) $\tan\dfrac{2\pi}{3}$ (k) $\tan\dfrac{11\pi}{6}$ (l) $\tan\dfrac{7}{6}\pi$

5. Draw graphs of $y = \sin\theta$, $y = \cos\theta$ and $y = \tan\theta$ to show that $\sin(-\theta) = -\sin\theta$, $\cos(-\theta) = \cos\theta$, $\tan(-\theta) = -\tan\theta$. Hence, write down the exact values of

   (a) $\sin\left(-\dfrac{\pi}{3}\right)$ (b) $\cos\left(-\dfrac{4}{3}\pi\right)$ (c) $\tan\left(-\dfrac{11\pi}{6}\right)$

3.3 Trigonometric Identities

An identity is a statement of equivalence. For example, $(x+1)^2 = x^2 + 2x + 1$ is an identity because it is true for all values of x, and so an identity cannot be solved like an equation. The following six identities are useful in a variety of situations. In each case, except for identity (2), a proof of the identity is given.

Identity (1): $\cos^2\theta + \sin^2\theta = 1$

Figure 3.17 shows a right $\triangle OPN$ in which $OP = 1$, $ON = \cos\theta$ and $PN = \sin\theta$. Using Pythagoras' theorem, $ON^2 + NP^2 = OP^2$. But, $ON = \cos\theta$, $NP = \sin\theta$ and $OP = 1 \Rightarrow (\cos\theta)^2 + (\sin\theta)^2 = 1^2$. It is customary to write $(\cos\theta)^2$ as $\cos^2\theta$ and similarly $(\sin\theta)^2$ as $\sin^2\theta$ so that $\cos^2\theta + \sin^2\theta = 1$.

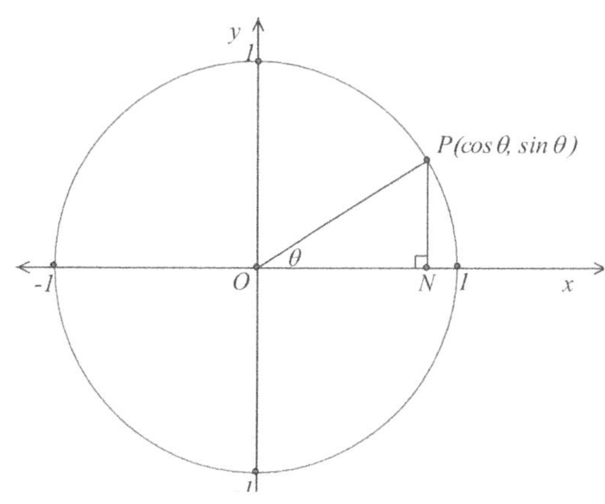

Identity (2): $\tan\theta = \dfrac{\sin\theta}{\cos\theta}$

This identity is simply the definition of $\tan\theta$.

Figure 3.17

Identities (3) and (4): $\cos 2\theta = 2\cos^2\theta - 1$ and $\sin 2\theta = 2\cos\theta\sin\theta$

Figure 3.18 shows a $\triangle ACD$ which is right-angled at C. $AB = BD = 1$ unit. $C\hat{A}D = \theta$. $\triangle ABD$ is isosceles so that its base angles are equal; therefore, $A\hat{D}B = \theta$. $C\hat{B}D$ is an exterior angle of $\triangle ABD$ and is equal to the sum of the two interior opposite angles, $B\hat{A}D$ and $A\hat{D}B$. Therefore, $C\hat{B}D = \theta + \theta = 2\theta$.

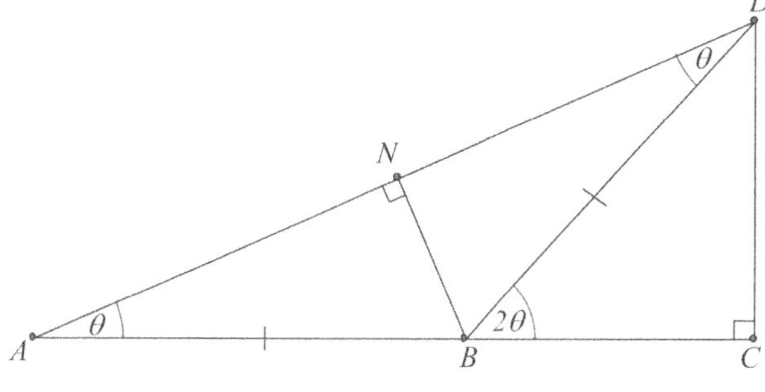

Now, $\cos 2\theta = \dfrac{BC}{BD} = \dfrac{BC}{1} = BC$. BN bisects

Figure 3.18

isosceles $\triangle ABD \Rightarrow \cos\theta = \dfrac{AN}{AB} = \dfrac{AN}{1} = AN$. Similarly, $ND = \cos\theta$ so $AD = 2\cos\theta$.

But $\cos\theta = \dfrac{AC}{AD} = \dfrac{AB + BC}{AD} = \dfrac{1 + BC}{AD} = \dfrac{1 + \cos 2\theta}{2\cos\theta} \Rightarrow 2(\cos\theta)^2 = 1 + \cos 2\theta \Rightarrow \cos 2\theta = 2(\cos\theta)^2 - 1$

$\Rightarrow \cos 2\theta = 2\cos^2\theta - 1$ (Identity (3)).

Also, $\sin 2\theta = \dfrac{DC}{BD} = \dfrac{DC}{1} = DC$ and $\sin\theta = \dfrac{DC}{AD}$. Therefore, $\sin\theta = \dfrac{\sin 2\theta}{2\cos\theta} \Rightarrow \sin 2\theta = 2\cos\theta\sin\theta$ (Identity (4)).

Identity (5): $\cos 2\theta = 1 - 2\sin^2\theta$

$\cos 2\theta = 1 - 2\sin^2\theta$ can be obtained by combining identity (3) with identity (1):

$$\cos^2\theta + \sin^2\theta = 1 \qquad \Rightarrow \cos^2\theta = 1 - \sin^2\theta$$

$$\cos 2\theta = 2\cos^2\theta - 1 \qquad \Rightarrow \cos 2\theta = 2(1 - \sin^2\theta) - 1 = 2 - 2\sin^2\theta - 1$$

$$\Rightarrow \cos 2\theta = 1 - 2\sin^2\theta \quad \text{(Identity (5))}$$

Identity (6): $\cos 2\theta = \cos^2\theta - \sin^2\theta$

$$\cos 2\theta = 2\cos^2\theta - 1 = \cos^2\theta + \cos^2\theta - 1$$

$$\Rightarrow \cos 2\theta = \cos^2\theta + (1 - \sin^2\theta) - 1 = \cos^2\theta - \sin^2\theta$$

$$\Rightarrow \cos 2\theta = \cos^2\theta - \sin^2\theta \quad \text{(Identity (6))}$$

Example 3.2: If θ is acute and $\sin\theta = \dfrac{2}{3}$, use the identities to find the exact value(s) of

    (a) $\cos\theta$ (b) $\cos 2\theta$

Solution 3.2: (a) Using identity (1), $\cos^2\theta + \sin^2\theta = 1 \Rightarrow \cos^2\theta + \left(\dfrac{2}{3}\right)^2 = 1 \Rightarrow \cos^2\theta + \dfrac{4}{9} = 1$.

Therefore, $\cos^2\theta = 1 - \dfrac{4}{9} = \dfrac{5}{9} \Rightarrow \cos\theta = \pm\sqrt{\dfrac{5}{9}}$. However, as θ is acute, $\cos\theta > 0$ so that

$$\cos\theta = \frac{\sqrt{5}}{3}.$$

(b) Using identity (5), $\cos 2\theta = 1 - 2\sin^2\theta \Rightarrow \cos 2\theta = 1 - 2\left(\dfrac{2}{3}\right)^2 = 1 - \dfrac{8}{9} = \dfrac{1}{9}$.

Example 3.3: Use the identities described in this section to show that
$$1 - 2\sin\theta - \cos\theta + \sin 2\theta = (1 - \cos\theta)(1 - 2\sin\theta).$$

Solution 3.3: Expand the right hand side: $(1 - \cos\theta)(1 - 2\sin\theta) = 1 - \cos\theta - 2\sin\theta + 2\sin\theta\cos\theta$.

But $2\sin\theta\cos\theta = \sin 2\theta$; hence $(1 - \cos\theta)(1 - 2\sin\theta) = 1 - \cos\theta - 2\sin\theta + \sin 2\theta$

$$= 1 - 2\sin\theta - \cos\theta + \sin 2\theta$$

Exercise 3.3

1. $\triangle ABC$ is right-angled at B, $AB = x$, $BC = 5$, $CA = 8$ and $\hat{A} = \theta$.

    (a) Write down the value of $\sin\theta$.   (b) Use Pythagoras' theorem to find the exact value of x.

    (c) Write down the exact value of $\cos\theta$.   (d) Show that $\cos^2\theta + \sin^2\theta = 1$

2. Given that $\sin\theta = \dfrac{2}{5}$, find the exact values of $\cos\theta$ if (a) θ is acute (b) θ is obtuse.

3. (a) If $\cos\theta = \dfrac{2}{3}$, find the value of k, given that $\sin\theta = \dfrac{\sqrt{k}}{3}$ and that $0 < \theta < \dfrac{\pi}{2}$.

(b) If $\pi < \theta < 2\pi$ and $\cos\theta = \dfrac{9}{41}$ find the exact value of $\sin\theta$.

4. If $\cos\theta = -\dfrac{5}{13}$, $0 < \theta < \pi$ find the exact value of

(a) $\sin\theta$ (b) $\sin 2\theta$ (c) $\cos 2\theta$

5. Find the possible values of $\cos\theta$ and $\cos 2\theta$, given that $\sin\theta = \dfrac{3}{5}$.

6. If $\sin\theta = \dfrac{7}{25}$, find all possible values of

(a) $\cos\theta$ (b) $\tan\theta$ (c) $\cos 2\theta$ (d) $\sin 2\theta$

7. Use the identities established in section 3.3 to show that

(a) $1 + \tan\theta = \dfrac{\cos\theta + \sin\theta}{\cos\theta}$, $-\dfrac{\pi}{2} < \theta < \dfrac{\pi}{2}$ (b) $\left(\cos\theta + \sin\theta\right)^2 = 1 + \sin 2\theta$

(c) $1 - \tan^2\theta = \dfrac{\left(\cos\theta - \sin\theta\right)\left(\cos\theta + \sin\theta\right)}{\left(1 - \sin\theta\right)\left(1 + \sin\theta\right)}$, $-\dfrac{\pi}{2} < \theta < \dfrac{\pi}{2}$

(d) $2\sin^2 3\theta = 1 - \cos 6\theta$ (e) $1 + \cos\theta + \cos 2\theta = \cos\theta\left(1 + 2\cos\theta\right)$

3.4 Composite Trigonometric Functions

Transformations can be used to obtain composite trigonometric functions.

3.4.1 Transformation (1) : Translation

A translation of c parallel to the x-axis and k parallel to the y-axis

transforms $y = \sin x$ to $y = \sin(x - c) + k$. For example, if $c = \dfrac{\pi}{3}$ and

$k = 2$ then $y = \sin x$ is transformed to $y = \sin\left(x - \dfrac{\pi}{3}\right) + 2$, as shown in

figure 3.19 (note that the thick line describes the transformed function).

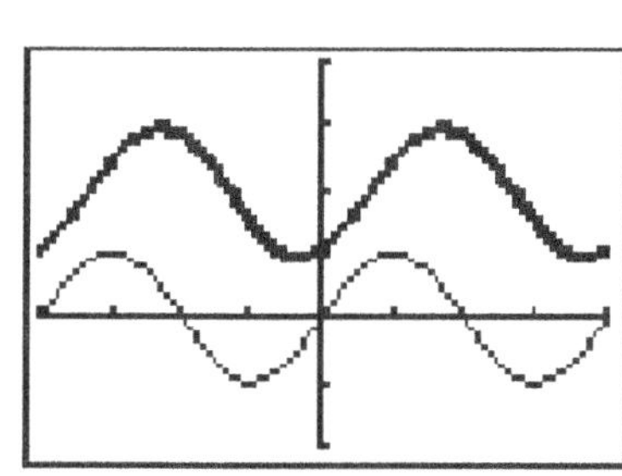

Figure 3.19

A translation of $\dfrac{\pi}{2}$ parallel to the x-axis transforms $y = \cos x$ to

$y = \cos\left(x - \dfrac{\pi}{2}\right)$ as shown in figure 3.20. It is clear that the image of

this translation is $y = \sin x$ so that $\cos\left(x - \dfrac{\pi}{2}\right) = \sin x$.

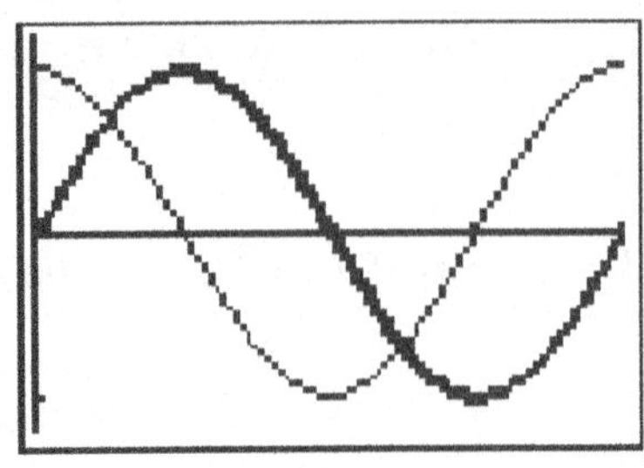

Figure 3.20

3.4.2 Transformation (2) : Stretch

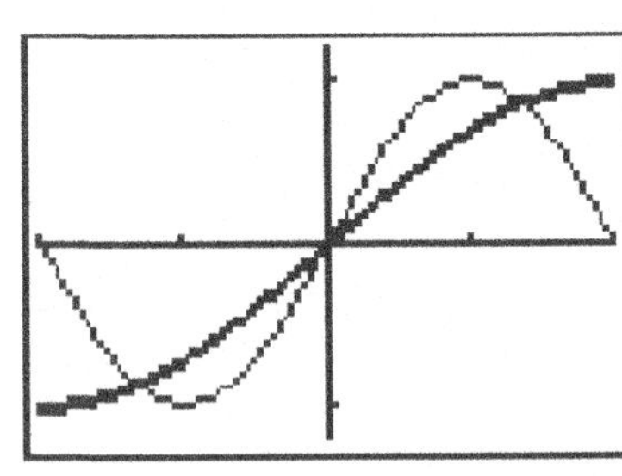

Figure 3.21

A stretch of factor a $(a \neq 0)$ parallel to the x-axis transforms

$y = \sin x$ to $y = \sin\left(\dfrac{x}{a}\right)$. This has the effect of changing the period

of the function. Now, the period of $y = \sin x$ is 2π, and so the effect

of a stretch of factor a parallel to the x-axis is to increase the period

by a factor of a so the period of $y = \sin\left(\dfrac{x}{a}\right)$ is $2\pi a$. For example,

if $a = 2$ then the period of $y = \sin\left(\dfrac{x}{2}\right)$ is 4π, as shown in figure 3.21.

If $a = \dfrac{1}{2}$, then $y = \sin x$ is transformed to $y = \sin\left(\dfrac{x}{1/2}\right) = \sin 2x$ and

the period of $y = \sin 2x$ is π, as shown in figure 3.22. In general, the

period of the functions $y = \sin kx$ and $y = \cos kx$ is $\dfrac{2\pi}{k}$.

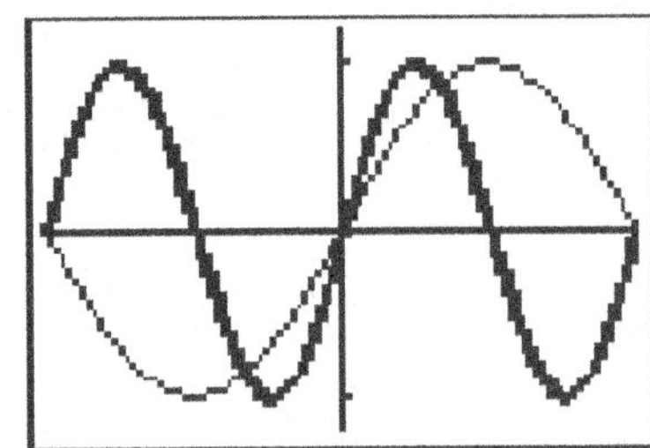

Figure 3.22

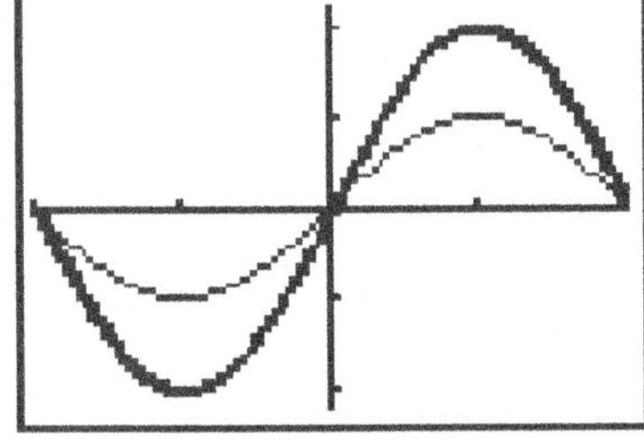

Figure 3.23

A stretch of factor a parallel to the y-axis
transforms $y = \sin x$ to $y = a \sin x$. This
has the effect of increasing the magnitude or *amplitude* of the wave by a
factor of a. If $a = 2$, then $y = \sin x$ is transformed to $y = 2 \sin x$, and,
although the period of the function remains at 2π, the amplitude is
doubled, as shown in figure 3.23.

Exercise 3.4

1.   In each case, state the transformation which maps the function $y = \sin x$ to

(a) $y = 5 \sin x$ 　　　　(b) $y = \sin \dfrac{1}{3} x$ 　　　　(c) $y = \sin\left(x - \dfrac{\pi}{6}\right)$

2. In each case, state the transformation which maps the function $y = \cos x$ to

(a) $y = \dfrac{2}{5}\cos x$ 　　(b) $y = \cos(x + \pi)$ 　　(c) $y = \cos\left(\dfrac{x}{4}\right)$

3. In each case, state the transformations which map f_1 to f_2.

(a) $f_1 : x \mapsto \tan x$, $f_2 : x \mapsto 1 + \tan 2x$ 　　(b) $f_1 : x \mapsto \sin x$, $f_2 : x \mapsto 2\sin 3x$

(c) $f_1 : x \mapsto \cos x$, $f_2 : x \mapsto 2\sin x$

4. Find the period of the following functions.

(a) $f(x) = \sin\dfrac{1}{2}x$ 　　(b) $f(x) = 4\sin 2x$ 　　(c) $f(x) = 3\cos 2x$ 　　(d) $f(x) = \tan 3x$

(e) $f(x) = \cos\left(\dfrac{\pi x}{3}\right)$ 　　(f) $f(x) = \sin^2 x$ 　　(g) $f(x) = \cos x + \sin x$

5. In each case, state the function which is obtained when the given transformation is applied to the given function.

(a) $y = \sin x$; stretch of factor 2 parallel to the y-axis

(b) $y = \cos x$; translation of 1 unit in the positive y direction

(c) $y = \tan x$; translation of $\dfrac{\pi}{3}$ units in the positive x direction

(d) $y = \sin x$; stretch of factor 2 parallel to the x-axis

(e) $y = \tan x$; stretch of factor $\dfrac{1}{3}$ parallel to the x-axis

(f) $y = \cos x$; reflection in the x-axis

6. f is a function of x such that $f(x) = 5\sin\left(3x - \dfrac{2\pi}{3}\right)$. Find

(a) the period of f 　　(b) 　(i) the maximum value of f 　(ii) the minimum value of f

(c) the exact value of 　　(i) $f(0)$ 　　(ii) $f\left(\dfrac{\pi}{2}\right)$

7. A trigonometric function is of the form $y = a\sin bx + c$ and has a period of $\dfrac{2\pi}{3}$, find

(a) the value of b.

If also the points with coordinates $(0,\ 5)$ and $\left(\dfrac{\pi}{2},\ 9\right)$ lie on the graph of f find

(b) the values of c and a

8.   Figure 3.24 shows the graph of $y = a\sin(bx+c)+d, \quad 0 \le x \le 2\pi$

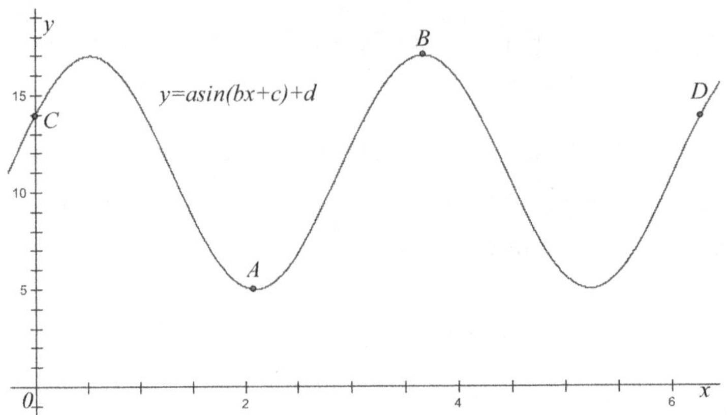

Figure 3.24

The point A has a minimum value of 5 and B has a maximum value of 17 on the graph of y.

The point C has coordinates $(0,\ 14)$ and the point D has coordinates $(2\pi,\ 14)$.

(a) Find the value of (i) a (ii) d.
(b) Write down the period of the function and hence find the value of b.
(c) Find the value of c.

9.   Figure 3.25 shows the graph of $f(x) = a\sin bx, \quad 0 \le x \le \pi$

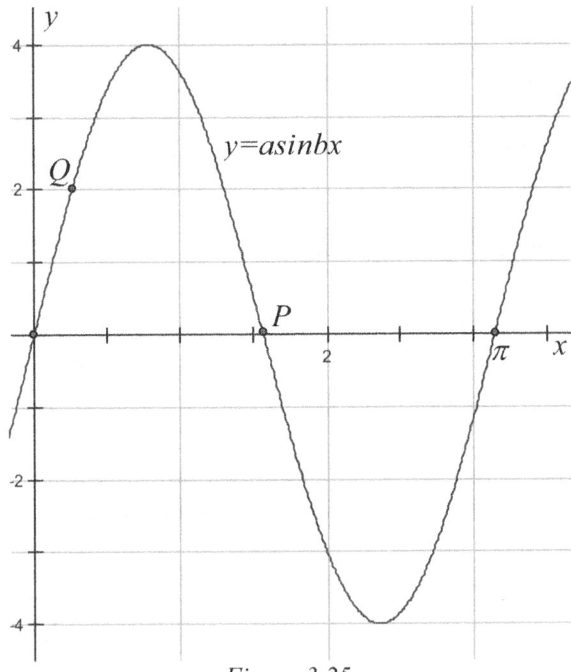

Figure 3.25

(a) Write down (i) the period of f
                      (ii) the amplitude of f
(b) Write down the values of (i) a
                          (ii) b

The point P has coordinates $(p,\ 0)$ and Q has coordinates $(q,\ 2)$

(c) Find the exact value of p.
(d) Find the exact value of q.

10. The magnitude of an alternating current is a function of time, t and is defined to be
$i(t) = 5\sin 120\pi t$, $t \geq 0$ where i is measured in ohms and t in seconds. Find the
(a) period of the function in seconds.
(b) number of cycles of the current per second.
(c) magnitude of the current when (i) $t = 0.2$ (ii) $t = 5.703$.

11. The time, t (in hours), of sunset after 18:00 at a specific location is a function of d, the
number of days after the beginning of the year, where $t = -2\cos\left(\dfrac{2\pi}{365}(d+11)\right)$, $1 \leq d \leq 365$.

Find the time of sunset on
(a) January 1   (b) February 15   (c) June 25

3.5 Solution of Trigonometric Equations

3.5.1 Solution of Equations of the Forms $\sin\theta = k$, $\cos\theta = k$ and $\tan\theta = k$ in the Interval $0 \leq \theta < 2\pi$

In general, equations of the form $\sin\theta = k$, $\cos\theta = k$ and
$\tan\theta = k$ have two solutions in the interval $0 \leq \theta < 2\pi$.

If $k > 0$ then $\sin\theta = k$ has solutions in quadrants 1 and 2.
$\cos\theta = k$ has solutions in quadrants 1 and 4.
$\tan\theta = k$ has solutions in quadrants 1 and 3.

If $k < 0$ then $\sin\theta = k$ has solutions in quadrants 3 and 4.
$\cos\theta = k$ has solutions in quadrants 2 and 3.
$\tan\theta = k$ has solutions in quadrants 2 and 4.

Figure 3.26 summarizes this situation.

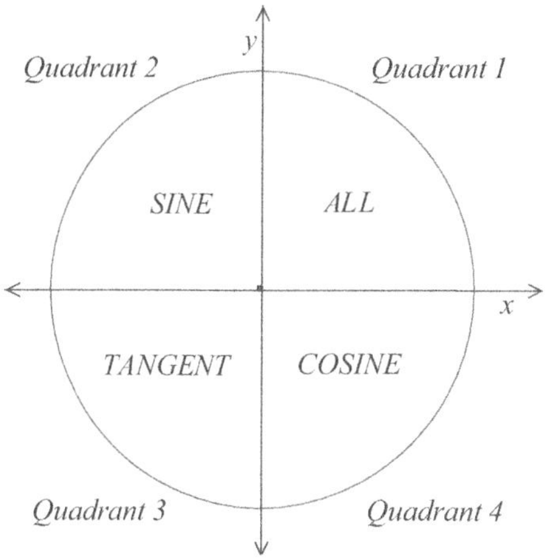

Figure 3.26

If the solution in the first quadrant is α, then the solution in the
second quadrant is $\pi - \alpha$, the solution in the third quadrant is $\pi + \alpha$ and the solution in the fourth
quadrant is $2\pi - \alpha$ (section 3.2.2). As the solutions in the second, third and fourth quadrants are
dependent on α, it is a good idea to find α even if $k < 0$.

Example 3.4: Solve $\sin\theta = -\dfrac{\sqrt{3}}{2}$ in the interval $0 \le \theta < 2\pi$.

Solution 3.4: As $k < 0$, the equation has solutions in the third and fourth quadrants. Now $\sin\dfrac{\pi}{3} = \dfrac{\sqrt{3}}{2}$ (table 3.2), so $\alpha = \dfrac{\pi}{3}$ and therefore $\theta = \pi + \dfrac{\pi}{3} = \dfrac{4\pi}{3}$ and

$$\theta = 2\pi - \dfrac{\pi}{3} = \dfrac{5\pi}{3}.$$

Example 3.5: Solve $\sqrt{2} - 2\cos\theta = 0$ in the interval $0 \le \theta < 2\pi$.

Solution 3.5: $\sqrt{2} - 2\cos\theta = 0 \Rightarrow \cos\theta = \dfrac{\sqrt{2}}{2} = \dfrac{1}{\sqrt{2}}$. As $k > 0$, the equation has solutions in the first and fourth quadrants. $\cos\dfrac{\pi}{4} = \dfrac{1}{\sqrt{2}}$ (table 3.2), so $\alpha = \dfrac{\pi}{4}$ and therefore the solutions are $\theta = \dfrac{\pi}{4}$ and $\theta = 2\pi - \dfrac{\pi}{4} = \dfrac{7\pi}{4}$.

If solutions are required in a different interval, then solutions can be found in the interval $0 \le \theta < 2\pi$ and the periodic nature of trigonometric functions is used to obtain solutions in the required interval. For example, suppose solutions of $\cos\theta = -0.5$ are required in the interval, $-\pi < \theta \le \pi$. The solutions of this equation in the interval $0 \le \theta < 2\pi$ are $\dfrac{2\pi}{3}$ and $\dfrac{4\pi}{3}$. $\dfrac{2\pi}{3}$ is already in the required interval, but $\dfrac{4\pi}{3} > \pi$, so this solution lies outside the required interval. However, as the period of $y = \cos\theta$ is 2π, if $\dfrac{4\pi}{3}$ is a solution, so is $\dfrac{4\pi}{3} - 2\pi = -\dfrac{2\pi}{3}$ and this solution is in the interval $-\pi < \theta \le \pi$. Therefore, the required solutions are $-\dfrac{2\pi}{3}$ and $\dfrac{2\pi}{3}$.

Example 3.6: Find all the solutions of $\sqrt{2}\sin\theta + 1 = 0$ in the interval, $-\pi < \theta \le \pi$.

Solution 3.6: $\sqrt{2}\sin\theta + 1 = 0 \Rightarrow \sin\theta = -\dfrac{1}{\sqrt{2}}$. The solutions will be in the third and fourth quadrants, and as $\alpha = \dfrac{\pi}{4}$, $\theta = \pi + \dfrac{\pi}{4} = \dfrac{5\pi}{4}$ and $\theta = 2\pi - \dfrac{\pi}{4} = \dfrac{7\pi}{4}$.

Both of these solutions are outside the interval $-\pi < \theta \le \pi$, but, if the period 2π is subtracted from each solution, the two solutions in the required interval are obtained. Hence the solutions are $\theta = \dfrac{5\pi}{4} - 2\pi = -\dfrac{3\pi}{4}$

and $\theta = \dfrac{7\pi}{4} - 2\pi = -\dfrac{\pi}{4}$.

Exercise 3.5

1. In each case, solve the equation in the interval $0 \le x < 2\pi$, giving all solutions.

(a) $\sin x = \dfrac{1}{2}$ (b) $\tan x = 1$ (c) $\cos x = 1$ (d) $\sin x = 0$

(e) $\cos x = -\dfrac{1}{\sqrt{2}}$ (f) $\tan x = -\dfrac{1}{\sqrt{3}}$ (g) $\cos x = -\dfrac{1}{2}$ (h) $\sin x = \dfrac{1}{\sqrt{2}}$

(i) $\tan x = 0$

2. In each case, find all the solutions of the equation in the interval $-\pi < x \le \pi$.

(a) $\cos x = 0$ (b) $\tan x = -1$ (c) $\sin x = -\dfrac{1}{2}$ (d) $\cos x = \dfrac{\sqrt{3}}{2}$

(e) $\sin x + 1 = 0$ (f) $\tan x = \sqrt{3}$ (g) $\sqrt{2}\cos x - 1 = 0$ (h) $\sin x = \sin\dfrac{\pi}{6}$

3. In each case, solve the given trigonometric equation in the interval $0 \le x < 2\pi$.

(a) $\sin\left(x + \dfrac{\pi}{3}\right) = 1$ (b) $\tan\left(x - \dfrac{\pi}{2}\right) = 1$ (c) $1 + \sin x = \dfrac{1}{2}$ (d) $\sqrt{2}\cos x - 1 = 0$

(e) $\sin^2 x = \dfrac{1}{2}$ (f) $\cos^2 x = \dfrac{1}{4}$ (g) $\tan^2 x = \dfrac{1}{3}$

4. (a) Given that $\cos x = \dfrac{1}{\sqrt{2}}$ and $\sin x = -\dfrac{1}{\sqrt{2}}$ for $0 \le x < 2\pi$, find the exact value of $\tan x$ and the exact value of x.

(b) If $\cos 2x = \dfrac{1}{2}$ and $-\dfrac{\pi}{2} < x \le \dfrac{\pi}{2}$, find the exact value of $\cos x$ and the exact values of x.

5. Consider the function $f(x) = 3\cos\left(\dfrac{\pi x}{2}\right) + 2, \quad -2 \le x \le 2$

(a) Sketch the graph of the function f.
(b) Write down (i) the period and (ii) the amplitude of f.
(c) Find $f(2)$

(d) Find the values of x for which $f(x) = \dfrac{1}{2}$.

(e) Show that $f(x) = 5 - 6\sin^2\dfrac{\pi x}{4}$

3.5.2 Solution of More Advanced Trigonometric Equations

In this section, the solution of trigonometric equations will be expected either in radians or in degrees. If the interval in which solutions are required is given in radians, then the solutions should be given in radians. If the interval in which the solutions are required is given in degrees, then the solutions should be given in degrees.

Unless exact solutions, in terms of multiples of π, are required, trigonometric equations can be solved to any reasonable degree of accuracy by using a graphing calculator. Example 3.7 uses this method as a means of checking the answer obtained using the standard trigonometric method. Some of the questions in Exercise 3.7 offer a similar opportunity for you to use your graphing calculator.

Trigonometric equations may occur which are quadratic or may reduce to quadratic equations, in terms of sine, cosine or tangent functions. In this case, it may be helpful to use a substitution thereby changing the trigonometric equation into an algebraic one. When the quadratic equation has been solved, you will be left with two simple trigonometric equations, and you can solve those as in section 3.5.1.

Example 3.7: Solve $6\sin^2\theta - \sin\theta - 1 = 0$ in the interval $-180° < \theta \le 180°$. Use your graphing calculator to check your solutions.

Solution 3.7: Let $s = \sin\theta$, then $6s^2 - s - 1 = 0 \Rightarrow 6s^2 - 3s + 2s - 1 = 0 \Rightarrow 3s(2s-1) + 1(2s-1) = 0$

$\Rightarrow (2s-1)(3s+1) = 0 \Rightarrow 2s-1 = 0, \ 3s+1 = 0 \Rightarrow s = \dfrac{1}{2}, \ s = -\dfrac{1}{3}$. So, $\sin\theta = \dfrac{1}{2}$ and

$\sin\theta = -\dfrac{1}{3}$. For $\sin\theta = \dfrac{1}{2}$, solutions are in the first and second quadrants,

$\theta = 30°, \ 150°$. For $\sin\theta = -\dfrac{1}{3}$, solutions are in the third and fourth quadrants,

$\theta = 199.5°, \ 340.5° \Rightarrow \theta = 30°, \ 150°, \ -19.5°, \ -160.5°$.

Check using graphing calculator:

Figure 3.27   *Figure 3.28*

Figure 3.27 shows the way in which the function needs to be entered into the calculator, and figure 3.28 shows the graph of $y = 6\sin^2\theta - \sin\theta - 1$ and the solution $x = -19.5°$ obtained using the graphing calculator. The other solutions can be found in a similar way.

If the equation contains two different trigonometric functions then trigonometric identities can be used to express the equation in terms of a single trigonometric function.

Example 3.8: Solve $2\sin^2\theta = 2 + \cos\theta$ in the interval $-\pi < \theta \le \pi$.

Solution 3.8: As $\cos^2\theta + \sin^2\theta = 1$ then $\sin^2\theta = 1 - \cos^2\theta$.

So $2(1 - \cos^2\theta) = 2 + \cos\theta \Rightarrow 2 - 2\cos^2\theta = 2 + \cos\theta$

$\Rightarrow -2\cos^2\theta = \cos\theta \Rightarrow 2\cos^2\theta + \cos\theta = 0 \Rightarrow \cos\theta(2\cos\theta + 1) = 0$

$\Rightarrow \cos\theta = 0$ and $\cos\theta = -\frac{1}{2}$.

For $\cos\theta = 0$, solutions in the interval $0 \le \theta < 2\pi$ are $\frac{\pi}{2}$ and $\frac{3\pi}{2}$. $\frac{\pi}{2}$ is already in the interval $-\pi < \theta \le \pi$, and if 2π is subtracted from $\frac{3\pi}{2}$, the resulting solution, $-\frac{\pi}{2}$, will also be in the interval $-\pi < \theta \le \pi$.

For $\cos\theta = -\frac{1}{2}$, solutions in the interval $0 \le \theta < 2\pi$ are $\pi + \frac{\pi}{3} = \frac{4\pi}{3}$ and $\pi - \frac{\pi}{3} = \frac{2\pi}{3}$. $\frac{2\pi}{3}$ is already in the interval $-\pi < \theta \le \pi$, and, if 2π is subtracted from $\frac{4\pi}{3}$, the resulting solution, $-\frac{2\pi}{3}$ will be in the interval $-\pi < \theta \le \pi$. Therefore,

$$\theta = -\frac{\pi}{2}, \ \frac{\pi}{2}, \ -\frac{2\pi}{3}, \ \frac{2\pi}{3}.$$

Example 3.9: Solve $3\sin\theta - 5\cos\theta = 0$ in the interval $0° \le \theta < 360°$.

Solution 3.9: This equation can be solved by dividing both sides of the equation by $\cos\theta$ provided that $\cos\theta \neq 0$. $\dfrac{3\sin\theta - 5\cos\theta}{\cos\theta} = \dfrac{0}{\cos\theta} \Rightarrow 3\left(\dfrac{\sin\theta}{\cos\theta}\right) - 5\left(\dfrac{\cos\theta}{\cos\theta}\right) = 0, \ \cos\theta \neq 0$.

Therefore, $3\tan\theta - 5 = 0 \Rightarrow \tan\theta = \dfrac{5}{3} \Rightarrow \theta = 59.03624...° = 59.0°$ and

$\theta = 180° + 59.0° = 239.0°$ so $\theta = 59.0°, \ 239°$.

If the trigonometric equation involves functions whose unknown angles differ, for example, θ and 2θ, then identities can be used to rewrite the equation so that the angles are equal.

Example 3.10: Solve $\cos 2\theta = \cos\theta$ in the interval $0 \le \theta < \pi$.

Solution 3.10: Using the identity $\cos 2\theta = 2\cos^2\theta - 1$ we obtain $2\cos^2\theta - 1 = \cos\theta$.

Let $c = \cos\theta$. $2c^2 - 1 = c \Rightarrow 2c^2 - c - 1 = 0 \Rightarrow 2c^2 - 2c + c - 1 = 0$

$\Rightarrow 2c(c-1) + 1(c-1) = 0 \Rightarrow (c-1)(2c+1) = 0 \Rightarrow c - 1 = 0,\ 2c + 1 = 0 \Rightarrow c = 1,\ c = -\frac{1}{2}$.

So $\cos\theta = 1$ and $\cos\theta = -\frac{1}{2}$.

For $\cos\theta = 1,\ \theta = 0$.

For $\cos\theta = -\frac{1}{2},\ \theta = \pi - \frac{\pi}{3} = \frac{2\pi}{3}$.

So $\theta = 0,\ \frac{2\pi}{3}$.

Exercise 3.6

1. In each case, solve the equation in the interval $0 \le \theta < 2\pi$, giving your answers exactly.

   (a) $\sin^2\theta + \sin\theta = 0$ (b) $2\cos^2\theta + \cos\theta = 1$ (c) $2\sin^2\theta - 1 = 0$

2. Find the solutions of the following equations in the interval $-180° < x \le 180°$.

   (a) $6\sin^2 x + \sin x - 1 = 0$ (b) $3\cos^2 x + 2\cos x - 1 = 0$ (c) $4\tan^2 x - \tan x - 3 = 0$

3. Show that the equation $2\sin^2 x = 1 - \cos x$ may be written in the form $2\cos^2 x - \cos x - 1 = 0$.

   Hence show that in the interval $-\pi < x \le \pi$, two of the solutions are $x = 0$ and $x = \frac{2\pi}{3}$, and

   find the third solution.

4. Use the identity $\sin 2x = 2\sin x \cos x$ to show that the equation $\sin 2x = \cos x$ can be written in

   the form $\cos x(2\sin x - 1) = 0$. Show that $\frac{\pi}{6}$ is a solution of this equation, and find two more

   solutions in the interval $0 \le x < \pi$.

5. In each case, find all the solutions of the following equations in the interval $0° \le x < 360°$. Give
   your answers correct to the nearest degree.

   (a) $\sin x = 4\cos x$ (b) $3\cos x + 2\sin x = 0$ (c) $\sin 2x = 3\cos^2 x$

6. Draw the graphs of the functions $f : x \mapsto 1 - \sin 2x$ and $g : x \mapsto 2\cos x$ for the domain

   $0° \le x \le 360°$. Use your graph to state how many solutions of the equation $1 - \sin 2x = 2\cos x$

   exist in the interval $0° \le x < 360°$. Find, correct to four decimal places, the least value of x in
   this interval which satisfies the equation.

7. The time, t (in hours), of sunset after 18:00 at a specific location is a function of d, the

   number of days after the beginning of the year, where $t = -2\cos\left(\frac{2\pi}{365}(d+11)\right),\ 1 \le d \le 365$.

   Find the dates
   (a) on which sunset is at 18:00   (b) when the sun sets at 17:00
   (c) when the sun sets at 19:31

8.   Sea level in a harbor rises and falls due to the tide so that its height, h measured in meters above the sea bed (assumed to be horizontal), is given by $h(t) = a + b\sin kt$ where t is time, measured in hours after midnight. Two high tides occur in any 25 hour period. The depth of the sea water at midnight $(t = 0)$ is 16m. At high tide the depth of the sea water is 25m. Find the

(a) values of a and b (b) period of the tide and show that $k = \dfrac{4\pi}{25}$

(c) depth of sea water in the harbor at   (i) 06:00   (ii) 14:00

(d) proportion of time during which the depth of water in the harbor is less then 10m

3.5.3 Solution of Trigonometric Equations Using the Graphing Calculator

The graphing calculator can be used to solve trigonometric equations in a manner similar to that described in section 2.2 of Unit 2.

Example 3.11: Find all solutions of $3\sin x + 2\cos x = 1$ in the interval, $0° \le x < 360°$. Give your answers to the nearest degree.

Solution 3.11: Firstly, ensure that your calculator is in 'degree' mode and then obtain a graph of the functions

$$f(x) = 3\sin x + 2\cos x \text{ and } g(x) = 1 \text{ for the domain}$$

$0° \le x \le 360°$. The solutions occur at the x - coordinates of the intersections of the graphs, $x = 130.2078...$ and $x = 342.4129...$, so, correct to the nearest degree, $x = 130°$ and $x = 342°$. Figure 3.29 shows the final screen display for the lower intersection point.

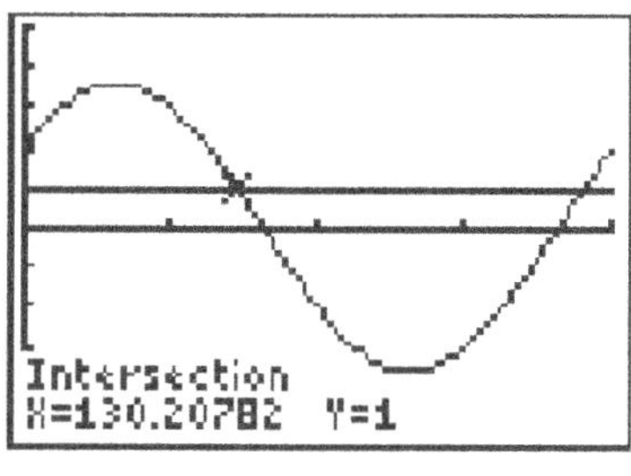

Figure 3.29

Example 3.12: Find all the solutions of $x + 7\sin x = 0$ in the interval $0 \le x < 2\pi$. Give your answers correct to three significant figures.

Solution 3.12: Firstly, ensure that your calculator is in 'radian' mode and then obtain a graph of the function $y = x + 7\sin x$ for the domain $0 \le x \le 2\pi$. The solutions are the zeros of the function which are $x = 0$, $x = 3.698707...$ and $x = 5.401716...$ so, correct to three significant figures $x = 0$, $x = 3.70$ and $x = 5.40$. Figure 3.30 shows the lower zero.

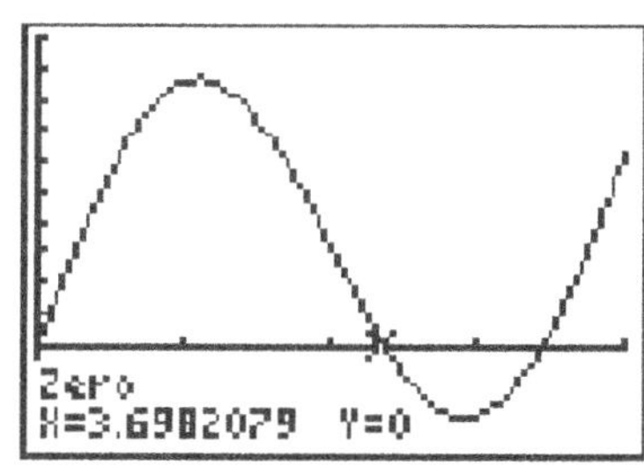

Figure 3.30

Exercise 3.7

1. (a) Sketch the graph of $y = x^2 \sin x - 2$, for $0 \le x \le \pi$.

   (b) State how many solutions of the equation $x^2 \sin x - 2 = 0$ there are in the interval $0 \le x < \pi$.

   (c) Find the solution of $x^2 \sin x - 2 = 0$ closest to the origin, correct to four decimal places.

2. Functions f and g are defined by $f : x \mapsto 3\sin x - 2$ and $g : x \mapsto \sin 2x$.

   (a) On the same axes, sketch the graphs of $y = f(x)$ and $y = g(x)$ for $0° \le x < 180°$.

   (b) Find both solutions of the equation $\sin 2x = 3\sin x - 2$ in the interval $0° \le x < 180°$.

3. Let $f(x) = \sin(3\sin x)$ with domain $0 \le x < 2\pi$.

   (a) Sketch the graph of $y = f(x)$ for the given domain.

   (b) In each case, state the number of solutions that exist in the interval $0 \le x < 2\pi$.

      (i) $\sin(3\sin x) = 0$ (ii) $\sin(3\sin x) = x - 1$

      (iii) $2\sin(3\sin x) = 1$ (iv) $\sin(3\sin x) = 1 - \dfrac{x}{3}$

4. A car with worn shock absorbers hits an obstruction in the road at time $t = 0$. The height, h centimeters, of the front bumper above the road during the next t seconds is given by
   $h(t) = 20 + 18e^{-0.4t} \sin 2t$, $t \ge 0$.

   (a) Sketch the graph of the height of the front bumper above the road for the first 10 seconds after the car hits the obstruction.

   (b) Find the minimum clearance between the front bumper and the road and the value of t when it occurs.

   (c) Find how much time elapses before the vibration caused by the car hitting the obstruction is reduced to ± 2cm.

3.6 The Solution of Triangles

In this section, all angles will be given in degrees and all answers will be expressed correct to three significant figures. It is customary to denote the three vertices of a triangle by the letters A, B and C and the sides opposite those vertices by the letters a, b and c respectively (figure 3.31).

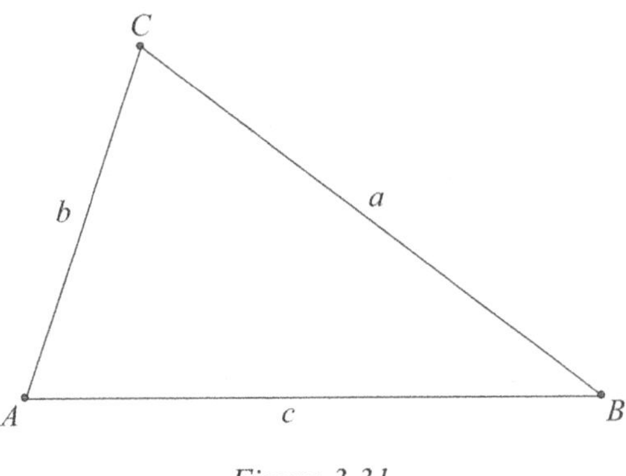

Figure 3.31

Solving a triangle means finding the magnitudes of any unknown angles or sides of a uniquely defined triangle. By 'uniquely defined', we mean that sufficient data is given about the lengths of the sides and the magnitudes of the angles so that only one triangle can be formed. For example, if you only know the length of two sides of a triangle, it has not been defined because there are an infinite number of triangles which can be drawn with that data. Therefore, it is necessary to establish the minimum data of a triangle which enables it to be defined.

Minimum requirements for the definition of a unique triangle:

- Three sides only (SSS). In this case, solving a triangle involves finding the magnitudes of the three unknown angles.

- Two angles and one side only (AAS). In this case, solving a triangle involves finding the magnitude of the third angle and the lengths of the remaining two sides.

- Two sides and the included angle (SAS). The 'included angle' is the angle between the two known sides. In this case, solving a triangle means finding the length of the third side and the magnitudes of the remaining two angles.

- One right angle, the hypotenuse and one other side (RHS). In this case, solving the triangle means finding the length of the third side and the magnitude of the remaining two angles.

3.6.1 Solving Right Triangles and Isosceles Triangles

An isosceles triangle can be bisected to produce two congruent right triangles. Simple trigonometry is all that is required to solve a right triangle.

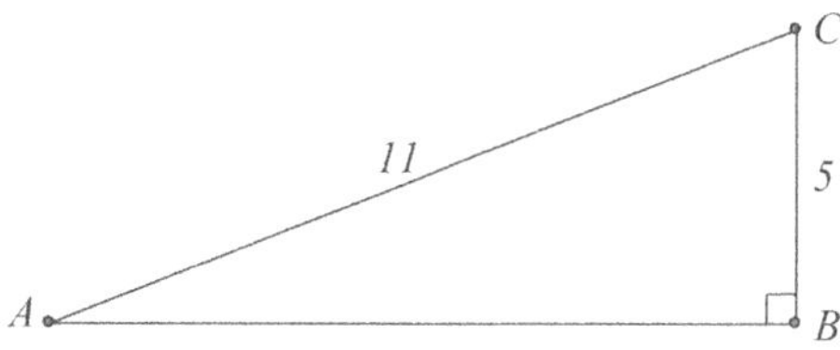

Example 3.13: Solve the triangle for which $a = 5$, $b = 11$ and $\hat{B} = 90°$.

Solution 3.13: You need to find $\hat{A}$ and $\hat{C}$ (i.e., the magnitudes of angle A and angle C) and the length of c.

Figure 3.32

Referring to figure 3.32, $\cos\hat{C} = \dfrac{5}{11}$, $\left(\hat{C} < 90°\right) \Rightarrow \hat{C} = 62.96430...°$.

Therefore $\hat{A} = 90° - 62.96430° = 27.03569...°$ and $\sin\hat{C} = \dfrac{c}{11} \Rightarrow c = 11\sin\hat{C}$

$= 9.797958....$ Therefore $\hat{A} = 27.0°$, $\hat{C} = 63.0°$ and $c = 9.80$.

Example 3.14: ΔABC is isosceles with $AB = BC = 15$ and $\hat{A} = 72°$. Find AC.

Solution 3.14: Bisect ΔABC through $[BN]$ where N is the midpoint of $[AC]$. Now ΔABN is right-angled at N (Figure 3.33). $\cos\hat{A} = \dfrac{AN}{AB} \Rightarrow AN = AB\cos\hat{A}$

$= 15\cos 72° = 4.63525$. $AC = 2AN = 2 \times 15 \times \cos 72°$

$= 9.270509....$ Therefore, $AC = 9.27$

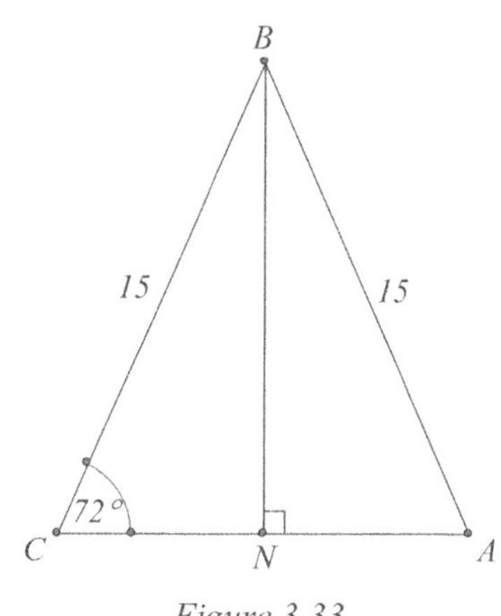

Figure 3.33

3.6.2 Solving Scalene Triangles

Scalene triangles are those which are neither right-angled nor isosceles, and, in order to solve them, it will be necessary to use two more trigonometric tools: *the cosine rule* and *the sine rule.*

The Cosine Rule

The cosine rule is an extension of Pythagoras' theorem to non-right-angled triangles. Figure 3.34 shows $\triangle ABC$ in which $CN = h$, $AN = n$. Proving the cosine rule:

Step 1: Draw $\triangle ABC$ and draw $[CN]$, the perpendicular from C to $[AB]$.

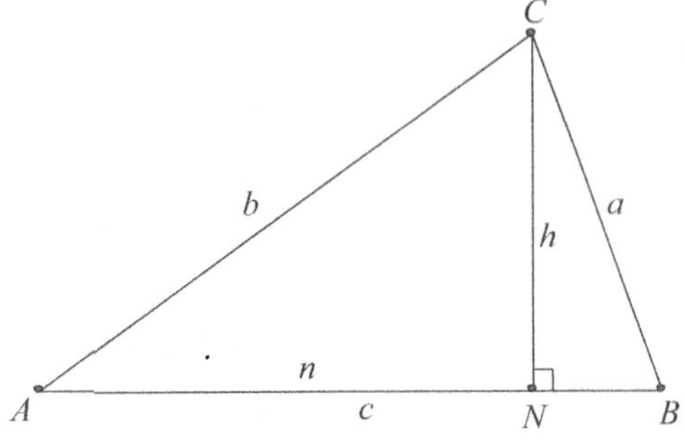

Figure 3.34

Step 2: Use Pythagoras' theorem in $\triangle ANC$ and $\triangle BNC$ to

eliminate h : $\quad (c-n)^2 + h^2 = a^2 \ \ldots\ldots (1)$

$$n^2 + h^2 = b^2 \ \ldots\ldots\ldots (2)$$

Subtract (2) from (1): $\left((c-n)^2 + h^2\right) - \left(n^2 + h^2\right) = a^2 - b^2$

$$\Rightarrow (c-n)^2 - n^2 = a^2 - b^2$$

$$\Rightarrow c^2 - 2cn + n^2 - n^2 = a^2 - b^2$$

$$\Rightarrow c^2 - 2cn = a^2 - b^2 \ \ldots\ldots (3)$$

Step 3: Use simple trigonometry on $\triangle ANC$ to eliminate n : $\cos \hat{A} = \dfrac{n}{b} \Rightarrow n = b\cos \hat{A} \ \ldots\ldots (4)$

Substitute (4) into (3): $\quad c^2 - 2cb\cos \hat{A} = a^2 - b^2$

Step 4: Rearrange and extend: $\quad a^2 = b^2 + c^2 - 2bc\cos \hat{A}$

As the cosine rule is used to find angles as well as sides, it is useful to rearrange these formulae

so that $\cos \hat{A}$ is the subject: $2bc\cos \hat{A} = b^2 + c^2 - a^2 \Rightarrow \cos \hat{A} = \dfrac{b^2 + c^2 - a^2}{2bc}$.

Because the vertices can be labeled in three ways, there are three labels for each of these arrangements of the cosine rule:

$$a^2 = b^2 + c^2 - 2bc\cos \hat{A} \ , \quad b^2 = c^2 + a^2 - 2ca\cos \hat{B} \ , \quad c^2 = a^2 + b^2 - 2ab\cos \hat{C}$$

$$\text{and } \cos \hat{A} = \frac{b^2 + c^2 - a^2}{2bc} \ , \quad \cos \hat{B} = \frac{c^2 + a^2 - b^2}{2ca} \ , \quad \cos \hat{C} = \frac{a^2 + b^2 - c^2}{2ab} \ .$$

The cosine rule is used when the data provided is either SSS or SAS.

Example 3.15: If, in ΔABC, $a = 17$, $b = 20$, $c = 25$ find A.

Solution 3.15: $\cos \hat{A} = \dfrac{b^2 + c^2 - a^2}{2bc} = \dfrac{20^2 + 25^2 - 17^2}{2 \times 20 \times 25} = 0.736$, $\left(A < 180^\circ\right) \Rightarrow \hat{A} = 42.60822\ldots = 42.6^\circ$.

Example 3.16: In ΔABC, $a = 8$, $c = 7$, $\hat{B} = 131^\circ$. Find b and $\hat{C}$.

Solution 3.16: $b^2 = c^2 + a^2 - 2ca \cos \hat{B} \Rightarrow b^2 = 7^2 + 8^2 - 2 \times 7 \times 8 \cos 131^\circ \Rightarrow b = 13.65572\ldots = 13.7$

$\cos \hat{C} = \dfrac{a^2 + b^2 - c^2}{2ab} = \dfrac{8^2 + 13.65572^2 - 7^2}{2 \times 8 \times 13.65572}$ $\left(\hat{C} < 180^\circ\right) \Rightarrow \hat{C} = 22.75976\ldots^\circ = 22.8^\circ$

The Sine Rule

Proving the sine rule:

Step 1: Draw ΔABC and a line perpendicular to $[AB]$ from C to intersect $[AB]$ at N. $CN = h$ (figure 3.35).

Step 2: Use simple trigonometry to write angles A and B in terms of h. From ΔANC, we deduce that

$$\sin \hat{A} = \frac{h}{b} \quad\ldots\ldots\ldots\ldots\ldots\ldots \quad (1)$$

From ΔBNC, we deduce that $\sin \hat{B} = \dfrac{h}{a}$ ... (2)

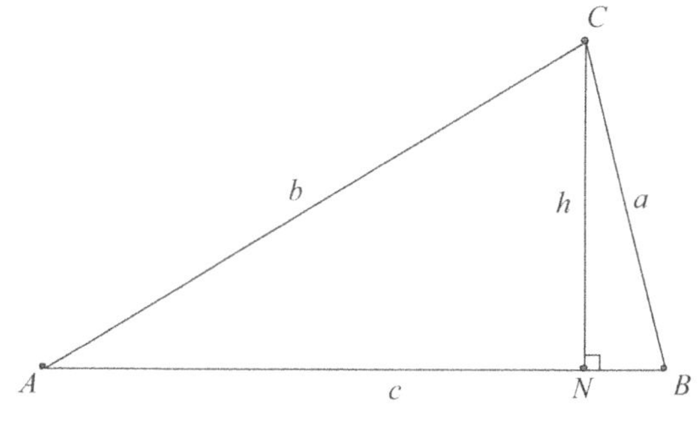

Figure 3.35

Step 3: Use (1) and (2) to eliminate h: $\quad h = b \sin \hat{A}$ and $h = a \sin \hat{B}$

$$\Rightarrow a \sin \hat{B} = b \sin \hat{A} \Rightarrow \frac{a}{\sin \hat{A}} = \frac{b}{\sin \hat{B}} \quad\ldots\ldots\ldots\ldots\ldots\ldots\ldots (3)$$

Step 4: Extending (3). Because the vertices can be labeled in three ways, the sine rule can be written in three ways: $\quad \dfrac{a}{\sin \hat{A}} = \dfrac{b}{\sin \hat{B}}$, $\quad \dfrac{b}{\sin \hat{B}} = \dfrac{c}{\sin \hat{C}}$, $\quad \dfrac{c}{\sin \hat{C}} = \dfrac{a}{\sin \hat{A}}$

This can be more conveniently expressed by $\dfrac{a}{\sin \hat{A}} = \dfrac{b}{\sin \hat{B}} = \dfrac{c}{\sin \hat{C}}$.

The sine rule is used when the data provided is AAS.

Example 3.17: Solve ΔABC if $a = 45$, $\hat{A} = 97^\circ$ and $\hat{B} = 24^\circ$.

Solution 3.17: $\hat{C} = 180^\circ - \left(97^\circ + 24^\circ\right) = 59^\circ$. $\quad \dfrac{b}{\sin \hat{B}} = \dfrac{a}{\sin \hat{A}} \Rightarrow \dfrac{b}{\sin 24^\circ} = \dfrac{45}{\sin 97^\circ}$

105

$$\Rightarrow b = \frac{45\sin 24^\circ}{\sin 97^\circ} = 18.4406 \Rightarrow b = 18.4 \text{ . Similarly, } \frac{c}{\sin \hat{C}} = \frac{a}{\sin \hat{A}} \Rightarrow \frac{c}{\sin 59^\circ} = \frac{45}{\sin 97^\circ}$$

$$\Rightarrow c = \frac{45\sin 59^\circ}{\sin 97^\circ} = 38.86220... = 38.9 \text{ .}$$

In general, it is unwise to use the sine rule to find angles. The following shows why:

Suppose you want to solve $\triangle ABC$ which has sides $a=8$, $b=11$, $c=14$ and $\hat{A} = 34.772^\circ$. If you use the sine rule to find $\hat{C}$, then $\dfrac{c}{\sin \hat{C}} = \dfrac{a}{\sin \hat{A}} \Rightarrow \sin \hat{C} = \dfrac{c\sin \hat{A}}{a} \Rightarrow \sin \hat{C} = \dfrac{14\sin 34.772}{8}$

$\left(\hat{C} < 180^\circ \right) \Rightarrow \hat{C} = 86.41796...^\circ$ or $180^\circ - 86.41796...^\circ = 93.58203...^\circ = 93.6^\circ \Rightarrow \hat{C} = 86.4^\circ, 93.6^\circ$.
Unfortunately, it is not clear which of these two angles is the correct one. If, however, the cosine rule is used to find $\hat{C}$, there is no ambiguity:

$$\cos \hat{C} = \frac{a^2 + b^2 - c^2}{2ab} = \frac{8^2 + 11^2 - 14^2}{2\times 8 \times 11} = -0.0625 \left(\hat{C} < 180^\circ \right) \Rightarrow \hat{C} = 93.58332...^\circ = 93.6^\circ \text{ .}$$

Exercise 3.8

1. In each case, solve $\triangle ABC$.

   (a) $a=3$, $b=5$, $c=6$ (b) $a=17$, $b=13$, $\hat{C}=53^\circ$

   (c) $b=11$, $\hat{A}=72^\circ$, $\hat{B}=29^\circ$ (d) $a=14.7$, $b=19.3$, $c=25.1$

   (e) $c=8.6$, $\hat{A}=38^\circ$, $\hat{B}=24^\circ$ (f) $b=1.92$, $c=0.778$, $\hat{A}=115^\circ$

2. $\triangle ABC$ is isosceles with $AB = AC = 15$cm and $B\hat{A}C = 20^\circ$. M is the midpoint of $[AB]$.

   (a) Draw a diagram of $\triangle ABC$ and mark point M on the triangle.

   (b) Find BC and $A\hat{B}C$.

   (c) Find MC.

3. Figure 3.36 shows $\triangle ABC$ in which $\hat{A}=30^\circ$, $\hat{B}=50^\circ$.
   N lies on AB such that $[CN]$ is perpendicular to $[AB]$
   and $CN=5$m. Find AB.

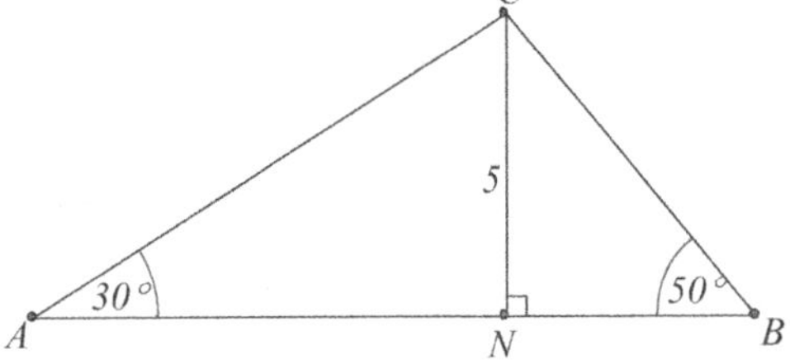

4. In $\triangle ABC$, $AB=7.6$cm, $BC=11.2$cm and $A\hat{B}C=68^\circ$.

   P lies on $[BC]$ such that $[AP]$ bisects $B\hat{A}C$.

   (a) Draw a diagram of $\triangle ABC$ and mark on it the point P.
   Find   (b) AC (c) AP (d) PC

Figure 3.36

5. $\triangle ABC$ has sides $BC=38$, $CA=33$, $AB=30$. P lies on $[AB]$ such that $[CP]$ is perpendicular to $[AB]$. Q lies on $[AC]$ such that $[BQ]$ is perpendicular to $[AC]$.

(a) Draw a diagram of $\triangle ABC$ showing the positions of P and Q.

(b) Find PQ.

6. In Section 3.6.2, the cosine and sine rules were proven for acute angled triangles. Use $\triangle ABC$, shown in figure 3.37, to prove that the cosine and sine rules are both true for an obtuse angled triangle. (An obtuse angle is one between $90°$ and $180°$.)

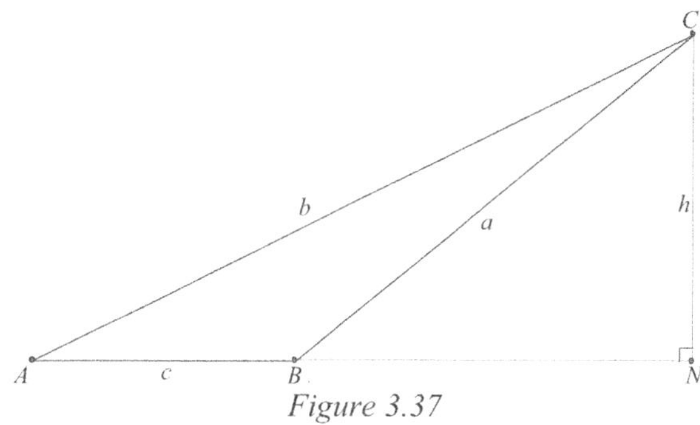

Figure 3.37

3.6.3 The Ambiguous Case

If the given data consists of two sides and an angle which is not the included angle, then two possible triangles can be drawn. In other words, the given data is ambiguous because it could apply to either one of the two triangles. Figure 3.38 shows an example of the ambiguous case in which $c = 10$, $a = 6$, $\hat{A} = 30°$. The two possible triangles are $\triangle ABC_1$ and $\triangle ABC_2$. Both triangles can be solved by using either the cosine rule or the sine rule.

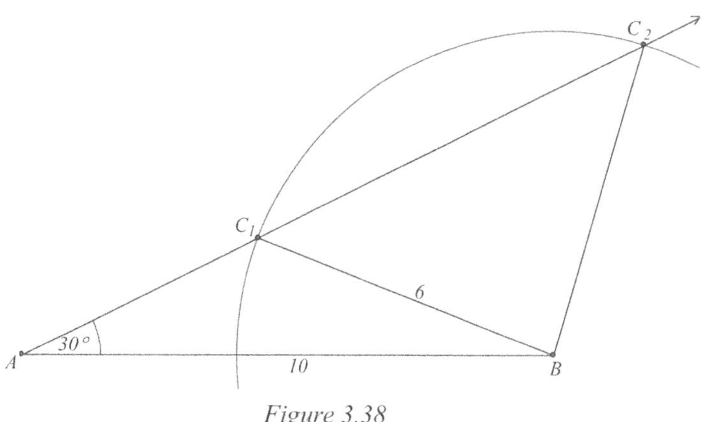

Figure 3.38

The following example uses the cosine rule, and the next exercise provides an opportunity to use the sine rule.

Example 3.18: $\triangle ABC$ has $AB = 3$cm, $C\hat{A}B = 37°$ and $CB = 2$cm.

        (a) Draw a diagram of both triangles which satisfy this data.

        (b) Find AC for both triangles.

Solution 3.18: (a) See figure 3.39.

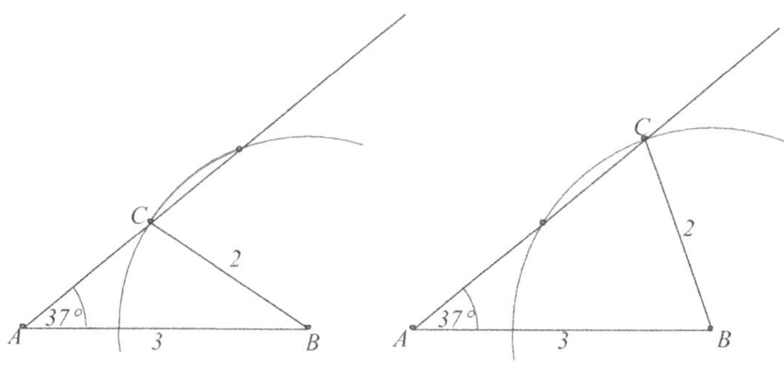

Figure 3.39

(b) Use the cosine rule to find b :

$$a^2 = b^2 + c^2 - 2bc \cos \hat{A}$$
$$\Rightarrow 2^2 = b^2 + 3^2 - 2 \times b \times 3 \cos 37°$$
$$\Rightarrow 4 = b^2 + 9 - 6b \cos 37° \Rightarrow b^2 - 6 \cos 37° b + 5 = 0$$
$$\Rightarrow b = \frac{6 \cos 37° \pm \sqrt{36 \cos^2 37° - 20}}{2}$$
$$= 3.256352..., \ 1.535460... \ .$$

So, $AC = 3.26$ or 1.54.

Exercise 3.9

1. $\triangle ABC$ has $AC = 8$, $AB = 11$ and $\hat{B} = 44°$.

(a) Draw a line of length 11cm and label it AB. Draw another line L through B at an angle of $44°$ to $[AB]$. Use a compass, centered at A, and draw a circular arc with radius 8cm to intersect L in two places C_1 and C_2.

(b) Measure BC_1 and BC_2.

(c) Use the sine rule to find, correct to three significant figures, BC_1 and BC_2.

2. $\triangle ABC$ has $a = 33$, $b = 27$ and $\hat{B} = 19°$.

(a) Draw two different diagrams each showing a triangle which satisfies the given data.

(b) Use the cosine rule to find both possible values of c.

3. $\triangle PQR$ has $PQ = 51$m, $QR = 15$m and $\hat{P} = 11°$. Find both possible values of $\hat{R}$.

4. Use the sine rule to solve both forms of $\triangle ABC$ which satisfy the conditions:
$AC = 16.6$, $CB = 13.9$ and $\hat{A} = 33°$.

3.6.4 The Area of a Triangle

We shall show that the area of triangle ABC, shown in Figure 3.40, is $\frac{1}{2}bh$, where b is the base and h the height of the triangle.

Figure 3.41 shows rectangle $ADEC$ in which the base $[CA]$ coincides with the base of $\triangle ABC$ and the height $[AD]$ is equal to the height h of $\triangle ABC$. Therefore, rectangle $ADEC$ has area bh.

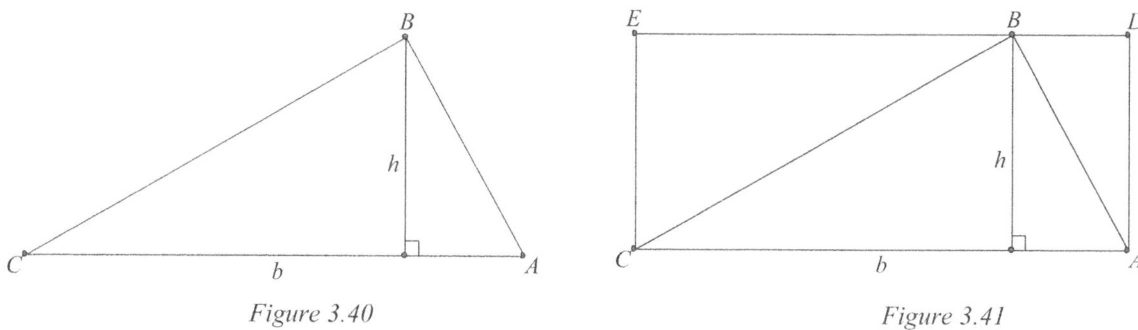

Figure 3.40   Figure 3.41

Figure 3.42 shows figure 3.41 extended by drawing $[AF]$ parallel to $[CB]$ forming a parallelogram $AFBC$. In addition, $[CA]$ is extended to G so that $[GF]$ is perpendicular to $[CA]$.

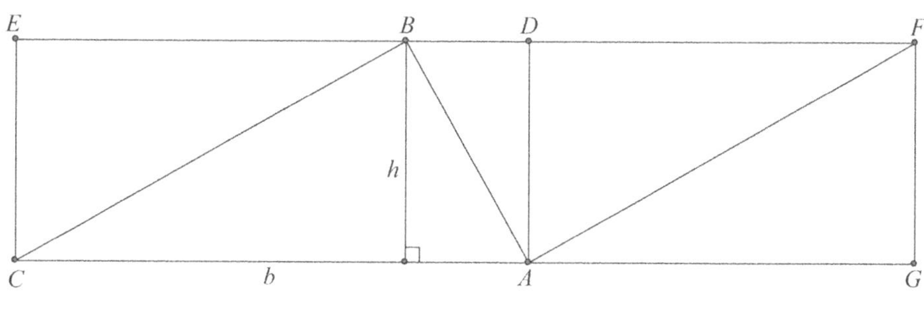

Figure 3.42

Then $\triangle AGF$ is congruent to $\triangle BEC$ from which we may deduce that the area of parallelogram $AFBC$ is equal to the area of rectangle $ADEC = bh$. Finally, we note that AB bisects parallelogram $AFBC$ so that the area of $\triangle ABC = \frac{1}{2}bh = \frac{1}{2} \times (\text{length of base})(\text{perpendicular height})$. This is a well-known result but not always immediately useful for finding the area of a scalene triangle.

A more useful formula can be obtained if the lengths of two sides of the triangle are known, and the magnitude of the angle between these sides is also known. Figure 3.43 shows $\triangle ABC$ in which it is assumed that both sides b and c, and the magnitude of angle A are known.

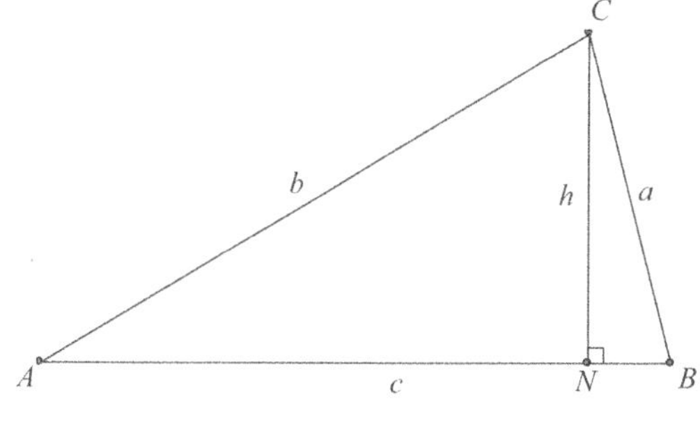

Figure 3.43

109

The area of $\triangle ABC = \dfrac{1}{2}ch$, and $\sin \hat{A} = \dfrac{h}{b} \Rightarrow h = b\sin\hat{A} \Rightarrow$ Area $\triangle ABC = \dfrac{1}{2}cb\sin\hat{A} = \dfrac{1}{2}bc\sin\hat{A}$.

Similarly, of course, the area of $\triangle ABC = \dfrac{1}{2}ab\sin\hat{C} = \dfrac{1}{2}ca\sin\hat{B}$.

Example 3.19: Find the area of $\triangle ABC$ in which $b = 18$, $c = 13$ and $\hat{A} = 32°$

Solution 3.19: The area of $\triangle ABC = \dfrac{1}{2}bc\sin\hat{A} = \dfrac{1}{2} \times 18 \times 13 \times \sin 32° = 62.00055... = 62.0$

so the area of $\triangle ABC = 62.0$

Exercise 3.10

1. In each case find the area of $\triangle ABC$.

   (a) $a = 17$, $b = 31$, $\hat{C} = 55°$ (b) $b = 3.28$, $c = 2.88$, $\hat{A} = 18.3°$ (c) $c = 173$, $a = 103$, $\hat{B} = 147°$

2. $\triangle ABC$ has sides $AB = 3.86\text{cm}$, $BC = 2.18\text{cm}$, $CA = 4.03\text{cm}$. Find the area of $\triangle ABC$.

3. In $\triangle ABC$, $AB = 15$, $BC = 9$ and $\hat{A} = 30°$. Find the area of both triangles which satisfy this data.

4. $\triangle PQR$ has $\hat{P} = 75°$, $\hat{Q} = 69°$ and $PQ = 8\text{m}$. Find the area of $\triangle PQR$.

5. In $\triangle ABC$, $AB = 8$, $AC = 5$ and the area of $\triangle ABC$ is $10\sqrt{3}$. Find both possible exact values of BC.

6. Figure 3.44 shows an equilateral $\triangle A_1 B_1 C_1$ with an inscribed $\triangle A_2 B_2 C_2$ constructed inside it whose vertices are the midpoints of the sides of $\triangle A_1 B_1 C_1$. More triangles are inscribed in a similar manner.

   (a) If $A_1 B_1 = B_1 C_1 = C_1 A_1 = 1$, find the exact area of $\triangle A_1 B_1 C_1$.

   (b) Show that the area of $\triangle A_2 B_2 C_2$ is $\dfrac{\sqrt{3}}{16}$.

   (c) Find the area of $\triangle A_3 B_3 C_3$.

   (d) The area of $\triangle A_{12} B_{12} C_{12} = \dfrac{\sqrt{3}}{2^n}$. Find the value of n.

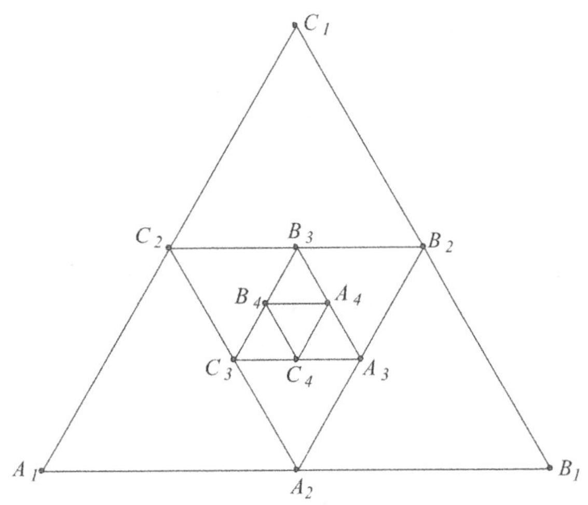

Figure 3.44

7. Figure 3.45 shows a tetrahedron (triangular based pyramid) in which $[OX]$ and $[OY]$ lie in a horizontal plane while $[OZ]$ is vertical. $OX = 16\text{cm}$, $OY = 20\text{cm}$, $OZ = 9\text{cm}$ and $X\hat{O}Y = 125°$. Find XY, YZ and ZX, and show that the area of $\triangle XYZ = 195\text{cm}^2$.

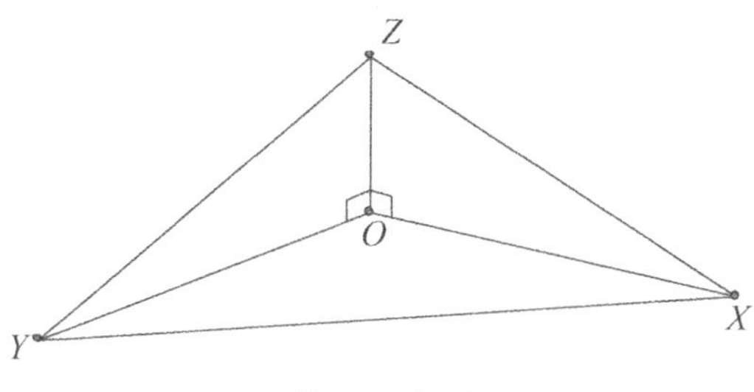

Figure 3.45

Unit 3 Review Exercises

Part 1

No graphing calculators should be used to answer questions in Part 1.

1. Let $\sin 50° = p$, $\cos 20° = q$. Find, in terms of p or q

    (a) $\sin 130°$ (b) $\cos 160°$ (c) $\cos 40°$

Show that

    (d) $p = 2q^2 - 1$

2. Given that $\cos 30° = \dfrac{\sqrt{3}}{2}$, $\cos 45° = \dfrac{1}{\sqrt{2}}$, $\sin 30° = \dfrac{1}{2}$, $\cos 60° = \dfrac{1}{2}$,

Find (a) $\cos 210°$ (b) $\cos 300°$ (c) $\cos 135°$ (d) $\sin 150°$

3. Let $a = \cos 10°$, $b = \sin 20°$

(a) Show that $\sin 10° = \sqrt{1 - a^2}$

(b) Find $\sin 170°$ in terms of a.

(c) Find $\sin 20°$ in terms of a and hence show that $4a^4 = 4a^2 - b^2$.

4. Let $u = \sin 65°$.

(a) Use the identity, $\cos^2\theta + \sin^2\theta = 1$ to write $\cos 65°$ in terms of u.

If, in addition, $v = \cos 130°$

(b) Find v in terms of u.

(c) Find $\sin 130°$ in terms of u and hence show that $\tan 130° = \dfrac{2u\sqrt{1 - u^2}}{1 - 2u^2}$.

5. Consider the function, $h(x) = 2\sin 3x$.

(a) Write down the period of h.

(b) Sketch the curve of h for $0 \le x \le \pi$.

6. Figure 3.46 shows triangle ABC in which $AC = 5$, $BC = 1$ and the size of $A\hat{B}C = 90°$. Let the size of $B\hat{A}C = \theta$.

Find the value of AB, giving your answer in the form $a\sqrt{b}$ where $a, b \in \mathbb{Z}^+$

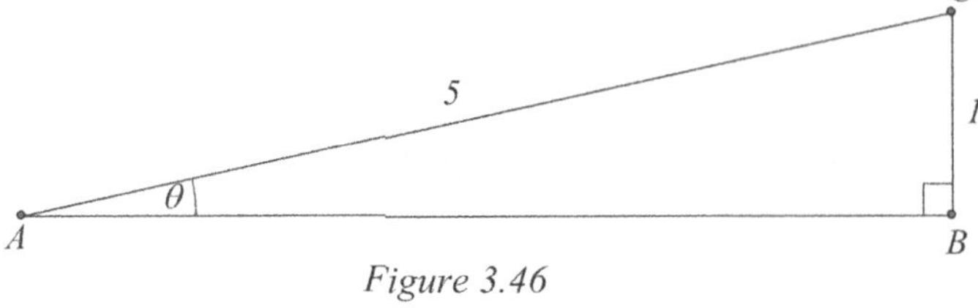

Figure 3.46

(a) Show that $\cos\theta = \dfrac{2\sqrt{6}}{5}$.

(b) Find the value of $\sin 2\theta$.

7. Consider the function $f(x) = 8x^2 - 6x + 1$

(a) Solve the equation $f(x) = 0$

Now consider the function $g(\theta) = 8\cos^4\theta - 6\cos^2\theta + 1$.

(b) Use your answers to (a) to find all four solutions of the equation $g(\theta) = 0$ in the interval $0 \le \theta < \pi$.

8. Figure 3.47 shows a circle with center O and radius 6 cm. Points A and B lie on the circle and the size of $A\hat{O}B$ is θ radians.

(a) Find, in terms of θ
    (i) the area of sector OAB
    (ii) the length of arc AB.

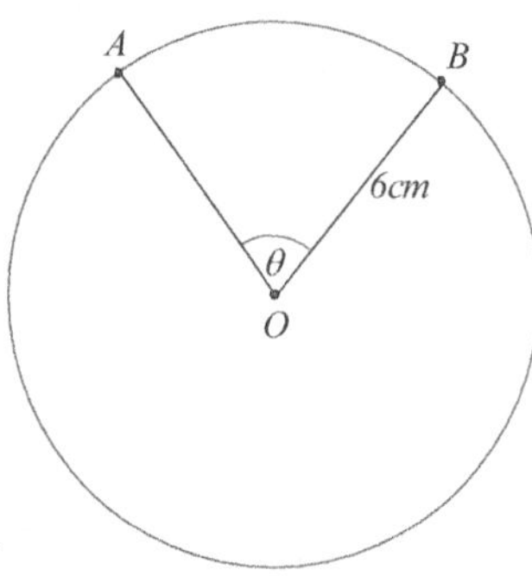

The area of sector OAB is 10 cm^2

(b) Calculate the value of θ, giving your answer in both radians and

    degrees.

Figure 3.47

9. Figure 3.48 shows a circle, C_1 of unit radius, with two circles of equal radius, C_2 and C_3 , touching C_1 internally and intersecting each other at points A and B inside C_1 . O is the center of circle C_2 .

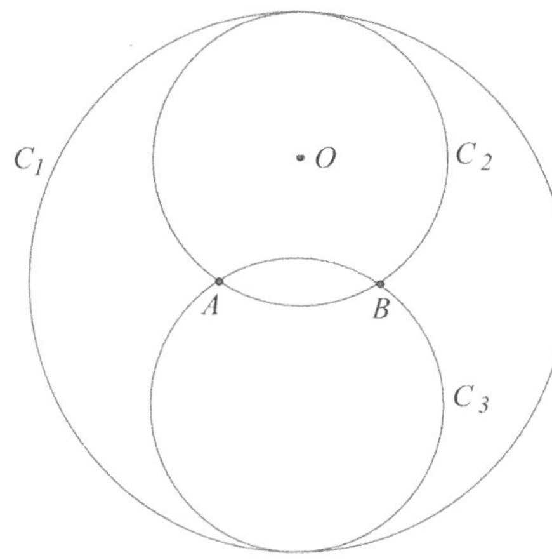

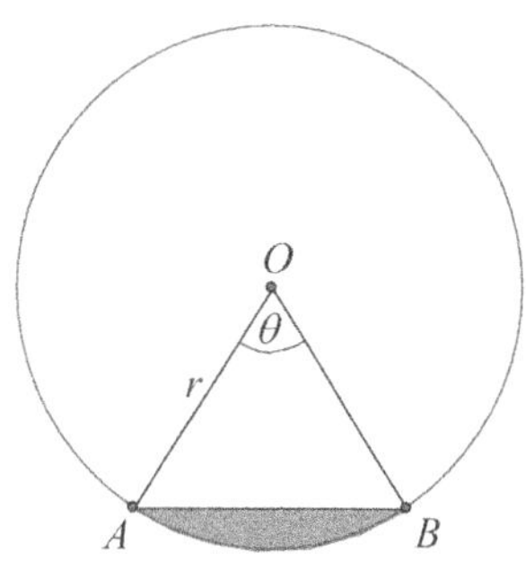

Figure 3.48   *Figure 3.49*

(a) Figure 3.49 shows circle C_2 . If the radius of C_2 is r and the size of $A\hat{O}B$ is θ radians, show that the shaded region has area $\dfrac{1}{2}r^2(\theta - \sin\theta)$.

(b) Given that $\theta = \dfrac{2\pi}{3}$ show, by considering the diameter of the unit circle C_1 , that $r = \dfrac{2}{3}$.

(c) Use your answers to (a) and (b) to show that the area of overlap between circles C_2 and C_3 is $\dfrac{4}{9}\left(\dfrac{2\pi}{3} - \dfrac{\sqrt{3}}{2}\right)$.

10. Consider the function $f(x) = a + b\sin kx$

(a) Given that $f(0) = 3$ write down the value of a.

The minimum value of f is 1 and the period of f is $\dfrac{\pi}{3}$

(b) Write down the value of b.
(c) Find the value of k.
(d) Sketch the graph of $y = f(x)$ for the domain $x \in [0, \pi]$
(e) Find the number of solutions of the equation $f(x) = 4$.

11. A sector of a circle has angle θ , radians and radius r.
(a) Write down, in terms of r and θ , the area of the sector
(b) Find, in terms of r and θ , the perimeter of the sector.

The perimeter of the sector is numerically equal to the area of the sector.

(c)   (i) Show that $\theta = \dfrac{4}{r-2}$ (ii) If the perimeter is 16 cm find the radius of the sector.

Part 2

Graphing calculators will usually be needed to answer questions in Part 2.

12. Figure 3.50 shows a circle with center O and radius 4 cm. The tangents at points P and Q meet at S. OS is 5cm.

    Calculate
    (a) PS
    (b) the size of $P\hat{O}S$, giving your answer in radians.
    (c) the area of quadrilateral $OPSQ$
    (d) the area of sector OPQ
    (e) the area of the shaded region.

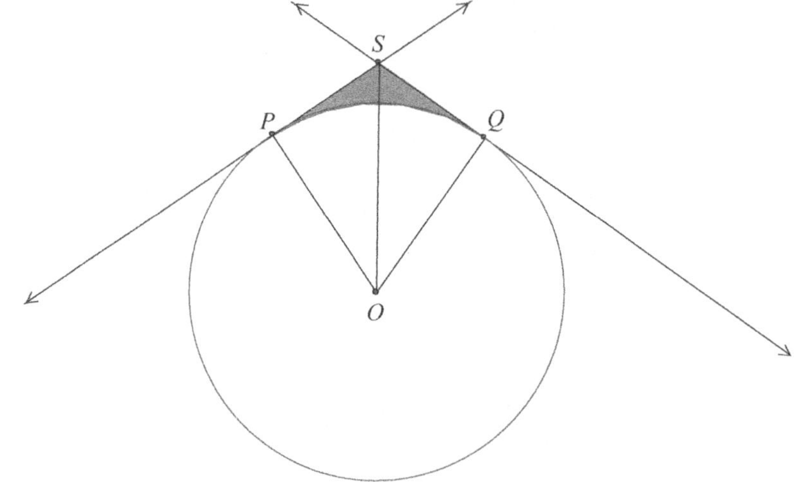

Figure 3.50

13. Consider the functions

    $$f(x) = \tan\frac{1}{2}x - 2\sin x \ , \ g(x) = \cos x \ , \ 0 \le x \le \pi .$$

    (a) On the same axes, sketch the graph of $y = f(x)$ and $y = g(x)$

    (b) Solve the equation $\tan\frac{1}{2}x - 2\sin x = \cos x$ in the interval $0 \le x < \pi$

14. In figure 3.51 $AB = BC = 10$, the size of $D\hat{A}C = 18°$ and the size of $D\hat{B}C = 35°$
    (a) Write down the size of $A\hat{B}D$ and hence find the size of $A\hat{D}B$.
    (b) Calculate BD.
    (c) Find the area of ΔBCD.

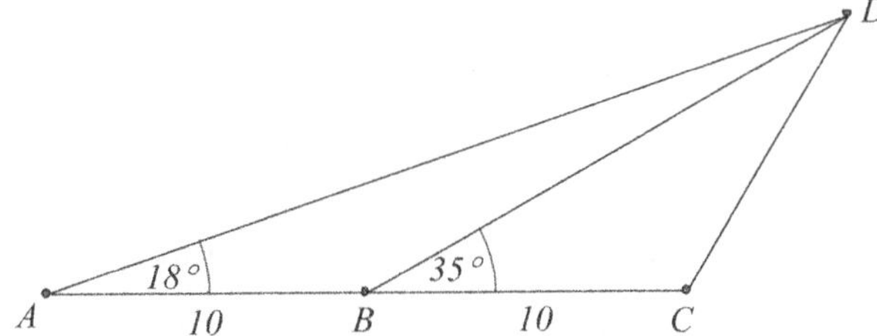

Figure 3.51

15. (a) Factorize $5c^2 - 2c - 3$
    (b) Use your answer to part (a) to solve the equation $5\cos^2\theta - 2\cos\theta - 3 = 0$ in the interval $0° \le \theta < 360°$.

16. Figure 3.52 shows $\triangle BCD$ in which $AB = 8$ cm, $AD = 6$ cm, the size of $A\hat{B}D = 30°$, the point D lies on $[AC]$ and $D\hat{A}B$ is obtuse. The size of $D\hat{B}C = 15°$. Let $BD = x$.

Use the cosine rule in $\triangle ABD$ to show that
$$x^2 - 8\sqrt{3}x + 28 = 0$$

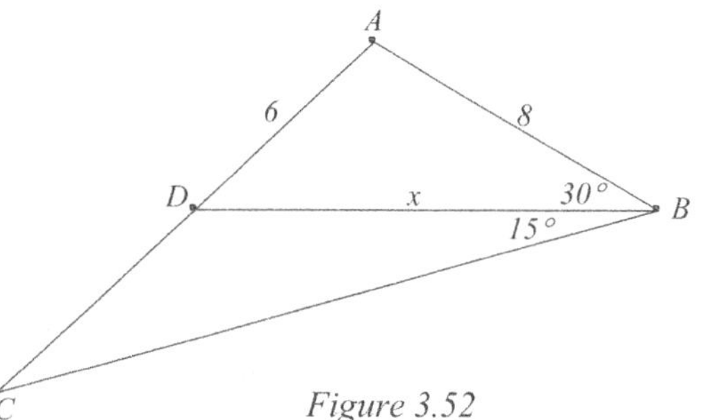

Figure 3.52

(a) Solve the equation shown in (a) in order to find the value of x. Explain why one of the two solutions you obtained was inappropriate.

(b) Find the size of $D\hat{A}B$.

(c) Hence calculate DC.

(d) Show that the area of triangle DBC is 24.8 cm^2

17. Let $f(\theta) = \sin\theta + 2\cos 2\theta + 1$, $\quad -90° < \theta \le 90°$.

(a) Use the double angle formulae to show that $f(\theta) = 3 + \sin\theta - 4\sin^2\theta$.

Consider the function $g(x) = 3 + x - 4x^2$.

(b) Solve the equation $g(x) = 0$

(c) Use your answers to (b) to show that $\theta = 90°$ is a solution of $f(\theta) = 0$ and find the other solution of this equation, correct to the nearest degree.

18. During her vacation at the seaside, Jennifer measures the height h, in meters, of the tide relative to a fixed point on the beach at four times during the day. The times, t are measured in hours after noon. The results are shown in the table below.

Time, t (hours)	0	2	7
Height, h (meters)	4	5.5	3.134

She uses the function $h(t) = a + b\sin(k\pi t)$ to model the height of the tide.

(a) Write down the value of a.

(b) Given that there are two high tides during a 24 hour period show that $k = \dfrac{1}{6}$.

(c) Find the value of b.

Estimate
(d) the height of the tide at 17:00
(e) the times when high tide occurs.

19.	Let $f(x)=12\cos x+5\sin x,\ 0\le x\le 2\pi$

(a) Sketch the graph of $y=f(x)$

(b) Find the coordinates of the maximum and minimum points on the graph.

Given that f can be written in the form $f(x)=a\sin(x+b)$ find

(c)	(i) the value of a	(ii) the value of b by using your answer to (b).

(d) Solve the equation $a\sin(x+b)=0$ in the interval $0\le x<2\pi$ using the values of a and b found in part (c).

(e) Verify that your solutions in (d) satisfy the equation $\tan x=-\dfrac{12}{5}$ and explain why

this is so.

20.	Let $f(x)=1-4\sin^2 x$.

(a) Show that $f(x)=f(-x)$ and explain what significance this has on the graph of $y=f(x)$

(b) Show that $f(x)=1-4\sin^2 x$ can be written in the form $f(x)=a+b\cos 2x$ and find the values of a, and b.

(c) Sketch the graph of $y=f(x)$ for the domain $-\pi\le x\le\pi$.

(d) Write down the number of solutions of $f(x)=0,\ \ -\pi\le x\le\pi$

(e) Find these solutions in terms of π .

UNIT 4: VECTORS

4.1 Elementary Vectors

Figure 4.01 shows a river, flowing from west to east, with a current of $2\,\text{ms}^{-1}$ (meters per second). The river is 500m wide. A boat is traveling across the river from point A on the south bank towards point B on the north bank, 200m downstream of point A.

If the boat starts at A and finishes at B then it has undergone a *displacement* which has translated the boat 200m east and 500m north.

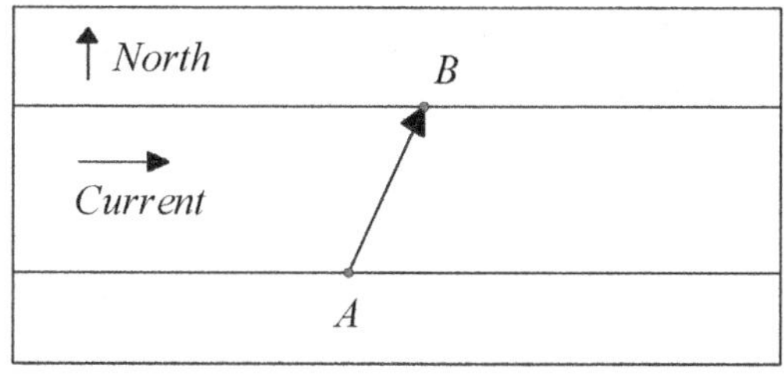

Figure 4.01

This displacement is an example of a *vector*. All vectors have a *magnitude* (length) and a direction. A vector can be represented by an arrow because an arrow has both magnitude (the length of the arrow) and direction (where the arrow is pointing). The arrow joining A to B in figure 4.01 is a vector.

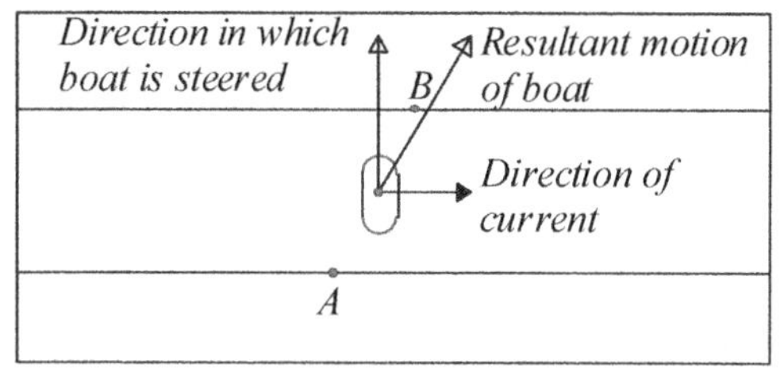

Figure 4.02

Suppose that the boat is capable of traveling in still water at $5\,\text{ms}^{-1}$ and that it is steered in a northerly direction (perpendicular to the bank). Figure 4.02 shows the way in which the boat would move across the river.

Note that, although the boat is directed at right angles to the bank, its actual path, relative to the bank, is not at right angles to the bank because of the effect of the current. We can use Pythagoras' theorem and simple trigonometry to show that, relative to the bank, the boat travels at a speed of $5.39\,\text{ms}^{-1}$ in a direction $21.8°$ east of north.

The *velocity* of an object refers to both its speed and its direction, so that the velocity of the boat is $5.39\,\text{ms}^{-1}$ in a direction $21.8°$ east of north.

Velocity is another example of a vector. As stated above, all vectors have a magnitude and a direction and may be represented by an arrow; however, a vector does not usually have position, so all arrows in figure 4.03 are identical vectors.

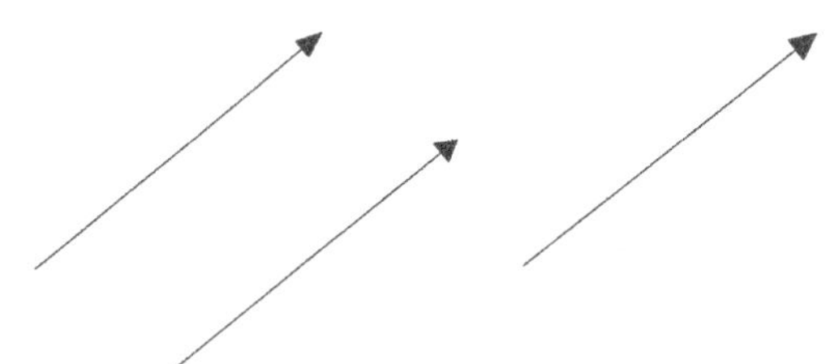

Figure 4.03

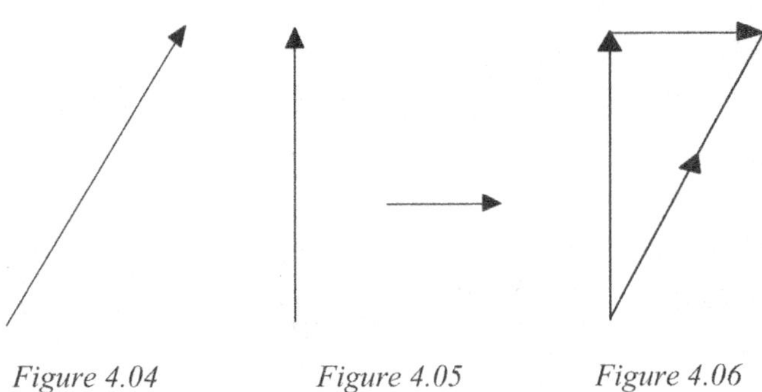

Figure 4.04 shows the vector which represents the velocity of the boat relative to the bank.

Figures 4.05 shows the vector which represents the velocity of the boat in still water and also the velocity of the current of the river.

The combined effect of the vectors representing the boat's velocity in still water and the velocity of the river's current, gives the velocity of the boat relative to the bank.

Figure 4.04   *Figure 4.05*   *Figure 4.06*

Figure 4.06 shows a vector triangle in which the vector representing the velocity of the river is added to the vector representing the velocity of the boat relative to the river. This gives the vector representing the velocity of the boat relative to the bank. This is an example of vector addition.

One vector can be added to a second vector by placing the back end of the second vector against the pointed end of the first vector. The result is a single vector starting at the back of the first vector and ending at the pointed end of the second vector as shown in figure 4.06.

The addition of three vectors can be carried out by drawing a vector quadrilateral, as shown in figure 4.07.

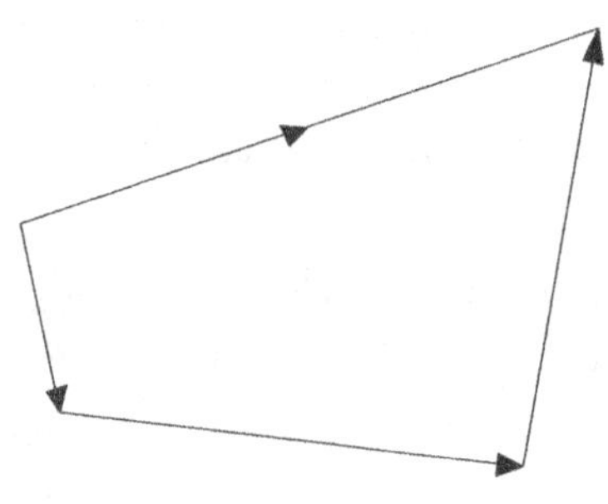

Figure 4.07

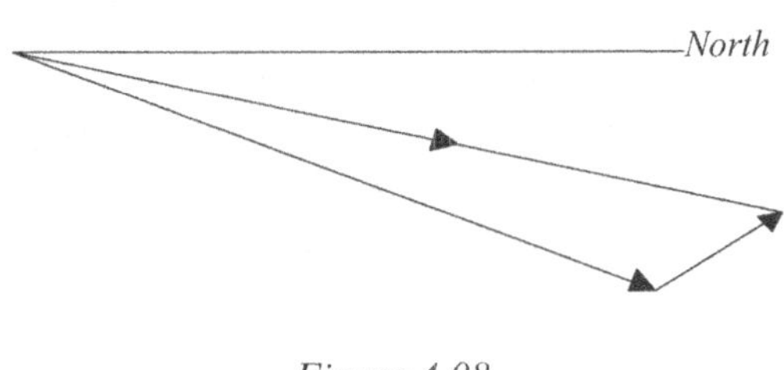

Figure 4.08

Now, consider a small aircraft whose velocity, relative to the air around it is $70\,\text{ms}^{-1}$ in a direction $20°$ east of north. The wind velocity is $15\,\text{ms}^{-1}$ in a direction of $30°$ west of north. Figure 4.08 shows the corresponding vector triangle.

The velocity vector of the aircraft relative to the ground can be found by using triangle trigonometry, as in figure 4.09. The vector from P to Q can be denoted by $\overrightarrow{PQ}$, the vector from R to Q by $\overrightarrow{RQ}$, and so on. Figure 4.09 shows ΔPQR which corresponds to the velocity vector triangle in figure 4.08.

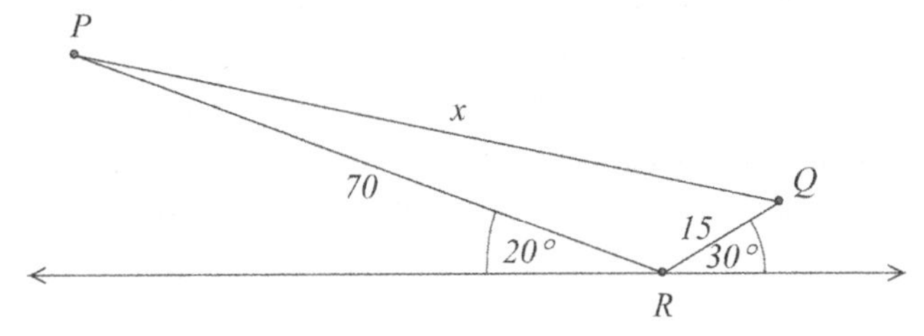

Figure 4.09

Note that $P\hat{R}Q = 180° - (20° + 30°) = 130°$.

The distance x from P to Q can be found using the cosine rule (Unit 3):

$$x = \sqrt{70^2 + 15^2 - 2 \times 70 \times 15 \times \cos 130°} = \sqrt{6474.85398} = 80.46647... = 80.5$$

The direction of $\overrightarrow{PQ}$ can be found by first calculating $Q\hat{P}R$, using the sine rule (Unit 3):

$$\frac{\sin Q\hat{P}R}{15} = \frac{\sin 130^\circ}{80.4665} \Rightarrow \sin Q\hat{P}R = \frac{15\sin 130^\circ}{80.4665} \Rightarrow Q\hat{P}R = 8.209942... = 8.21^\circ$$

Therefore, the direction of $\overrightarrow{PQ}$ is $20^\circ - 8.2^\circ = 11.8^\circ$ east of north. Hence, the velocity vector of the aircraft relative to the ground has magnitude 80.5ms^{-1} and direction 11.8° east of north.

Exercise 4.1

1.   In each case, draw vector triangles to show the vector sum of the vector pair.

(a)   (b)   (c)

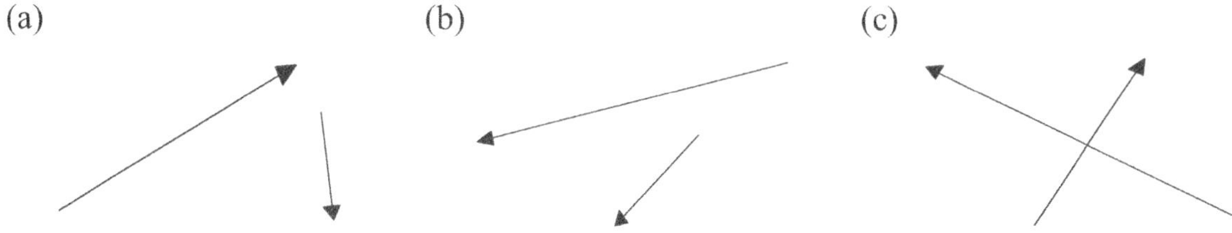

2.   Draw vector polygons to show the vector sum of the following sets of vectors.

(a)   (b)   (c)

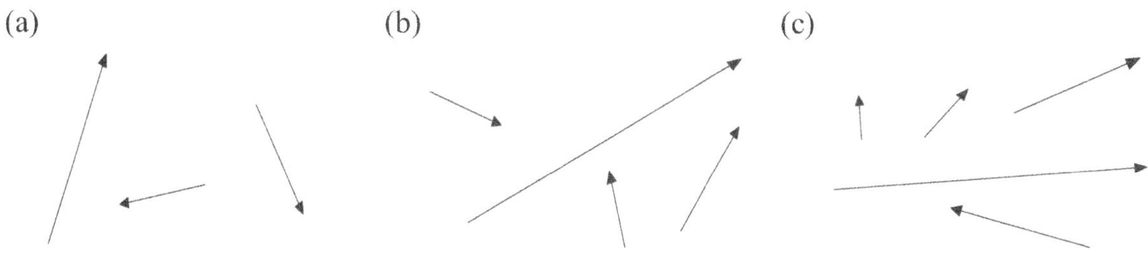

3.   A boat is moving due west in the ocean with speed 10ms^{-1}. A man, standing on deck, fires a harpoon due north. The speed of the harpoon is 36ms^{-1}. Find the velocity of the harpoon relative to the ocean.

4.   Starting at point P in the desert, a truck driver is given the following commands:
      (1)   Drive in a direction 18° east of north for 25km.
      (2)   Drive in a direction 100° east of north for 38km.

If these commands are carried out in the order (1) and then (2), the truck arrives at point Q.
(a) Draw a large, neat diagram of the truck's route.
(b) Use the sine and/or cosine rule to find the displacement vector $\overrightarrow{PQ}$.

5. Sheila can swim in still water at a speed of 0.83ms^{-1}. She is trying to swim across a river, the width of which is 35m, to a point directly opposite. The river flows with a constant current of 0.48ms^{-1}.

(a) Draw a velocity vector diagram showing Sheila's velocity vector relative to the river bank.
(b) Find Sheila's speed relative to the bank.
(c) Show that it takes Sheila approximately 52 seconds to cross the river.

4.2 Vector Algebra

Figures 4.10 and 4.11 show two kinds of notation for describing a vector algebraically.

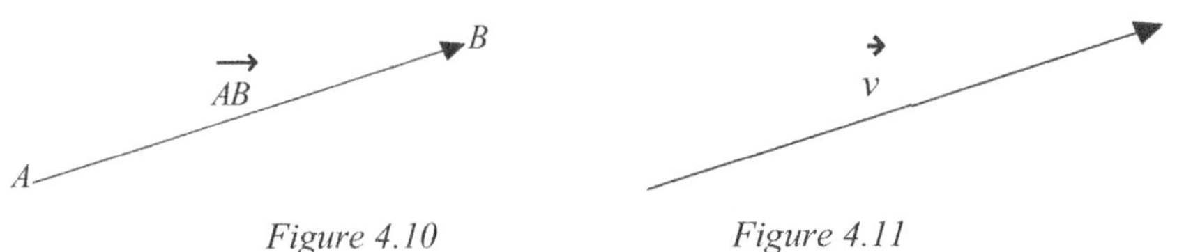

Figure 4.10   Figure 4.11

Vectors are often written algebraically in bold italic script, *v*, for example, but, because it is difficult to hand-write in this way, it is strongly recommended that you write vectors as they are written in this text, with a letter with an arrow pointing to the right drawn above it, like this: $\vec{v}$.

4.2.1 Vector Addition

Section 4.1 introduced the concept of vector addition in an informal manner. Vector addition is now developed in a more general way.

The sum of vectors $\vec{a}$ and $\vec{b}$ (figure 4.12) is shown in figure 4.13.

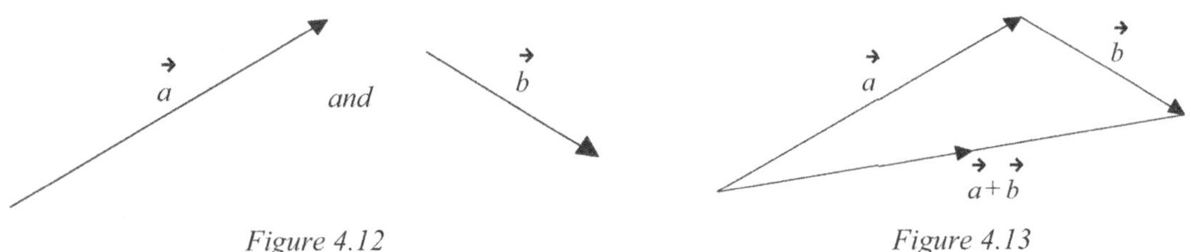

Figure 4.12   Figure 4.13

On the other hand, the sum of the vectors $\vec{b}$ and $\vec{a}$ (figure 4.14) is shown in figure 4.15.

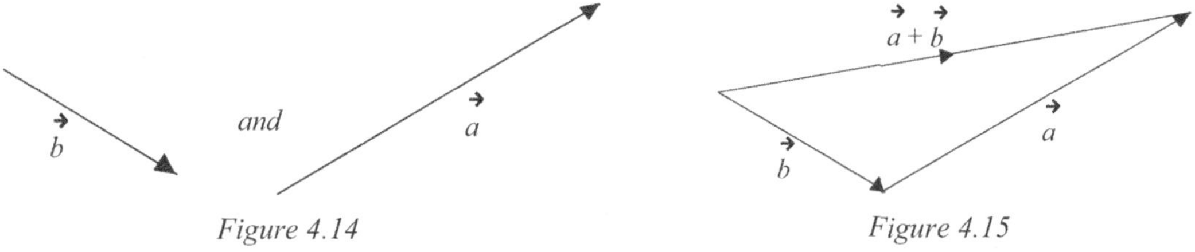

Figure 4.14   Figure 4.15

Figure 4.16 shows that $\vec{a}+\vec{b}=\vec{b}+\vec{a}$. Therefore, the order in which vectors are added together is unimportant, or, to put it in more mathematical terminology, vector addition is *commutative*.

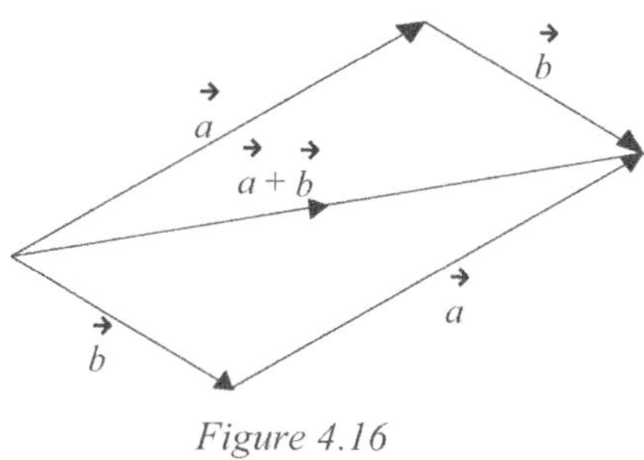

Figure 4.16

Figure 4.17 shows the addition of three vectors, $\vec{a}$, $\vec{b}$ and $\vec{c}$, where $\vec{u}=\vec{a}+\vec{b}$ and $\vec{u}+\vec{c}=\vec{v}$. Therefore, $\left(\vec{a}+\vec{b}\right)+\vec{c}=\vec{v}$.

Figure 4.18 shows the addition of $\vec{a}$, $\vec{b}$ and $\vec{c}$, where $\vec{u}=\vec{a}+\vec{b}$ and $\vec{w}=\vec{b}+\vec{c}$. Now $\vec{a}+\vec{w}=\vec{v}$ and $\vec{u}+\vec{c}=\vec{v}$. But $\vec{v}=\vec{a}+\vec{w}=\vec{u}+\vec{c}=\vec{a}+\left(\vec{b}+\vec{c}\right)=\left(\vec{a}+\vec{b}\right)+\vec{c}$. Therefore, whether $\vec{c}$ is added to $\vec{a}+\vec{b}$ or $\vec{b}+\vec{c}$ is added to $\vec{a}$ is unimportant. We say, then, that vector addition is *associative*.

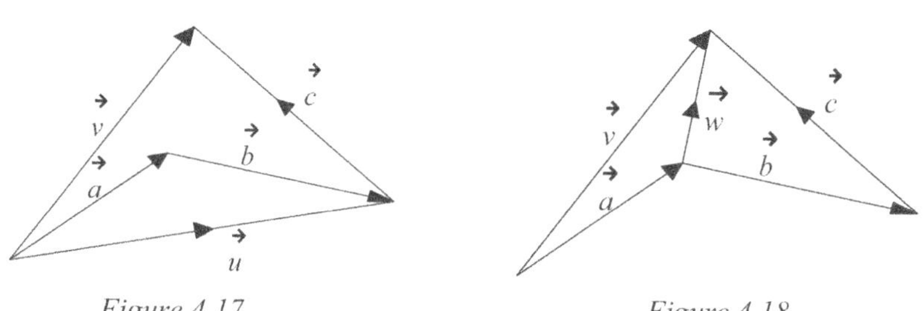

Figure 4.17   *Figure 4.18*

4.2.2 The Difference of Two Vectors

If vector $\vec{v}$ is ⟍⟍⟍, then vector $-\vec{v}$ is ⟍⟍⟍. So that $-\vec{v}$ is equal in magnitude but opposite in direction to $\vec{v}$. This makes possible the subtraction of vectors.

If $\vec{a}$ is ⟍⟍⟍ and $\vec{b}$ is ⟍⟍⟍, then $-\vec{b}$ is ⟍⟍⟍ so that $\vec{u}=\vec{a}+\left(-\vec{b}\right)=\vec{a}-\vec{b}$, as shown by figure 4.19. Therefore, vector $\vec{b}$ can be subtracted from vector $\vec{a}$ by the addition of vector $-\vec{b}$ to vector $\vec{a}$.

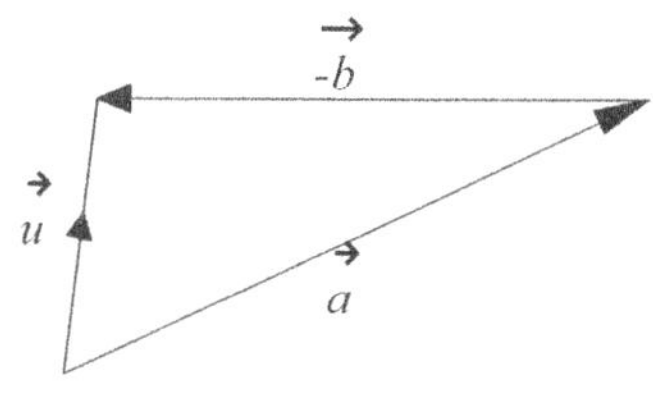

Figure 4.19

121

4.2.3 The Multiplication of a Vector by a Scalar

Real numbers, $\mathbb{R}$, are referred to as *scalars* when they are used together with vectors. It is important that the notation used to represent vectors is clearly distinguishable from that used to represent real numbers.

If $\vec{a}$ is the vector   , then $2\vec{a}$ is the vector   , which is parallel to $\vec{a}$ but whose magnitude is twice that of $\vec{a}$. More generally, if vector $\vec{a}$ is parallel to vector $\vec{b}$, there is a scalar value k, such that $\vec{b} = k\vec{a}$ and $\vec{b}$ has magnitude k times the magnitude of $\vec{a}$. If k is negative, the vector $k\vec{a}$ is still parallel to $\vec{a}$ but points in the opposite direction.

4.2.4 The Zero Vector

For the sake of completeness, it is necessary to define the zero vector, $\vec{0}$, the magnitude of which is 0 and the direction of which remains undefined. It can occur in numerous situations, one of which is when a closed polygon of vectors, as shown in figure 4.20, is expressed algebraically.

Then $\overrightarrow{AB} + \overrightarrow{BC} + \overrightarrow{CD} + \overrightarrow{DE} + \overrightarrow{EF} + \overrightarrow{FA} = \vec{0}$.

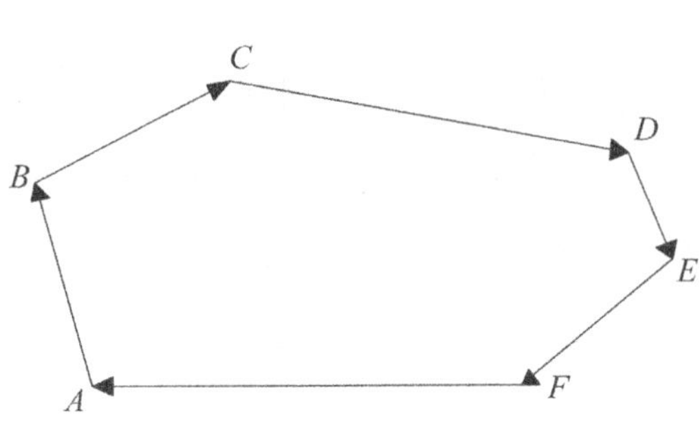

Figure 4.20

Example 4.1: Figure 4.21 shows $\triangle AOB$ in which $\overrightarrow{OA} = \vec{a}$ and $\overrightarrow{OB} = \vec{b}$. M is the midpoint of $\overrightarrow{AB}$. Find, in terms of $\vec{a}$ and $\vec{b}$, the vector $\overrightarrow{OM}$.

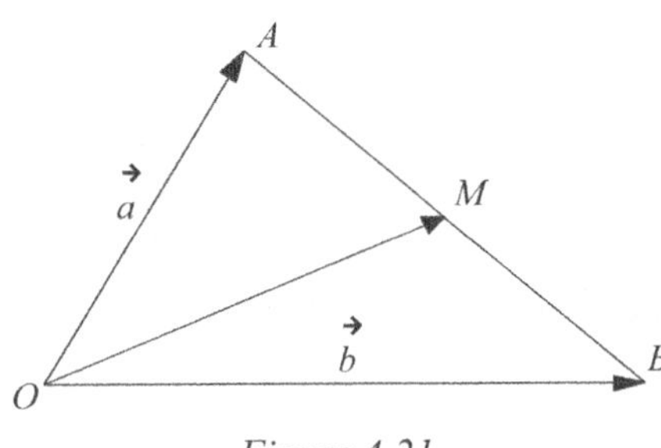

Figure 4.21

Solution 4.1: $\overrightarrow{OA} + \overrightarrow{AB} = \overrightarrow{OB} \Rightarrow \overrightarrow{AB} = \overrightarrow{OB} - \overrightarrow{OA} = \vec{b} - \vec{a}$.

$$\overrightarrow{AM} = \frac{1}{2}\overrightarrow{AB} \Rightarrow \overrightarrow{AM} = \frac{1}{2}\left(\vec{b} - \vec{a}\right) = \frac{1}{2}\vec{b} - \frac{1}{2}\vec{a}.$$

$$\overrightarrow{OM} = \vec{a} + \overrightarrow{AM} = \vec{a} + \left(\frac{1}{2}\vec{b} - \frac{1}{2}\vec{a}\right) = \frac{1}{2}\vec{a} + \frac{1}{2}\vec{b}.$$

Therefore, $\overrightarrow{OM} = \frac{1}{2}\vec{a} + \frac{1}{2}\vec{b}$.

Figure 4.22 shows a quicker method of obtaining this result, but it requires some knowledge of the geometry of a parallelogram.

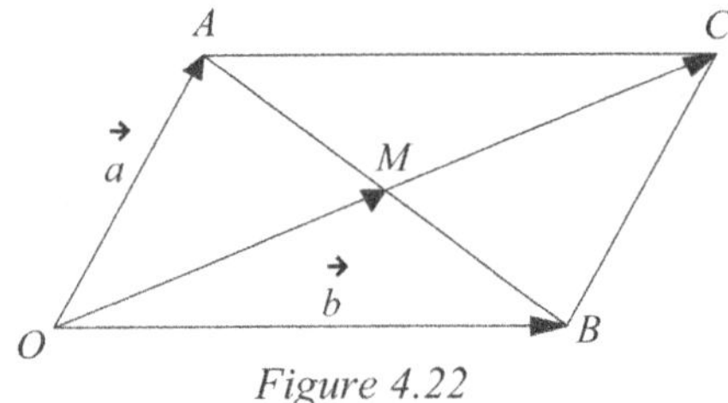

Figure 4.22

$\overrightarrow{OB} = \overrightarrow{AC} = \vec{b}$ so that $\overrightarrow{OC} = \vec{a} + \vec{b}$, and, because the diagonals bisect each other, $\overrightarrow{OM} = \frac{1}{2}\vec{a} + \frac{1}{2}\vec{b}$.

Exercise 4.2

1. Figure 4.23 shows quadrilateral $OPQR$. Copy the equations shown below and fill in the blanks.

   (a) $\overrightarrow{OQ} = \overrightarrow{OP} + \underline{\quad}$ (b) $\overrightarrow{RQ} = \overrightarrow{RO} + \underline{\quad}$

   (c) $\overrightarrow{OP} - \overrightarrow{OQ} = \underline{\quad}$ (d) $\overrightarrow{PR} + \underline{\quad} + \overrightarrow{OQ} = \overrightarrow{OR}$

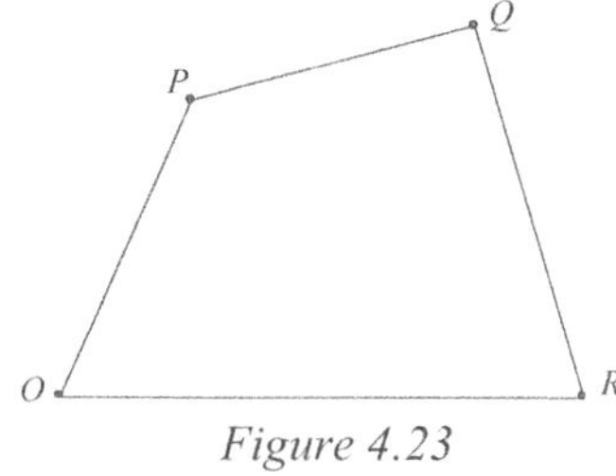
Figure 4.23

2. Simplify

   (a) $\overrightarrow{AB} + \overrightarrow{BC} + \overrightarrow{CD}$ (b) $\overrightarrow{OA} + \overrightarrow{AB} - \overrightarrow{CB}$ (c) $\overrightarrow{AB} + \overrightarrow{BA}$

3. Figure 4.24 shows a parallelogram $OACB$ in which $\overrightarrow{OA} = \vec{a}$ and $\overrightarrow{OB} = \vec{b}$. Write the following in terms of $\vec{a}$ and $\vec{b}$.

   (a) $\overrightarrow{AC}$ (b) $\overrightarrow{BC}$ (c) $\overrightarrow{OC}$ (d) $\overrightarrow{AB}$

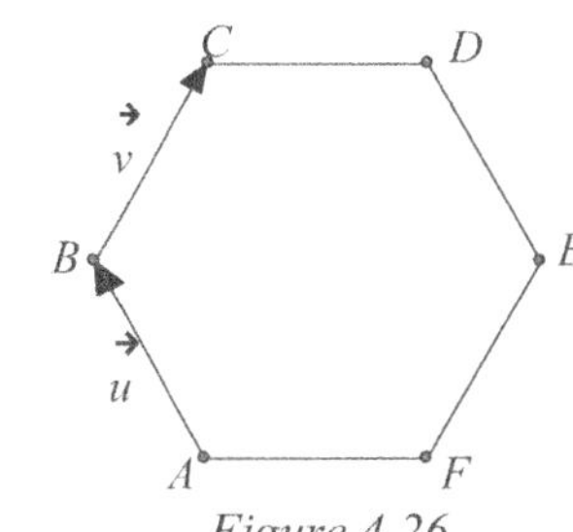
Figure 4.24

4. Figure 4.25 shows points O, A, B and C. If $\overrightarrow{OA} = \vec{a}$, $\overrightarrow{OB} = \vec{b}$ and B is the midpoint of $[AC]$, write the following in terms of $\vec{a}$ and $\vec{b}$.

   (a) $\overrightarrow{AB}$ (b) $\overrightarrow{BA}$ (c) $\overrightarrow{BC}$
   (d) $\overrightarrow{AC}$ (e) $\overrightarrow{OC}$

Figure 4.25

5. Figure 4.26 shows a regular hexagon, $ABCDEF$. If $\overrightarrow{AB} = \vec{u}$ and $\overrightarrow{BC} = \vec{v}$, write the following vectors in terms of $\vec{u}$ and $\vec{v}$

   (a) $\overrightarrow{AC}$ (b) $\overrightarrow{CD}$ (c) $\overrightarrow{AD}$ (d) $\overrightarrow{BF}$

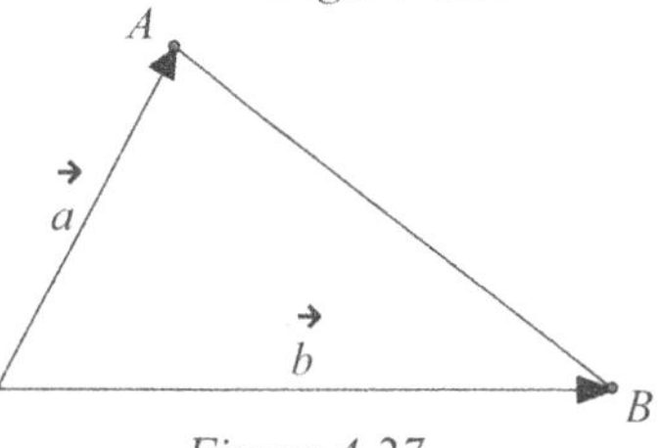
Figure 4.26

6. Figure 4.27 shows a ΔOAB. $\overrightarrow{OA} = \vec{a}$, $\overrightarrow{OB} = \vec{b}$ and P lies on $\overrightarrow{OA}$ such that $\overrightarrow{OP} = \dfrac{2}{3}\overrightarrow{OA}$. Q lies on $\overrightarrow{OB}$ such that $\overrightarrow{OQ} = \dfrac{1}{3}\overrightarrow{OB}$.

   (a) Copy the vector diagram shown and mark on it the positions of P and Q.

   (b) Find, in terms of $\vec{a}$ and $\vec{b}$, (i) $\overrightarrow{BA}$ (ii) $\overrightarrow{PB}$ (iii) $\overrightarrow{AQ}$
   (iv) $\overrightarrow{PQ}$.

Figure 4.27

7. Figure 4.28 shows a quadrilateral $OABC$.
   $\overrightarrow{OA} = \vec{a}$, $\overrightarrow{OB} = \vec{b}$, $\overrightarrow{OC} = \vec{c}$ and $\vec{a} + \dfrac{1}{2}\vec{c} = \dfrac{4}{5}\vec{b}$.

   Find, in terms of $\vec{a}$ and $\vec{b}$,

   (a) $\overrightarrow{AB}$ (b) $\overrightarrow{BC}$ (c) $\overrightarrow{CA}$

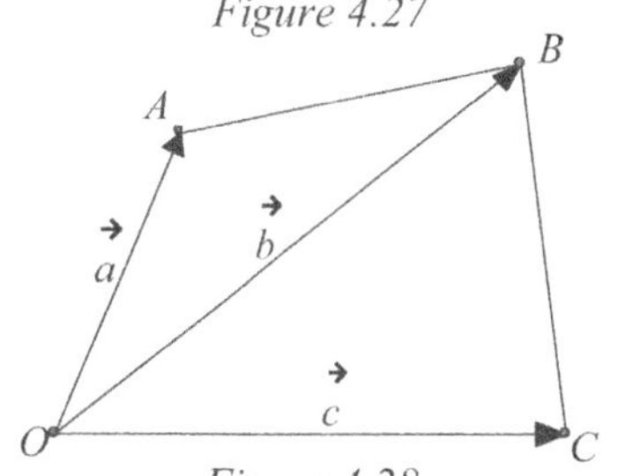
Figure 4.28

Use the results you obtained in parts (a) and (b) to check that (d) $\overrightarrow{OA}+\overrightarrow{AB}+\overrightarrow{BC}+\overrightarrow{CO}=\vec{0}$

8.   Figure 4.29 shows triangle OAB. $\overrightarrow{OA}=\vec{a}$, $\overrightarrow{OB}=\vec{b}$.

P is the midpoint of $[AB]$. R is the midpoint of $\overrightarrow{OA}$.

$\overrightarrow{OQ}=\dfrac{7}{8}\overrightarrow{OB}$. Find the following vectors in terms of $\vec{a}$ and $\vec{b}$.

(a) $\overrightarrow{AB}$ (b) $\overrightarrow{OP}$ (c) $\overrightarrow{AQ}$ (d) $\overrightarrow{QR}$

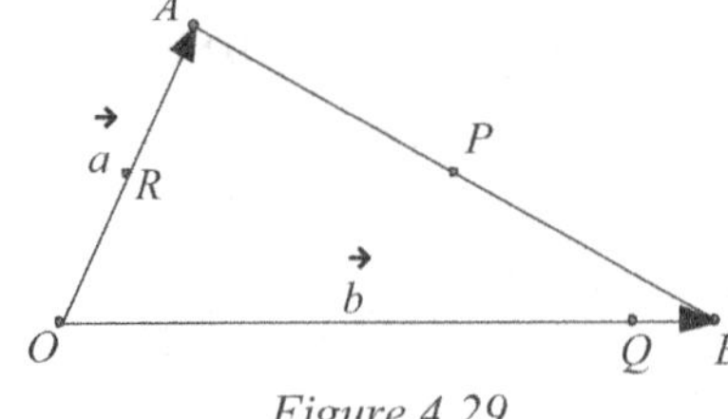

Figure 4.29

4.2.5 Base Vectors

Base vectors are a convenient way of expressing vectors when using a Cartesian coordinate system. Base vectors $\vec{i}$ and $\vec{j}$ are used in two dimensions. $\vec{i}$ is a vector of length one unit parallel to the x-axis, and $\vec{j}$ is a vector of length one unit parallel to the y-axis. Figure 4.30 shows how $\vec{u}$ can be written in terms of base vectors $\vec{i}$ and $\vec{j}$.

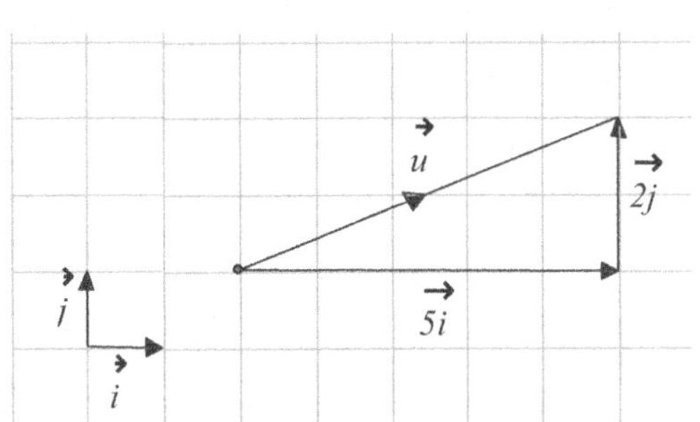

Figure 4.30

The *component* of $\vec{u}$ parallel to the x-axis is 5, and the component of $\vec{u}$ parallel to the y-axis is 2. Therefore, $\vec{u}=5\vec{i}+2\vec{j}$.

Base vectors $\vec{i}$, $\vec{j}$ and $\vec{k}$ are used in three dimensions, where $\vec{i}$ and $\vec{j}$ are defined as for two dimensions, and $\vec{k}$ is a vector of length one unit parallel to the z-axis. Figure 4.31 shows a set of mutually perpendicular

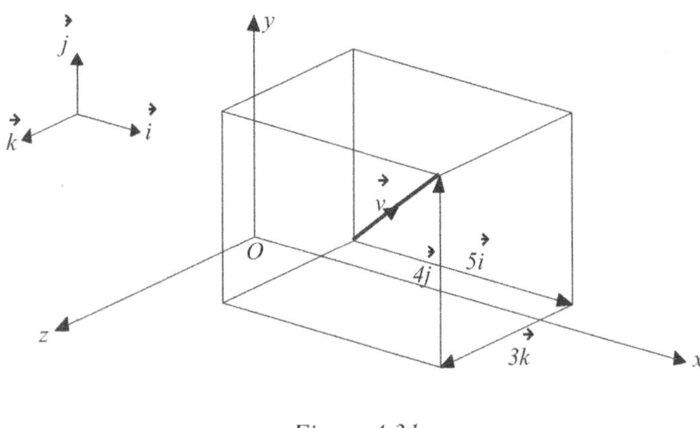

Figure 4.31

axes x, y and z, where the x and z axes lie in the horizontal plane and the y-axis is vertical. It also shows the vector, $\vec{v}$. The component of $\vec{v}$ parallel to the x-axis is 5, the component parallel to the y-axis is 4 and the component parallel to the z-axis is 3 so that $\vec{v}=5\vec{i}+4\vec{j}+3\vec{k}$.

An alternative way of expressing a vector is to use a column in which the x-component is the top number in the column and the second number in the column is the y-component. For example, in two dimensions, $\vec{i}=\begin{pmatrix}1\\0\end{pmatrix}$ and $\vec{j}=\begin{pmatrix}0\\1\end{pmatrix}$ and $\vec{u}=\begin{pmatrix}5\\2\end{pmatrix}$. In three dimensions, the z-component appears as the third number at the bottom of the column: $\vec{i}=\begin{pmatrix}1\\0\\0\end{pmatrix}$, $\vec{j}=\begin{pmatrix}0\\1\\0\end{pmatrix}$, $\vec{k}=\begin{pmatrix}0\\0\\1\end{pmatrix}$ and $\vec{v}=\begin{pmatrix}5\\4\\3\end{pmatrix}$. Vectors written in this notation are known as *column vectors*.

In general, a two-dimensional vector is written either as $\vec{v} = v_1\vec{i} + v_2\vec{j}$ or as $\vec{v} = \begin{pmatrix} v_1 \\ v_2 \end{pmatrix}$. A three-dimensional vector is written either as $\vec{v} = v_1\vec{i} + v_2\vec{j} + v_3\vec{k}$ or as $\vec{v} = \begin{pmatrix} v_1 \\ v_2 \\ v_3 \end{pmatrix}$, where v_1, v_2, $v_3 \in \mathbb{R}$.

You should be familiar with both notations.

The following example shows the use of base vector notation involving vector addition, vector subtraction and also multiplication of a vector by a scalar.

Example 4.2: (a) If $\vec{u} = \begin{pmatrix} 1 \\ -4 \end{pmatrix}$ and $\vec{v} = \begin{pmatrix} 5 \\ 2 \end{pmatrix}$, show, with the aid of a diagram, that $\vec{u} + \vec{v} = \begin{pmatrix} 6 \\ -2 \end{pmatrix}$.

(b) If $\vec{a} = 4\vec{i} - 3\vec{j} + 5\vec{k}$ and $\vec{b} = 2\vec{i} + 4\vec{j} - \vec{k}$, find the vector $2\vec{a} - 3\vec{b}$.

Solution 4.2: (a)

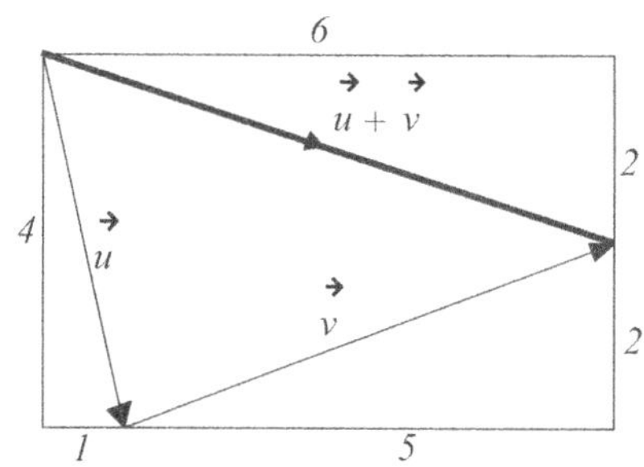

$$\vec{u} + \vec{v} = \begin{pmatrix} 1 \\ -4 \end{pmatrix} + \begin{pmatrix} 5 \\ 2 \end{pmatrix} = \begin{pmatrix} 1+5 \\ -4+2 \end{pmatrix} = \begin{pmatrix} 6 \\ -2 \end{pmatrix}$$

(b) $2\vec{a} - 3\vec{b} = 2\left(4\vec{i} - 3\vec{j} + 5\vec{k}\right) - 3\left(2\vec{i} + 4\vec{j} - \vec{k}\right) = 8\vec{i} - 6\vec{j} + 10\vec{k} - 6\vec{i} - 12\vec{j} + 3\vec{k}$

$= 2\vec{i} - 18\vec{j} + 13\vec{k}$

4.2.6 The Magnitude of a Vector

The magnitude of a vector has already been described in section 4.1 as the length of the arrow which represents the vector. This magnitude can be denoted using the absolute value symbol so that a vector $\vec{v}$ has magnitude $\left|\vec{v}\right|$.

For a two-dimensional vector, $\vec{v} = 5\vec{i} + \vec{j}$, $\left|\vec{v}\right|$ can be found using Pythagoras' theorem. Figure 4.32 shows the components of $\vec{v}$.

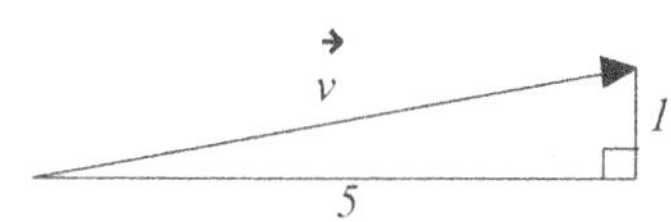

By Pythagoras' theorem, $\left|\vec{v}\right|^2 = 5^2 + 1^2 = 26 \Rightarrow \left|\vec{v}\right| = \sqrt{26}$.

Figure 4.32

Although it is not quite so obvious, Pythagoras' theorem works just as well for a three-dimensional vector. Figure 4.33 shows the vector $\vec{v}=5\vec{i}+\vec{j}+3\vec{k}$. ΔBDC is right-angled at D so that, by Pythagoras' theorem, $|\vec{u}|^2=3^2+1^2$. Moreover, ΔABC is right-angled at C so that

$$|\vec{v}|^2=5^2+|\vec{u}|^2=5^2+1^2+3^2=35 \text{ and } |\vec{v}|=\sqrt{35} \text{ .}$$

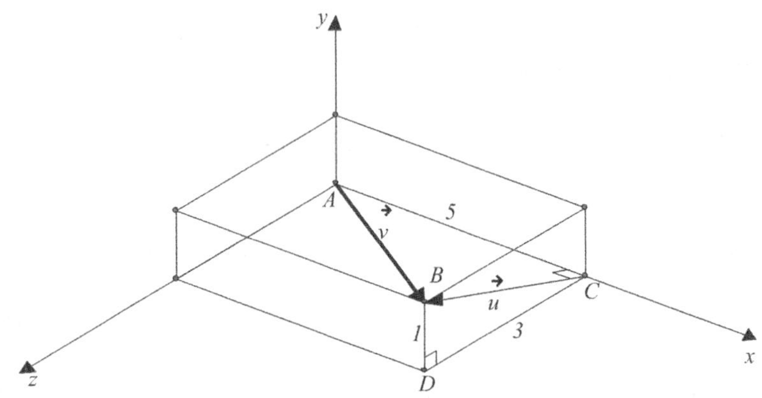

Figure 4.33

More generally, if $\vec{v}=v_1\vec{i}+v_2\vec{j}+v_3\vec{k}$, $|\vec{v}|=\sqrt{v_1^2+v_2^2+v_3^2}$.

4.2.7 Unit Vectors

A *unit vector* has unit magnitude. So, for example, $\dfrac{2}{3}\vec{i}-\dfrac{1}{3}\vec{j}+\dfrac{2}{3}\vec{k}$, $0.6\vec{i}+0.8\vec{j}$ and $\dfrac{6}{7}\vec{i}+\dfrac{3}{7}\vec{j}+\dfrac{2}{7}\vec{k}$

are all unit vectors because $\sqrt{\left(\frac{2}{3}\right)^2+\left(-\frac{1}{3}\right)^2+\left(\frac{2}{3}\right)^2}=\sqrt{0.6^2+0.8^2}=\sqrt{\left(\frac{6}{7}\right)^2+\left(\frac{3}{7}\right)^2+\left(\frac{2}{7}\right)^2}=1.$

Given any vector, $\vec{v}$, it is possible to find a unit vector parallel to $\vec{v}$ by dividing it by its magnitude, $|\vec{v}|$. For example, if $\vec{v}=6\vec{i}+9\vec{j}+2\vec{k}$ then $|\vec{v}|=\sqrt{6^2+9^2+2^2}=11$. Therefore, a unit vector parallel to $\vec{v}$ is $\dfrac{1}{11}\left(6\vec{i}+9\vec{j}+2\vec{k}\right)=\frac{6}{11}\vec{i}+\frac{9}{11}\vec{j}+\frac{2}{11}\vec{k}$.

Example 4.3: If $\vec{a}=\begin{pmatrix}1\\0\\2\end{pmatrix}$ and $\vec{b}=\begin{pmatrix}-1\\2\\-5\end{pmatrix}$ find   (a) $\left|3\vec{a}+\vec{b}\right|$, the magnitude of vector $3\vec{a}+\vec{b}$

(b) a unit vector parallel to $\vec{a}$.

Solution 4.3: (a) $3\vec{a}+\vec{b}=3\begin{pmatrix}1\\0\\2\end{pmatrix}+\begin{pmatrix}-1\\2\\-5\end{pmatrix}=\begin{pmatrix}3+-1\\0+2\\6-5\end{pmatrix}=\begin{pmatrix}2\\2\\1\end{pmatrix}$. So $\left|3\vec{a}+\vec{b}\right|=\sqrt{2^2+2^2+1^2}=\sqrt{4+4+1}=3$.

(b) $|\vec{a}| = \sqrt{1^2 + 0^2 + 2^2} = \sqrt{5}$. So a unit vector, $\vec{u}$ parallel to $\vec{a}$, is $\vec{u} = \dfrac{1}{\sqrt{5}}\begin{pmatrix} 1 \\ 0 \\ 2 \end{pmatrix} = \begin{pmatrix} 0.447 \\ 0 \\ 0.894 \end{pmatrix}$.

4.2.8 Position Vectors

A position vector is a vector with the additional property that it is fixed at its back end to the origin, O.

This being the case, its pointed end clearly has a position. More specifically, any point P, with coordinates (x, y) in the two-dimensional Cartesian plane, has associated with it a position

vector $\overrightarrow{OP} = \begin{pmatrix} x \\ y \end{pmatrix}$, as shown in figure 4.34.

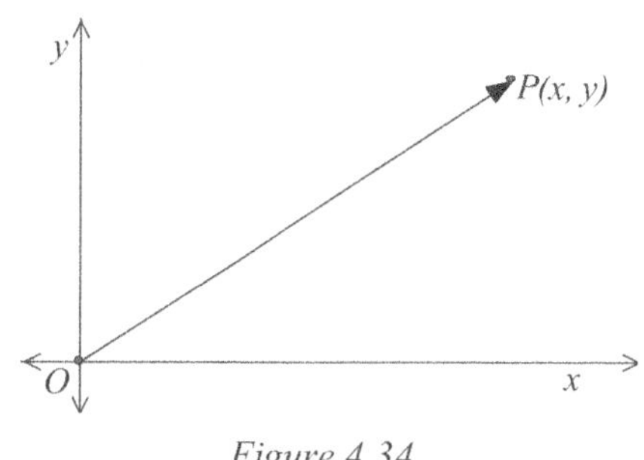

Figure 4.34

In a similar way, any point P with coordinates (x, y, z) in three-dimensional Cartesian space, has

associated with it a position vector $\overrightarrow{OP} = \begin{pmatrix} x \\ y \\ z \end{pmatrix}$.

Exercise 4.3

1.   Using a suitable scale, draw on squared paper, the vectors.

    (a) $3\vec{i} + 5\vec{j}$ (b) $-2\vec{i}$ (c) $4\vec{i} - 3\vec{j}$ (d) $3\vec{j} - \vec{i}$

2.   Using a suitable scale, draw vector triangles on squared paper, showing the following results, where $\vec{a} = \begin{pmatrix} 2 \\ 1 \end{pmatrix}$ and $\vec{b} = \begin{pmatrix} 3 \\ -2 \end{pmatrix}$.

    (a) $\vec{a} + \vec{b} = \begin{pmatrix} 5 \\ -1 \end{pmatrix}$ (b) $2\vec{a} + \vec{b} = \begin{pmatrix} 7 \\ 0 \end{pmatrix}$ (c) $\vec{a} - 2\vec{b} = \begin{pmatrix} -4 \\ 5 \end{pmatrix}$

3.   (a) If $\vec{u} = \begin{pmatrix} 4 \\ 3 \end{pmatrix}$, $\vec{v} = \begin{pmatrix} -2 \\ 7 \end{pmatrix}$, find   (i) $\vec{u} + \vec{v}$ (ii) $2\vec{u} + 3\vec{v}$ (iii) $\dfrac{1}{2}(\vec{u} - \vec{v})$

    (b) If $\vec{u} = 5\vec{i} - 2\vec{j} + \vec{k}$, $\vec{v} = 3\vec{i} + \vec{j} + \vec{k}$, find   (i) $2\vec{u} + \vec{v}$ (ii) $-\vec{u} + 2\vec{v}$ (iii) $\dfrac{1}{3}\vec{u} - \dfrac{1}{2}\vec{v}$

4.   If $\vec{a}$ and $\vec{b}$ are parallel and $\vec{a} = \begin{pmatrix} -2 \\ 0 \\ 3 \end{pmatrix}$, show that there are two vectors, $\vec{b}$, which satisfy the

    condition $|\vec{a}| = 3|\vec{b}|$. State these two vectors.

127

5.   Find unit vectors parallel to the following vectors.

(a) $\begin{pmatrix} 8 \\ 15 \end{pmatrix}$ (b) $\begin{pmatrix} -7 \\ 3 \end{pmatrix}$ (c) $\begin{pmatrix} 3 \\ 0 \end{pmatrix}$ (d) $6\vec{i} - 5\vec{j}$ (e) $-0.43\vec{i} + 0.37\vec{j}$

6.   Let $\vec{a} = \begin{pmatrix} 2 \\ 9 \end{pmatrix}$, $\vec{b} = \begin{pmatrix} 7 \\ -3 \end{pmatrix}$.

(a) Find   (i) $|\vec{a} + \vec{b}|$ (ii) $|\vec{a}| + |\vec{b}|$

(b) Show that $|\vec{a} + \vec{b}| < |\vec{a}| + |\vec{b}|$.

(c) Investigate whether $|\vec{a} + \vec{b}| < |\vec{a}| + |\vec{b}|$ for all vectors $\vec{a}$ and $\vec{b}$.

7.   Find the values of the constants a and b if $a\begin{pmatrix} 2b \\ -3 \end{pmatrix} + \begin{pmatrix} 1 \\ 4 \end{pmatrix} = \begin{pmatrix} 7 \\ 2 \end{pmatrix}$.

8.   If $\vec{a} = 5\vec{i} + \vec{j}$, $\vec{b} = 3\vec{i} + 3\vec{j}$ and $\vec{c} = -\vec{i} - 2\vec{j}$, find the values of the constants m and n such that $m\vec{a} + n\vec{b} = 12\vec{c}$.

9.   If $A(1,\ 4,\ 2)$, $B(-3,\ 1,\ 6)$ and $C(5,\ 5,\ -6)$ are points whose coordinates are given with respect to an origin, O,

(a) write down position vectors $\overrightarrow{OA}$, $\overrightarrow{OB}$ and $\overrightarrow{OC}$.

(b) show that $|\overrightarrow{OA}| = \sqrt{21}$ and find $|\overrightarrow{OB}|$ and $|\overrightarrow{OC}|$.

10.   The position vectors of a triangle ABC are $\overrightarrow{OA} = \begin{pmatrix} 4 \\ 5 \end{pmatrix}$, $\overrightarrow{OB} = \begin{pmatrix} -3 \\ 6 \end{pmatrix}$, and $\overrightarrow{OC} = \begin{pmatrix} 7 \\ -2 \end{pmatrix}$.

(a) Find $\overrightarrow{AB}$, $\overrightarrow{BC}$ and $\overrightarrow{CA}$ and show that $\overrightarrow{AB} + \overrightarrow{BC} + \overrightarrow{CA} = \vec{0}$.

(b) Find $|\overrightarrow{OA}|$, $|\overrightarrow{OB}|$ and $|\overrightarrow{AB} + 2\overrightarrow{BC}|$.

4.3 The Scalar (or Dot) Product

4.3.1 The Scalar Product of Two Vectors in Two Dimensions

Before defining the scalar product and showing how it is used to find angles, it may be valuable to ensure that the meaning of an angle between two vectors is clear. How is it possible to talk about the angle between the two vectors $\vec{a}$ and $\vec{b}$ shown in figure 4.35?

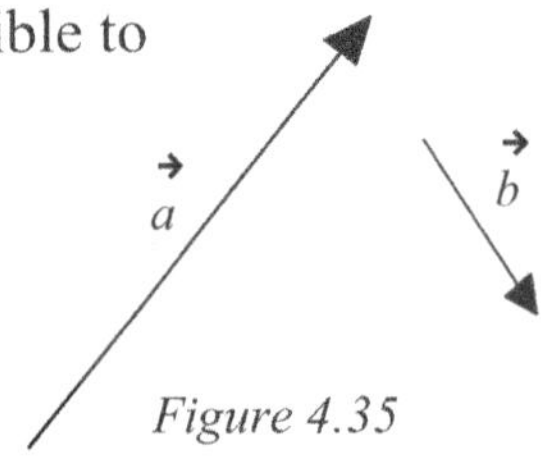

As they have no position, they can be translated so that their back ends are together.

Figure 4.35

Then the angle θ, shown in figure 4.36, represents the angle between $\vec{a}$ and $\vec{b}$. This makes it possible to talk about the angle between two vectors without worrying about actually sliding them together in the manner just described.

The scalar product, or dot product, provides an efficient method of finding angles between vectors. For the vectors $\vec{a}$ and $\vec{b}$, the scalar product is written $\vec{a} \bullet \vec{b}$, and if $\vec{a} = a_1\vec{i} + a_2\vec{j}$, $\vec{b} = b_1\vec{i} + b_2\vec{j}$, then $\vec{a} \bullet \vec{b} = a_1 b_1 + a_2 b_2$.

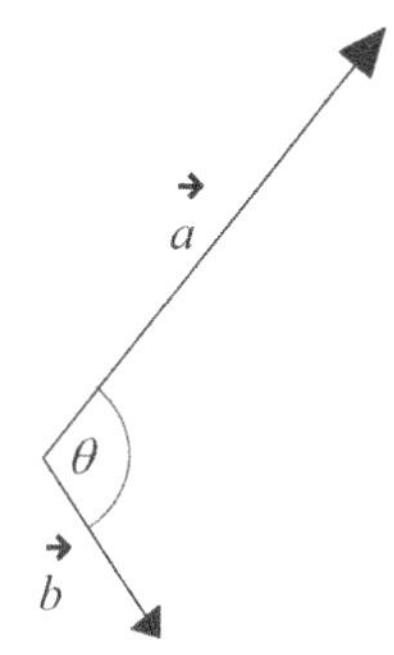

Figure 4.36

The significance of this last expression will soon become evident. In the rather heavy algebra that follows, the cosine rule is being used in $\triangle OAB$, shown in figure 4.37, to find an expression for the cosine of the angle θ between vectors $\overrightarrow{OA} = \vec{a}$ and $\overrightarrow{OB} = \vec{b}$, expressed as general column vectors.

Suppose that $\overrightarrow{OA} = \vec{a} = \begin{pmatrix} a_1 \\ a_2 \end{pmatrix}$, $\overrightarrow{OB} = \vec{b} = \begin{pmatrix} b_1 \\ b_2 \end{pmatrix}$.

Then, as $\overrightarrow{AB} = \overrightarrow{AO} + \overrightarrow{OB} = \overrightarrow{OB} - \overrightarrow{OA}$,

$$\overrightarrow{AB} = \begin{pmatrix} b_1 \\ b_2 \end{pmatrix} - \begin{pmatrix} a_1 \\ a_2 \end{pmatrix} = \begin{pmatrix} b_1 - a_1 \\ b_2 - a_2 \end{pmatrix}$$

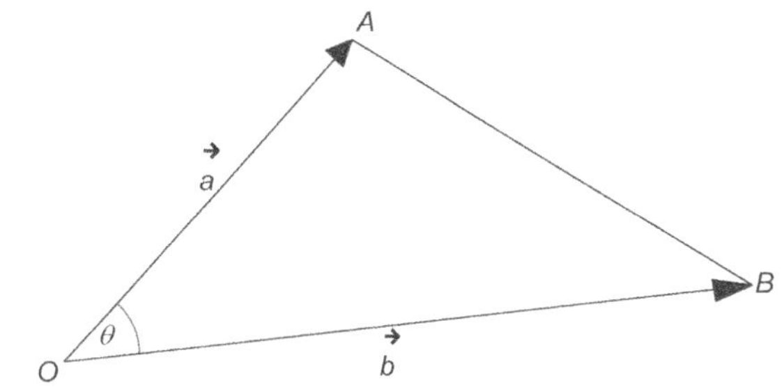

Figure 4.37

In order to use the cosine rule in $\triangle OAB$, it will be necessary to find magnitudes of vectors $\overrightarrow{OA}$, $\overrightarrow{OB}$ and $\overrightarrow{AB}$

By Pythagoras' theorem, $\left|\overrightarrow{OA}\right|^2 = \left|\vec{a}\right|^2 = a_1^2 + a_2^2$, $\left|\overrightarrow{OB}\right|^2 = \left|\vec{b}\right|^2 = b_1^2 + b_2^2$ and

$$\left|\overrightarrow{AB}\right|^2 = \left|\vec{b} - \vec{a}\right|^2 = (b_1 - a_1)^2 + (b_2 - a_2)^2 . \quad \text{But, by the cosine rule, } \cos\theta = \frac{\left|\vec{a}\right|^2 + \left|\vec{b}\right|^2 - \left|\overrightarrow{AB}\right|^2}{2\left|\vec{a}\right|\left|\vec{b}\right|}$$

$$\Rightarrow \cos\theta = \frac{\left(a_1^2 + a_2^2\right) + \left(b_1^2 + b_2^2\right) - \left[(b_1 - a_1)^2 + (b_2 - a_2)^2\right]}{2\left|\vec{a}\right|\left|\vec{b}\right|}$$

$$= \frac{a_1^2 + a_2^2 + b_1^2 + b_2^2 - \left[b_1^2 - 2a_1 b_1 + a_1^2 + b_2^2 - 2a_2 b_2 + a_2^2\right]}{2\left|\vec{a}\right|\left|\vec{b}\right|}$$

$$= \frac{2\left(a_1 b_1 + a_2 b_2\right)}{2\left|\vec{a}\right|\left|\vec{b}\right|} = \frac{a_1 b_1 + a_2 b_2}{\left|\vec{a}\right|\left|\vec{b}\right|}$$

The expression in the numerator is called the *scalar* (or *dot*) *product* and is written $\vec{a} \cdot \vec{b}$. Therefore,

$\vec{a} \cdot \vec{b} = a_1 b_1 + a_2 b_2$. In addition, $\cos\theta = \dfrac{\vec{a} \cdot \vec{b}}{\left|\vec{a}\right|\left|\vec{b}\right|} \Rightarrow \vec{a} \cdot \vec{b} = \left|\vec{a}\right|\left|\vec{b}\right|\cos\theta$.

Therefore, $\vec{a} \cdot \vec{b} = \left|\vec{a}\right|\left|\vec{b}\right|\cos\theta$ and $\vec{a} \cdot \vec{b} = a_1 b_1 + a_2 b_2$.

It may be helpful to think of the scalar product as an area. $\vec{a} \cdot \vec{b} = \left|\vec{a}\right|\left|\vec{b}\right|\cos\theta = \left|\vec{a}\right|\left(\left|\vec{b}\cos\theta\right|\right)$, so that $\vec{a} \cdot \vec{b}$ represents the area of a rectangle with sides whose lengths are $\left|\vec{a}\right|$ and $\left|\vec{b}\right|\cos\theta$. Alternatively, $\vec{a} \cdot \vec{b}$ represents the area of the rectangle whose lengths are $\left|\vec{b}\right|$ and $\left|\vec{a}\right|\cos\theta$. Figure 4.38 shows the two different ways in which $\vec{a} \cdot \vec{b}$ can be visualized as an area.

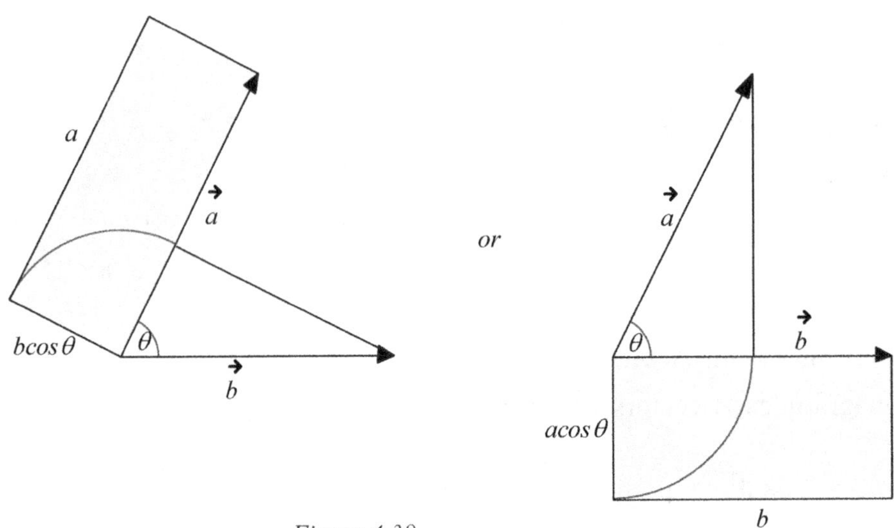

Figure 4.38

4.3.2 The Scalar Product of Two Vectors in Three Dimensions

Consider any two vectors $\vec{a} = a_1\vec{i} + a_2\vec{j} + a_3\vec{k}$ and $\vec{b} = b_1\vec{i} + b_2\vec{j} + b_3\vec{k}$ in three dimensions. As they have no fixed position, they can be translated until their back ends are together. The vectors $\vec{a}$ and $\vec{b}$ lie in a plane, so there is no problem about defining the angle between them. Moreover, everything that was done is section 4.3.1 remains valid except that, in three dimensions, the vectors have a third component. Therefore, the formulae for the scalar product in three dimensions are $\vec{a} \cdot \vec{b} = \left|\vec{a}\right|\left|\vec{b}\right|\cos\theta$ and $a \cdot b = a_1 b_1 + a_2 b_2 + a_3 b_3$.

4.3.3 The Angle between Two Vectors

The scalar product can be used to find the angle between two vectors, and the calculation, although slightly longer for three-dimensional vectors than it is for two-dimensions, is no more difficult.

Example 4.4:   If $\vec{a} = \begin{pmatrix} 4 \\ 1 \end{pmatrix}$, $\vec{b} = \begin{pmatrix} 3 \\ 4 \end{pmatrix}$, find the angle between $\vec{a}$ and $\vec{b}$.

Solution 4.4: $\vec{a} \cdot \vec{b} = a_1 b_1 + a_2 b_2 = 4 \times 3 + 1 \times 4 = 12 + 4 = 16$. $|\vec{a}| = \sqrt{4^2 + 1^2} = \sqrt{17}$,

$|\vec{b}| = \sqrt{3^2 + 4^2} = 5$. $\cos\theta = \dfrac{\vec{a} \cdot \vec{b}}{|\vec{a}||\vec{b}|}$, where θ is the angle between $\vec{a}$ and $\vec{b}$.

Therefore, $\cos\theta = \dfrac{16}{5\sqrt{17}} \Rightarrow \theta = 39.09385\ldots = 39.1°$.

Example 4.5:   Find the angle between the vectors $\vec{u} = 4\vec{i} + 6\vec{j} - 3\vec{k}$ and $\vec{v} = \vec{i} - \vec{j} - 2\vec{k}$.

Solution 4.5: $\vec{u} \cdot \vec{v} = 4 \times 1 + 6 \times (-1) + (-3) \times (-2) = 4 - 6 + 6 = 4$.

$\vec{u} \cdot \vec{v} = |\vec{u}||\vec{v}|\cos\theta$, where θ is the angle between the two vectors.

$|\vec{u}| = \sqrt{4^2 + 6^2 + (-3)^2} = \sqrt{61}$, $|\vec{v}| = \sqrt{1^2 + (-1)^2 + (-2)^2} = \sqrt{6}$

Therefore, $4 = \sqrt{61}\sqrt{6}\cos\theta \Rightarrow \cos\theta = \dfrac{4}{\sqrt{61}\sqrt{6}} \Rightarrow \theta = 77.93136\ldots = 77.9°$.

4.3.4 Perpendicular Vectors

Figure 4.39 shows two perpendicular vectors, $\vec{a}$ and $\vec{b}$. As $\vec{a}$ and $\vec{b}$ are perpendicular, non-zero vectors, the angle between them is a right angle and so
$\vec{a} \cdot \vec{b} = |\vec{a}||\vec{b}|\cos 90° = 0$.

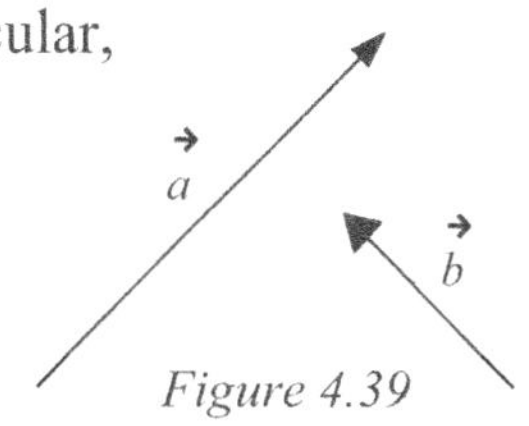

In addition, $\vec{a} \cdot \vec{b} = 0 \Rightarrow |\vec{a}||\vec{b}|\cos\theta = 0 \Rightarrow \cos\theta = 0 \Rightarrow \theta = 90°,\ 270°$. Therefore,

(a) if vectors $\vec{a}$ and $\vec{b}$ are perpendicular then $\vec{a} \cdot \vec{b} = 0$.

(b) if $\vec{a} \cdot \vec{b} = 0$ then $\vec{a}$ and $\vec{b}$ are perpendicular.

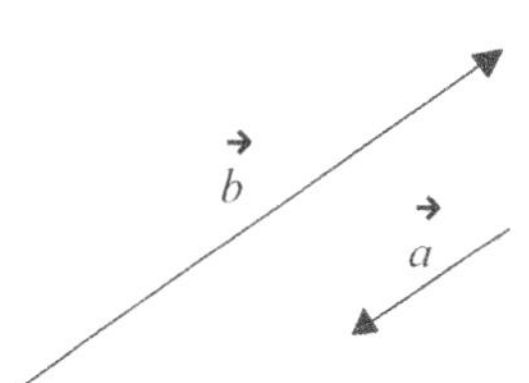

Figure 4.39

4.3.5 Parallel Vectors

As was shown in section 4.2.3, if vectors $\vec{a}$ and $\vec{b}$ are parallel, then $\vec{b} = k\vec{a}$, for some k, $k \in \mathbb{R}$. Figure 4.40 shows two parallel vectors $\vec{a}$ and $\vec{b}$ in which $k = -3$.

Figure 4.40

It is also true that if $\vec{b} = k\vec{a}$ then $\vec{a}$ and $\vec{b}$ are parallel. This makes it easy to see that the vectors $\vec{a} = 3\vec{i} + 4\vec{j}$ and $\vec{b} = 9\vec{i} + 12\vec{j}$ are parallel because $\vec{b} = 3\vec{a}$. Similarly, in three dimensions, if $\vec{a} = \vec{i} - 2\vec{j} + 3\vec{k}$ and $\vec{b} = -2\vec{i} + 4\vec{j} - 6\vec{k}$ then $\vec{b} = -2\vec{a}$ and so $\vec{a}$ and $\vec{b}$ are parallel.

In addition, if $\vec{a}$ and $\vec{b}$ are parallel then the angle θ between them is either $0°$ or $180°$, and, because $\cos 0° = 1$ and $\cos 180° = -1$, $\cos\theta = \pm 1$. Now $\vec{a} \cdot \vec{b} = |\vec{a}||\vec{b}|\cos\theta$ so that if $\vec{a}$ and $\vec{b}$ are parallel then $\vec{a} \cdot \vec{b} = \pm|\vec{a}||\vec{b}|$.

Example 4.6: If $\vec{u} = 2\vec{i} + b\vec{j} - \vec{k}$ and $\vec{v} = 6\vec{i} + \vec{j} - 3\vec{k}$, find the value of b which makes

    (a) $\vec{u}$ perpendicular to $\vec{v}$ (b) $\vec{u}$ parallel to $\vec{v}$

Solution 4.6: (a) $(2\vec{i} + b\vec{j} - \vec{k}) \cdot (6\vec{i} + \vec{j} + -3\vec{k}) = 2\times 6 + b\times 1 + (-1)\times(-3) = 15 + b$, but if $\vec{u}$ and $\vec{v}$ are perpendicular then $\vec{u} \cdot \vec{v} = 0 \Rightarrow 15 + b = 0 \Rightarrow b = -15$.

(b) Method 1: If $\vec{u}$ and $\vec{v}$ are parallel then $\vec{v} = a\vec{u}$ for some value of a. Comparing the x-components (or z-components) of $\vec{u}$ and $\vec{v}$, $a = 3$. Therefore, the y-component of $\vec{v}$ is 3 times the y-component of $\vec{u}$ $\Rightarrow 1 = 3b \Rightarrow b = \dfrac{1}{3}$.

Method 2: $\vec{u} \cdot \vec{v} = 15 + b$, $|\vec{u}| = \sqrt{2^2 + b^2 + (-1)^2} = \sqrt{5 + b^2}$ and

$|\vec{v}| = \sqrt{6^2 + 1^2 + (-3)^2} = \sqrt{46}$. Because $\vec{u}$ and $\vec{v}$ are parallel,

$\vec{u} \cdot \vec{v} = \pm|\vec{u}||\vec{v}|$ and so $15 + b = \pm\sqrt{5 + b^2}\sqrt{46} \Rightarrow (15 + b)^2 = 46(5 + b^2)$

$\Rightarrow 9b^2 - 6b + 1 = 0 \Rightarrow (3b - 1)^2 = 0 \Rightarrow b = \dfrac{1}{3}$.

Exercise 4.4

1. In each case, find the angle between the pair of vectors.

  (a) $\begin{pmatrix} 1 \\ 5 \end{pmatrix}, \begin{pmatrix} 2 \\ 2 \end{pmatrix}$ (b) $\begin{pmatrix} -3 \\ 1 \end{pmatrix}, \begin{pmatrix} 4 \\ -2 \end{pmatrix}$ (c) $\begin{pmatrix} 5 \\ 3 \end{pmatrix}, \begin{pmatrix} 1 \\ -2 \end{pmatrix}$ (d) $\begin{pmatrix} 7 \\ 4 \end{pmatrix}, \begin{pmatrix} 0 \\ 3 \end{pmatrix}$

2. In each case, find the angle between the pair of vectors.

  (a) $\begin{pmatrix} 1 \\ 1 \\ -1 \end{pmatrix}, \begin{pmatrix} 2 \\ 3 \\ 2 \end{pmatrix}$ (b) $\begin{pmatrix} 6 \\ -2 \\ 1 \end{pmatrix}, \begin{pmatrix} -2 \\ -5 \\ 4 \end{pmatrix}$ (c) $\begin{pmatrix} -4 \\ 3 \\ -3 \end{pmatrix}, \begin{pmatrix} 5 \\ 1 \\ -6 \end{pmatrix}$ (d) $\begin{pmatrix} 4 \\ 5 \\ 6 \end{pmatrix}, \begin{pmatrix} 3 \\ 4 \\ 6 \end{pmatrix}$

3. If A, B, C and D have coordinates $(1, 2)$, $(0, 2)$, $(-2, 4)$ and $(3, 5)$ respectively, find the angle between the following pairs of vectors.

  (a) $\overrightarrow{AB}$, $\overrightarrow{CD}$ (b) $\overrightarrow{AC}$, $\overrightarrow{CD}$ (c) $\overrightarrow{AD}$, $\overrightarrow{BC}$

4. If A, B, C and D have position vectors $\begin{pmatrix} 0 \\ 1 \\ 3 \end{pmatrix}$, $\begin{pmatrix} 2 \\ 1 \\ 5 \end{pmatrix}$, $\begin{pmatrix} -3 \\ 2 \\ 0 \end{pmatrix}$ and $\begin{pmatrix} 1 \\ 4 \\ 1 \end{pmatrix}$ respectively, find the angle between the following pairs of vectors.

(a) $\overrightarrow{AB}$, $\overrightarrow{DC}$ (b) $\overrightarrow{CB}$, $\overrightarrow{AD}$ (c) $\overrightarrow{BC}$, $\overrightarrow{BD}$ (d) $\overrightarrow{DC}$, $\overrightarrow{DA}$

5. (a) If $\vec{a}$ and $\vec{b}$ are perpendicular vectors and $\vec{a} = \begin{pmatrix} 10 \\ 5 \end{pmatrix}$ and $\vec{b} = \begin{pmatrix} -8 \\ k \end{pmatrix}$, find the value of k.

(b) If $\vec{u}$ and $\vec{v}$ are perpendicular vectors and $\vec{u} = 3\vec{i} + \vec{j} + c\vec{k}$, $\vec{v} = -5\vec{i} - 9\vec{j} + 8\vec{k}$, find the value of c.

6. Given that $\vec{a} = \begin{pmatrix} 3 \\ 2 \end{pmatrix}$, $\vec{b} = \begin{pmatrix} 1 \\ -4 \end{pmatrix}$, $\vec{c} = \begin{pmatrix} 1 \\ m \end{pmatrix}$ and $\vec{d} = \begin{pmatrix} -10 \\ n \end{pmatrix}$ and that $\vec{d} = (\vec{a} \cdot \vec{c})\vec{b} - (\vec{a} \cdot \vec{b})\vec{c}$, find the values of the constants m and n.

7. Write down unit vectors which are perpendicular to the following vectors.

(a) $\begin{pmatrix} 5 \\ 3 \end{pmatrix}$ (b) $\begin{pmatrix} 8 \\ -11 \end{pmatrix}$ (c) $\begin{pmatrix} 3.29 \\ 7.16 \end{pmatrix}$ (d) $\begin{pmatrix} -1 \\ 0 \end{pmatrix}$ (e) $4\vec{i} + 3\vec{j}$ (f) $-1.5\vec{j}$

8. Each vector in column one is parallel to exactly one vector in column two. Find which vector in column two is parallel to each vector in column one.

	Column One	Column Two
(a)	$3\vec{i} - \vec{j} + \vec{k}$	$15\vec{i} + 15\vec{j} + 60\vec{k}$
(b)	$5\vec{i} + 5\vec{j} + 20\vec{k}$	$-3\vec{i} + 4\vec{j} + 6\vec{k}$
(c)	$\frac{1}{2}\vec{i} - \frac{2}{3}\vec{j} - \vec{k}$	$-9\vec{i} + 3\vec{j} - 3\vec{k}$
(d)	$\vec{i} + \vec{k}$	$\frac{4}{3}\vec{i} - \frac{5}{6}\vec{j} + \frac{5}{9}\vec{k}$
(e)	$-\frac{2}{5}\vec{i} + \frac{1}{4}\vec{j} - \frac{1}{6}\vec{k}$	$-4\vec{i} - 4\vec{k}$

9. Show that the vectors $\begin{pmatrix} 1 \\ 1 \\ -3 \end{pmatrix}$, $\begin{pmatrix} 1 \\ -10 \\ -3 \end{pmatrix}$ and $\begin{pmatrix} 3 \\ 0 \\ 1 \end{pmatrix}$ are mutually perpendicular.

10.   In each case, find the value of c such that $\vec{u}$ and $\vec{v}$ are perpendicular.

(a) $\vec{u} = 3\vec{i} + 5\vec{j} - \vec{k}$; $\vec{v} = 7\vec{i} - 4\vec{j} + c\vec{k}$

(b) $\vec{u} = \vec{i} - 2\vec{j} + 8\vec{k}$; $\vec{v} = c\vec{i} + 3\vec{j} + \vec{k}$

(c) $\vec{u} = 2\vec{i} + 5\vec{j} + 9\vec{k}$; $\vec{v} = 4\vec{i} + cj - 6\vec{k}$

(d) $\vec{u} = c\vec{i} - 3\vec{j} + 2\vec{k}$; $\vec{v} = 2\vec{i} + 4j + c\vec{k}$

11.   If $\vec{u} = \vec{i} + 2\vec{j} - 3\vec{k}$ and $\vec{v} = 3\vec{i} + 6\vec{j} - 9\vec{k}$,

(a) show that $\vec{u} \bullet \vec{v} = |\vec{u}||\vec{v}|$.

(b) find a vector $\vec{w}$ such that $\vec{u} \bullet \vec{w} = -|\vec{u}||\vec{w}|$.

4.4 The Vector Equation of a Line

4.4.1 The Derivation of the Vector Equation of a Line

Perhaps it is a good idea to be reminded of how a vector differs from a straight line. A straight line has a direction, and it continues indefinitely. A vector, on the other hand, can be represented by an arrow so it has a finite magnitude and a direction. It also has a 'directional sense' indicated by the arrow head. Nevertheless, it is possible, and very useful, to represent a straight line using vectors.

Figure 4.41 shows line L joining points A and B , where $\vec{a}$ and $\vec{b}$ are the position vectors of A and B respectively.

The straight line L can be represented by a position vector, $\vec{r}$, from the origin O to an arbitrary point, P, on the line. Now $\vec{r} = \vec{a} + \overrightarrow{AP}$, but, as $\overrightarrow{AP}$ is parallel to $\overrightarrow{AB}$, it can be expressed as a scalar multiple of $\overrightarrow{AB}$. Therefore, $\vec{r} = \vec{a} + t\overrightarrow{AB}$ where t is a scalar value.

Since $\overrightarrow{AB} = -\vec{a} + \vec{b}$, $\vec{r} = \vec{a} + t\left(\vec{b} - \vec{a}\right)$. Therefore, the vector equation of a line which passes through two points A and B , whose position vectors are $\vec{a}$ and $\vec{b}$ respectively, is $\vec{r} = \vec{a} + t\left(\vec{b} - \vec{a}\right)$.

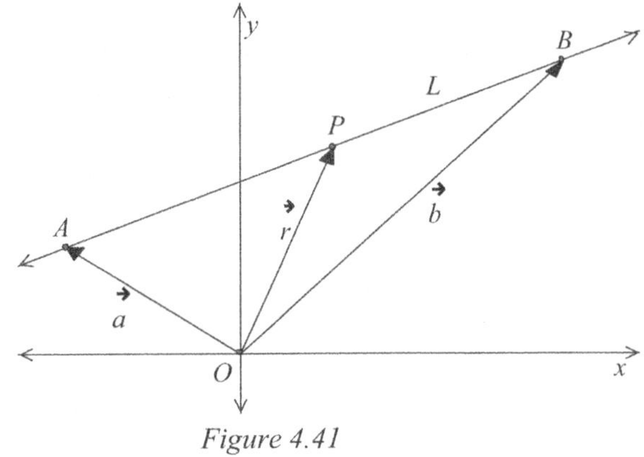

Figure 4.41

As t varies, so P moves along the line, and $\vec{r}$, the position vector of P , varies accordingly. For example, when $t = 0$, P is located at A ; when $t = 1$, P is located at B ; when $t < 0$, P is located on the line to the left of A , and so on.

4.4.2 The Vector Equation of a Line in Two-Dimensional Space

Suppose that A has coordinates $(-3, 4)$ and B has coordinates $(5, 11)$. Note that, as P is an arbitrary point on L, it can be represented by (x, y). Therefore, $\overrightarrow{OP} = \vec{r} = \begin{pmatrix} x \\ y \end{pmatrix}$ and $\begin{pmatrix} x \\ y \end{pmatrix} = \begin{pmatrix} -3 \\ 4 \end{pmatrix} + t \begin{pmatrix} 5-(-3) \\ 11-4 \end{pmatrix}$

$$\Rightarrow \begin{pmatrix} x \\ y \end{pmatrix} = \begin{pmatrix} -3 \\ 4 \end{pmatrix} + t \begin{pmatrix} 8 \\ 7 \end{pmatrix}.$$

It is useful to think of the equation of a line as consisting of two parts: a position vector of a point on the line, in this case $\begin{pmatrix} -3 \\ 4 \end{pmatrix}$, and a vector indicating direction, in this case, $t \begin{pmatrix} 8 \\ 7 \end{pmatrix}$. The position vector can be the position vector of *any* point <u>on the line</u> and the 'direction' vector can be *any* vector <u>parallel to the line</u>.

Therefore, $\begin{pmatrix} x \\ y \end{pmatrix} = \begin{pmatrix} -3 \\ 4 \end{pmatrix} + t \begin{pmatrix} 8 \\ 7 \end{pmatrix}$ and $\begin{pmatrix} x \\ y \end{pmatrix} = \begin{pmatrix} 5 \\ 11 \end{pmatrix} + t \begin{pmatrix} 16 \\ 14 \end{pmatrix}$ are different vector equations of the same line. There are an infinite number of such vector equations representing the same line.

4.4.3 Finding a Vector Parallel to a Given Line in a Plane

The vector equation of a line contains a vector parallel to that line. For example, $\begin{pmatrix} 8 \\ 7 \end{pmatrix}$ is a vector parallel to the line with vector equation $\begin{pmatrix} x \\ y \end{pmatrix} = \begin{pmatrix} -3 \\ 4 \end{pmatrix} + t \begin{pmatrix} 8 \\ 7 \end{pmatrix}$. However, if the line is written in Cartesian form, it is necessary to do some algebra in order to find a vector parallel to the line; $ax + by + c = 0$ can be rearranged to $y = -\dfrac{a}{b}x - \dfrac{c}{b}$ where the gradient of the line is $\dfrac{-a}{b}$ or $\dfrac{a}{-b}$.

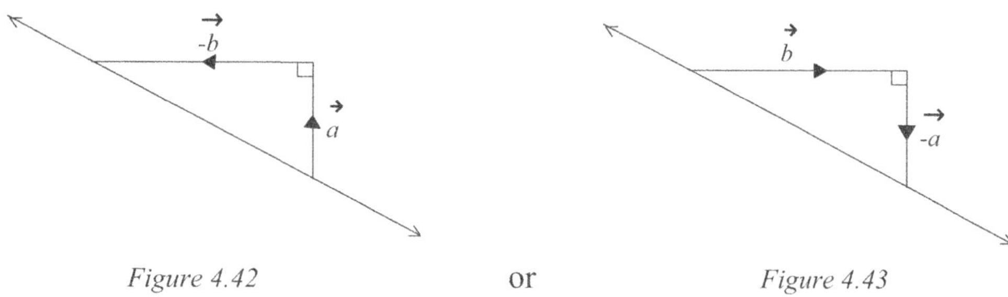

Figure 4.42   or   Figure 4.43

Figures 4.42 and 4.43 show that a vector parallel to the line is $\begin{pmatrix} -b \\ a \end{pmatrix}$ or $\begin{pmatrix} b \\ -a \end{pmatrix}$. There is an infinite number of vectors parallel to the line, but the ones shown are the easiest to identify.

4.4.4 The Angle between Two Lines in Two-Dimensional Space

The angle between two lines, L_1 and L_2, in a plane is simply the angle between two vectors, $\vec{v}_1$ and $\vec{v}_2$, in which $\vec{v}_1$ is parallel to L_1 and $\vec{v}_2$ is parallel to L_2. The appropriate method, described in section 4.4.3, can be used to find $\vec{v}_1$ and $\vec{v}_2$ from the given lines L_1 and L_2.

Example 4.7: (a) Find the acute angle between the lines with vector equations $L_1 : \vec{r} = \begin{pmatrix} 1 \\ 3 \end{pmatrix} + s \begin{pmatrix} -2 \\ 1 \end{pmatrix}$

and $L_2 : \vec{r} = \begin{pmatrix} 2 \\ 1 \end{pmatrix} + t \begin{pmatrix} 3 \\ 2 \end{pmatrix}$.

(b) Find the acute angle between the lines whose Cartesian equations are
$L_1 : x + y - 3 = 0$ and $: L_2 : 2x - 3y + 1 = 0$.

Solution 4.7: (a) L_1 has 'direction' vector $\vec{v}_1 = \begin{pmatrix} -2 \\ 1 \end{pmatrix}$, L_2 has 'direction' vector $\vec{v}_2 = \begin{pmatrix} 3 \\ 2 \end{pmatrix}$.

$\vec{v}_1 \bullet \vec{v}_2 = -2 \times 3 + 1 \times 2 = -6 + 2 = -4$, $|\vec{v}_1| = \sqrt{5}$, $|\vec{v}_2| = \sqrt{13}$. If θ is the angle between

the lines L_1 and L_2, $\cos\theta = \dfrac{-4}{\sqrt{13}\sqrt{5}} \Rightarrow \theta = 119.7448... = 119.7°$. Since θ is obtuse, it

is necessary to take the supplementary angle in order to obtain the acute angle
between the lines. So, the acute angle between L_1 and L_2 is $180° - 119.7° = 60.3°$.

(b) Figure 4.44 shows the lines
represented by $L_1 : x + y - 3 = 0$ and
$L_2 : x + y - 3 = 0$. A vector parallel to
$x + y - 3 = 0$ is $\vec{u} = \begin{pmatrix} 1 \\ -1 \end{pmatrix}$ and one

parallel to $2x - 3y + 1 = 0$ is $\vec{v} = \begin{pmatrix} 3 \\ 2 \end{pmatrix}$.

Now, $\vec{u} \bullet \vec{v} = 1 \times 3 + (-1) \times 2 = 3 - 2 = 1$,

$|\vec{u}| = \sqrt{1^2 + (-1^2)} = \sqrt{2}$ and

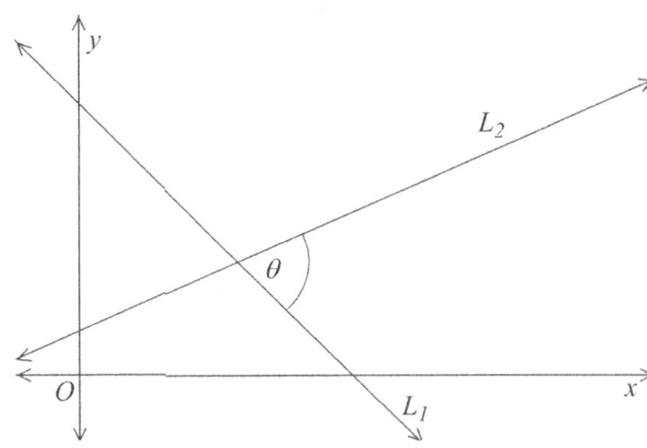

Figure 4.44

$|\vec{v}| = \sqrt{3^2 + 2^2} = \sqrt{13}$. So, $\cos\theta = \dfrac{\vec{u} \bullet \vec{v}}{|\vec{u}||\vec{v}|} = \dfrac{1}{\left(\sqrt{2} \times \sqrt{13}\right)}$, $\theta = 78.69006... = 78.7°$.

4.4.5 An Application to Constant Velocity Motion in Two-Dimensional Space

The vector equation of a line has a useful application: for an object moving with *constant* velocity, displacement, $\vec{s}$, velocity, $\vec{v}$, and time, t, are related by the formula $\vec{s} = t\vec{v}$, where $\vec{s}$ and $\vec{v}$ are vectors and t is a scalar.

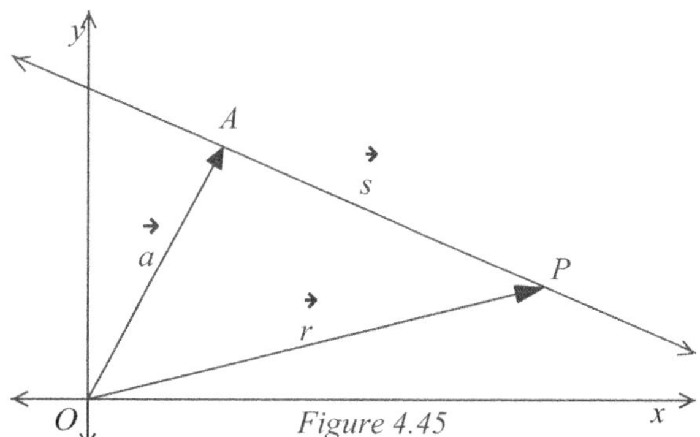

Figure 4.45 is a diagram of a cyclist who begins a journey at point A with position vector $\vec{a}$, relative to an origin O. The cyclist continues with velocity $\vec{v}$ so that after time t he is at point P.

Therefore, $\vec{s} = \overrightarrow{AP} = -\vec{a} + \vec{r} = \vec{r} - \vec{a}$, but, because $\vec{s} = t\vec{v}$, $\vec{r} - \vec{a} = t\vec{v} \Rightarrow \vec{r} = \vec{a} + t\vec{v}$.

Suppose that $\vec{a} = \begin{pmatrix} 3 \\ 5 \end{pmatrix}$ km and the cyclist's velocity is $\begin{pmatrix} 10 \\ -2 \end{pmatrix}$ kmh^{-1}. P is an arbitrary position on the cyclist's path, so $\overrightarrow{OP} = \vec{r}$ can be written as $\begin{pmatrix} x \\ y \end{pmatrix}$ so that

$$\begin{pmatrix} x \\ y \end{pmatrix} = \begin{pmatrix} 3 \\ 5 \end{pmatrix} + t \begin{pmatrix} 10 \\ -2 \end{pmatrix}$$ for any value of $t \geq 0$. The position vector of the cyclist can be found at any time by replacing t with the appropriate numerical value.

Example 4.8: An object, P, moves in a straight line with constant velocity. Its position vector, relative to an origin, O, is given by $\begin{pmatrix} x \\ y \end{pmatrix} = \begin{pmatrix} 5 \\ 9 \end{pmatrix} + t \begin{pmatrix} 4 \\ -3 \end{pmatrix}$, $t \geq 0$, where t is measured in hours and displacement is measured in kilometers.

(a) Find the coordinates of P (i) initially   (ii) after 5 hours.

Another object, Q, also moves in a straight line with constant velocity so that its position vector at time t is given by $\begin{pmatrix} x \\ y \end{pmatrix} = \begin{pmatrix} -3 \\ 4 \end{pmatrix} + t \begin{pmatrix} 5 \\ -1 \end{pmatrix}$, $t \geq 0$.

(b) Show that the paths of P and Q intersect, but P and Q do not collide.

Solution 4.8: (a) (i) $t = 0 \Rightarrow \begin{pmatrix} x \\ y \end{pmatrix} = \begin{pmatrix} 5 \\ 9 \end{pmatrix} + 0 \begin{pmatrix} 4 \\ -3 \end{pmatrix} = \begin{pmatrix} 5 \\ 9 \end{pmatrix}$, so the coordinates of P are $(5, 9)$.

(ii) $t = 5 \Rightarrow \begin{pmatrix} x \\ y \end{pmatrix} = \begin{pmatrix} 5 \\ 9 \end{pmatrix} + 5 \begin{pmatrix} 4 \\ -3 \end{pmatrix} = \begin{pmatrix} 5 \\ 9 \end{pmatrix} + \begin{pmatrix} 20 \\ -15 \end{pmatrix} = \begin{pmatrix} 25 \\ -6 \end{pmatrix}$, so the coordinates of P are $(25, -6)$.

(b) Because $\begin{pmatrix} 4 \\ -3 \end{pmatrix}$ is not parallel to $\begin{pmatrix} 5 \\ -1 \end{pmatrix}$, the paths of P and Q are not parallel.

Therefore, they intersect. Let the intersection point be N and suppose that P reaches N at time t_1 and Q reaches N at time t_2.

Equating x components of P and Q, we obtain $5 + 4t_1 = -3 + 5t_2$.

Equating y components of P and Q, we obtain $9 - 3t_1 = 4 - t_2$.

Solving for t_1 and t_2 gives $t_1 = 3$ and $t_2 = 4$. So P reaches the intersection point after three hours and Q reaches the intersection after four hours. As P and Q pass the intersection point at different values of t, they clearly do not collide.

The next example shows how clever use of the scalar product simplifies what, appears to be, a tricky problem.

Example 4.9: A cyclist travels at a speed of 26kmh^{-1} in a direction $\begin{pmatrix} 12 \\ -5 \end{pmatrix}$ relative to an origin, O.

She starts at point $A\ (-2,\ 10)$, and, after one hour, she has reached point B.

(a) Write down a unit vector parallel to the cyclist's velocity and use it to find her velocity as a column vector.

(b) Find   (i) $\overrightarrow{OA}$ (ii) $\overrightarrow{AB}$ (iii) $\overrightarrow{OB}$

(c) Find the coordinates of B.

After t hours she is at point P.

(d) On Cartesian axes, show the cyclist's path and the points O, A, B and an arbitrary point P.

(e) Find   (i) $\overrightarrow{AP}$ (ii) $\overrightarrow{OP}$ (iii) $\overrightarrow{OP}\bullet\overrightarrow{AP}$ in terms of t

(f) Hence, find the time, to the nearest minute, for the cyclist to be closest to O and the distance $\left|\overrightarrow{OP}\right|$ at this time.

Solution 4.9: (a) The magnitude of $\begin{pmatrix} 12 \\ -5 \end{pmatrix}$ is $\sqrt{12^2 + (-5)^2} = \sqrt{169} = 13$. Therefore, a unit vector

parallel to her velocity is $\begin{pmatrix} \frac{12}{13} \\ -\frac{5}{13} \end{pmatrix}$. If she travels at a speed of 26kmh^{-1}, then her

velocity is $26\begin{pmatrix} \frac{12}{13} \\ -\frac{5}{13} \end{pmatrix} = \begin{pmatrix} 24 \\ -10 \end{pmatrix}\text{kmh}^{-1}$.

(b)   (i) $\overrightarrow{OA} = \begin{pmatrix} -2 \\ 10 \end{pmatrix}\text{km}$ (ii) $\overrightarrow{AB} = \begin{pmatrix} 24 \\ -10 \end{pmatrix}\text{km}$

(iii) $\overrightarrow{OB} = \overrightarrow{OA} + \overrightarrow{AB} = \begin{pmatrix} -2 \\ 10 \end{pmatrix} + \begin{pmatrix} 24 \\ -10 \end{pmatrix} = \begin{pmatrix} 22 \\ 0 \end{pmatrix}$ km.

(c) The coordinates of B are $(22, 0)$.

(d)

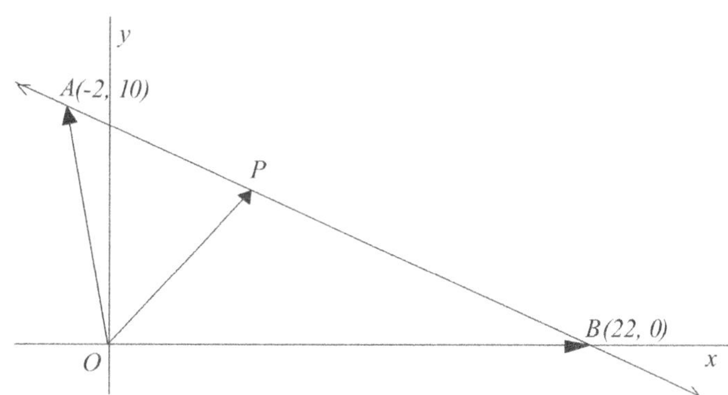

(e) (i) $\overrightarrow{AP} = t\begin{pmatrix} 24 \\ -10 \end{pmatrix}$ (ii) $\overrightarrow{OP} = \overrightarrow{OA} + \overrightarrow{AP} = \begin{pmatrix} -2 \\ 10 \end{pmatrix} + t\begin{pmatrix} 24 \\ -10 \end{pmatrix} = \begin{pmatrix} -2 + 24t \\ 10 - 10t \end{pmatrix}$

(iii) $\overrightarrow{OP} \cdot \overrightarrow{AP} = \begin{pmatrix} -2 + 24t \\ 10 - 10t \end{pmatrix} \cdot \begin{pmatrix} 24t \\ -10t \end{pmatrix} = (-2 + 24t)24t + (10 - 10t)(-10t)$

$$= -48t + 576t^2 - 100t + 100t^2 = 676t^2 - 148t$$

(f) The cyclist is closest to O when $\overrightarrow{OP}$ is perpendicular to $\overrightarrow{AP}$, and, when this is so,

$\overrightarrow{OP} \cdot \overrightarrow{AP} = 0 \Rightarrow 676t^2 - 148t = 0 \Rightarrow t(676t - 148) = 0$.

Therefore, $t = \dfrac{148}{676}$ which is $60 \times \dfrac{148}{676} = 13.13609... = 13\,\text{minutes} = 13$ minutes to the

nearest minute. Substituting $t = \dfrac{148}{676} = 0.2189349...$ into $\overrightarrow{OP}$,

$$\overrightarrow{OP} = \begin{pmatrix} -2 + 24t \\ 10 - 10t \end{pmatrix} = \begin{pmatrix} -2 + 24 \times 0.2189349... \\ 10 - 10 \times 0.2189349... \end{pmatrix} = \begin{pmatrix} 3.254437... \\ 7.810651... \end{pmatrix}$$

$$\Rightarrow |\overrightarrow{OP}| = \sqrt{3.254437...^2 + 7.810651...^2} = 8.461538... = 8.46 \text{ km.}$$

Exercise 4.5

1. Write the following straight lines, whose vector equations are given, in Cartesian form.

(a) $\begin{pmatrix} x \\ y \end{pmatrix} = \begin{pmatrix} 1 \\ 2 \end{pmatrix} + t\begin{pmatrix} 2 \\ -1 \end{pmatrix}$ (b) $\begin{pmatrix} x \\ y \end{pmatrix} = \begin{pmatrix} -5 \\ -2 \end{pmatrix} + t\begin{pmatrix} 4 \\ 9 \end{pmatrix}$ (c) $\begin{pmatrix} x \\ y \end{pmatrix} = \begin{pmatrix} 13 \\ 5 \end{pmatrix} + (t-1)\begin{pmatrix} 8 \\ -3 \end{pmatrix}$

(d) $\vec{r} = (3\vec{i} - 4\vec{j}) + t(5\vec{i} + \vec{j})$ (e) $\vec{r} = (-7\vec{i} + 2\vec{j}) + 3t\vec{i}$

2. Find, correct to the nearest degree, the non-obtuse angle between the following pairs of lines expressed in Cartesian form.

(a) $x + y + 1 = 0$; $y = x + 3$ (b) $2x + 3y - 9 = 0$; $4x - y + 3 = 0$

(c) $5x - 6y + 7 = 0$; $3x + 4y + 23 = 0$ (d) $y = 4x - 11$; $\dfrac{1}{3}x + \dfrac{1}{2}y = 8$

3.	Figure 4.46 shows the vertices of parallelogram $OABC$, in which $\overrightarrow{OA} = \begin{pmatrix} -2 \\ 5 \end{pmatrix}$, $\overrightarrow{OC} = \begin{pmatrix} 6 \\ -3 \end{pmatrix}$.

(a) Write down $\overrightarrow{AB}$ and hence find $\overrightarrow{OB}$.

(b) L_1 is the line through (AC). If a vector equation for L_1 is $\begin{pmatrix} x \\ y \end{pmatrix} = \begin{pmatrix} -2 \\ 5 \end{pmatrix} + t \begin{pmatrix} 1 \\ k \end{pmatrix}$, find the value of k.

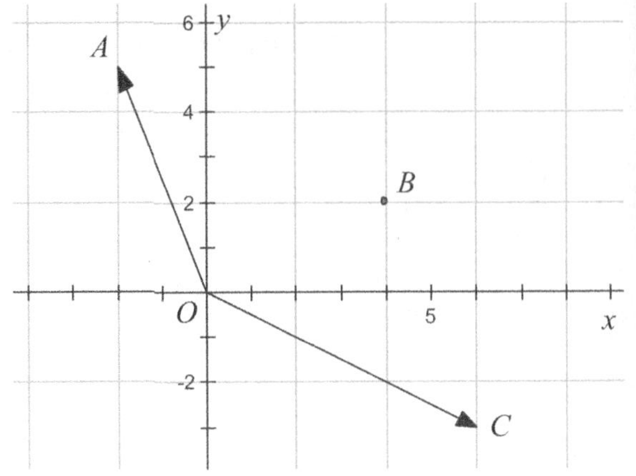

Figure 4.46

A particle P travels with constant velocity from A to C in 8 seconds.

(c) Find	(i) the coordinates of P 1 second after leaving A.

(ii) the value of t when P has coordinates $(1, 2)$.

(d) L_2 is the line through B perpendicular to (AC). State a vector equation of L_2.

(e) Find the minimum distance from P to B and show that it occurs 4.5 seconds after P leaves A.

4.	A has coordinates $(3, 7)$ and B has coordinates $(9, 1)$ relative to an origin, O. The line L_1 passes through A and B.

(a) Find a vector equation for L_1.

(b) Show that the point P with coordinates $(1, 9)$ lies on L_1.

(c) Write down a vector equation of the line L_2 which passes through P and is perpendicular to L_1.

(d) Find the coordinates of the point Q where L_2 cuts the x-axis.

5.	In this question, all position vectors are given relative to an origin, O. Distances are measured in meters, and time, t, is measured in seconds.

An antelope runs across flat ground, and his path is given by the vector equation $\begin{pmatrix} x \\ y \end{pmatrix} = \begin{pmatrix} -9 \\ 11 \end{pmatrix} + t \begin{pmatrix} 4 \\ 3 \end{pmatrix}$, $t \geq 0$.

(a) Find the speed of the antelope.
(b) Find the coordinates of the point where the antelope starts.
(c) Find the coordinates of the antelope's position after four seconds.

A cheetah chases after the antelope, starting at $t = 0$, from a point with coordinates $(0, -25)$ and runs in a straight line with constant velocity. The cheetah catches the antelope after 11 seconds.

(d) Find the coordinates of the point where the cheetah catches the antelope.
(e) Find the speed of the cheetah.
(f) Explain why this question offers an inadequate model of a realistic chase.

140

6.   Triangle ABC is such that the vertex A has coordinates $(-3,\ -1)$, $\overrightarrow{AB}=\begin{pmatrix}4\\2\end{pmatrix}$ and $\overrightarrow{AC}=\begin{pmatrix}5\\5\end{pmatrix}$.

(a) Find $\overrightarrow{BC}$

(b) Find $\overrightarrow{OB}$

(c) Hence, find, in the form $\vec{r}=\vec{a}+t\vec{b}$, a vector equation for the line, L which passes through the points B and C.

The point D, lies on L and has coordinates $(h,\ k)$

(d) (i) Find $\overrightarrow{AD}$, in terms of h and k.

  (ii) Given that $\overrightarrow{AD}$ is perpendicular to L, show that $h+3k+6=0$

  (iii) Use your answer to (c) to find h and k.

(e) Hence find the distance of A from L.

4.4.6 The Vector Equation of a Line in Three-Dimensional Space

Suppose that A has coordinates $(2,\ 5,-2)$ and B has coordinates $(3,\ 6,-2)$. As in section 4.4.2, P is

an arbitrary point on L such that $\overrightarrow{OP}=\vec{r}=\begin{pmatrix}x\\y\\z\end{pmatrix}$, and $\begin{pmatrix}x\\y\\z\end{pmatrix}=\begin{pmatrix}2\\5\\-2\end{pmatrix}+t\begin{pmatrix}3-2\\6-5\\-2+2\end{pmatrix}\Rightarrow\begin{pmatrix}x\\y\\z\end{pmatrix}=\begin{pmatrix}2\\5\\-2\end{pmatrix}+t\begin{pmatrix}1\\1\\0\end{pmatrix}$

represents a line L through A and B.

As is the case for the vector equation of a line in a plane there is an infinite number of vector equations to represent the same line. For example, the following three vector equations all represent L :

$$\begin{pmatrix}x\\y\\z\end{pmatrix}=\begin{pmatrix}0\\3\\-2\end{pmatrix}+t\begin{pmatrix}3\\3\\0\end{pmatrix} \qquad \begin{pmatrix}x\\y\\z\end{pmatrix}=\begin{pmatrix}5\\8\\-2\end{pmatrix}+t\begin{pmatrix}1\\1\\0\end{pmatrix} \qquad \begin{pmatrix}x\\y\\z\end{pmatrix}=\begin{pmatrix}2\\5\\-2\end{pmatrix}+t\begin{pmatrix}-2\\-2\\0\end{pmatrix}$$

These equations represent L because $(0,\ 3,-2)$, $(5,\ 8,-2)$ and $(2,\ 5,-2)$ are coordinates of points

lying on L and $\begin{pmatrix}3\\3\\0\end{pmatrix}$, $\begin{pmatrix}1\\1\\0\end{pmatrix}$ and $\begin{pmatrix}-2\\-2\\0\end{pmatrix}$ are all vectors parallel to L.

4.4.7 The Angle between Two Lines in Three-Dimensional Space

It is useful to consider the angle between two lines in three–dimensions even if the lines do not necessarily intersect. Consider lines L_1 and L_2 where L_1 is $\vec{r_1}=\vec{a_1}+t_1\vec{b_1}$ and L_2 is $\vec{r_2}=\vec{a_2}+t_2\vec{b_2}$.

Therefore, $\vec{b_1}$ is parallel to L_1 and $\vec{b_2}$ is parallel to L_2. Hence, the angle between L_1 and L_2 is equal to the angle between $\vec{b_1}$ and $\vec{b_2}$. This provides a convenient way of finding the angle between two lines.

Suppose that L_1 has vector equation $\vec{r}_1 = \left(\vec{i} - 2\vec{j} + \vec{k}\right) + t_1\left(3\vec{i} + \vec{j} - \vec{k}\right)$ and L_2 has vector equation $\vec{r}_2 = \left(2\vec{i} + 3\vec{j} + 4\vec{k}\right) + t_2\left(\vec{i} - \vec{k}\right)$. The angle, θ, between L_1 and L_2 is the angle between the vectors $\vec{b}_1 = 3\vec{i} + \vec{j} - \vec{k}$ and $\vec{b}_2 = \vec{i} - \vec{k}$ which is given by $\cos\theta = \dfrac{\vec{b}_1 \cdot \vec{b}_2}{\left|\vec{b}_1\right|\left|\vec{b}_2\right|}$. We now calculate $\vec{b}_1 \cdot \vec{b}_2$, $\left|\vec{b}_1\right|$ and $\left|\vec{b}_2\right|$:

$\vec{b}_1 \cdot \vec{b}_2 = 3 \times 1 + 1 \times 0 + (-1)(-1) = 4$, $\left|\vec{b}_1\right| = \sqrt{3^2 + 1^2 + (-1)^2} = \sqrt{11}$, $\left|\vec{b}_2\right| = \sqrt{1^2 + 0^2 + (-1)^2} = \sqrt{2}$. So, if $\cos\theta = \dfrac{4}{\sqrt{11}\sqrt{2}}$ then $\theta = 31.48215... = 31.5°$.

Example 4.10: Find the acute angle between the lines whose vector equations are $\begin{pmatrix} x \\ y \\ z \end{pmatrix} = \begin{pmatrix} 1 \\ 0 \\ 3 \end{pmatrix} + s\begin{pmatrix} 2 \\ -1 \\ 5 \end{pmatrix}$

and $\begin{pmatrix} x \\ y \\ z \end{pmatrix} = \begin{pmatrix} 2 \\ 1 \\ -1 \end{pmatrix} + t\begin{pmatrix} 1 \\ 3 \\ -2 \end{pmatrix}$.

Solution 4.10: Let $\vec{v}_1 = \begin{pmatrix} 2 \\ -1 \\ 5 \end{pmatrix}$, $\vec{v}_2 = \begin{pmatrix} 1 \\ 3 \\ -2 \end{pmatrix}$ and the angle between the lines be θ.

$$\vec{v}_1 \cdot \vec{v}_2 = 2 \times 1 + (-1) \times 3 + 5 \times (-2) = -11$$

$$\left|\vec{v}_1\right| = \sqrt{2^2 + (-1)^2 + 5^2} = \sqrt{30}, \quad \left|\vec{v}_2\right| = \sqrt{1^2 + 3^2 + (-2)^2} = \sqrt{14}$$

$$\cos\theta = \frac{-11}{\sqrt{30}\sqrt{14}} \Rightarrow \theta = 122.4623...°$$

Therefore, the acute angle between the lines is $180° - 122.5° = 57.5°$.

Example 4.11: Line L passes through $P(1, -5, 3)$ and $Q(4, 1, -1)$.

    (a) Find the vector equation of line L through (PQ).

    (b) Show that the line passes through point $R(-5, -17, 11)$ but not through point $S(4, 1, 0)$.

    (c) Show that $P\hat{Q}S \approx 59.2°$.

Solution 4.11: (a) $\overrightarrow{PQ} = \overrightarrow{OQ} - \overrightarrow{OP} = \begin{pmatrix} 3 \\ 6 \\ -4 \end{pmatrix}$. The equation of the line is $\vec{r} = \overrightarrow{OP} + t\overrightarrow{PQ}$. Therefore the equation of L is $\begin{pmatrix} x \\ y \\ z \end{pmatrix} = \begin{pmatrix} 1 \\ -5 \\ 3 \end{pmatrix} + t \begin{pmatrix} 3 \\ 6 \\ -4 \end{pmatrix}$.

(b) At the point on L where $x = -5$, $-5 = 1 + 3t \Rightarrow t = -2$. Then $y = -5 + 6(-2) = -17$ and $z = 3 - 4(-2) = 11$. Therefore, the point $R(-5, -17, 11)$ lies on L. However, at a point on L where $x = 4$, $4 = 1 + 3t \Rightarrow t = 1$. Then $y = -5 + 6 \times 1 = 1$ and $z = 3 - 4 \times 1 = -1$ (not 0), so $S(4, 1, 0)$ cannot lie on L.

(c) $\overrightarrow{SQ} = \overrightarrow{OQ} - \overrightarrow{OS} = \begin{pmatrix} 0 \\ 0 \\ -1 \end{pmatrix}$ and $\overrightarrow{PQ} = \begin{pmatrix} 3 \\ 6 \\ -4 \end{pmatrix} \Rightarrow \overrightarrow{PQ} \cdot \overrightarrow{SQ} = 4$. $\left| \overrightarrow{SQ} \right| = 1$,

$$\left| \overrightarrow{PQ} \right| = \sqrt{3^2 + 6^2 + (-4)^2} = \sqrt{61} \Rightarrow \cos P\hat{Q}S = \frac{4}{\sqrt{61}} \Rightarrow P\hat{Q}S = 59.19301... = 59.2°$$

4.4.8 An Application to Constant Velocity Motion in Three-Dimensional Space

In this section, we explore the application of the vector equation of a line to a problem involving motion in a straight line with constant velocity.

Example 4.12: In this question, distances are measured in meters and velocities in meters per second. An aircraft is flying in a straight line. At a given moment, when $t = 0$, it has position vector $\left(8000\vec{i} + 2050\vec{j} + 13000\vec{k} \right)$ and velocity vector $\left(-35\vec{i} - 10\vec{j} - 50\vec{k} \right)$, relative to a set of axes in which the x-axis is due north, the y-axis is vertically upwards and the z-axis is due east. An airstrip is modeled by the line segment with vector equation $\vec{r} = \left(895\vec{i} + 2850\vec{k} \right) - s\left(7\vec{i} + 10\vec{k} \right)$, $0 \leq s \leq 80$.

(a) Find the speed of the aircraft.
(b) Write down, in terms of t, the equation of the flight path of the aircraft for $t \geq 0$.
(c) Find the value of t when the aircraft touches down.
(d) Show that the aircraft lands on the airstrip and find the distance, after it touches down, in which it must stop.

Solution 4.12: (a) The speed of the aircraft is $\sqrt{(-35)^2 + (-10)^2 + (-50)^2} = 61.84658... = 61.8 \text{ ms}^{-1}$.

(b) $\vec{r} = \left(8000\vec{i} + 2050\vec{j} + 13000\vec{k} \right) + t\left(-35\vec{i} - 10\vec{j} - 50\vec{k} \right)$

(c) The aircraft touches down when the y component of its position vector is zero. Therefore, $2050 - 10t = 0 \Rightarrow t = 205$.

(d) At $t = 205$, $\vec{r} = \left(8000\vec{i} - 2050\vec{j} + 13000\vec{k} \right) + 205\left(-35\vec{i} - 10\vec{j} - 50\vec{k} \right)$

$\Rightarrow \vec{r} = \left(8000 - 7175 \right)\vec{i} + 0\vec{j} + \left(13000 - 10250 \right)\vec{k} = 825\vec{i} + 2750\vec{k}$. Therefore, the coordinates of the point of touch down are $\left(825, 0, 2750 \right)$. If the aircraft lands on the airstrip, $(825,\ 0,\ 2750)$ will satisfy the equation of the airstrip. Equating the x component, $825 = 895 - 7s \Rightarrow s = 10$. Equating the z component, $2750 = 2850 - 10s \Rightarrow s = 10$. As the values of s are consistent, the point $(825,\ 0,\ 2750)$ lies on the airstrip. The position vector of the end of the airstrip is $\left(895\vec{i} + 2850\vec{k}\right) - 80\left(7\vec{i} + 10\vec{k}\right) = 335\vec{i} + 2050\vec{k}$. The position vector of the touch-down point is $\left(825\vec{i} + 2750\vec{k}\right)$. Therefore, the distance in which the aircraft must stop is $\sqrt{(825 - 335)^2 + (2750 - 2050)^2} = 854.4588... = 854$ m.

Exercise 4.6

1. In each case, find the acute angle between the given lines.

(a) $\vec{r} = \left(2\vec{i} + 6\vec{j} - \vec{k}\right) + p\left(\vec{i} - 2\vec{j} + 5\vec{k}\right),\quad \vec{r} = \left(3\vec{i} - \vec{k}\right) + q\left(3\vec{i} + 3\vec{j} + 2\vec{k}\right)$

(b) $\vec{r} = \left(4\vec{i} - \vec{j} + \vec{k}\right) + p\left(6\vec{i} + \vec{j} - 3\vec{k}\right),\quad \vec{r} = \left(-\vec{i} - 2\vec{j} + 3\vec{k}\right) + q\left(5\vec{i} + 4\vec{j} + 3\vec{k}\right)$

(c) $\vec{r} = p\left(-\vec{i} - 8\vec{j} + 2\vec{k}\right),\quad \vec{r} = \left(4\vec{i} - \vec{j} + 3\vec{k}\right) + q\left(-3\vec{i} + 8\vec{j} - 7\vec{k}\right)$

2. In each case, find the acute angle between the given lines.

(a) $\begin{pmatrix} x \\ y \\ z \end{pmatrix} = \begin{pmatrix} 1 \\ 1 \\ 1 \end{pmatrix} + r\begin{pmatrix} 3 \\ 5 \\ 1 \end{pmatrix},\qquad \begin{pmatrix} x \\ y \\ z \end{pmatrix} = \begin{pmatrix} 0 \\ 2 \\ 4 \end{pmatrix} + s\begin{pmatrix} 2 \\ -1 \\ 1 \end{pmatrix}$

(b) $\begin{pmatrix} x \\ y \\ z \end{pmatrix} = \begin{pmatrix} 6 \\ -7 \\ 2 \end{pmatrix} + r\begin{pmatrix} -3 \\ -10 \\ 5 \end{pmatrix},\qquad \begin{pmatrix} x \\ y \\ z \end{pmatrix} = \begin{pmatrix} 6 \\ 2 \\ -2 \end{pmatrix} + s\begin{pmatrix} 0 \\ 1 \\ 0 \end{pmatrix}$

(c) $\begin{pmatrix} x \\ y \\ z \end{pmatrix} = \begin{pmatrix} 1 \\ 0 \\ 4 \end{pmatrix} + r\begin{pmatrix} -3 \\ 4 \\ -2 \end{pmatrix},\qquad \begin{pmatrix} x \\ y \\ z \end{pmatrix} = \begin{pmatrix} 0 \\ 3 \\ 5 \end{pmatrix} + s\begin{pmatrix} 0.7 \\ 1.2 \\ -0.3 \end{pmatrix}$

3. (i) A line L passes through a point $A(3, -1,\ 1)$ and is parallel to the line whose vector equation is $\vec{r} = \left(\vec{i} - 2\vec{j} + \vec{k}\right) + c\left(2\vec{i} + 3\vec{j} - 2\vec{k}\right)$. Write down, in the form $\vec{r} = \vec{a} + t\vec{b}$, the vector equation of L and show that it passes through the point with coordinates $(7,\ 5, -3)$.

(ii) A line l passes through two points whose coordinates are $(1, -3, -2)$ and $(2,\ 1,\ 4)$. Find, in the form $\vec{r} = \vec{a} + t\vec{b}$, the vector equation of l and test whether or not the point with coordinates $(0, -7, -8)$ lies on l.

4. A has coordinates $(1,\ 3, -4)$ and B has coordinates $(5, -1,\ 0)$

(a) Write down the coordinates of M, the midpoint of $[AB]$

C has coordinates $(6, -3, 5)$

(b) Write down, in the form $\vec{r} = \vec{a} + t\vec{b}$, an equation of the line which passes through C and M.

(c) Find $\left| \overrightarrow{AB} \right|$ and $\left| \overrightarrow{CB} \right|$

(d) Show that $A\hat{B}C \simeq 147°$

(e) Find the area of triangle ABC.

5.   In this question, distances are measured in meters and velocities in meters per second.

A submarine is rising to the surface of the sea in a straight path. At time $t = 0$, the submarine is at a point P with coordinates $(350,\ -217,\ 785)$ relative to a set of axes whose origin, O, lies on the surface of the sea. The x-axis is due north, the y-axis is vertically upwards and the z-axis is due east. The velocity of the submarine is given by the vector $\vec{v} = -9\vec{i} + 7\vec{j} - 4\vec{k}$.

(a) Write down the position vector, $\overrightarrow{OP}$.

(b) Find the speed of the submarine.

(c) The submarine's path can be modeled by a straight line. Find, in terms of t, the vector equation of this line.

(d) A stationary fishing boat is located at point Q with coordinates $(10,\ 0,\ 600)$. How far is the submarine from the fishing boat when it surfaces?

6.   In this question, units of distance are in meters and units of velocity are in meters per second.

The path of a skier traveling down a slope is modeled by a straight line whose vector equation is $\vec{r} = \vec{a} + t\vec{v}$. $\vec{a} = \begin{pmatrix} 5.83 \\ 2.11 \\ 3.74 \end{pmatrix}$ is the position vector of his path at time $t = 0$, and $\vec{v} = \begin{pmatrix} 0.711 \\ -0.283 \\ 1.782 \end{pmatrix}$ is the skier's velocity for $t \geq 0$. The components of both $\vec{a}$ and $\vec{v}$ are relative to a fixed set of axes.

(a) Find the skier's speed.

(b) Find the skier's position vector after 10 seconds.

(c) Find the time taken for the skier to reach the point whose coordinates are $(43.51, -12.89, 98.19)$.

7.   In this question, units of distance are shown in meters.

Juan has lost his watch, which is lying at the bottom of a swimming pool. Figure 4.47 shows the swimming pool with a system of Cartesian axes superimposed.

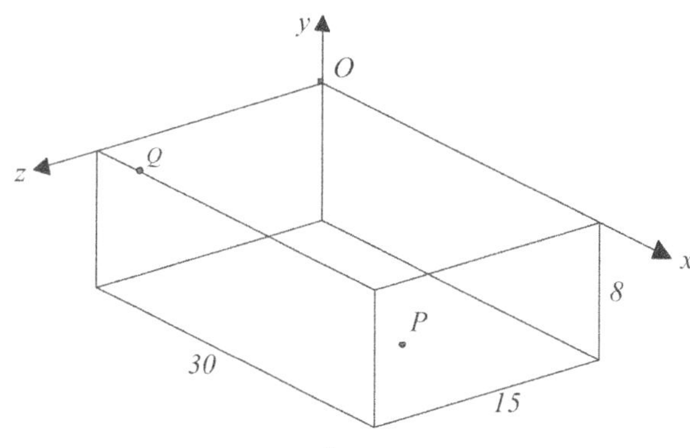

Figure 4.47

The watch is located at point P whose coordinates are $(21, -8,\ 8)$. Juan is standing on the edge of the pool at point Q with coordinates $(4,\ 0,\ 15)$ relative to the origin, O.

(a) Find $\overrightarrow{QP}$.

(b) Juan dives into the pool at Q and swims in a straight line to P. Swimming at a speed of $1\ \mathrm{ms^{-1}}$, show that his velocity is approximately $0.848\vec{i} - 0.399\vec{j} - 0.349\vec{k}$.

(c) If Juan can maintain this speed while picking up the watch and then rise vertically to the surface, how long will he need to hold his breath?

4.5 Parallel, Coincident, Intersecting and Non-Intersecting Lines

4.5.1 Parallel and Coincident Lines

By now you are aware that a line cannot be expressed by a unique vector equation. For example,

$$\begin{pmatrix} x \\ y \\ z \end{pmatrix} = \begin{pmatrix} 1 \\ 0 \\ 1 \end{pmatrix} + p \begin{pmatrix} 1 \\ -2 \\ 1 \end{pmatrix} \text{ and } \begin{pmatrix} x \\ y \\ z \end{pmatrix} = \begin{pmatrix} 7 \\ -12 \\ 7 \end{pmatrix} + q \begin{pmatrix} -3 \\ 6 \\ -3 \end{pmatrix}$$ represent the same line, l. You can test this by

assuming that the two equations represent <u>different</u> lines, l_1 and l_2. Then choose an arbitrary value of p. For example, suppose that we put $p = 2$; then $x = 1 + 2 \times 1 = 3$, $y = 0 - 2 \times 2 = -4$ and $z = 1 + 2 = 3$. Therefore, the point $(3, -4,\ 3)$ lies on l_1.

Now test to see if the point $(3, -4,\ 3)$ lies on l_2 by equating x components: $3 = 7 - 3q \Rightarrow q = \dfrac{4}{3}$; by equating y components: $-4 = -12 + 6q \Rightarrow q = \dfrac{4}{3}$; by equating z components: $3 = 7 - 3q \Rightarrow q = \dfrac{4}{3}$.

As the results are consistent, it is clear that the point $(3, -4,\ 3)$ lies on l_1 and l_2, and, as you know that the lines are parallel, then l_1 and l_2 are coincident.

On the other hand, the line L, with vector equation $\begin{pmatrix} x \\ y \\ z \end{pmatrix} = \begin{pmatrix} 1 \\ 2 \\ 3 \end{pmatrix} + r \begin{pmatrix} -1 \\ 2 \\ -1 \end{pmatrix}$, although parallel to l, is not

coincident with it. An easy way to decide is to consider an arbitrary point, R on L. You can choose R by taking an arbitrary value for r. Let $r = 1$, for example; then $\overrightarrow{OR} = \begin{pmatrix} 1 \\ 2 \\ 3 \end{pmatrix} + \begin{pmatrix} -1 \\ 2 \\ -1 \end{pmatrix} = \begin{pmatrix} 0 \\ 4 \\ 2 \end{pmatrix}$ and

therefore R has coordinates $(0,\ 4,\ 2)$.

Now test to see whether R lies on l by equating components of l with coordinates of R :

$$1+p=0 \Rightarrow p=-1$$
$$0-2p=4 \Rightarrow p=-2$$
$$1+p=2 \Rightarrow p=1$$

The test fails because the values of p are inconsistent, so you know that L and l are not coincident lines.

A similar argument may be used to test whether parallel lines in a plane are distinct (different) or coincident.

Example 4.13: Decide whether the line L_1 , with equation $\vec{r}=\left(3\vec{i}-2\vec{j}\right)+p\left(2\vec{i}+5\vec{j}\right)$, and the line L_2 , with equation $\vec{r}=\left(\vec{i}+8\vec{j}\right)+q\left(4\vec{i}+10\vec{j}\right)$, are distinct or coincident lines.

Solution 4.13: L_1 and L_2 are parallel because $4\vec{i}+10\vec{j}=2\left(2\vec{i}+5\vec{j}\right)$. Let $p=0$ then the point $\left(3,-2\right)$ lies on L_1 . Now, if L_2 is coincident with L_1 , the point $\left(3,-2\right)$ will also lie on L_2 . $x=3$, so $1+4q=3 \Rightarrow q=\dfrac{1}{2}$ and $y=-2$, so $8+10q=-2 \Rightarrow q=-1 \neq \dfrac{1}{2}$, so the values of q are inconsistent and therefore $\left(3,-2\right)$ does not lie on L_2 . Hence, L_1 and L_2 are distinct lines.

4.5.2 Intersecting and Non-Intersecting Lines

Consider two lines L_1 and L_2 with vector equations:

$$L_1 : \vec{r_1}=\left(2\vec{i}+2\vec{j}-\vec{k}\right)+p\left(\vec{i}-3\vec{j}+\vec{k}\right) \qquad L_2 : \vec{r_2}=\left(3\vec{i}-5\vec{j}+\vec{k}\right)+q\left(-3\vec{i}+5\vec{j}-2\vec{k}\right)$$

These lines intersect at a point P . Because P lies on both lines, the components of L_1 and L_2 at P are equal. Therefore, equating each component of L_1 and L_2 gives

$$2+p=3-3q \quad \ldots\ldots\ldots\ldots\ldots\ldots (1)$$
$$2-3p=-5+5q \quad \ldots\ldots\ldots\ldots\ldots .(2)$$
$$-1+p=1-2q \quad \ldots\ldots\ldots\ldots\ldots .(3)$$

Using (1) and (2) to solve for p and q gives $p+3q=1 \Rightarrow 3p+9q=3$, and $3p+5q=7$. Hence $q=-1$ and $p=4$. Now, if these values are substituted into (3), the left hand side is $-1+4=3$ and the right hand side is $1-2\left(-1\right)=3$ so that a consistent result is obtained. This indicates that, indeed, L_1 and L_2 intersect at P . The coordinates of P are found by substituting the value of p into the equation of L_1 : $x=2+p=2+4=6$, $y=2-3\times4=-10$, $z=-1+4=3$. Therefore, the coordinates of

P are $(6, -10, 3)$. Alternatively, P could be found by substituting the value of q into the equation of L_2.

Now consider the lines L_3 and L_4 with vector equations:

$$L_3 : \vec{r_3} = \left(\vec{i} + 3\vec{j} + 4\vec{k}\right) + p\left(4\vec{i} - 2\vec{j} + \vec{k}\right) \qquad L_4 : \vec{r_4} = \left(2\vec{i} + \vec{k}\right) + q\left(3\vec{i} + \vec{j} - 4\vec{k}\right)$$

We <u>assume</u> that these lines intersect at point Q. Then, at Q, the components of each line are equal and

$$1 + 4p = 2 + 3q \ldots \ldots \ldots \ldots (4)$$
$$3 - 2p = 0 + q \ldots \ldots \ldots \ldots (5)$$
$$4 + p = 1 - 4q \ldots \ldots \ldots \ldots (6)$$

Using (4) and (5) to solve for p and q gives $4p - 3q = 1$ and $2p + q = 3 \Rightarrow 6p + 3q = 9$. Hence $p = 1$, $q = 1$. Now, if these values are substituted into (6), the left hand side is $4 + 1 = 5$ but the right hand side is $1 - 4 = -3$ and an inconsistent result is obtained. Because $5 \neq -3$, the initial assumption that the lines intersect is false, and, therefore, L_3 and L_4 do not intersect.

Example 4.14: Lines l and m have vector equations $\vec{r} = \begin{pmatrix} 1 \\ 3 \\ 2 \end{pmatrix} + p\begin{pmatrix} -2 \\ c \\ 1 \end{pmatrix}$; $\vec{r} = \begin{pmatrix} 0 \\ 1 \\ 1 \end{pmatrix} + q\begin{pmatrix} 1 \\ 2 \\ -2 \end{pmatrix}$, where c

is a constant. Find the value of c such that l and m intersect.

Solution 4.14: Let the intersection point be P. Then, at P

$$1 - 2p = q \ldots \ldots \ldots \ldots (1)$$
$$3 + cp = 1 + 2q \ldots \ldots \ldots (2)$$
$$2 + p = 1 - 2q \ldots \ldots \ldots (3)$$

Using (1) and (3) to solve for p and q gives $2p + q = 1 \Rightarrow 4p + 2q = 2$. From (3), $p + 2q = -1$ and subtracting gives $3p = 3 \Rightarrow p = 1 \Rightarrow 2 \times 1 + q = 1 \Rightarrow q = -1$. Because l and m intersect, at the point of intersection, equation (1), (2) and (3) are consistent. Therefore, $3 + cp = 1 + 2q \Rightarrow 3 + c = 1 - 2 = -1 \Rightarrow c = -4$.

Exercise 4.7

1.   In each case, decide whether the given pair of lines intersects. If so, find the coordinates of the point of intersection.

(a) $\vec{r} = \left(\vec{i} + \vec{j} - 4\vec{k}\right) + p\left(-3\vec{i} + 2\vec{j} + \vec{k}\right)$; $\vec{r} = \left(7\vec{i} + 5\vec{j} + 10\vec{k}\right) + q\left(\vec{i} - \vec{j} - \vec{k}\right)$

(b) $\vec{r} = \left(2\vec{j} + \vec{k}\right) + p\left(\vec{i} + \vec{j} + \vec{k}\right)$; $\vec{r} = \left(-3\vec{i} - \vec{j} + 2\vec{k}\right) + q\left(2\vec{i} + 3\vec{k}\right)$

(c) $\vec{r} = \left(7\vec{i} + 2\vec{j} + \vec{k}\right) + 2p\vec{i}$; $\vec{r} = \left(2\vec{i} + 2\vec{j} + \vec{k}\right) + q\left(2\vec{i} + 2\vec{j} + \vec{k}\right)$

(d) $\vec{r} = p\left(3\vec{i} - 5\vec{j} + 8\vec{k}\right)$; $\vec{r} = \left(6\vec{i} - 11\vec{j}\right) + q\left(3\vec{i} - \vec{k}\right)$

2. In each case, find the value of c such that the given pair of lines intersects.

(a) $\vec{r} = \begin{pmatrix} 1 \\ -1 \\ 0 \end{pmatrix} + p\begin{pmatrix} c \\ 0 \\ 1 \end{pmatrix}$; $\vec{r} = \begin{pmatrix} -2 \\ -1 \\ 1 \end{pmatrix} + q\begin{pmatrix} 5 \\ 1 \\ 3 \end{pmatrix}$
(b) $\vec{r} = \begin{pmatrix} 0 \\ 0 \\ -1 \end{pmatrix} + p\begin{pmatrix} 1 \\ -1 \\ 1 \end{pmatrix}$; $\vec{r} = \begin{pmatrix} 3 \\ c \\ -3 \end{pmatrix} + q\begin{pmatrix} 2 \\ -1 \\ -3 \end{pmatrix}$

(c) $\vec{r} = \begin{pmatrix} 1 \\ 1 \\ 1 \end{pmatrix} + p\begin{pmatrix} 2 \\ 1 \\ 1 \end{pmatrix}$; $\vec{r} = \begin{pmatrix} 2 \\ 0 \\ 5 \end{pmatrix} + q\begin{pmatrix} 6 \\ c \\ 1 \end{pmatrix}$

3. In each case, test whether the given pair of parallel lines are distinct or coincident.

(a) $\begin{pmatrix} x \\ y \end{pmatrix} = \begin{pmatrix} 1 \\ 3 \end{pmatrix} + s\begin{pmatrix} -1 \\ 2 \end{pmatrix}$; $\begin{pmatrix} x \\ y \end{pmatrix} = \begin{pmatrix} 3 \\ 3 \end{pmatrix} + t\begin{pmatrix} -1 \\ 2 \end{pmatrix}$

(b) $\vec{r} = \left(2\vec{i} + 5\vec{k}\right) + s\left(3\vec{i} + 2\vec{j}\right)$; $\vec{r} = \left(-7\vec{i} - 6\vec{j} + 5\vec{k}\right) + t\left(3\vec{i} + 2\vec{j}\right)$

(c) $\vec{r} = \left(2\vec{i} - 5\vec{j} - \vec{k}\right) + s\left(\vec{i} - 2\vec{j} + 4\vec{k}\right)$; $\vec{r} = \left(-\vec{i} + \vec{j} - 13\vec{k}\right) + t\left(3\vec{i} - 6\vec{j} + 12\vec{k}\right)$

(d) $\vec{r} = 4\vec{j} + s\left(\vec{i} - \vec{j}\right)$; $\vec{r} = \left(2\vec{i} - 11\vec{j}\right) + t\left(-\vec{i} + \vec{j}\right)$

(e) $\vec{r} = \begin{pmatrix} 3 \\ 10 \\ -1 \end{pmatrix} + s\begin{pmatrix} -3 \\ 0 \\ 1 \end{pmatrix}$; $\vec{r} = \begin{pmatrix} -5 \\ 10 \\ 17 \end{pmatrix} + t\begin{pmatrix} 3 \\ 0 \\ -1 \end{pmatrix}$

4. P has coordinates $\left(1, 5, -2\right)$ and Q has coordinates $\left(-2, 8, -4\right)$.

(a) Write down $\overrightarrow{PQ}$

The line L_1 passes through points P and Q and the line L_2 has equation $\vec{r} = \begin{pmatrix} 1 \\ 5 \\ 3 \end{pmatrix} + t\begin{pmatrix} 1 \\ k \\ 1 \end{pmatrix}$

(b) Write down, in the form $\vec{r} = \vec{a} + s\vec{b}$, an equation for L_1.
(c) If L_1 and L_2 intersect at the point R find
    (i) the value of k.
    (ii) the coordinates of R.

5. In this question, units of distance are measured in kilometers.

An aircraft is flying in a straight line with vector equation $\vec{r} = \begin{pmatrix} 7 \\ 6 \\ 15 \end{pmatrix} + t\begin{pmatrix} 6 \\ 0 \\ 8 \end{pmatrix}$, $t \geq 0$ relative to a

set of Cartesian axes with an origin at sea level. The x-axis is due north, the y-axis is vertically upwards and the z-axis is due east. A missile is fired at the aircraft from a point P

whose coordinates are $(9.2,\ 0,\ 17)$ at time $t=0$, where t is measured in minutes. The

velocity of the missile, measured in kilometers per minute, is $\begin{pmatrix} 2 \\ 15 \\ 5 \end{pmatrix}$.

(a) Write down the vector equation of the path of the missile in terms of t.

(b) Show that the path of the missile and the path of the aircraft intersect at Q, whose coordinates are $(10,\ 6,\ 19)$.

(c) Find the time when the aircraft reaches Q and the time when the missile reaches Q. Hence show that the missile does not hit the aircraft.

Unit 4 Review Exercises

Part 1

No graphing calculators should be used to answer questions in Part 1.

1. Vectors $\vec{a}$, $\vec{b}$ and $\vec{c}$ are defined by $\vec{a} = \begin{pmatrix} 1 \\ 1 \\ 0 \end{pmatrix}$, $\vec{b} = \begin{pmatrix} 4 \\ 1 \\ 2 \end{pmatrix}$ and $\vec{c} = \begin{pmatrix} -3 \\ 2 \\ 6 \end{pmatrix}$. Given that $\vec{a}+t\vec{b}$ is

perpendicular to $\vec{c}$ find the value of t.

2. The line L passes through $A(1,\ 3)$ and $B(2,-2)$.

(a) Find the vector $\overrightarrow{AB}$

(b) Show that $\begin{pmatrix} 5 \\ 1 \end{pmatrix}$ is perpendicular to $\overrightarrow{AB}$.

L passes through A and is perpendicular to the line joining A and B.

(c) Write down a vector equation of the line L in the form $\vec{r}=\vec{a}+t\vec{b}$.

3. Let $\vec{a} = 6\vec{i} - 8\vec{j} + 2\vec{k}$ and $\vec{b} = \vec{i} + 4\vec{j} + p\vec{k}$.

(a) Find the value of p if $\vec{a}$ and $\vec{b}$ are perpendicular.

(b) Find the value of p and of q if $\vec{a}+q\vec{b}$ is parallel to the x-axis

4. The line L_1 is represented by $\vec{r}_1 = \begin{pmatrix} 1 \\ 1 \\ -1 \end{pmatrix} + s\begin{pmatrix} 2 \\ 0 \\ 3 \end{pmatrix}$ and the line L_2 by $\vec{r}_2 = \begin{pmatrix} 2 \\ -2 \\ 7 \end{pmatrix} + t\begin{pmatrix} 4 \\ 0 \\ c \end{pmatrix}$.

(a) Show that L_1 and L_2 do not intersect whatever the value of c.

(b) Find the value of c for which L_1 and L_2 are parallel.

5. The line L is parallel to the vector $\vec{v} = -3\vec{i} + \vec{j} + \vec{k}$ and passes through the point $A(4,\ 0,\ 5)$.

(a) Write down, in the form $\vec{r} = \vec{a} + t\vec{b}$, a vector equation of L.

(b) The point $B(2, -2,\ c)$ lies on L. Find the value of c.

(c) Determine whether the point $C(13, -3,\ 2)$ lies on L.

6. A vector equation for the line L is $\vec{r} = \begin{pmatrix} 1 \\ 2 \\ -3 \end{pmatrix} + s\begin{pmatrix} 4 \\ -1 \\ 2 \end{pmatrix}$. Which of the following lines

$$\text{A:}\ \vec{r} = \begin{pmatrix} 1 \\ 1 \\ 0 \end{pmatrix} + t\begin{pmatrix} 0 \\ 2 \\ 1 \end{pmatrix} \qquad \text{B:}\ \vec{r} = \begin{pmatrix} -3 \\ 3 \\ -5 \end{pmatrix} + t\begin{pmatrix} 8 \\ -2 \\ 4 \end{pmatrix} \qquad \text{C:}\ \vec{r} = \begin{pmatrix} 1 \\ 2 \\ -3 \end{pmatrix} + t\begin{pmatrix} 2 \\ -1 \\ 4 \end{pmatrix}$$

$$\text{D:}\ \vec{r} = \begin{pmatrix} 1 \\ 0 \\ -3 \end{pmatrix} + t\begin{pmatrix} 4 \\ -1 \\ 2 \end{pmatrix} \qquad \text{E:}\ \vec{r} = \begin{pmatrix} -7 \\ 3 \\ -4 \end{pmatrix} + t\begin{pmatrix} 2 \\ 0 \\ 1 \end{pmatrix}$$

is (i) distinct but parallel to (ii) coincident with (iii) perpendicular to (iv) intersects the line L.

7. Figure 4.48 shows a square $ABCD$. The line L_1 through A and C has vector equation

$\vec{r} = \begin{pmatrix} 4 \\ 4 \end{pmatrix} + s\begin{pmatrix} 4 \\ 3 \end{pmatrix}$. The point A has coordinates $(0,\ 1)$. The line L_2 through B and C has vector

equation $\vec{r} = \begin{pmatrix} 7 \\ 0 \end{pmatrix} + t\begin{pmatrix} 1 \\ 7 \end{pmatrix}$.

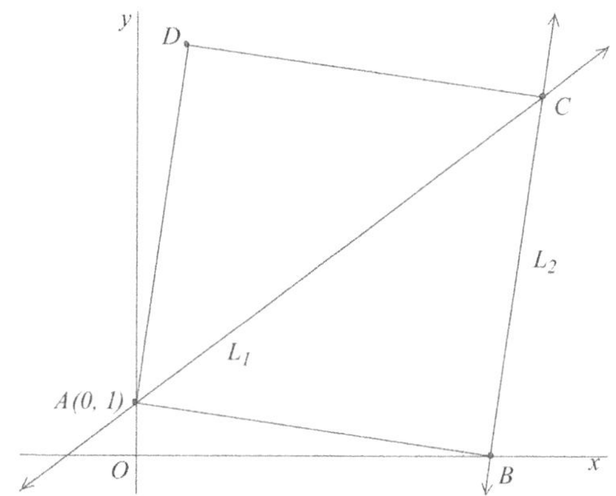

Figure 4.48

(a) Find the coordinates of C, the point where L_1 and L_2 intersect.

(b) Write down a vector parallel to the side $[AB]$ and hence find, in the form $\vec{r} = \vec{a} + p\vec{b}$, a vector equation of L_3, the line passing through A and B.

(c) Find the coordinates of B.

(d) Find the coordinates of D.

8. The line L_1 has equation $\vec{r} = \left(\vec{i} - 2\vec{k}\right) + s\left(2\vec{i} + \vec{j} + \vec{k}\right)$ and the line L_2 has equation

$$\vec{r} = \left(2\vec{i} - \vec{j}\right) + t\left(\vec{i} - \vec{j} - \vec{k}\right).$$

(a) Show that L_1 and L_2 are perpendicular

(b) Show that L_1 and L_2 do not intersect.

9. (a) If $\vec{a} = \begin{pmatrix} 1 \\ 2 \\ n \end{pmatrix}$ and $|\vec{a}| = 3$ find the possible values of n.

(b) If vector $\vec{p} = \begin{pmatrix} 1 \\ 3 \\ 1 \end{pmatrix}$, vector $\vec{q} = \begin{pmatrix} 2 \\ 1 \\ -5 \end{pmatrix}$ and vector $\vec{r} = \begin{pmatrix} 16 \\ m \\ n \end{pmatrix}$, find the values of m and n given

that $\vec{p}$, $\vec{q}$ and $\vec{r}$ are mutually perpendicular.

10. $\vec{a} = \vec{i} + 2\vec{j} - 2\vec{k}$, $\vec{b} = 3\vec{i} - 4\vec{j} + 6\vec{k}$ and $\vec{c} = -2\vec{i} + \vec{j} - 2\vec{k}$

(a) Find $\left(\vec{a} + \vec{b}\right) \cdot \vec{c}$

(b) Find $|\vec{a} + \vec{b}||\vec{c}|$

(c) Use your answers to (a) and (b) to show that $\vec{a} + \vec{b}$ is parallel to $\vec{c}$

Part 2

Graphing calculators will usually be needed to answer questions in Part 2.

11. $A(1,\ 3,\ 5)$ and $B(2, -1,\ 4)$ are points in three-dimensional space given relative to axes with origin O.

(a) Write down $\overrightarrow{AB}$

(b) Show that $\left|\overrightarrow{AB}\right| = 3\sqrt{2}$

Let $\overrightarrow{AC} = -\vec{i} + 2\vec{j} - 3\vec{k}$

(c) Find the coordinates of C

(d) Find the angle between $\overrightarrow{AB}$ and $\overrightarrow{AC}$.

12. The line L_1 passes through the point A with coordinates $(1,\ 2,\ -4)$ and the point B with coordinates $(3,\ 2,\ -5)$.

(a) Write down $\overrightarrow{AB}$

(b) Write a vector equation for L_1 in the form $\vec{r} = \vec{a} + s\vec{b}$

The line L_2 has vector equation $\vec{r} = \begin{pmatrix} -8 \\ 3 \\ -7 \end{pmatrix} + t \begin{pmatrix} -2 \\ 0 \\ 1 \end{pmatrix}$

(c) Explain why L_1 and L_2 are parallel.

(d) Show that the point C with coordinates $(-2, 3, -10)$ lies on L_2.

(e) Find $\overrightarrow{AC}$

(f) Show that $\overrightarrow{AC}$ is perpendicular to L_1 and L_2.

(g) Find the distance between L_1 and L_2.

13. Figure 4.49 shows the points $B(1, 8, 2)$ and $C(3, 3, 1)$ form two vertices of a parallelogram $OABC$, where O is the origin.

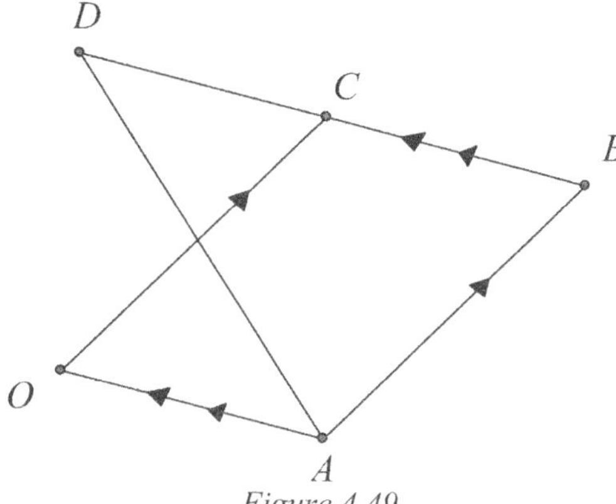

Figure 4.49

(a) Find the vector $\overrightarrow{BC}$

(b) Find (i) $\overrightarrow{OA}$ (ii) $B\hat{O}A$

The point D lies on the line passing through points B and C so that $\overrightarrow{BD} = 2\overrightarrow{BC}$.

(c) Find $\overrightarrow{AD}$.

14. Figure 4.50 shows a parallelogram $OABC$ in which A has coordinates $(1, 5)$ and C has coordinates $(8, 2)$.

(a) Write down $\overrightarrow{OA}$ and $\overrightarrow{OC}$

(b) Write down $\overrightarrow{OA} + \overrightarrow{OC}$ and hence find $\overrightarrow{OB}$.

$[OB]$ and $[AC]$ intersect at M.

(c) Find $A\hat{M}B$

(d) Show that the area of parallelogram $OABC$ is 38.

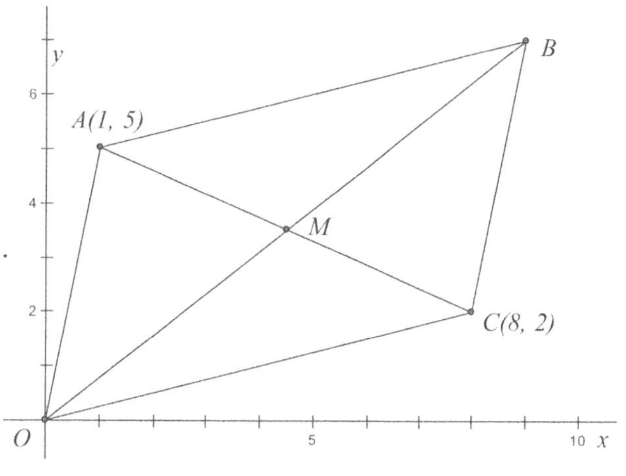

Figure 4.50

15. Three lines L_1, L_2, L_3 are defined as follows:

$$L_1 : \vec{r} = \begin{pmatrix} -2 \\ 3 \\ 4 \end{pmatrix} + p \begin{pmatrix} -4 \\ a \\ 1 \end{pmatrix}; \quad L_2 : \vec{r} = \begin{pmatrix} 6 \\ -4 \\ b \end{pmatrix} + q \begin{pmatrix} 0 \\ 1 \\ 2 \end{pmatrix}; \quad L_3 : \vec{r} = \begin{pmatrix} 2 \\ 3 \\ c \end{pmatrix} + s \begin{pmatrix} -1 \\ 1 \\ 3 \end{pmatrix}$$

(a) Find the values of a, b and c so that L_1, L_2 and L_3 all meet in a point P.

(b) Find the coordinates of P.

(c) Find the acute angle between L_1 and L_2.

16. A balloon is located on a horizontal plane at a point A with coordinates $(150, 250, 0)$ relative to a set of Cartesian axes whose origin is O. The x-axis is due east, the y-axis due north and the z-axis vertically upwards. Units of distance are measured in meters and time is measured in minutes. The balloon moves in a straight line at constant speed so that its position vector

$\overrightarrow{OP}$, after t minutes is given by the vector equation $\vec{r} = \begin{pmatrix} 150 \\ 250 \\ 0 \end{pmatrix} + t \begin{pmatrix} 200 \\ 200 \\ 100 \end{pmatrix}$, $t \geq 0$.

(a) Find the speed of the balloon.

(b) Find the value of t when the balloon reaches a height of 300 meters.

When the balloon reaches a height of 500 meters its velocity changes so that its position vector

is now $\vec{r} = \begin{pmatrix} 1150 \\ 1250 \\ 500 \end{pmatrix} + (t-5) \begin{pmatrix} -10 \\ -20 \\ -20 \end{pmatrix}$, $t \geq 5$ and continues at this velocity until it reaches the

ground.

(c) Find the length of time that the balloon was in the air.

(d) How far is the balloon from its starting position at A?

17. A harbor is located at the origin of coordinates $O(0, 0)$, and a rock is located at $A(10, 13)$, distances being measured in kilometers. A sail boat leaves harbor at 09:00 and travels with

constant velocity $\begin{pmatrix} 4 \\ 3 \end{pmatrix}$, measured in kilometers per hour. P is the position of the ship at time t,

where t is measured in hours, after 09:00

(a) Find $\overrightarrow{OP}$ and $\overrightarrow{AP}$ when $t = 1$, $t = 2$, $t = 5$.

(b) Find $\overrightarrow{AP}$ at time t.

(c) Find, correct to the nearest minute, when the ship is, for the first time, 15 km from the rock.

(d) Find, as a function of t, the scalar product $\overrightarrow{AP} \cdot \begin{pmatrix} 4 \\ 3 \end{pmatrix}$.

(e) Use part (d) to find, correct to the nearest minute, when the ship is nearest to the rock.

UNIT 5: DIFFERENTIATION

5.1 Rates of Change

5.1.1 Mean Velocity and Instantaneous Velocity

No doubt you are already familiar with the formula $v = \dfrac{s}{t}$, where v is velocity, s is distance and t is time. It is important to realize that this relation is only meaningful for constant or average (mean) velocity. The first part of this unit will be concerned with refinements to this relation which will enable you to find instantaneous velocity.

Consider a small balloon which is released from ground level at $s = 0$ and $t = 0$. The balloon rises vertically with gradually decreasing speed until it comes to rest, after 1 minute, at a height of 120 meters, after which the balloon bursts. The height of the balloon, s in meters, as a function of time, t in seconds, is $s(t) = 4t - \dfrac{1}{30}t^2$. Figure 5.01 shows the distance-time graph of the balloon.

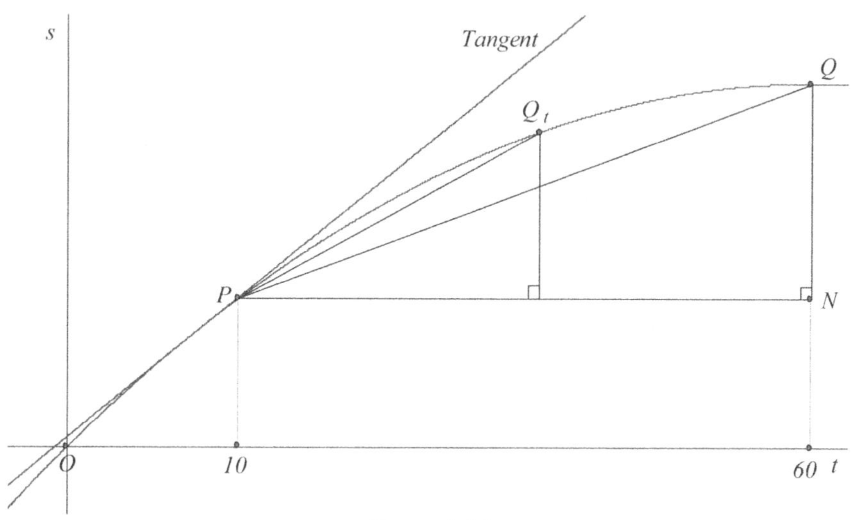

Figure 5.01

Suppose that we are interested in estimating the instantaneous velocity of the balloon at P when $t = 10$. In figure 5.01, the gradient of the chord $PQ = \dfrac{QN}{PN}$ represents the mean velocity of the balloon during the time interval from $t = 10$ to $t = 60$. Similarly, the gradient of the chord PQ_t represents the mean velocity during the time interval from $t = 10$ to an arbitrary time t. As $t \to 10$ so the chord PQ_t approaches the tangent to the curve at P, and the gradient of the tangent at P represents the instantaneous velocity at $t = 10$. This process will be explained in section 5.1.3. However, for now, we look at a more direct, if less illuminating, method of finding instantaneous velocity.

One way to get an accurate, but not necessarily exact, estimate of the gradient of a tangent is to use the graphing calculator. Graph the function $s(t) = 4t - \dfrac{1}{30}t^2$ for the domain $0 \le t \le 60$ and draw the tangent at $t = 10$. The equation of the tangent is shown at the bottom of the screen in the form

$y = mx + c$, so you can read off m, the gradient of the tangent, which is the instantaneous velocity at $t = 10$. In this case, $m = 3.33$, so the velocity at $t = 10$ is $3.33\,\text{ms}^{-1}$ (see figure 5.02).

t	v
0	4
10	3.333
20	2.667
30	2
40	1.333
50	0.667
60	0

Table 5.1

It is now possible to find the velocity function of the balloon by finding the instantaneous velocity at regular intervals along the domain and then by plotting these velocity values against the corresponding value of t. Table 5.1 shows v, the velocity of the balloon, at 10 second intervals.

Figure 5.02

This data can be entered into lists and plotted on a statistical plot. Then, a curve can be fitted to the points of the plot. The function, whose graph is this curve, represents the velocity function of the balloon. Figures 5.03 to 5.07 show how to find the approximate velocity function which is $v(t) = -0.0667t + 4.00$.

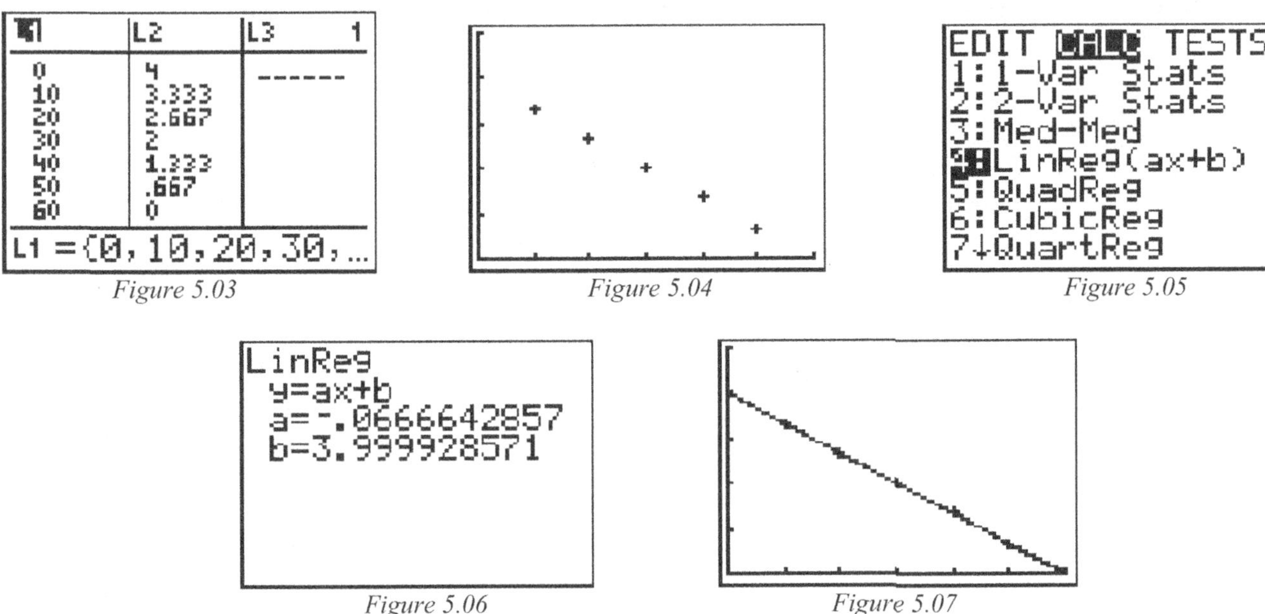

Figure 5.03

Figure 5.04

Figure 5.05

Figure 5.06

Figure 5.07

Velocity is an example of a *rate of change*. Other examples are acceleration (the rate of change of velocity with respect to time), gas consumption (the rate of consumption of gasoline with respect to distance traveled), and economic growth rate, (the rate at which the economy grows with respect to time). The velocity function is an example of a *gradient function* in the sense that it is a function which describes the gradient, or, rate of change of the displacement, as t varies.

5.1.2 Differentiation

The process of determining the gradient function from a function $f(x)$ is called *differentiation*. The gradient function of $f(x)$ is called the *derivative* and is denoted by $f'(x)$. We shall now find the derivatives of a number of simple functions.

If $f(x)=1$, then the graph of f is a line parallel to the x-axis. Therefore, the gradient of the graph is 0 for all values of x, so $f'(x)=0$. If $f(x)=x$, then the graph of f is a straight line, and the gradient is 1 so that $f'(x)=1$. Summarizing these two results, we obtain:

$$f(x)=1 \Rightarrow f'(x)=0$$
$$f(x)=x \Rightarrow f'(x)=1$$

For the function $f(x)=x^2$, it will be necessary to use the graphing calculator to find the gradient function in a manner similar to the one used to find the velocity function of the balloon in section 5.1.1 (figures 5.08 to 5.13):

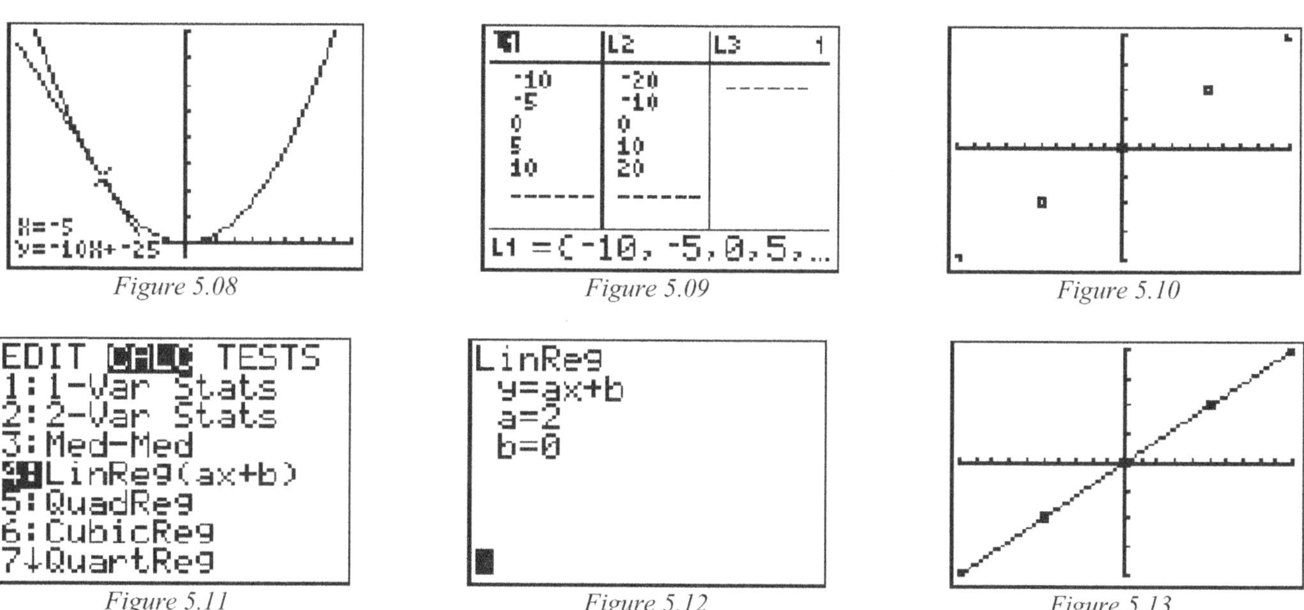

Figure 5.08   Figure 5.09   Figure 5.10

Figure 5.11   Figure 5.12   Figure 5.13

Therefore the derivative of $f(x)=x^2$ is $f'(x)=2x$.

Exercise 5.1

In questions $1-4$, use your graphing calculator to find the derivative of the following functions.

1. $f(x)=x^3$ (use quadratic regression)

2. $f(x)=3x^2+5x+9$ (use linear regression)

3. $f(x)=x^4$ (use cubic regression)

4. $f(x) = \sqrt{x}$, $x > 0$ (use power regression)

5.   Use your results from questions 1 and 2 to conjecture the derivative of
        $f(x) = ax^3 + bx^2 + cx + d$, where a, b, c and d are constants.

5.1.3 Limits

It is not always possible to find exact derivatives using your graphing calculator, and, in any case, this method does not really explain how the gradient of the tangent is calculated. This section shows how limits are used to find the exact gradient of a tangent by considering the gradient of a chord.

It is not possible to calculate the gradient of the tangent at a point on a curve using the same method that is used to calculate the gradient of a chord because you would end up with the fraction $\dfrac{0}{0}$, which is undefined. So, it is necessary to use a different technique to find the gradient of a tangent. This section investigates such a technique.

Enter the functions $f(x) = ((x+1)^2 - 1)/x$, $g(x) = (\sin x)/x$, $h(x) = 3(e^x - 1)/x$ into your graphing calculator and draw their graphs with a window of $-4 \leq x \leq 4$ and $-4 \leq y \leq 4$. All three graphs appear to be well-behaved and continuous over the domain shown. However, in all three cases, $f(0)$, $g(0)$ and $h(0)$ appear to be undefined because their denominators are zero, and yet the graphs seem to indicate that $f(0) = 2$, $g(0) = 1$ and $h(0) = 3$.

Instead of simply substituting $x = 0$ into these functions, let's see what happens to $f(x)$ if we move x towards 0 ($x \to 0$, in mathematical notation):

$$f(x) = \frac{(x+1)^2 - 1}{x} = \frac{x^2 + 2x + 1 - 1}{x} = \frac{x^2 + 2x}{x} = \frac{x(x+2)}{x} = x + 2$$

Then, as $x \to 0$, so $f(x) \to 2$. This is called a limiting process and is written $\lim_{x \to 0} f(x) = 2$.

Let's see how this works for a specific function, $f(x) = x^2$.

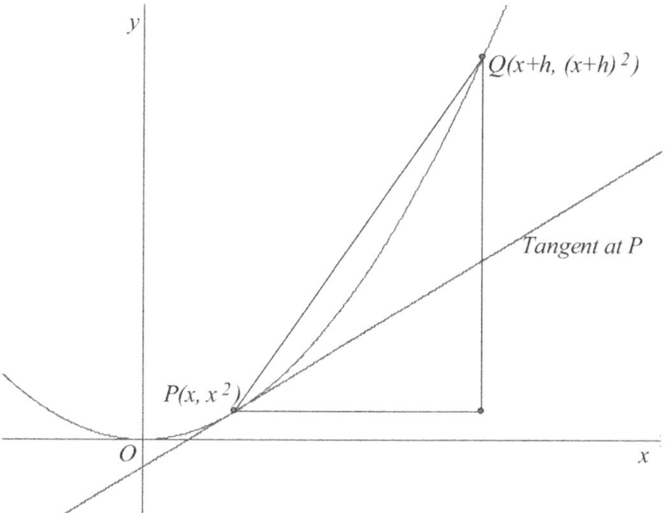

Figure 5.14

$$f(x) = x^2 \Rightarrow f'(x) = \lim_{h \to 0}\left(\frac{f(x+h) - f(x)}{(x+h) - x} \right) = \lim_{h \to 0}\left(\frac{(x+h)^2 - x^2}{(x+h) - x} \right)$$

$$= \lim_{h \to 0}\left(\frac{x^2 + 2xh + h^2 - x^2}{h} \right) = \lim_{h \to 0}\left(\frac{2xh + h^2}{h} \right)$$

$$= \lim_{h \to 0}\left(\frac{h(2x+h)}{h} \right) = \lim_{h \to 0}(2x+h) = 2x$$

The derivative of $f(x) = x^3$ can be found in a similar manner:

$$f'(x) = \lim_{h \to 0}\left(\frac{f(x+h) - f(x)}{h} \right) = \lim_{h \to 0}\left(\frac{(x+h)^3 - x^3}{h} \right) = \lim_{h \to 0}\left(\frac{x^3 + 3x^2h + 3xh^2 + h^3 - x^3}{h} \right)$$

$$= \lim_{h \to 0}\left(\frac{3x^2h + 3xh^2 + h^3}{h} \right) = \lim_{h \to 0}\left(\frac{h(3x^2 + 3xh + h^2)}{h} \right) = \lim_{h \to 0}(3x^2 + 3xh + h^2) = 3x^2$$

Exercise 5.2
Use the method of limits to find the derivative of the following functions:

1. $f(x) = x^2 + 3x$ 2. $f(x) = 3 - 2x$ 3. $f(x) = x^3 + 4x^2$

5.1.4 Differentiation of Polynomial Functions

If you have worked through sections 5.1.2 and 5.1.3, you will have noticed that a certain predictable pattern has emerged relating $f'(x)$ to $f(x)$, namely that

$$f(x) = x^n \Rightarrow f'(x) = nx^{n-1} \text{ for } n \in \mathbb{Q} \ldots\ldots\ldots\ldots (1)$$
$$f(x) = ax^n \Rightarrow f'(x) = a\left(nx^{n-1}\right) = nax^{n-1} \ldots\ldots\ldots\ldots (2)$$
$$f(x) = x^n + x^m \Rightarrow f'(x) = nx^{n-1} + mx^{m-1} \ldots\ldots\ldots\ldots (3)$$

Using the statements (1), (2) and (3), you are now able to simply write down the derivatives of polynomial functions.

Example 5.1: Differentiate $f(x) = 3x^5 + 7x^2 + 1$.

Solution 5.1: $f'(x) = 3(5x^4) + 7(2x) + 0 \Rightarrow f'(x) = 15x^4 + 14x$

Example 5.2: Differentiate $f(x) = \dfrac{3}{x^2} + \dfrac{5}{x} - 11x^3$, $x \neq 0$.

Solution 5.2: First it is necessary to write f in a form suitable for differentiating.
$$f(x) = 3x^{-2} + 5x^{-1} - 11x^3 \Rightarrow f'(x) = 3\left(-2x^{-3}\right) + 5\left(-1x^{-2}\right) - 11\left(3x^2\right)$$
$$\Rightarrow f'(x) = -6x^{-3} - 5x^{-2} - 33x^2 = -\frac{6}{x^3} - \frac{5}{x^2} - 33x^2, \ x \neq 0$$

Example 5.3: Differentiate $f(x) = \dfrac{2}{\sqrt{x}}$, $x > 0$.

Solution 5.3: $f(x) = 2x^{-\frac{1}{2}}$ so $f'(x) = 2\left(-\dfrac{1}{2}x^{-\frac{3}{2}}\right) = -\dfrac{1}{\sqrt{x^3}}$, $x > 0$

Example 5.4: Find the gradient of the tangent to the graph of $f(x) = x^4 - 2x^3 - 21x^2$ at the point whose coordinates are $(-4, \ 48)$.

Solution 5.4: Firstly it is necessary to find the derivative, $f'(x)$ and then the required gradient is $f'(-4)$. $f'(x) = 4x^3 - 2(3x^2) - 21(2x) = 4x^3 - 6x^2 - 42x$
$$\Rightarrow f'(-4) = 4(-4)^3 - 6(-4)^2 - 42(-4) = -184$$

Exercise 5.3
1.   Write down $f'(x)$ for each of the given functions.

(a) $f(x) = x^7$ 　　　　　　(b) $f(x) = 9x^5$ 　　　　　　(c) $f(x) = 2 + x - 3x^3$

(d) $f(x) = 3(x^2 - 2)$ 　　　(e) $f(x) = 2x^4 + 3x^2 - 6x + 13$ 　　　(f) $f(x) = \dfrac{1}{3}x^5 - \dfrac{2}{5}x^3 - \dfrac{1}{2}x$

(g) $f(x) = \dfrac{x^3 + 5}{3}$ (h) $f(x) = x - \dfrac{3}{x}$,   (i) $f(x) = x^5 - \dfrac{1}{x^5}$

(j) $f(x) = x^{2k+1}$

2.   Differentiate the following functions.

    (a) $g(x) = 3\sqrt{x}$ (b) $g(x) = x^4 + \dfrac{1}{x}$ (c) $g(x) = \dfrac{1}{x^3} - \dfrac{2}{x} + 3$ (d) $g(x) = \sqrt{x} + \dfrac{1}{\sqrt{x}}$

3.   Find the gradient of the tangent to the graphs of each of the given functions at the point where $x = 2$.

    (a) $f(x) = x^4$ (b) $f(x) = 3 + x^4$ (c) $f(x) = x^2 - 9x + 4$

    (d) $f(x) = \dfrac{1}{3x^2}$ (e) $f(x) = 4\sqrt{x}$

5.1.5 Differentiation Using the $\dfrac{dy}{dx}$ Notation

You are probably already familiar with three different notations for a function. We take, as an example, the function which maps x to $x^2 + 5x + 11$ and show the appropriate notation for the derivative:

$$f : x \mapsto x^2 + 5x + 11 \Rightarrow f' : x \mapsto 2x + 5$$

$$f(x) = x^2 + 5x + 11 \Rightarrow f'(x) = 2x + 5$$

$$y = x^2 + 5x + 11 \Rightarrow \frac{dy}{dx} = 2x + 5$$

If a function is written in the form '$y =$', then the derivative is written '$\dfrac{dy}{dx} =$'. What follows shows how this notation arises using the example, $y = x^2 + 5x + 11$. Δx means 'a small increase in x,' and Δy means 'a small corresponding increase in y.'

In figure 5.15, the gradient of $PQ = \dfrac{QN}{PN} = \dfrac{\Delta y}{\Delta x}$, but $QN = (x + \Delta x)^2 + 5(x + \Delta x) + 11 - (x^2 + 5x + 11)$

and $PN = (x + \Delta x) - x \Rightarrow \dfrac{\Delta y}{\Delta x} = \dfrac{\left((x + \Delta x)^2 + 5(x + \Delta x) + 11\right) - (x^2 + 5x + 11)}{(x + \Delta x) - x}$

$$= \frac{x^2 + 2x\Delta x + (\Delta x)^2 + 5x + 5\Delta x + 11 - x^2 - 5x - 11}{\Delta x}$$

$$= \frac{2x\Delta x + (\Delta x)^2 + 5\Delta x}{\Delta x} = \frac{\Delta x(2x + 5 + \Delta x)}{\Delta x} = 2x + 5 + \Delta x$$

As $\Delta x \to 0$ and Q moves down the curve towards P, so the chord PQ becomes, by the limiting process, the tangent at P. Therefore,

$$\lim_{\Delta x \to 0}\left(\frac{\Delta y}{\Delta x}\right) = \lim_{\Delta x \to 0}(2x+5+\Delta x) = 2x+5 \text{ and}$$

$$\lim_{\Delta x \to 0}\left(\frac{\Delta y}{\Delta x}\right) \text{ is abbreviated to } \frac{dy}{dx}.$$

Thus, the derivative of $y = x^2 + 5x + 11$ is

written $\dfrac{dy}{dx} = 2x+5$

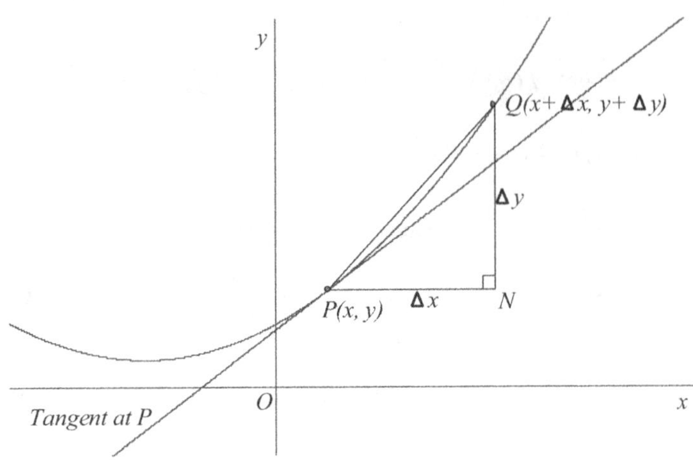

Figure 5.15

Example 5.5: Differentiate $y = 4x^3 - 5x^2 + 3x + 2$.

Solution 5.5: When a function is given in the ' $y =$ ' form, you should give the derivative in the ' $\dfrac{dy}{dx} =$ ' form so, in this case, $\dfrac{dy}{dx} = 4(3x^2) - 5(2x) + 3 = 12x^2 - 10x + 3$.

Example 5.6 Find the derivative of $y = (3x+1)(x^2+2)$.

Solution 5.6: As you have not yet learned how to differentiate the product of functions, it is necessary to multiply out the function before trying to differentiate it:

$$y = (3x+1)(x^2+2) = 3x^3 + x^2 + 6x + 2 \implies \frac{dy}{dx} = 3(3x^2) + 2x + 6 = 9x^2 + 2x + 6$$

Example 5.7 The line with equation $y = 2x - 10$ is tangent to the graph with equation $y = x^2 + 4x - 9$ at the point $(a,\ b)$. Find the values of a and b .

Solution 5.7: Firstly, you need to find the gradient of the curve at $x = a$ and put this equal to 2, which is the gradient of the tangent. $\dfrac{dy}{dx} = 2x + 4$, so that $2a + 4 = 2 \implies a = -1$. Then,

$$b = a^2 + 4a - 9 \implies b = (-1)^2 + 4(-1) - 9 = -12 .$$ Therefore, $a = -1$ and $b = -12$.

Exercise 5.4
1.   Differentiate the following functions.
   (a) $y = x^4$ (b) $y = 4x^6$ (c) $y = 3 - 7x^3$ (d) $y = 12x^4 - 7x^3 - 3x^2 + 5x - 5$

   (e) $y = \dfrac{4}{x^3} + \dfrac{3}{x^2} - \dfrac{2}{3x}$ (f) $y = 3\sqrt{x} + \sqrt{x^3} + \sqrt{x^5}$

2. Differentiate the following functions.

(a) $y = x^3(x+3)$ (b) $y = (2x^2 + 3)(x-2)$ (c) $y = x(x^2 + 1)^2$ (d) $y = \dfrac{x+3}{x}$

3. Differentiate the following functions.

(a) $f(x) = (1 - 3x)(1 + x)$ (b) $f(x) = x(x^3 + x^2 + x + 1)$ (c) $f(x) = (x+8)(3 - x^3)$

(d) $f(x) = \dfrac{x^3 + x^2 + 1}{\sqrt{x}}$ (e) $f(x) = \left(x + \dfrac{1}{2x}\right)^2$

4. Find the gradient of the tangent to the curve with the given equation at the given point.

(a) $y = x^2 + 5$, (3, 14)   (b) $f(x) = \dfrac{x+4}{x^3}$ (1, 5)   (c) $g(x) = 3\sqrt{x} - \dfrac{1}{x}$, $\left(\dfrac{1}{4}, \dfrac{5}{2}\right)$

(d) $y = x^3 + 5x + 1$, $(-2, -17)$

5. Find the coordinates of the point(s) on the curve with the given equation where the tangent has the given gradient.

(a) $y = x^3$; 48   (b) $f(x) = x^2 + 5x + 6$; 8   (c) $g(x) = \dfrac{4+x}{x}$; -1

6. A curve has equation $y = ax^2 + bx$, and $3x + y + 11 = 0$ is the equation of the tangent to the curve at the point $(-2, -5)$. Find the value of a and b.

7. $g(x) = mx + c$ is the equation of the tangent to the curve with equation $f(x) = 2x^2 - 7x + 3$ at the point whose coordinates are (5, 18). Find the value of m and c.

8. The line with equation $y = x + 4$ is tangent to the curve with equation $y = x^2 + k$. Find the value of k.

5.2 Graphical Behavior of Functions

5.2.1 The Second Derivative

The second derivative, written $f''(x)$, describes the way in which the gradient function, $f'(x)$ changes. Figure 5.16 shows a portion of the graph of $y = f(x)$. As x increases, the gradient becomes less negative; then, passing through zero the gradient becomes more and more positive.

Therefore, the rate of change of the gradient is positive, and so for a section of the graph

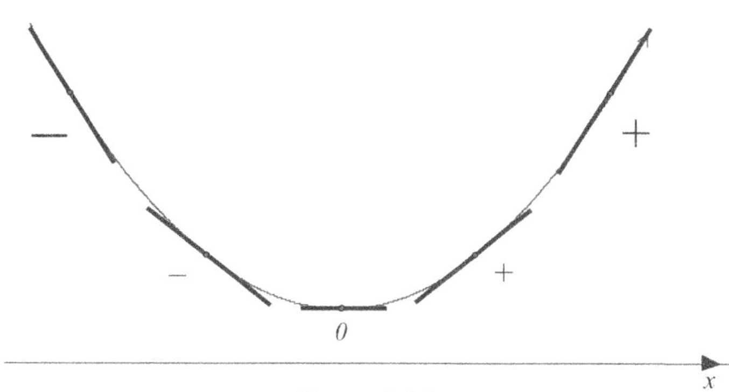

Figure 5.16

163

which has a 'concave up' shape (figure 5.17), the second derivative is positive and $f''(x) > 0$.

In figure 5.18, as x increases, the gradient becomes less positive. Passing through zero, the gradient becomes more negative.

The rate of change of the gradient is therefore negative, and, for a section of the graph which has a 'concave down' shape (figure 5.19), the second derivative is negative and $f''(x) < 0$.

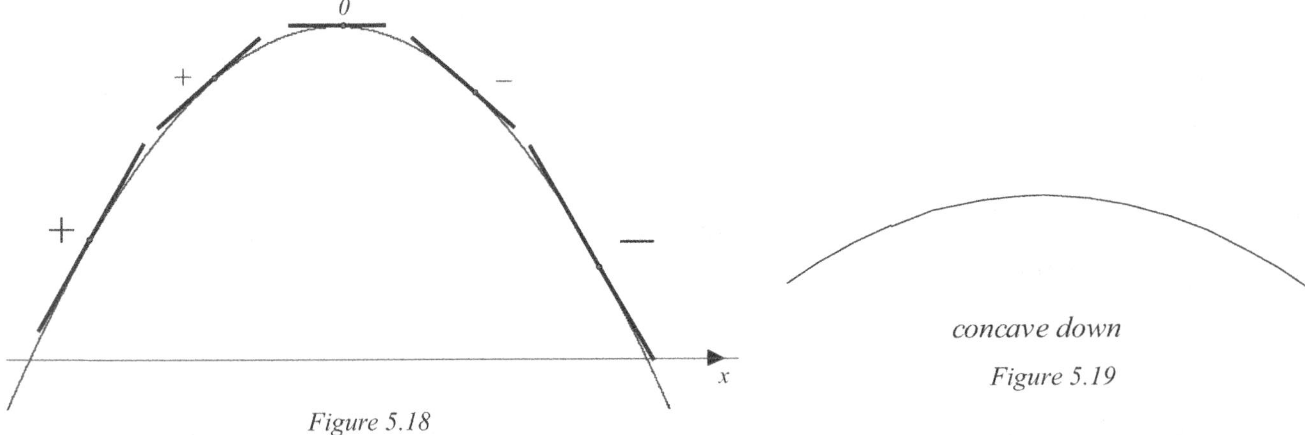

Figure 5.17

Figure 5.18

concave down

Figure 5.19

5.2.2 Local Maximum and Minimum Points

Figure 5.20 shows a point, P with coordinates $\left(x_1, f\left(x_1\right)\right)$ on the graph of a function. P is a local maximum point, in the sense that at that point the function has a larger value than all points nearby on the curve. P is called a 'local maximum' rather than a 'maximum' because there may be other points on the graph of the function where the function has a greater value, such as point R. It should be clear that the gradient of the tangent at P is zero, so that $f'(x_1) = 0$. The curve close to P is 'concave down', and therefore the value of $f''(x_1) < 0$.

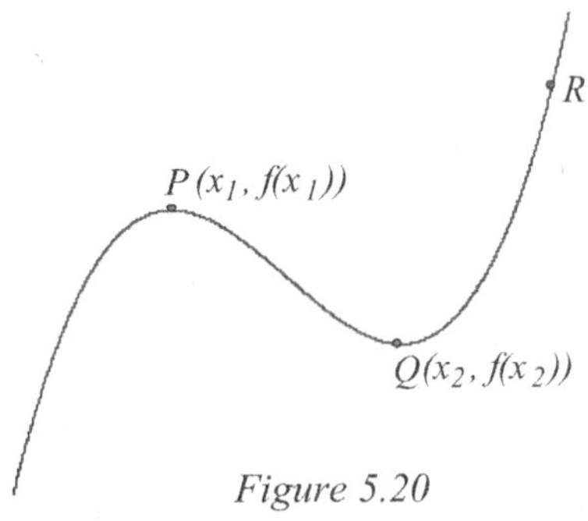

Figure 5.20

Similarly, Q is a local minimum point because the value of the function at Q is less than all other nearby points. Again, the gradient of the tangent at Q is zero so that $f'(x_2) = 0$. However, now the shape of the curve through Q is 'concave up', so that $f''(x_2) > 0$.

5.2.3 Points of Inflection

Figure 5.21 shows a section of the graph of $y = f(x)$.

Consider a point moving from left to right in the direction of increasing values of x. To begin with, as it moves through the 'concave down' section, the second derivative is negative. Then, further to the right it moves into the 'concave up' section of the curve, and the second derivative is positive. Because the graph is continuous as the point moves from left to right, it must pass through a point where the value of the second derivative changes from negative to positive. At that point, $f''(x) = 0$. This point is called a *point of inflection.*

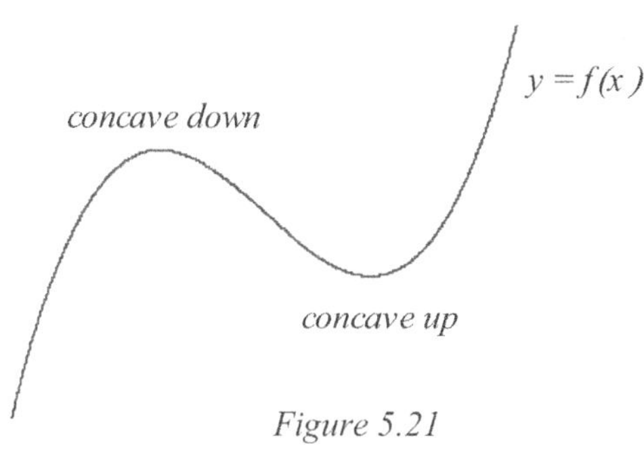

Figure 5.21

Figure 5.22 shows the position of point P, a point of inflection where $f''(x) = 0$. A point of inflection is where the curve changes from being 'concave down' to 'concave up' or from being 'concave up' to being 'concave down'.

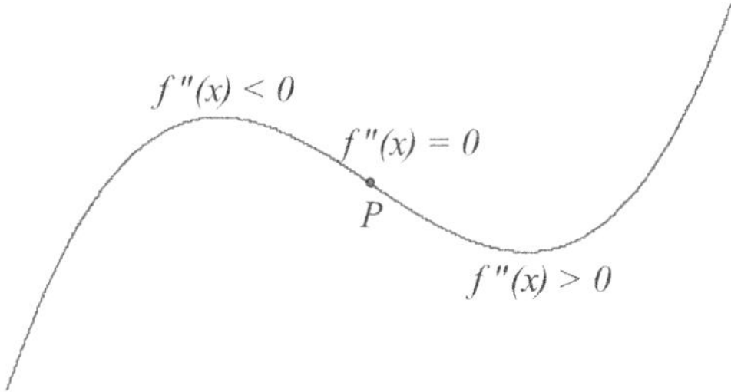

Figure 5.22

While it is true that if an inflection occurs at $x = a$ then $f''(a) = 0$, it is not necessarily true that if $f''(a) = 0$ then the graph of $y = f(x)$ has an inflection at $x = a$. A counter-example is the function $f(x) = x^4$, for which $f''(0) = 0$ but, at $x = 0$, there is a minimum value not an inflection.

Example 5.8: Consider figure 5.23 and then use the terms > 0, < 0, or $= 0$ to make a true statement about

(a) $f(a)$

(b) $f'(a)$

(c) $f''(a)$

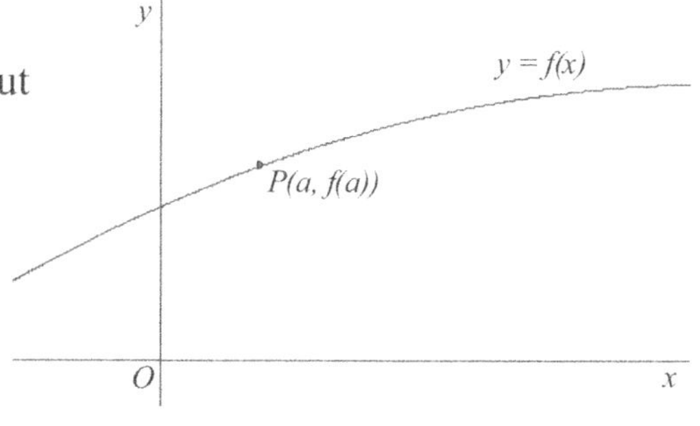

Figure 5.23

Solution 5.8: (a) As P is above the x-axis, $f(a) > 0$.

(b) The gradient of the tangent at P is positive; therefore, $f'(a) > 0$.

(c) The curve is 'concave down'; therefore, $f''(a) < 0$

5.2.4 The Derivative of a Graphically Defined Function

The derivative of the function $y = f(x)$, defined only in terms of its graph, can be drawn by considering the gradient of the tangent to $y = f(x)$ at points on the graph.

The graph of the derivative $y = f'(x)$ is such that its y-coordinate at $x = a$ is equal to the gradient of the curve of $y = f(x)$ at $x = a$. Figure 5.24 shows an example.

The gradient of the tangent at point P_i on the graph of $f(x)$ at the top of figure 5.24 is m_i, and m_i becomes the y-ordinate of the corresponding point T_i (x_i, m_i) in the graph at the bottom of figure 5.24. This need only be done for a few points, P_i and then they can be joined by a smooth curve to give an estimate of the graph of $f'(x)$.

Notice that at P_2 and P_4, the gradient is zero so that in the graph of $f'(x)$, the points T_2 and T_4 have a y-ordinate of zero and so are positioned on the x-axis.

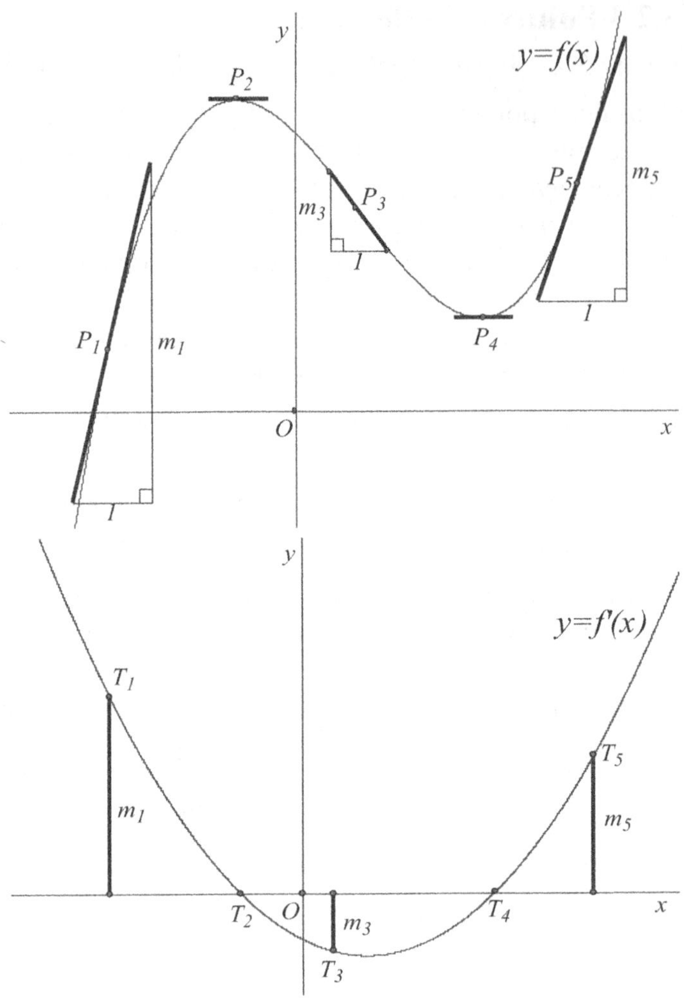

Figure 5.24

Exercise 5.5

1. In each case, use one of the terms, > 0, < 0, or $= 0$, to make a true statement about

   (a) $f(a)$, (b) $f'(a)$, and (c) $f''(a)$.

(i)

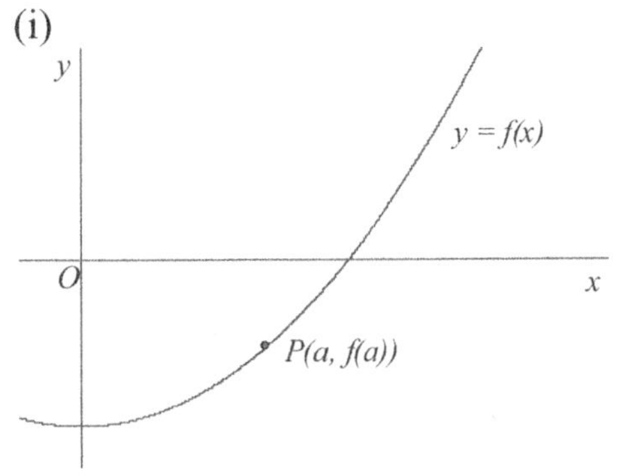

(ii)

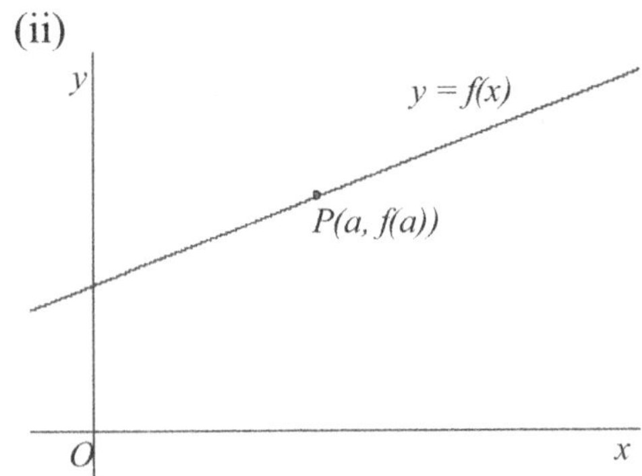

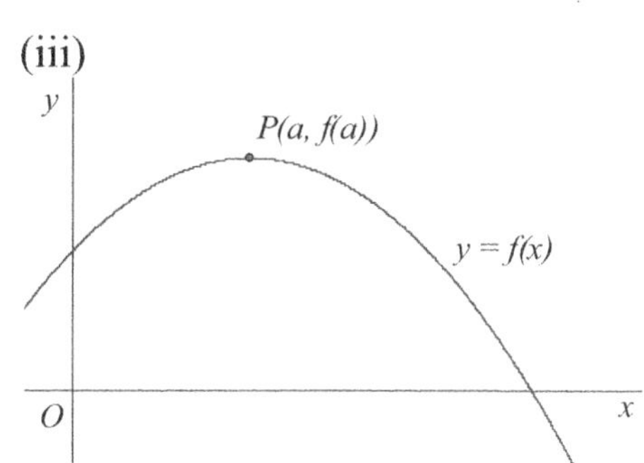

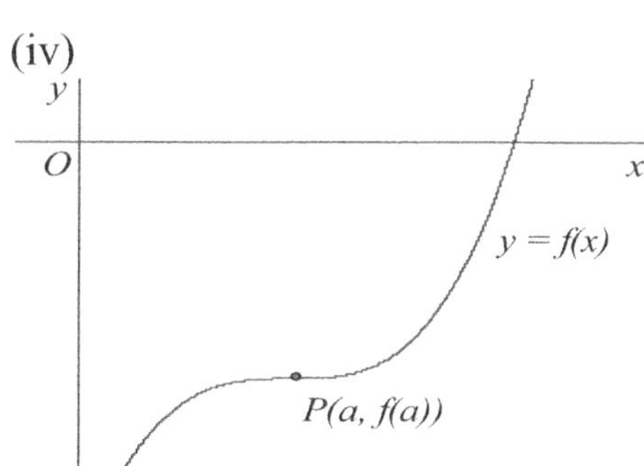

2. In each case, draw a section of the graph $y = f(x)$ close to the point $P(a, f(a))$ so that $f(a), f'(a)$ and $f''(a)$ have the given values.

(a) $f(a) > 0, f'(a) < 0, f''(a) = 0$ (b) $f(a) = 0, f'(a) = 0, f''(a) > 0$

(c) $f(a) < 0, f'(a) > 0, f''(a) > 0$ (d) $f(a) > 0, f'(a) = 0, f''(a) < 0$

3. The graphs in the right hand column are, not necessarily in the correct order, the derivatives of the graphs in the left hand column. Match each graph (a) to (d) with its derivative, (i) to (iv).

(a)

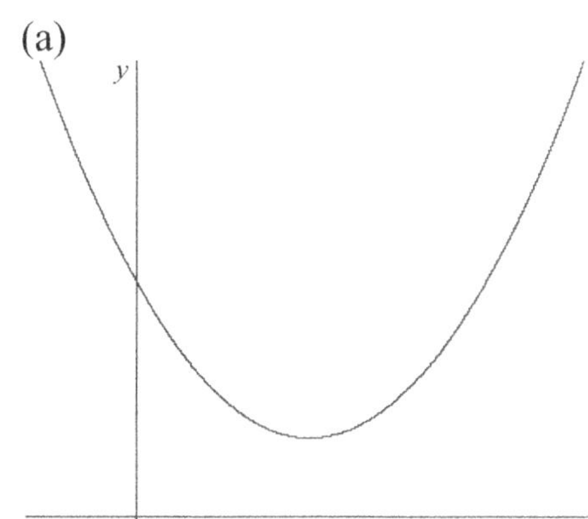

(i)

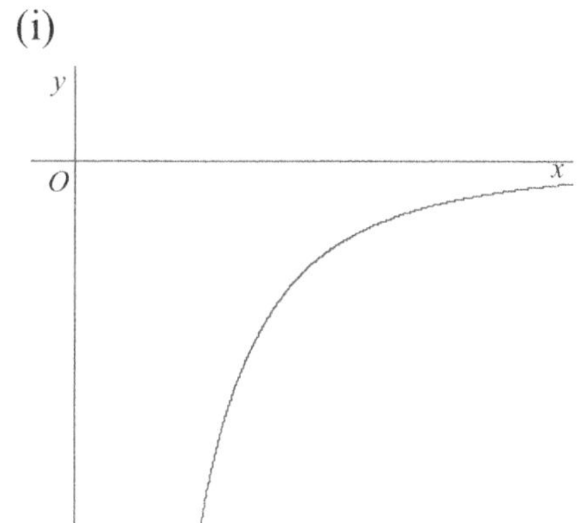

(b)

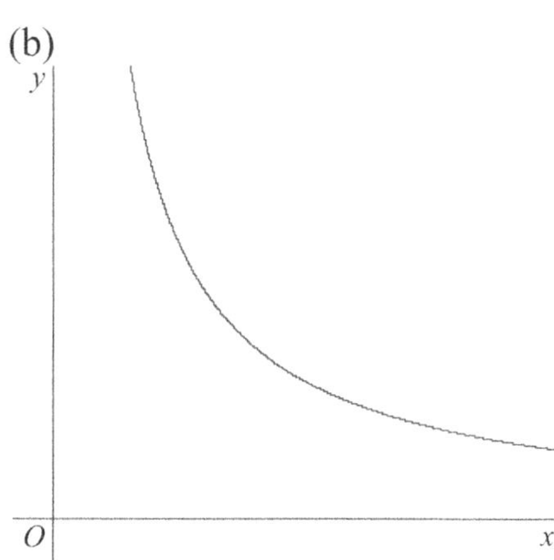

(ii)

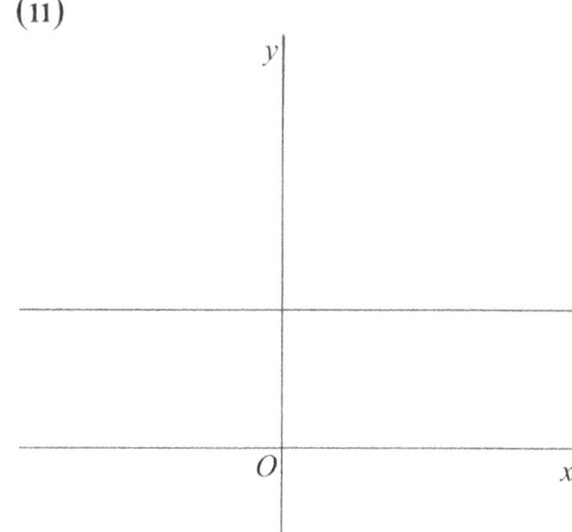

(c)

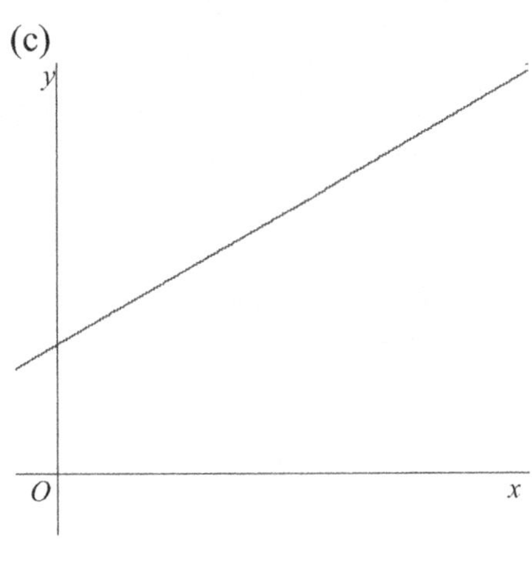

(iii)

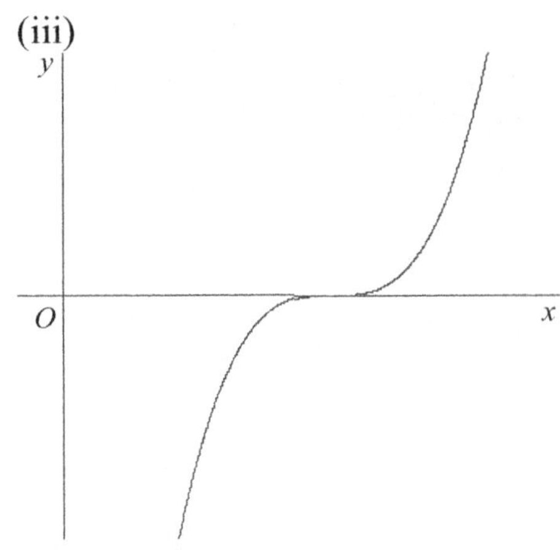

(d)

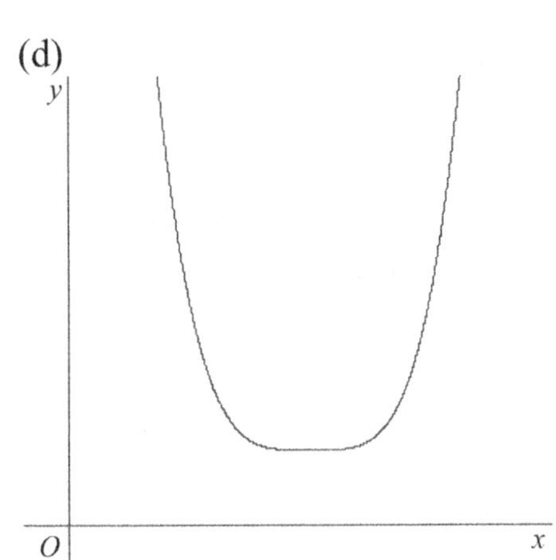

(iv)

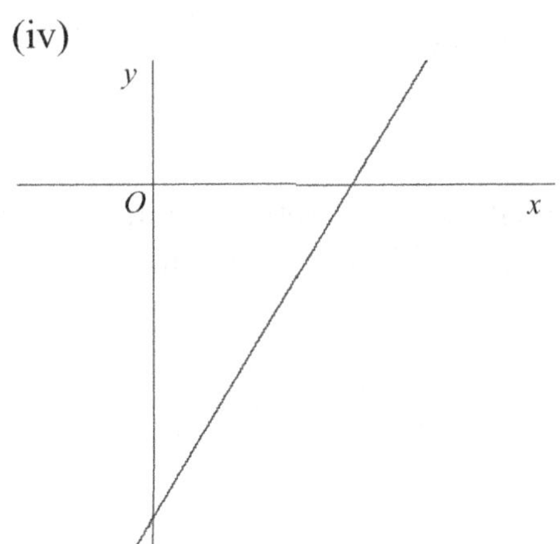

4.   Copy the graphs shown below and underneath each draw the derivative, as in figure 5.24.

(a)

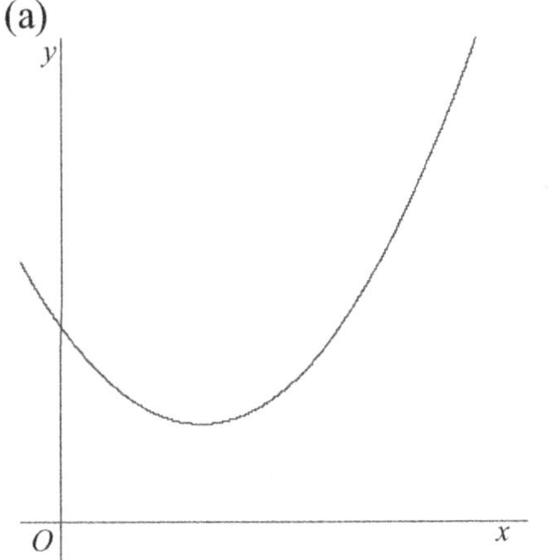

(b)

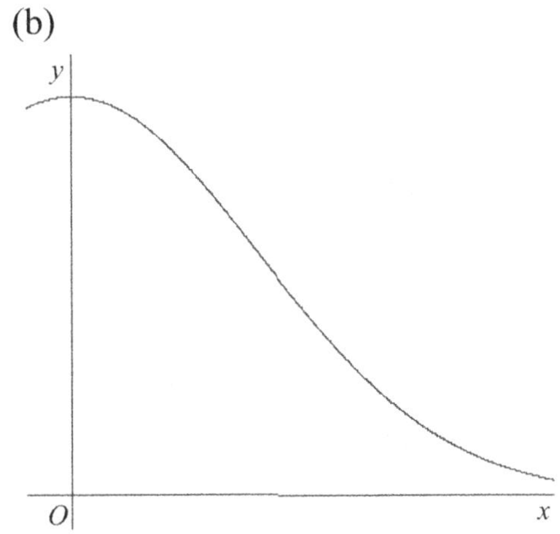

(c)

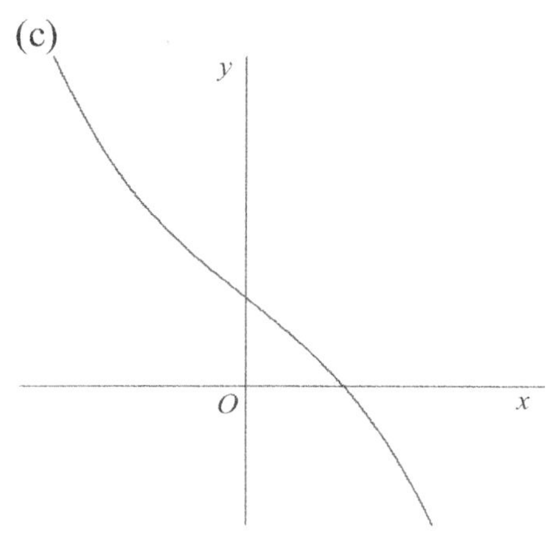

(d)

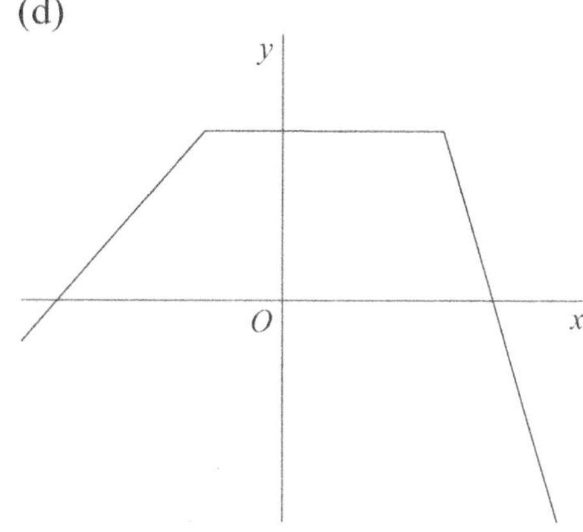

5.3 Applications of Differentiation 1

5.3.1 Maximum and Minimum Values of a Function

Figure 5.25 shows two parts of a graph of the function $y = f(x)$ containing a local maximum point at P and a local minimum point at Q. Maximum and minimum points on a curve are known collectively as *turning points*.

This section looks at two methods of identifying turning points as having either maximum values or minimum values.

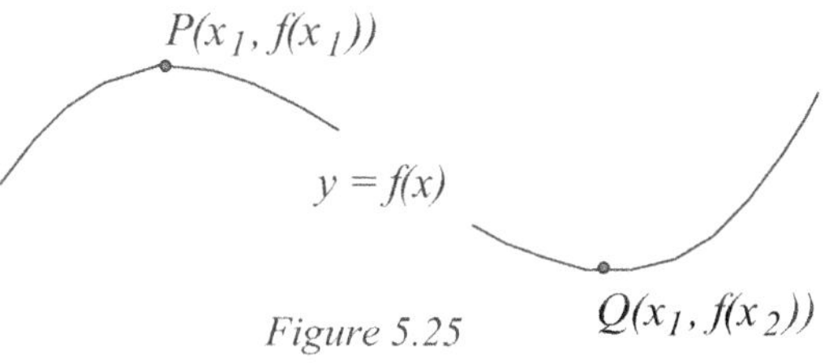

Figure 5.25

Method 1 - Using the second derivative to identify turning points:

In section 5.2 we showed that, at a local maximum point $P\left(x_1, f\left(x_1\right)\right)$ on the graph of $y = f(x)$, $f'(x_1) = 0$ and $f''(x_1) < 0$. And we also showed that, at a minimum point $Q\left(x_2, f\left(x_2\right)\right)$, $f'(x_2) = 0$ and $f''(x_2) > 0$ (figure 5.25).

This enables us to find the position of maximum and minimum values of a function. Consider the function $f(x) = x^3 - 3x^2 + 9 \Rightarrow f'(x) = 3x^2 - 6x \Rightarrow f''(x) = 6x - 6$.

At turning points, we know that $f'(x) = 0 \Rightarrow 3x^2 - 6x = 0 \Rightarrow 3x(x-2) = 0 \Rightarrow x = 0,\ x = 2$.

At $x = 0$, $f(0) = 0^3 - 3 \times 0^2 + 9 = 9$ and $f''(0) = 6 \times 0 - 6 = -6 < 0$; therefore, the function has a maximum value of 9 when $x = 0$.

At $x = 2$, $f(2) = 2^3 - 3(2)^2 + 9 = 5$ and $f''(2) = 6 \times 2 - 6 = 6 > 0$; therefore, the function has a minimum value of 5 when $x = 2$.

Method 2 - Using a table of first derivative signs to identify turning points

It is also useful sometimes to be able to identify maximum and minimum values of a function without having to find the second derivative. This can be done by investigating the gradient of the tangent at points close to, and on either side of, the turning point.

Consider any turning points of the function $y = x\left(\sqrt{x} - 6\right)$, $x > 0 \Rightarrow y = x^{\frac{3}{2}} - 6x \Rightarrow \dfrac{dy}{dx} = \dfrac{3}{2}x^{\frac{1}{2}} - 6$.

At turning points $\dfrac{dy}{dx} = 0 \Rightarrow \dfrac{3}{2}x^{\frac{1}{2}} - 6 = 0 \Rightarrow 3x^{\frac{1}{2}} - 12 = 0 \Rightarrow x^{\frac{1}{2}} = 4 \Rightarrow x = 16$. Therefore, a turning

point occurs at $x = 16$, and we investigate the gradient of the curve on either side of this point. Table 5.2 shows this investigation in tabular form.

x	15	16	17
$\dfrac{dy}{dx}$	−0.1905	0	0.1847
Tangent	↘	→	↗

Table 5.2

The tangents in the bottom row show the rough shape of the curve close to the turning point; therefore, y has a minimum value when $x = 16$.

At $x = 16$, $y = 16\left(\sqrt{16} - 6\right) = 16(4 - 6) = -32$; so, there is a minimum value of -32 when $x = 16$.

You may ask,

- why identify maximum and minimum points in the ways described when you could simply enter the function into your graphing calculator and observe the graph displayed on the screen?

- why distinguish a maximum from a minimum point using the sign of the derivative when it appears obvious that the point with greater y-ordinate is the maximum and the point with the lower y-ordinate the minimum?

These questions may best be answered by entering the function $f(x) = x + \dfrac{1}{100x}$, $x \neq 0$ into your graphing calculator with a domain of $-10 \leq x \leq 10$. The calculator image gives no hint of a maximum or minimum. Only when you change the window will you see the maximum and minimum. You then realize that the local minimum value of the function is greater than the local maximum value of the function.

Example 5.9:   The function f is defined by $f(x) = x^2 + \dfrac{2}{x}$, $x > 0$.

    (a) Find the derivative of f.

    (b) Solve the equation $f'(x) = 0$. Hence find the value of f where there is a turning point.

    (c) Use a table to show that this value is a minimum as opposed to a maximum value.

Solution 5.9:   (a) $f(x) = x^2 + 2x^{-1} \Rightarrow f'(x) = 2x - 2x^{-2} \Rightarrow 2x - \dfrac{2}{x^2}$, $x > 0$

(b) $f'x)=0 \Rightarrow 2x-\dfrac{2}{x^2}=0 \Rightarrow x^3=1 \Rightarrow x=1$. Therefore, the value of f where there is a

turning point is $f(1)=1^2+\dfrac{2}{1}=3$

(c)

x	0.5	1	2
$f'(x)$	-7	0	3.5
Tangent	$\searrow$	$\rightarrow$	$\nearrow$

The arrows in the third row of the table show that $f(1)=3$ is a minimum value of f.

Example 5.10: Consider the function $f(x)=x+\dfrac{1600}{x}$, $x>0$.

(a) Find $f'(x)$ and $f''(x)$.

(b) Show that a minimum or maximum value of f occurs when $x=40$ and find $f(40)$.

(c) Find $f''(40)$ and use it to investigate whether $f(40)$ is a maximum or a minimum value.

Solution 5.10: (a) $f(x)=x+1600x^{-1} \Rightarrow f'(x)=1-1600x^{-2}=1-\dfrac{1600}{x^2}$, $x>0$

$\Rightarrow f''(x)=2\left(1600x^{-3}\right)=\dfrac{3200}{x^3}$, $x>0$

(b) Maximum and minimum values occur when $f'(x)=0$

$\Rightarrow 1-\dfrac{1600}{x^2}=0 \Rightarrow x^2=1600 \Rightarrow x=\pm 40$ but $x>0$ so $x=40$.

$f(40)=40+\dfrac{1600}{40}=80$

(c) $f''(40)=\dfrac{3200}{64000}=\dfrac{1}{20}>0$. Therefore, $f(40)=80$ is a minimum value.

Exercise 5.6

1. Differentiate each of the given functions and hence find the maximum or minimum value(s) of each function. In each case, use a table to determine whether the values are maximums or minimums.

(a) $f(x)=6-2x-x^2$ (b) $f(x)=x^2+7x-4$ (c) $f(x)=x^3-5x^2-8x+11$

(d) $f(x)=2x^3+x^2-4x-5$ (e) $f(x)=x^4-256x+15$ (f) $f(x)=\sqrt{x}-2x$, $x>0$

2.   For each of the functions,

(a) $y = x^2 + 6x + 8$ (b) $y = x^3 - 12x + 3$ (c) $y = x^5 - 80x$

find (i) $\dfrac{dy}{dx}$

(ii) any values of x at which either a maximum or a minimum occur.

(iii) the coordinates of the turning point and, by investigating the value of the second derivative at the turning point(s) found in (ii), determine whether the turning point has a maximum or a minimum value.

3.   For each of the functions,

(a) $f(x) = x^3 + 9x - 7$ (b) $f(x) = x^2 - \dfrac{1}{x},\ x \neq 0$ (c) $f(x) = x^4 + 3x^3 + 3x^2 - x + 5$

find (i) $f''(x)$

(ii) the coordinates of any points of inflection.

5.3.2 Optimization Problems 1

Optimization problems are ones which involve maximums or minimums. Finding the dimensions of a metal container which minimizes the amount of metal used is an example of an optimization problem.

The methods described in section 5.3.1 can be used to solve optimization problems.

Example 5.11: A rectangle has length x and width y and its perimeter is $10\,\text{m}$.

(a) Write y as a function of x.

(b) Find an expression, in terms of x, for the area, A of the rectangle.

(c) Solve $\dfrac{dA}{dx} = 0$ and hence find the values of x and y that maximize A.

Solution 5.11: (a) $2x + 2y = 10 \Rightarrow x + y = 5 \Rightarrow y = 5 - x$ (b) $A = xy = x(5 - x) = 5x - x^2$

(c) $\dfrac{dA}{dx} = 5 - 2x = 0 \Rightarrow x = 2.5 \Rightarrow y = 5 - 2.5 = 2.5$

You should check that these values of x and y do, in fact, give a maximum value rather than a minimum value. This can be done by using a table.

x	2	2.5	3
$\dfrac{dA}{dx}$	1	0	−1
Tangent	↗	→	↘

Therefore, the maximum area occurs when $x = y = 2.5$.

Exercise 5.7

1.   A variable W is related to variables x, y by $W = xy$ and $x + 3y = 8$.

(a) Write y in terms of x.

(b) Substitute the answer from (a) into $W = xy$ and hence write W in terms of x.

(c) Differentiate W to obtain $\dfrac{dW}{dx}$.

(d) Solve $\dfrac{dW}{dx}=0$ and hence show that W has a turning point when $x=4$.

(e) By investigating how the sign of $\dfrac{dW}{dx}$ changes as x increases from 3 to 5, find whether W is a maximum or minimum value when $x=4$, and find this maximum or minimum value.

2.   A manufacturing process produces x digital cameras at a total cost of $\$\left(10000-400x+8x^2\right)$.

Each digital camera is sold for $\$\left(400-2x\right)$.

(a) Find P, the profit made by the sale of the total consignment, in terms of x.

(b) Show that the profit, P, is maximized if each digital camera is sold for \$320.

(c) Show that the profit is a maximum, rather than a minimum, for this sale price.

3.   A square, $PQRS$ of side y cm is fitted into a long channel 8cm wide (figure 5.26). The distance of S from the side of the channel touching the corner P is x cm.

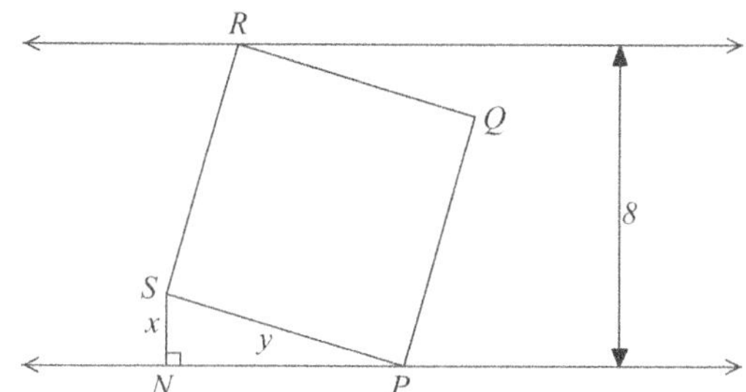

Figure 5.26

(a) Show that the area, A of $PQRS$ is $64-16x+2x^2$.

(b) Differentiate A and solve the equation $A'(x)=0$.

(c) Use a table to investigate whether A is maximized or minimized at this value of x.

4.   A closed tube has a square cross-section of side x m and length l m. The sum of the perimeter of the cross-section and the length of the tube is 6m. The volume of the tube is V m^3.

(a) Show that $4x+l=6$.   (b) Show that $V=6x^2-4x^3$.

(c) Differentiate V and solve the equation $V'(x)=0$.

(d) Show that V has a maximum value when $x=1$ and find this value.

5.   A rectangular box with an open top has volume 36 cm^3. The width of the base of the box is x cm. and the length of the base is twice the width. The height of the box is h cm.

(a) Find h in terms of x.

(b) Find the dimensions of the box which minimize the surface area of the box.

(c) Show that these dimensions give a minimum rather than a maximum surface area

6.   A cylindrical can has volume 2000π cm^3. Note that the height of the can, h and the radius of the can, x are measured in cm.

(a) Find h in terms of x.

(b) Find A, the surface area of the can, including both circular ends, in terms of x.

(c) Find the minimum value of A.

7. A rectangle $PQRS$ has base PQ on the x-axis and opposite vertices R and S lie on the parabola $f(x) = x - x^2$ and O is the origin of axes.

(a) By letting $OP = x$, show that if $A(x)$ is the area of rectangle $PQRS$, then
$$A(x) = 2x^3 - 3x^2 + x.$$

(b) Find $A'(x)$ and hence find the maximum area of the rectangle $PQRS$ and the value of x at which this maximum occurs.

5.4 Differentiation of Other Functions

The method used in section 5.1.2 will serve to find the derivative of other functions.

5.4.1 The Derivative of the Logarithmic Function $f(x) = \ln x, \ x > 0$

Enter the function $f(x) = \ln x, \ x > 0$ into your graphing calculator and, for various values of x, draw the tangent and read off the gradient, m, of the tangent at that value of x. Table 5.3 shows some values of x with corresponding values of m for $f(x) = \ln x$.

x	m
0.1	10.000
0.5	2.000
1	1.000
2	0.500
5	0.200
10	0.100
25	0.040

Table 5.3

It is not necessary, in this instance, to do a statistical plot of these points and fit a curve to them because you can probably spot a pattern in table 5.3: m is $\dfrac{1}{x}$. Therefore, the derivative of $f(x) = \ln x, \ x > 0$ is $f'(x) = \dfrac{1}{x}, \ x > 0$.

5.4.2 The Derivative of the Exponential Function $f(x) = e^x$

Table 5.4 shows the results when a similar procedure is carried out for the function $f(x) = e^x$. In this case, it is helpful to include also a column for the value of the function.

x	m	$f(x)$
−5	0.00674	0.00674
−1	0.36788	0.36788
0	1.00000	1.00000
1	2.71828	2.71828
2	7.38906	7.38906
5	148.41318	148.41316

Table 5.4

Again, it is unnecessary to do a statistical plot because it is clear from table 5.4 that the gradient at x is equal to the value of the function at x so that $f'(x) = e^x$.

5.4.3 The Derivative of the Trigonometric Function $f(x)=\sin x$

Figures 5.27 and 5.28 show, for $f(x)=\sin x$, values of x and corresponding values of m stored as lists and then plotted on a statistical plot.

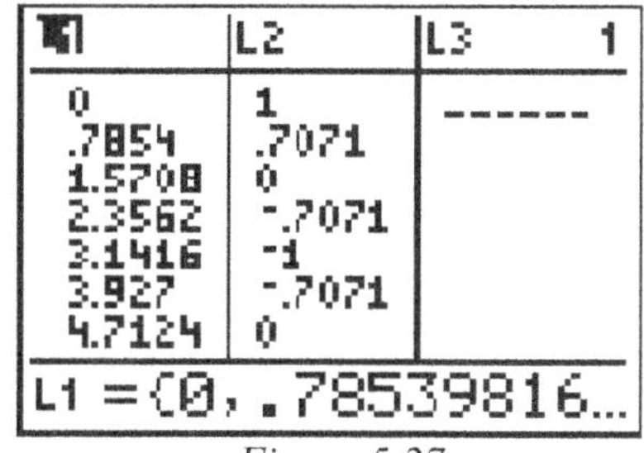

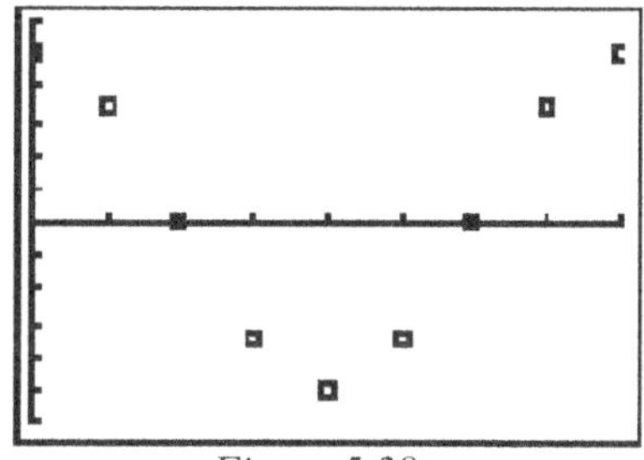

Figure 5.27   *Figure 5.28*

It seems reasonable to expect that the derivative of $f(x)=\sin x$ is a trigonometric function, and a glance at the statistical plot suggests that $f'(x)=\cos x$ — which is, in fact, the derivative of $f(x)=\sin x$. If a similar process is carried out with the function $f(x)=\cos x$, then the derivative is found to be $f'(x)=-\sin x$.

To summarize,

$$f(x)=x^n \qquad \Rightarrow \qquad f'(x)=nx^{n-1}$$

$$f(x)=\ln x,\ x>0 \ \Rightarrow \ f'(x)=\frac{1}{x},\ \ x>0$$

$$f(x)=e^x \qquad \Rightarrow \qquad f'(x)=e^x$$

$$f(x)=\sin x \qquad \Rightarrow \qquad f'(x)=\cos x$$

$$f(x)=\cos x \qquad \Rightarrow \qquad f'(x)=-\sin x$$

5.5 Techniques of Differentiation

5.5.1 Differentiation of Composite Functions

Since the derivative of $f(x)=e^x$ is $f'(x)=e^x$, it might seem reasonable that the derivative of $f(x)=e^{2x}$ would be $f'(x)=e^{2x}$. The method of section 5.1.2 is used to find the derivative of $f(x)=e^{2x}$. Table 5.5 shows, for $f(x)=e^{2x}$, the values of m and $f(x)$ for various values of x.

Because $m=f'(x)$, table 5.5 shows that $f'(x)=2f(x)$. Therefore $f'(x)=2e^{2x}$. We now see why this is so.

x	m	$f(x)$
-1	0.271	0.153
-0.5	0.736	0.368
0	2	1
0.5	5.437	2.718
1	14.778	7.389
2	109.1964	54.598
3	806.858	403.429
4	5961.920	2980.960

Table 5.5

A composite function is one in which one function is 'nested' inside another and can be written $y=f(g(x))$. So, $y=f(g(x))$ can be written in terms of two component functions: $y=f(u)$ in which $u=g(x)$ is nested.

175

The derivative, $\dfrac{dy}{dx}$, can be expressed in terms of the derivatives of the two component functions, $\dfrac{dy}{du}$

and $\dfrac{du}{dx}$. Bearing in mind that Δx, Δu and Δy are small but finite increments of x, u and y

respectively, then $\dfrac{\Delta y}{\Delta x} = \dfrac{\Delta y}{\Delta u} \times \dfrac{\Delta u}{\Delta x}$. If the limit is taken of both sides of this identity, as in the process

described in section 5.1.3, then $\dfrac{dy}{dx} = \dfrac{dy}{du} \times \dfrac{du}{dx}$. Using functional notation, this statement is

$f'(x) = f'(u) \times u'(x)$.

Composite functions may have more than two components nested inside one another.

Consider a function that has three components: $y = f(g(h(x)))$. Its components are

$y = f(v)$, $v = g(u)$ and $u = h(x)$, and its derivative is $\dfrac{dy}{dx} = \dfrac{dy}{dv} \times \dfrac{dv}{du} \times \dfrac{du}{dx}$, forming a kind of chain.

For this reason, this method of differentiation is called the *chain rule*.

Example 5.12: Use the chain rule to differentiate the function $y = \left(2x^2 + 3\right)^5$.

Solution 5.12: The components are $y = u^5$ and $u = 2x^2 + 3$. Therefore, $\dfrac{dy}{du} = 5u^4$, $\dfrac{du}{dx} = 4x$

$$\Rightarrow \frac{dy}{dx} = \frac{dy}{du} \times \frac{du}{dx} = \left(5u^4\right)(4x) = 20x\left(2x^2 + 3\right)^4.$$

Example 5.13: Find $\dfrac{dy}{dx}$ if $y = e^{3x}$.

Solution 5.13: The components are $y = e^u$ and $u = 3x$. Therefore, $\dfrac{dy}{du} = e^u$, $\dfrac{du}{dx} = 3$

$$\Rightarrow \frac{dy}{dx} = \frac{dy}{du} \times \frac{du}{dx} = \left(e^u\right)(3) = 3e^{3x}.$$

Example 5.14: Differentiate $f(x) = \cos 5x$.

Solution 5.14: The components are $f(u) = \cos u$ and $u(x) = 5x$. Therefore, $f'(u) = -\sin u$, $u'(x) = 5$

$$\Rightarrow f'(x) = f'(u) \times u'(x) = (-\sin u)(5) = -5\sin 5x.$$

Example 5.15: Differentiate $f(x) = \ln \sin x$.

Solution 5.15: The components are $f(u) = \ln u$ and $u(x) = \sin x$. Therefore, $f'(u) = \dfrac{1}{u}$, $u'(x) = \cos x$

$$\Rightarrow f'(x) = f'(u) \times u'(x) = \left(\frac{1}{u}\right)(\cos x) = \frac{\cos x}{\sin x} = \frac{1}{\tan x}.$$

Exercise 5.8

1. Differentiate the following functions using the chain rule.

   (a) $f(x) = (3x+1)^3$

   (b) $f(x) = (1-4x^2)^2$

   (c) $f(x) = \sqrt{x^2+5}$

   (d) $g(x) = e^{-x}$

   (e) $g(x) = 3e^{4x}$

   (f) $g(x) = e^{1-x^2}$

   (g) $h(x) = \sin 3x$

   (h) $h(x) = \cos\left(2x - \dfrac{\pi}{6}\right)$

   (i) $h(x) = \sin^2 x$

   (j) $f(x) = \ln(1+5x)$

   (k) $f(x) = 4\ln(x^2+3)$

   (l) $f(x) = \ln(1+e^x)$

2. Find $\dfrac{dy}{dx}$ for each of the given functions using the chain rule.

   (a) $y = (2x-5)^2$

   (b) $y = (x^2-4)^3$

   (c) $y = (3-5x)^4$

   (d) $y = e^{x^2}$

   (e) $y = e^{1-2x}$

   (f) $y = \cos^3 x$

   (g) $y = 3\sin 4x$

   (h) $y = \sin 3x^2$

   (i) $y = \ln\dfrac{1}{x}$, $x > 0$

   (j) $y = \ln(x^3-3)$

   (k) $y = \ln 4x$

   (l) $y = \dfrac{2}{\cos x}$

3. Write down the derivatives of the following functions. After differentiating a number of composite functions, you may notice a pattern emerging, and you will be able to write down the derivatives without going through all the steps of the chain rule.

   (a) $f(x) = (1-7x)^4$

   (b) $f(x) = 5e^{-4x}$

   (c) $y = \ln(x + \sin x)$

   (d) $y = \dfrac{1}{x^2-3}$

   (e) $g(x) = (2-\cos x)^2$

   (f) $f(x) = \cos(e^{-x})$

   (g) $f(x) = \sin\left(\dfrac{\pi}{2} - 6x\right)$

   (h) $y = (1+e^x)^3$

   (i) $y = \sqrt{\sin x + 1}$

   (j) $f(x) = e^{-\cos x}$

   (k) $y = \ln \cos 2x$

   (l) $g(x) = \sin^2 3x$

5.5.2 Differentiation of Product Functions

A product function can be written in the form $y(x) = u(x) \times v(x)$ and using the method of section 5.1.5

$$\frac{dy}{dx} = \lim_{\Delta x \to 0}\left(\frac{(u+\Delta u)(v+\Delta v)-uv}{\Delta x}\right) = \lim_{\Delta x \to 0}\left(\frac{uv+u\Delta v+v\Delta u+\Delta u\Delta v-uv}{\Delta x}\right)$$

$$= \lim_{\Delta x \to 0}\left(u\frac{\Delta v}{\Delta x}+v\frac{\Delta u}{\Delta x}+\Delta u\frac{\Delta v}{\Delta x}\right) \text{ and since } \lim_{\Delta x \to 0}\left(\Delta u\frac{\Delta v}{\Delta x}\right)=0,$$

So that $y = uv \Rightarrow \dfrac{dy}{dx} = u\dfrac{dv}{dx} + v\dfrac{du}{dx}$. This formula is known as the *product rule*.

The alternative notation gives the product rule as:

$$f(x) = u(x)v(x) \Rightarrow f'(x) = u(x)v'(x) + u'(x)v(x)$$

Example 5.16: Differentiate $f(x) = x^2 \cos x$.

Solution 5.16: Let $u(x) = \cos x$, $v(x) = x^2$. Therefore, $u'(x) = -\sin x$, $v'(x) = 2x$.

  Then, $f'(x) = u(x)v'(x) + v(x)u'(x) = (\cos x)2x + x^2(-\sin x)$.

  Therefore, $f'(x) = 2x\cos x - x^2 \sin x$.

The commutative property of addition and multiplication ensures that, had we put $u(x) = x^2$ and $v(x) = \cos x$, the final result would have been the same.

Example 5.17: Find the derivative of $y = (3x-1)(x+4)^3$.

Solution 5.17: Let $u = 3x-1$, $v = (x+4)^3$. Therefore, $\dfrac{du}{dx} = 3$, $\dfrac{dv}{dx} = 3(x+4)^2$.

  Then $\dfrac{dy}{dx} = u\dfrac{dv}{dx} + v\dfrac{du}{dx} = (3x-1)\left[3(x+4)^2\right] + (x+4)^3 3$.

  Factoring, $\dfrac{dy}{dx} = 3(x+4)^2\left[(3x-1)+(x+4)\right] = 3(x+4)^2(4x+3)$.

5.5.3 Differentiation of Quotient Functions

A quotient function is written in the form, $y = \dfrac{u}{v}$ and, using a method similar to the one in section

5.5.2, it can be shown that $\dfrac{dy}{dx} = \dfrac{v\dfrac{du}{dx} - u\dfrac{dv}{dx}}{v^2}$. With the alternative notation, if $y = \dfrac{u}{v}$ then

$f'(x) = \dfrac{v(x)u'(x) - u(x)v'(x)}{\left[v(x)\right]^2}$. This formula is known as the *quotient rule*.

Example 5.18: Find $f'(x)$ if $f(x) = \dfrac{x}{x^2 + 2}$.

Solution 5.18: Let $u(x) = x$, $v(x) = x^2 + 2$. Therefore, $u'(x) = 1$, $v'(x) = 2x$.

$$\text{Then } f'(x) = \frac{v(x)u'(x) - u(x)v'(x)}{\left[v(x)\right]^2} = \frac{(x^2 + 2)1 - x(2x)}{(x^2 + 2)^2} = \frac{2 - x^2}{(x^2 + 2)^2}.$$

Note that, because division is not commutative, $u(x)$ must always be chosen as the numerator and $v(x)$ as the denominator.

Example 5.19: Find the derivative of $y = \dfrac{e^x}{\sin x}$.

Solution 5.19: Let $u = e^x$, $v = \sin x$. Therefore, $\dfrac{du}{dx} = e^x$, $\dfrac{dv}{dx} = \cos x$.

$$\text{Then } \frac{dy}{dx} = \frac{v\frac{du}{dx} - u\frac{dv}{dx}}{v^2} = \frac{(\sin x)e^x - e^x(\cos x)}{(\sin x)^2} = \frac{e^x(\sin x - \cos x)}{\sin^2 x}, \quad 0 < x < \pi.$$

Exercise 5.9

1.   Find $f'(x)$ for each of the given functions.

    (a) $f(x) = x^2 \sin x$ (b) $f(x) = x^3(x+1)^2$ (c) $f(x) = e^x \cos x$

    (d) $f(x) = x \ln x$ (e) $f(x) = 4xe^x$ (f) $f(x) = x(\cos x + \sin x)$

2.   Find $f'(x)$ for each of the given functions.

    (a) $f(x) = \dfrac{x^2}{x^2 + 1}$ (b) $f(x) = \dfrac{x}{\ln x}$ (c) $f(x) = \dfrac{\sin x}{x}$

    (d) $f(x) = \dfrac{1 + \ln x}{x}$,   (e) $f(x) = \tan x$ (f) $f(x) = \dfrac{\cos x}{2 + \sin x}$

3. Consider the function $y = \dfrac{x}{2x+1}$, $x \neq -\dfrac{1}{2}$.

(a) Use the quotient rule to find $\dfrac{dy}{dx}$.

(b) By writing the function in the form $y = x(2x+1)^{-1}$, find $\dfrac{dy}{dx}$ by using the product rule.

4. (a) Explain why you would not use the quotient rule to differentiate $y = \dfrac{x + \sin x}{3}$.

(b) Consider the function $y = \dfrac{3}{x + \sin x}$, $x > 0$.

(i) Use the quotient rule to find $\dfrac{dy}{dx}$.   (ii) Use the chain rule to find $\dfrac{dy}{dx}$.

5. In each case, differentiate the given function using the most appropriate method.

(a) $f(x) = x^2 \cos 2x$ (b) $g(x) = \dfrac{2x}{x^2 + 1}$ (c) $y = \dfrac{\ln x}{x}$

(d) $y = x(1 - 4x)^3$ (e) $g(x) = \dfrac{x^2 + 4x}{2}$ (f) $f(x) = (x + 3)e^{-x}$

(g) $y = \dfrac{3x + 4}{2x + 3}$ (h) $g(x) = \ln\left(\dfrac{x^2 - 1}{x^2 + 1}\right)$ (i) $y = \dfrac{e^{2x}}{1 - 3x}$

(j) $f(x) = e^{-x} \sin 2x$ (k) $y = 2 \tan 3x$ (l) $g(x) = \sin 2x \cos x$

6. (a) If $y = \dfrac{e^x}{1 + e^x}$, show that $\dfrac{dy}{dx} = \dfrac{e^x}{\left(1 + e^x\right)^2}$.

(b) If $f(x) = \dfrac{\sin x - \cos x}{\sin x + \cos x}$, show that $f'(x) = \dfrac{2}{1 + \sin 2x}$.

(c) If $g(x) = \ln\left(\dfrac{x}{2x + 1}\right)$, show that $g'(x) = \dfrac{1}{x(2x + 1)}$.

5.6 Applications of Differentiation 2

5.6.1 Increasing and Decreasing Functions

An increasing function, $f(x)$, is one whose value always increases as x increases throughout its domain. A decreasing function is one whose value always decreases as x increases throughout its domain.

It is possible to identify an increasing function by showing that its gradient is always positive. In a similar way, it is possible to identify a decreasing function by showing that its gradient is always negative.

Example 5.20: Show that

(a) $f(x) = x^3 + 3x + 13$ increases for all values of x.

(b) $g(x) = \dfrac{1}{x}$, $x \neq 0$ decreases for all values of x.

Solution 5.20: (a) $f(x) = x^3 + 3x + 13 \Rightarrow f'(x) = 3x^2 + 3 = 3(x^2 + 1)$. For all values of x,

$x^2 \geq 0 \Rightarrow x^2 + 1 > 0 \Rightarrow 3(x^2 + 1) > 0$. Therefore, $f'(x) > 0$ for all values of x

and $f(x)$ is an increasing function.

(b) $g(x) = x^{-1} \Rightarrow g'(x) = -x^{-2} = -\dfrac{1}{x^2}$. For all values of x, $x^2 \geq 0 \Rightarrow \dfrac{1}{x^2} > 0$, $x \neq 0$.

Therefore, $-\dfrac{1}{x^2} < 0$ for all values of x, $x \neq 0$ and $g(x)$ is a decreasing function.

5.6.2 The Equation of a Tangent and a Normal at a Point on a Curve

The tangent at point P on a curve is the straight line which touches the curve at P. The normal at point P on a curve is the straight line passing through P which is perpendicular to the tangent at P (figure 5.29).

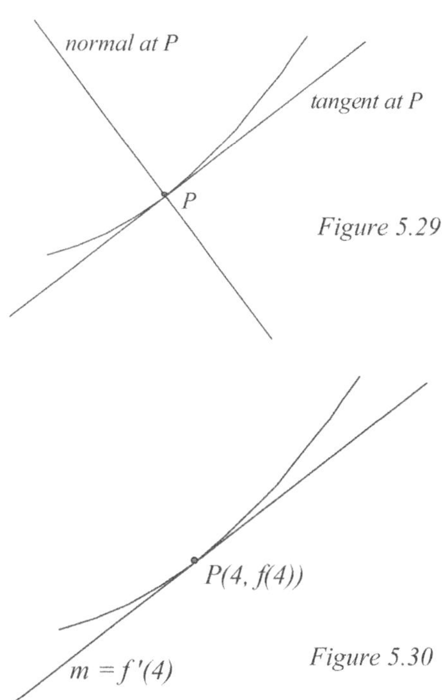

Consider the point P, with $x = 4$, on the curve of the function $f(x) = x^2 - 3x + 11$.

In order to find the equation of the tangent to the curve at P, it is necessary to find

- the gradient, m of the tangent to the curve at P
- the y-coordinate of P (figure 5.30)

The gradient at P is $f'(4)$. $f(x) = x^2 - 3x + 11 \Rightarrow f'(x) = 2x - 3$. Therefore, $f'(4) = 2(4) - 3 = 8 - 3 = 5$.

At P, $x = 4$ and $y = f(4)$. $f(4) = (4)^2 - 3(4) + 11 = 16 - 12 + 11 = 15$. Therefore the equation of the tangent at P is $\dfrac{y - 15}{x - 4} = 5 \Rightarrow y - 15 = 5(x - 4) \Rightarrow 5x - y - 5 = 0$

In order to find the equation of the normal at P, it is necessary to find

- The gradient of the normal to the curve at P
- The y-coordinate of P

Since the gradient of the tangent at P is 5, the gradient of the normal at P is $-\dfrac{1}{5}$ because the product of perpendicular gradients is -1.

The y-coordinate at P has already been calculated and is 15. Therefore the equation of the normal at P is $\dfrac{y - 15}{x - 4} = -\dfrac{1}{5} \Rightarrow 5(y - 15) = -1(x - 4) \Rightarrow x + 5y - 79 = 0$.

Example 5.21: (a) A curve has equation $y = 1 - \dfrac{1}{x^2}$, $x \neq 0$. Show that the equation of the tangent to this curve at the point where $x = -2$ is $x + 4y - 1 = 0$.

(b) Find the equation of the normal to the curve at the point where $x = -2$.

Solution 5.21: (a) The gradient of the tangent at the point on the curve where $x = -2$ is equal to the value of $\dfrac{dy}{dx}$ at that point. $y = 1 - \dfrac{1}{x^2} \Rightarrow y = 1 - x^{-2} \Rightarrow \dfrac{dy}{dx} = -(-2)(x)^{-3} = \dfrac{2}{x^3}$.

Therefore, at $x = -2$, $\dfrac{dy}{dx} = \dfrac{2}{(-2)^3} = \dfrac{2}{-8} = -\dfrac{1}{4}$. The y-coordinate at the point on the curve where $x = -2$ is $y = 1 - \dfrac{1}{(-2)^2} = 1 - \dfrac{1}{4} = \dfrac{3}{4}$. Therefore, the equation of the tangent at $x = -2$ is $\dfrac{y - \frac{3}{4}}{x - (-2)} = -\dfrac{1}{4} \Rightarrow 4y - 3 = -1(x + 2) \Rightarrow x + 4y - 1 = 0$.

(b) The gradient of the normal is 4, and, since the normal passes through the point with coordinates $\left(-2, \dfrac{3}{4}\right)$, the equation of the normal is $\dfrac{y - \frac{3}{4}}{x + 2} = 4$. Therefore,

$$y - \dfrac{3}{4} = 4(x + 2) \Rightarrow 4y - 3 = 16x + 32 \Rightarrow 16x - 4y + 35 = 0.$$

Exercise 5.10

1. In each case, use $f'(x)$ to decide whether $f(x)$ is an increasing or decreasing function.

(a) $f(x) = 5 - 2x$ (b) $f(x) = \dfrac{1}{x^3}$, $x \neq 0$ (c) $f(x) = \ln x$, $x > 0$

(d) $f(x) = e^{-x}$ (e) $f(x) = x^3 + x$

2. (a) If $y = xe^{-x}$, find $\dfrac{dy}{dx}$ and hence show that $y = xe^{-x}$ is decreasing, provided $x > 1$.

(b) If $y = kx - \sin x$, for what range of values of k is $y = kx - \sin x$ an increasing function?

3. In each case, find, in the form $ax + by + c = 0$, the equation of the tangent to the given curve at the given point.

(a) $y = x^3$; (2, 8)   (b) $f(x) = 2x - \dfrac{1}{x}$, $x \neq 0$; (1, 1)   (c) $f(x) = x^2 + 6x + 8$; (–2, 0)

(d) $y = x^3 - 6x^2 - 4x$; (4, -48)   (e) $f(x) = \sqrt{x}$; (4, 2)

4. Find, in the form $ax + by + c = 0$, the equation of the normal to the curves given in question 3, at the given points.

5. Find the equation of the tangent at $(0,\ 0)$ to the curve of the function $y = xe^{-x}$.

6. Given the curve of the function $y = x^2 + 2x + 5$, show that the equation of the normal to this curve at the point where $x = -2$ is $x - 2y + 12 = 0$.

7. Given the curve of the function $y = x^2 \ln x$, show that the equation of the tangent to this curve at the point where $x = e$ is $y = 3ex - 2e^2$.

8. Given the curve of the function $f(x) = x^2 e^{-x}$, show that the equation of the tangent to this curve at the point where $x = 1$ is $y = \dfrac{x}{e}$.

9. Given the curve of the function $f(x) = \dfrac{e^{2x}}{x}$, $x \neq 0$, show that the equation of the tangent to this curve at the point where $x = 1$ is $y = e^2 x$.

10. Given the curve of the function $f(x) = x \ln x$, show that the equation of the normal to this curve at the point $(e,\ e)$ is $x + 2y - 3e = 0$.

5.6.3 Optimization Problems 2

You should now be able to do optimization problems involving functions like those encountered in sections 5.4 and 5.5.

Example 5.22: Given the function $f(x) = \dfrac{e^x}{x}$, $x > 0$,

    (a) find $f'(x)$.   (b) solve the equation $f'(x) = 0$.

    (c) show that f has a minimum value when $x = 1$ and find the exact value of this minimum.

Solution 5.22: (a) Let $u(x) = e^x$, $v(x) = x \Rightarrow u'(x) = e^x$, $v'(x) = 1$

$$f'(x) = \frac{v(x)u'(x) - u(x)v'(x)}{[v(x)]^2} = \frac{xe^x - e^x 1}{x^2} = \frac{e^x(x-1)}{x^2}$$

Therefore, $f'(x) = \dfrac{e^x(x-1)}{x^2}$, $x > 0$.

(b) $f'(x) = 0 \Rightarrow \dfrac{e^x(x-1)}{x^2} = 0 \Rightarrow x - 1 = 0 \Rightarrow x = 1$

(c) f has a maximum or minimum value at $x = 1$. In order to show that f has a minimum rather than a maximum at $x = 1$, it is probably quicker to investigate the gradient on either side of $x = 1$

x	0.5	1	1.5
$f'(x)$	−3.30	0	0.996
Tangent	↘	→	↗

than to find $f''(1)$. The table shows that f has a minimum not a maximum at

$$x = 1 \text{ and the exact value of this maximum is } f(1) = \frac{e^1}{1} = e.$$

Example 5.23: A business manufactures t-shirts. The cost, in dollars, C of producing a t-shirt decreases as x, the number of t-shirts manufactured, increases. The function relating x and C is $C(x) = 3 + \dfrac{50}{x}$, $x > 0$. The sale price of a t-shirt, S also decreases as x increases. The function relating x and S is $S(x) = 15 - \dfrac{x^2}{6000}$, $x > 0$.

    (a) Show that the profit made on the sale of 200 t-shirts is $1016.67.

    (b) Find the profit, $P(x)$, on the sale of x t-shirts as a function of x, and show that

$$P'(x) = 12 - \frac{x^2}{2000}, \quad x > 0.$$

    (c) Hence show that when $x = 155$ the profit is maximized.

    (d) Show that $P''(155) < 0$ in order to verify that the profit is a maximum value rather than a minimum value.

Solution 5.23: (a) $C(200) = 3 + \dfrac{50}{200} = 3.25$, $S(200) = 15 - \dfrac{40000}{6000} = \dfrac{25}{3}$. Therefore, in dollars, the cost of producing 200 t-shirts is $200 \times 3.25 = 650$, and the income from the sale of 200 t-shirts is $200 \times \dfrac{25}{3} = 1666.67$. Hence, the profit is $\$(1666.67 - 650.00) = \$1016.67$.

(b) $P(x) = xS(x) - xC(x) = x\left(15 - \dfrac{x^2}{6000}\right) - x\left(3 + \dfrac{50}{x}\right) = 12x - \dfrac{x^3}{6000} - 50$.

Therefore, $P'(x) = 12 - \dfrac{3x^2}{6000} = 12 - \dfrac{x^2}{2000}$, $x > 0$.

(c) For optimum profit, $P'(x) = 0 \Rightarrow 12 - \dfrac{x^2}{2000} = 0 \Rightarrow x^2 = 24000$. Therefore, $x = \pm 154.9$ But as $x \in \mathbb{Z}^+$, $x = 155$.

(d) $P''(x) = -\dfrac{2x}{2000} = -\dfrac{x}{1000} \Rightarrow P''(155) = -\dfrac{155}{1000} = -0.155 < 0$ so that the profit is a maximum, rather than a minimum at $x = 155$.

Exercise 5.11

1. Given the function $f(x) = \dfrac{2x}{x^2 + 4}$,

(a) find $f'(x)$ and show that the solution of $f'(x) = 0$ is $x = \pm 2$.

(b) show that f has a maximum value when $x = 2$ by investigating the gradient at $x = 1$ and $x = 3$.

(c) Find the minimum value of f.

2. Given that $f(x) = xe^{-x}$,

(a) find $f'(x)$.

(b) solve the equation $f'(x) = 0$ and hence show that a maximum or minimum value occurs when $x = 1$.

(c) find the value of $f''(1)$ and hence show that f has a maximum value at $x = 1$.

(d) find the value of this maximum, giving your answer exactly.

3. Consider the function $y = \dfrac{\ln x}{x}$, $x > 0$.

(a) Sketch a graph of $y = \dfrac{\ln x}{x}$, $0 < x < 5$.

(b) Find $\dfrac{dy}{dx}$ and show that $\dfrac{d^2y}{dx^2} = \dfrac{2\ln x - 3}{x^3}$, $x > 0$.

(c) Show that y has a maximum, rather than a minimum, value when $x = e$.

(d) Find the exact value of this maximum.

4. Consider the function $f(x)=e^{-x}\sin x,\ 0\le x<\pi$.

   (a) Find $f'(x)$

   (b) Show that the only solution of $f'(x)=0$ in the given domain is $x=\dfrac{\pi}{4}$.

   (c) Show that $f''(x)=-2e^{-x}\cos x$, and investigate whether $f(x)$ has a maximum or a

   minimum value $x=\dfrac{\pi}{4}$.

5. Consider the function $g(x)=x-\sin 2x,\ 0\le x<\pi$.

   (a) Sketch a graph of $y=g(x)$, marking the minimum point P and the maximum point Q.

   (b) Write down $g'(x)$ and solve the equation $g'(x)=0$, giving solutions in the interval
   $0<x<\pi$.

   (c) Show that the coordinates of P are $\left(\dfrac{\pi}{6},\dfrac{\pi-3\sqrt{3}}{6}\right)$.

   (d) Find, as multiples of π, the coordinates of Q and use $g''(x)$ to show that Q has, in fact, a
   local maximum.

6. A wire of length 10cm is cut into two pieces. One piece is made into a circle and the other into
   a square. Assume that the circle has radius r cm and the square has side x cm.
   (a) Write down a formula relating r and x.

   (b) If A is the sum of the areas of the circle and the square, show that $A(x)=x^2+\dfrac{(5-2x)^2}{\pi}$.

   (c) Show that, correct to three significant figures, $x=1.40$ is a solution of the equation
   $A'(x)=0$ and find out whether $A(1.40)$ has a maximum or a minimum value.

7. Figure 5.31 shows the graph of $y=x^2$. Point
   P has coordinates $(0,\ 9.5)$ and point Q is a
   general point on the curve of $y=x^2$ and has
   coordinates $\left(x,\ x^2\right)$. Let L be the distance PQ.

   Show that $L=\sqrt{x^4-18x^2+90.25}$.

   (a) Find $\dfrac{dL}{dx}$.

   (b) Show that $x=0$, $x=3$ are solutions of the
   equation $\dfrac{dL}{dx}=0$ and find one other solution
   of this equation.
   (c) Hence find the coordinates of the two points
   on the curve which are closest to P.
   (d) Explain the significance of the solution, $x=0$.

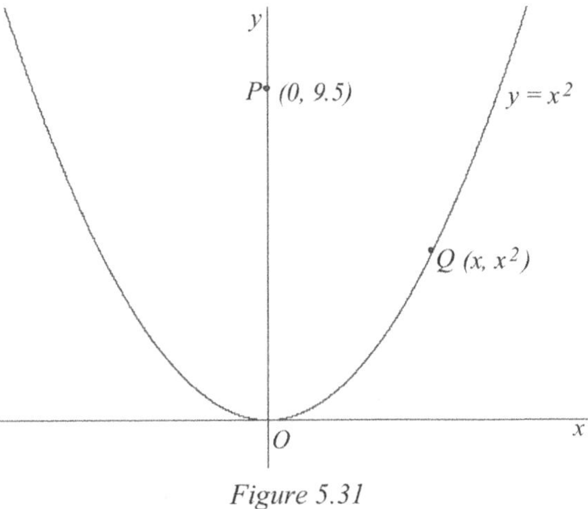

Figure 5.31

8.	A productivity function, f relates the amount of sugar, in kilograms, produced per hour by a sugar mill to the number of workers, x employed, where $f(x) = 500\left(1 - e^{-0.2x}\right)$, $x \in \mathbb{Z}^+$. In addition, the hourly production costs, in dollars, are related to the number of workers employed by the function $g(x) = 8x + 30$, $x \in \mathbb{Z}^+$. The mill sells 1 kg of sugar for $0.85.

(a) Show that $P(x)$, the hourly profit made by the mill, is $P(x) = 425\left(1 - e^{-0.2x}\right) - 8x - 30$.

(b) Find $P'(x)$ and, by solving the equation $P'(x) = 0$, find the number of workers who should be employed for maximum profit. Find this maximum profit.

(c) Use your graphing calculator to verify your answers to part (b).

9.	The cost, $\$C$, of producing x liters of wine in a given period is given by the function,

$C = 3x + \dfrac{x(x - 400)^2}{80000}$. The sale price of wine per liter is $8.

(a) Find, in terms of x, the profit, $\$P$, made on the sale of x liters of wine.

(b) Show that the sale of 300 liters brings a profit of $1462.50.

(c) Find $\dfrac{dP}{dx}$ and solve the equation $\dfrac{dP}{dx} = 0$ to show that the maximum profit occurs when approximately 655 liters of wine is produced. Find this maximum profit.

(d) Use your graphing calculator to verify your answers to part (c).

5.6.4 Kinematics

Kinematics is the name given to the study of motion without regard to its cause. The investigation of the balloon at the beginning of Unit 5 is an example of kinematics in one dimension. In this section, we are concerned only with kinematics in one dimension, i.e., motion in a straight line.

In Unit 4, a displacement was defined as a vector, describing the translation of an object from one point to another. For motion in a straight line, the *displacement* of an object, $s(t)$ at time t is its displacement from a fixed point, O on the straight line.

Displacements on the line on one side of O may be regarded as positive, and those on the line on the opposite side of O may be regarded as negative.

Velocity, $v(t)$, is the rate of change of displacement with respect to time, so $v = \dfrac{ds}{dt}$. *Acceleration*, $a(t)$ is the rate of change of velocity with respect to time, so $a = \dfrac{dv}{dt}$.

Example 5.24: A particle moves vertically with displacement function $s(t) = 56t - 6t^2$, $t \geq 0$, where s is measured in meters and t in seconds. Find
(a) $v(3)$, the velocity of the particle after three seconds.
(b) the maximum height reached by the particle.
(c) the acceleration of the particle.

Solution 5.24: (a) $s(t) = 56t - 6t^2 \Rightarrow s'(t) = 56 - 12t \Rightarrow v(t) = 56 - 12t$

$\Rightarrow v(3) = 56 - 12 \times 3 = 56 - 36 = 20$. Therefore the velocity of the particle after three seconds is $20\,\text{ms}^{-1}$.

(b) The maximum height of the particle occurs when the velocity is instantaneously zero. So, at maximum height, $v(t) = 0 \Rightarrow 56 - 12t = 0 \Rightarrow t = \dfrac{56}{12} = \dfrac{14}{3}$. Now

$$s\left(\frac{14}{3}\right) = 56 \times \frac{14}{3} - 6\left(\frac{14}{3}\right)^2 = 130\tfrac{2}{3}\,,$$ and therefore the maximum height is $130\tfrac{2}{3}$ m.

(c) $a(t) = v'(t) = -12$, so the acceleration of the particle is $-12\,\text{ms}^{-2}$.

Example 5.25: If y is the displacement of a particle which moves in a straight line relative to a fixed point O and $y(t) = 4\left(1 - e^{-0.5t}\right)$, $t \geq 0$.

(a) Find the distance of the particle from its starting position when $t = 2$.
(b) Find $v(t)$, the velocity of the particle as a function of time.
(c) Find the mean velocity during the first 2 seconds.
(d) Find the acceleration of the particle when $t = 5$.
(e) Describe, in words, the motion of the particle.

Solution 5.25: (a) $y(0) = 4\left(1 - e^0\right) = 0$, $y(2) = 4\left(-e^{-1}\right) = 2.5285$. Therefore the particle is

$2.53 - 0 = 2.53$ meters from its starting position.

(b) $s(t) = 4 - 4e^{-0.5t}$. Use the chain rule to differentiate $s(t)$ and find $v(t)$.

The components of $s(t)$ are $s(u) = 4 - 4e^u$ and $u(t) = -0.5t$

$\Rightarrow s'(u) = -4e^u$, $u'(t) = -0.5$. $v(t) = s'(t) = s'(u) \times u'(t)$

$\Rightarrow v(t) = \left(-4e^u\right)(-0.5) = 2e^u = 2e^{-0.5t}$.

(c) Mean velocity $= \dfrac{\text{distance moved}}{\text{time}} = \dfrac{2.5285}{2} = 1.26425$. Therefore, the mean velocity

is $1.26\,\text{ms}^{-1}$.

(d) $a(t) = v'(t)$. Now, $v(t) = 2e^{-0.5t}$ and, again, use the chain rule to differentiate $v(t)$ to find $a(t)$. The components are $v(u) = 2e^u$ and $u(t) = -0.5t$

$\Rightarrow a(t) = v'(u) = 2e^u$, $u'(t) = -0.5$. $a(t) = v'(t) = v'(u) \times u'(t)$

$= \left(2e^u\right)(-0.5) = -e^u = -e^{-0.5t} \Rightarrow a(5) = -e^{-2.5} = -0.8208499....$ Therefore, the

acceleration after 5 seconds is $-0.0821\,\text{ms}^{-2}$.

(e) The particle moves forward from O with initial velocity of $2\,\text{ms}^{-1}$, but it gradually slows down until its velocity is imperceptible. After it has gone $4m$, it appears to stop.

Exercise 5.12

In questions $1-3$, t is measured in seconds and displacement in meters.

1. A particle moves in a straight line so that its displacement from a fixed point, O is
   $y(t) = 5 + 12t - t^2$, $t \geq 0$. Find

   (a) $y(0)$, the initial displacement of the particle from O.

   (b) $v(t)$, the velocity of the particle in terms of t.

   (c) the value of t when the particle is instantaneously at rest and the displacement of the
   particle at this time.

   (e) the time, t when the velocity is 10.

2. The position of a particle, moving in a straight line at time t, is $x(t) = t^3 - 5t^2 + 3t$, for $t \geq 0$.
   Find

   (a) the initial displacement of the particle.

   (b) $v(t)$, the velocity of the particle in terms of t and show that $v(0) = 3$

   (c) the times when the particle is instantaneously at rest and the distance the particle
   travels between these two times.

3. The velocity of a particle at time t, moving in a straight line relative to a fixed origin, is
   $v(t) = \sin \dfrac{\pi t}{3}$, $t \geq 0$. Find

   (a) the value of t, at the first time that the particle reaches a velocity of $\dfrac{1}{2}$ ms^{-1}.

   (b) $a(t)$, the acceleration of the particle as a function of t.

   (c) the acceleration of the particle when $t = 1$.

4. A man drops a stone from a height of 1.4m above ground level into an old mine shaft. The
   height, s, of the stone above ground level is given by $s(t) = 1.4 - 5t^2$. The man hears the stone
   splash into water at the bottom of the shaft 5.2 seconds after the instant that he releases it.

   (a) Estimate the depth of the mine shaft. Assume that the time taken for the sound of the
   splash to reach the man's ear is negligible.

   (b) Use differentiation to find the velocity v ms^{-1} of the stone in terms of time t and
   show that the speed of the stone as it hits the water is 52 ms^{-1}. Neglect the effect of
   air resistance.

5. A particle moves along a straight line and has position at time t given by $x(t) = \ln\left(1 + 2t^2\right)$ for
   $t \geq 0$, where x is measured in meters and t in seconds.

   (a) Find the velocity function, $v(t)$, of the particle.

   (b) Show that the acceleration function is $a(t) = \dfrac{4\left(1 - 2t^2\right)}{\left(1 + 2t^2\right)^2}$.

   (c) Show that when the acceleration is zero the velocity is $\sqrt{2}$ ms^{-1}

   (d) Using the fact that the average speed is distance travelled divided by time taken, find the
   average velocity of the particle between $t = 1$ and $t = 3$.

UNIT 6: STATISTICS

6.1 Statistics: Basic Concepts

Statistics involve the collection, display and analysis of data. The collection of data and the display of non-numerical data, although very important, fall outside the scope of this course. The focus will be on the introduction of some ideas and techniques which are useful in the analysis of data.

6.1.1 Populations and Samples

A *population* is the group of items, known as *data points*, to which the statistical process is to be applied. Although there is no restriction on the number of items in a population, it will generally contain a large number, or possibly an infinite number, of items. Examples of a population are:

- the number of plants occurring in a given area of land
- the heights of students in a school
- the times required by different people to complete a given task

In order to investigate a population, it is possible to carry out a census in which each item of the population is measured. However, it is usually more practical, more economical and, possibly, more accurate to measure a selection of items taken from the population. This selection is called a *sample*. A statistical analysis of the data obtained from this sample is then carried out in order to gain information about the entire population.

6.1.2 Discrete Data

Discrete data $\{x_i\}$, $i \in \{1, 2, 3, \ldots, n\}$ consists of only exact data points. In general, $x_i \in \mathbb{Z}$. Examples of *discrete data* are:

- the number of cars parked in a parking lot
- the cost of an item of merchandise
- IQ scores
- the numbers on a six-sided die
- the number of balls in a bag

Unprocessed data is called raw data. Data Set 1 shows the raw data of IB grades obtained by a group of 49 Mathematical Studies candidates.

This data may be conveniently processed by arranging them into a frequency table as shown in table 6.1.

4	3	4	5	3	6	2	4	3	6
4	6	2	4	5	3	4	3	7	7
5	6	4	5	4	6	7	6	7	5
5	6	5	5	6	4	4	5	3	3
3	4	5	3	3	4	4	7	7	

Data Set 1

Table 6.1 shows the frequency distribution which is the function $f : x_i \to f_i$ for the data points $i \in \{1, 2, 3, \ldots, 7\}$. In this example, $x_1 = 1, f_1 = 0, x_2 = 2, f_2 = 2 \ldots, x_7 = 7, f_7 = 6$.

Grade, x_i	Frequency, f_i
1	0
2	2
3	10
4	13
5	10
6	8
7	6

Table 6.1

6.1.3 Continuous Data

Continuous data $\{x_i\}$, $i \in \{1, 2, 3, \ldots, n\}$ can only be measured approximately and will usually take values correct to a specified accuracy. The degree of accuracy will depend on the practical difficulties of measuring the data and the use to which the data will be put. In general, $x_i \in \mathbb{R}$. Examples of *continuous data* are:

- the height of apple trees in an orchard
- the times taken by individual students in a group to solve a problem
- the temperatures at a specific point at different times
- the masses of rabbits in a population of rabbits

It is important to realize that it is not the data itself but the nature of what is being measured that makes the data discrete or continuous. For example, the heights of people in a group may be measured to the nearest centimeter so that the data is a set of whole numbers; thus, they may appear to represent discrete data. However, the heights of the people are nevertheless approximations of a value which comes from a continuous number scale.

6.1.4 Grouped Data

It is often convenient to group data into intervals. Both discrete and continuous data can be grouped. Suppose 35 students do an examination and are awarded a percentage grade. Data Set 2 shows the data.

84	73	51	72	66
61	88	46	60	59
37	69	64	59	31
83	55	75	73	67
44	57	77	29	81
60	74	51	52	65
68	63	40	68	65

Data Set 2

The data can be analyzed by placing each data point into a particular interval, known as a *class interval*. The width of the class interval will depend on the number of data points, the accuracy of the data and the purpose for which the data is being collected and analyzed. In general, narrower class intervals will preserve more of the accuracy of the raw data but will involve more work.

The raw data can be transferred to the frequency table by using tally marks, as shown in the second column of table 6.2. This table shows the frequency table for class intervals of $0-9$, $10-19$, $20-29$, $\ldots$

Class Interval	Tally Marks	Frequency
0-9		0
10-19		0
20-29	1	1
30-39	11	2
40-49	111	3
50-59	11111, 11	7
60-69	11111, 11111, 11	12
70-79	11111, 1	6
80-89	1111	4
90-99		0

Table 6.2

The following example shows that continuous data can be grouped in the same way that discrete data is grouped. Data Set 3 shows the weights of 24 tomatoes measured to the nearest 0.1 of a gram. Table 6.3 shows this data grouped into the intervals $25.0 - 29.9,\ 30.0 - 34.9,\ \ldots$

33.5	39.1	29.8	35.5	32.3	37.6
29.4	30.6	35.0	41.2	33.4	34.9
37.1	36.8	28.7	32.7	38.0	30.7
31.9	35.8	37.2	34.4	39.5	36.8

Data Set 3

Class Interval	Tally Marks	Frequency
25.0-29.9	111	3
30.0-34.9	11111, 1111	9
35.0-39.9	11111, 11111, 1	11
40.0-44.9	1	1

Table 6.3

Exercise 6.1

1. Throw a die 60 times and record the score on the die. You may prefer to use the random number function on your graphing calculator to obtain the results more quickly and conveniently. Construct a frequency table of your results.

2   Throw two dice 60 times and record the sum of the scores on both dice. Construct a frequency table of your results.

3. Count the number of tosses of a coin until a head is obtained. Repeat this procedure 30 times and draw a frequency table of your results.

4. Use your graphing calculator to produce 80 random numbers between 0 and 99 inclusive. Arrange your raw data in class intervals $0 - 9,\ 10 - 19,\ 20 - 29,\ \ldots$

5. A traffic survey involves counting the number of cars passing a particular point between $16:00$ and $19:00$ for 50 days. Data Set 4 shows this data. Table 6.4 shows a frequency table for this data. Copy and complete the table.

492	574	541	315	476	594	618	603	405	537
349	638	426	597	531	524	819	430	641	715
390	563	664	495	473	568	507	584	741	722
327	633	447	490	488	587	603	728	426	459
530	556	649	691	388	492	546	702	513	584

Data Set 4

Class Interval	Tally Marks	Frequency
300-349		
350-399	11	2
400-449		
450-499		
500-549	11111, 111	8
550-599		
600-649		
650-699		
700-749		
750-799		
800-850		

Table 6.4

6. Suppose a student plays 50 games of computer solitaire and records the times to complete each game. The results, measured in minutes and seconds, to the nearest second, are shown in Data Set 5. Make a frequency table of the times, t using the 10 intervals, $0:30 \le t < 1:00,\ 1:00 \le t < 1:30,\ \dots\ 5:00 \le t < 5:30,\ 5:30 \le t < 6:00$.

3:23	4:12	1:56	1:45	1:56	1:22	2:30	1:43	2:49	1:38
3:05	2:45	1:12	1:31	2:10	1:41	1:52	1:15	1:39	1:55
0.55	1:23	1:57	1:11	2:16	5:32	1:40	1:03	1:29	1:34
2:51	1:14	1:54	1:36	2:48	0.59	1:57	3:28	1:27	2:21
1:44	2:30	4:37	0:48	2:08	1:39	4:51	1:36	1:07	2:18

Data Set 5

6.2 Frequency Histograms

The distribution of frequencies of continuous data can be represented by a frequency histogram. A *frequency histogram* is a diagrammatic representation of continuous data in which the height of a rectangle represents the frequency of the data.

175	166	172	181	174	171	169	158	166	170
177	182	164	171	179	157	174	164	179	185
187	163	158	170	179	159	176	161	170	183
162	179	182	189	191	184	171	166	168	155
169	166	164	195	163	167	163	170	191	182
179	178	179							

Data Set 6

Data Set 6 shows the heights, measured to the nearest centimeter, of 53 male employees of a hospital. Table 6.5 shows the raw data put into a frequency table with class intervals 155-159, 160-164, 165-169 . . . .

Figure 6.01 shows the frequency histogram which represents this data. Note that the boundaries of the frequency columns of the histogram represent the true limits of the data in each class interval. For example, the data point 165cm represents a height somewhere between 164.5cm and 165.5cm so that the boundaries of the rectangles of the frequency histogram are 154.5 – 159.5 ; 159.5 – 164.5 ; . . . ; 194.5 – 199.5

Class interval	Frequency
155 – 159	5
160 – 164	8
165 – 169	8
170 – 174	10
175 – 179	10
180 – 184	6
185 – 189	3
190 – 194	2
195 – 199	1

Table 6.5

Figure 6.01

Note that the horizontal scale does not start at 0, but at 155. In order to emphasize this fact, a short jagged line is inserted in the horizontal axis as shown.

Note also how the scale is drawn. The scale shown in figure 6.02 is inappropriate because the markings are located at inconvenient points; 154.5, 159.5, 164.5 . . . instead of convenient points 155, 160, 165 . . . .

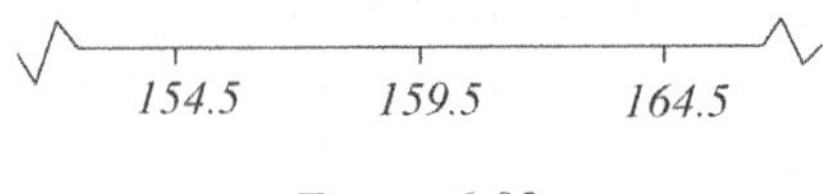

Figure 6.02

The scale shown in figure 6.03 is inappropriate because there are no numbers adjacent to the scale markings and the numerical intervals shown do not correspond to the markings on the scale.

Figure 6.03

Most graphing calculators are able to produce a frequency histogram without the need for the user to first arrange the raw data into a frequency table. Figures 6.04 to 6.06 show how a frequency histogram for data set 6 can be produced using a TI-84 Plus. The raw data is entered into a list (L_1 in this case), and a statistical plot and suitable window are selected.

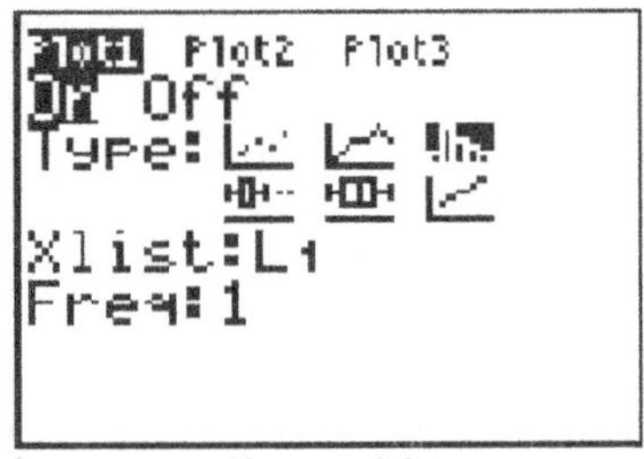

Figure 6.04

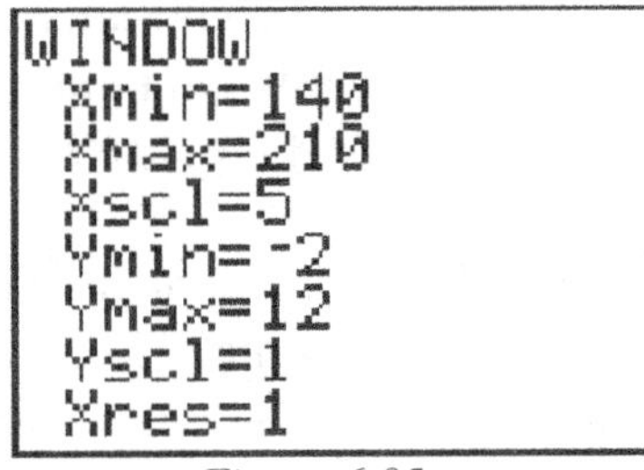

Figure 6.05

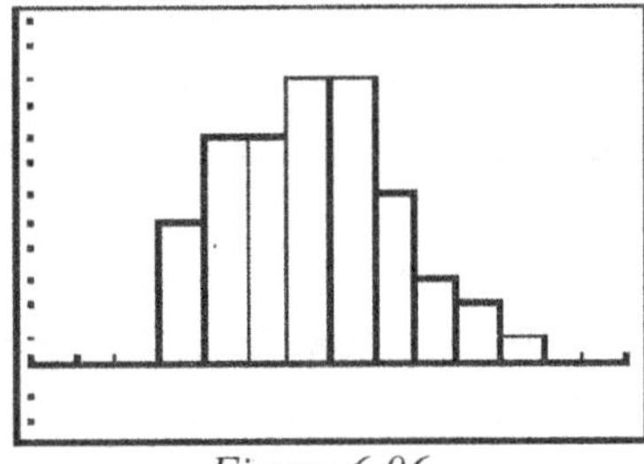

Figure 6.06

The horizontal scale of 5 (Xscl of figure 6.05) is critical, as it is this value which determines how the data is separated into the correct class intervals and the frequency histogram obtained.

This process can also be used, if necessary, to arrange the raw data into a frequency table by using the trace on the frequency histogram to show the frequency in each class interval as shown in figure 6.07.

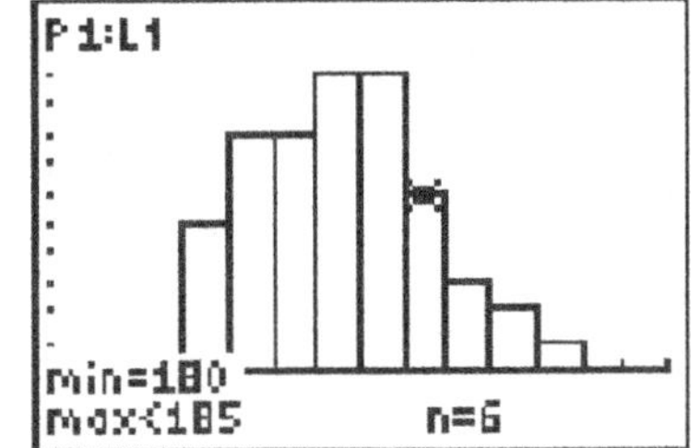

Figure 6.07

Exercise 6.2

1.	33 Granny Smith apples are weighed, and their masses, in grams, are recorded, correct to the nearest gram, as shown in data set 7.

181	188	176	182	185	180	194	177	180	183	183
190	179	181	182	187	173	182	186	187	189	177
192	185	184	188	177	187	190	181	176	189	191

Data Set 7

(a) Group the data into class intervals: 170-174, 175-179 . . . ,   190-194 and construct a frequency table.

(b) Draw a frequency histogram of the data.

2. Use a graphing calculator to generate 100 random numbers between 0 and 99.

(a) Make a frequency table in which the data is grouped into class intervals of width 10 units.

(b) Draw a frequency histogram of the grouped data.

(c) Comment on the expected shape of the frequency histogram. Does your frequency histogram conform to this shape?

3. Draw a frequency histogram of the data from data set 6 with class intervals of

(a) width 4

(b) width 10.

6.3 Cumulative Frequency

6.3.1 The Cumulative Frequency Table

Cumulative frequency is the total frequency of data points up to and including the data point under consideration. Table 6.6, which shows the IB grades of 49 students, is a copy of table 6.1 with a cumulative frequency column added. For grouped data, such as that in table 6.6, the cumulative frequency is the total frequency of all data points less than or equal to the upper class boundary of the class interval under consideration.

Grade, x_i	Frequency, f_i	Cumulative Frequency, F_i
1	0	0
2	2	2
3	10	12
4	13	25
5	10	35
6	8	43
7	6	49

Table 6.6

Table 6.7 shows a cumulative frequency table for the heights of the hospital employees whose frequency table was shown in table 6.5. Note that the upper class boundary is set to reflect the actual maximum height that an employee could be for his particular class interval. Note also that an additional row has been added at the top of the table to include the height below which there are no employees.

Upper Class Boundary	Frequency	Cumulative Frequency
≤ 154.5	0	0
≤ 159.5	5	5
≤ 164.5	8	13
≤ 169.5	8	21
≤ 174.5	10	31
≤ 179.5	10	41
≤ 184.5	6	47
≤ 189.5	3	50
≤ 194.5	2	52
≤ 199.5	1	53

Table 6.7

6.3.2 The Cumulative Frequency Curve

A cumulative frequency curve can be
drawn for grouped data. It is obtained
by plotting the cumulative frequency
against the corresponding upper class
boundary and joining the points with a
smooth curve. Figure 6.08 shows the
cumulative frequency curve drawn from
table 6.7

A cumulative frequency curve can be
drawn on a graphing calculator. Figures
6.09 – 6.12 show how a TI-84 Plus can
obtain a cumulative frequency curve.

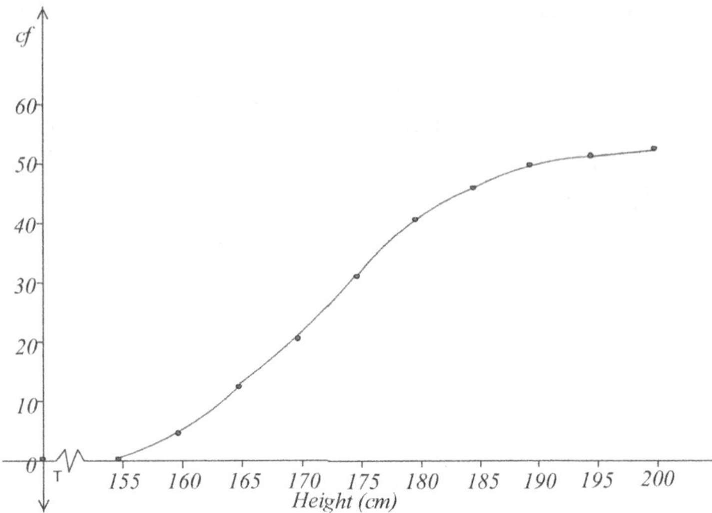

Figure 6.08

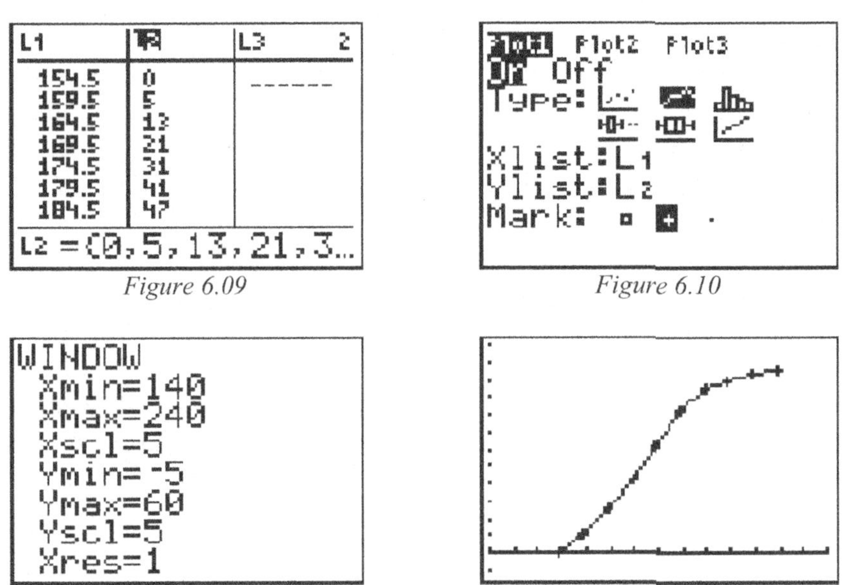

Figure 6.09

Figure 6.10

Figure 6.11

Figure 6.12

The graph obtained in figure 6.12 is not strictly a cumulative frequency curve because adjacent points
are joined by a straight line rather than by a smooth curve. However, in practice, the difference is not
large, provided that the class intervals are not too wide.

6.3.3 Percentiles and Quartiles

The cumulative frequency curve is a way of arranging and displaying the data in numerical order. A
percentile measures the position, in terms of a percentage, of a data point with respect to the rest of the
data. For example, the 90th percentile is that position which has 90% of the data less than or equal to it.
A *quartile* is a specific percentile. The *lower quartile* is the 25th percentile and the *upper quartile* is
the 75th percentile.

Example 6.1: (a) Find the 90th percentile height of the hospital employees (data set 6 and figure 6.08).

(b) Find what proportion of male employees are taller than 182cm.

Solution 6.1: (a) $90\% = 0.9$. $0.9 \times 53 = 47.7$. Therefore, the 90th percentile corresponds to the position 47.7 on the cumulative frequency scale. A construction line is drawn from 47.7 on the vertical axis onto the curve, and then is drawn at right angles to the horizontal axis where it intersects at 186, as shown in figure 6.13. So the 90th percentile height is 186cm.

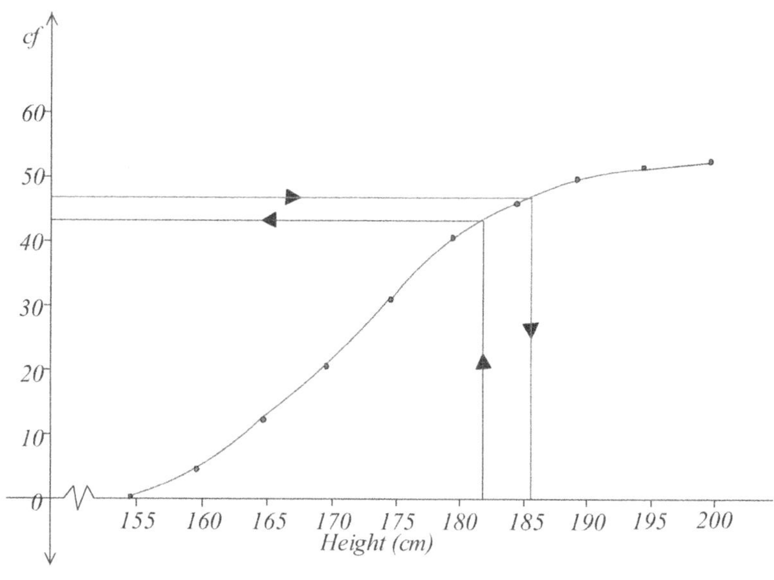

Figure 6.13

(b) A construction line is drawn vertically from 182 on the horizontal axis to the curve and then at right angles to intersect the vertical axis at 44. Therefore, the proportion of male employees who are taller then 182cm is $\dfrac{53-44}{53} = 0.170$ or 17.0%.

Exercise 6.3

1. Table 6.8 shows a frequency table of IB grades. Use table 6.8 to construct a cumulative frequency table.

Grade	Frequency
1	2
2	5
3	7
4	11
5	14
6	6
7	2

Table 6.8

2. Table 6.9 shows a cumulative frequency table of marks obtained in an examination by a group of 115 students.
   (a) Use table 6.9 to construct a cumulative frequency curve .
   (b) Find the (i) 90th percentile mark.
           (ii) 50th percentile mark.

Mark	Cumulative Frequency
≤ 10	0
≤ 20	2
≤ 30	7
≤ 40	13
≤ 50	24
≤ 60	41
≤ 70	75
≤ 80	94
≤ 90	112
≤ 100	115

Table 6.9

3. Table 6.10 shows the area, in hectares, of 87 farms, measured to the nearest 0.1ha (1 ha is $10\,000\,\text{m}^2$).
   (a) Construct a cumulative frequency table with two columns: one showing the upper class boundary and the other showing the cumulative frequency. The initial entry in each column should be 0.
   (b) Use the cumulative frequency table to draw a cumulative frequency curve.

4. The maximum temperature, measured in $^\circ\text{C}$ to the nearest 0.1°C, of 46 children suffering from a fever was recorded as shown in Data Set 8.
   (a) Group the data into appropriate class intervals.
   (b) Construct a cumulative frequency table.
   (c) Draw a cumulative frequency curve.
   (d) Use this curve to find the
       (i)   90th percentile temperature.
       (ii)   temperature above which 60% of the children's temperatures lie.

Area(ha)	Frequency
≤ 9.9	2
10-19.9	11
20-29.9	17
30-39.9	7
40-49.9	11
50-59.9	7
60-69.9	6
70-79.9	9
80-89.9	4
90-99.9	8
100 – 109.9	5

Table 6.10

39.2	39.2	39.5	39.0	39.4	39.3	38.8	39.6	39.4	39.1
39.2	39.3	38.2	38.9	39.0	39.4	39.3	39.1	39.7	39.0
39.2	39.5	38.7	38.1	38.8	39.3	39.0	38.6	39.6	39.3
39.2	39.1	39.4	39.1	39.0	39.1	39.7	39.1	39.2	38.9
38.8	39.0	38.4	38.7	39.1	39.1				

Data Set 8

5. Table 6.11 shows the number of earthquakes experienced at a particular location over a 50-year period. The intensity of the earthquakes is given by the Richter Scale.
   (a) Construct a cumulative frequency table.
   (b) Draw a cumulative frequency curve.
   (c) Find   (i) the 95th   (ii) the 60th   (iii) the 90th percentile.
   (d) Find the lower quartile and the upper quartile.

Earthquake Intensity	<3	3-3.5	3.5-4	4-4.5	4.5-5	5-6	6-7	>7
Frequency	74	55	39	11	7	3	1	0

Table 6.11

6.4 Measures of Central Tendency

'Measures of central tendency' is just a rather verbose term for 'averages' – a single number which is representative of a body of data.

6.4.1 The Arithmetic Mean

The *arithmetic mean* represents a set of data in that it is the sum of that data divided by the number of data points. If data represents an entire population, the arithmetic mean is denoted by μ (pronounced 'mew'). If the data represents a sample, then the arithmetic mean is denoted by $\bar{x}$. In general, the mean of a population, μ, is unknown, and the sample mean, $\bar{x}$, is used as an estimate of the population mean. It is therefore more likely that you will be confronted by sample data rather than data comprising a complete population, so it will generally be assumed that the data presented in an example or an exercise is from a sample.

Other kinds of mean are sometimes used, such as the harmonic mean and the geometric mean. However, the arithmetic mean is by far the most important and widely used, and it will, from now on, be referred to simply as the *mean*.

Ungrouped data: For an ungrouped set of raw data, $\bar{x} = x_1 + x_2 + x_3 + \ldots + x_n = \dfrac{1}{n}\sum_{i=1}^{n} x_i$. For

example, the mean $\bar{x}$ of the following data $93.1, \quad 74.8, \quad 88.1, \quad 80.9, \quad 77.3, \quad 83.7, \quad 90.9, \quad 95.6$

is $\bar{x} = \dfrac{93.1 + 74.8 + 88.1 + 80.9 + 77.3 + 83.7 + 90.9 + 95.6}{8} = \dfrac{684.4}{8} = 85.55$.

When data is presented in a frequency table with N class intervals whose frequencies are

$f_1, f_2, f_3, \ldots, f_N$, $\quad \bar{x} = \dfrac{f_1 x_1 + f_2 x_2 + \ldots + f_N x_N}{n} = \dfrac{1}{n}\sum_{i=1}^{N} f_i x_i$, where $n = f_1 + f_2 + \ldots + f_N = \sum_{i=1}^{N} f_i$.

The mean $\bar{x}$ of the data in data set 1 (see section 6.1.2) can be obtained by adding an extra column to the corresponding frequency table (table 6.1) as shown in table 6.12:

$$\bar{x} = \frac{0 + 4 + 30 + 52 + 50 + 48 + 42}{49} = \frac{226}{49} = 4.61$$

Notice that although the data are integers, that is $x_i \in \mathbb{Z}$, it is acceptable to give the mean as a decimal, that is $\bar{x} \in \mathbb{Q}$.

Grade, x_i	Frequency, f_i	$f_i x_i$
1	0	0
2	2	4
3	10	30
4	13	52
5	10	50
6	8	48
7	6	42

Table 6.12

Grouped data: For grouped data, the mid-interval value of each class interval is used as a representative for the data within that class interval. For data which is grouped into N class intervals, it is assumed that the f_i data points in the ith interval are uniformly spread throughout that interval. Although this is not usually the case, the error which results from this assumption is quite small, particularly when the number of data points per interval is large. The sum of all these f_i data points is $f_i x_i$, where x_i is the mid-interval value of the ith interval.

Example 6.2:   Table 6.5 shows the frequency table of the heights of the 53 hospital employees taken from data set 6. Copy this table and add to it two more columns: one for mid-interval values x_i and the other for the products $f_i x_i$. Use this table to find the mean height of the hospital employees, correct to two decimal places.

Solution 6.2:

Class Interval	Mid-Interval Value, x_i	f_i	$f_i x_i$
155 – 159	157	5	785
160 – 164	162	8	1296
165 – 169	167	8	1336
170 – 174	172	10	1720
175 – 179	177	10	1770
180 – 184	182	6	1092
185 – 189	187	3	561
190 – 194	192	2	384
195 – 199	197	1	197

Table 6.13

Then, as $\sum_{i=1}^{9} f_i = 53$, $\sum_{i=1}^{9} f_i x_i = 9141$, $\bar{x} = \frac{1}{n}\sum_{i=1}^{N} f_i x_i = \frac{1}{53}(9141) = 172.472$. Therefore, the mean height of the 53 employees is 172.47cm.

Although it is important to understand the process of obtaining the mean from a set of grouped data such as this, it will usually be expected that the mean will be obtained by appropriate use of a graphing calculator.

Example 6.2 can be solved by using a TI-84Plus. The mid-interval value for each class interval is entered into a list and the corresponding frequencies into another list. Figures 6.14 – 6.17 show the steps necessary to obtain the mean.

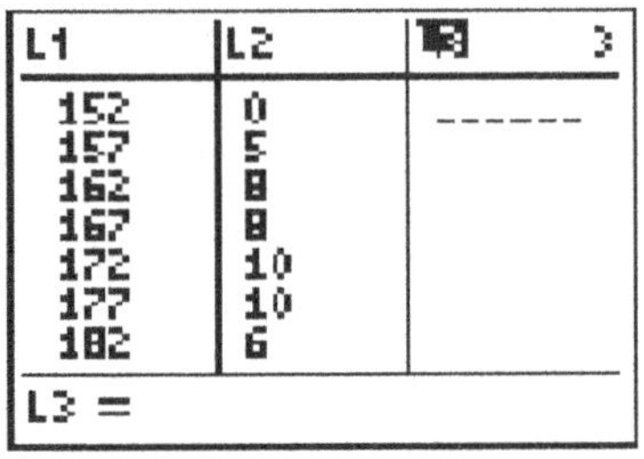

Figure 6.14

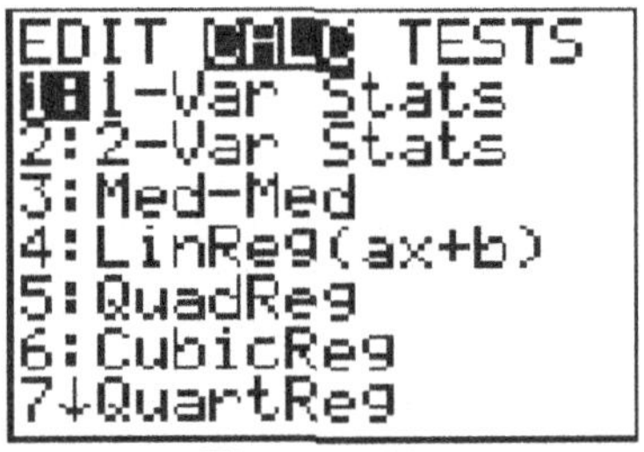

Figure 6.15

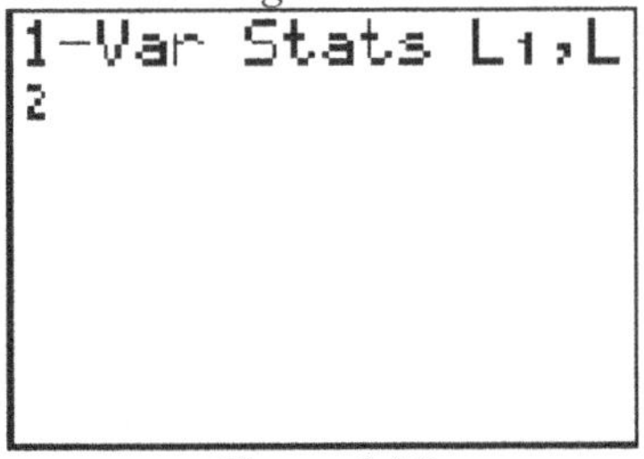

Figure 6.16

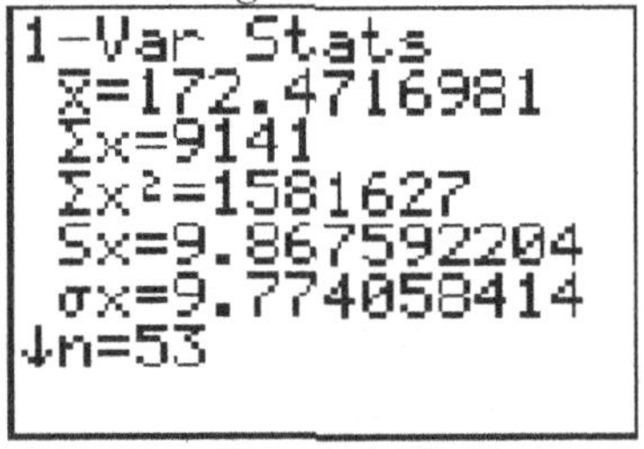

Figure 6.17

Exercise 6.4

1. Find the mean of each of the following sets of numbers.
   (a) 15.7, 18.9, 17.1, 17.5, 19.0, 14.8, 16.2
   (b) 0.519, 0.489, 0.530, 0.493, 0.522, 0.537, 0.481, 0.503
   (c) 1.27×10^{-8}, 1.09×10^{-8}, 9.88×10^{-9}, 1.31×10^{-8}, 1.17×10^{-8}, 1.08×10^{-8}

2. Table 6.14 shows a frequency table of the number of absences during one semester for a class of 23 students.

   Copy this table and add to it a column with values of $f_i x_i$ as in Example 6.2, and hence find the mean number of absences per student for the semester.

Number of Absences, x_i	Frequency, f_i
0	8
1	5
2	4
3	4
4	0
5	1
6	0
7	1
>7	0

Table 6.14

3. The data in table 6.15 shows the weight, in kilograms, of 45 babies at six months of age. Use table 6.15 to construct a table with one column for mid-interval values, a second column for frequencies and a third column for the product of mid-interval values and frequencies. Use it to find the mean weight of the babies.

Weight	Frequency
3 – 5	3
5 – 7	12
7 - 9	15
9 -11	11
11 - 13	4

Table 6.15

In the following questions, you should use your graphing calculator in the most efficient manner possible to find the mean of the data given.

4. In each case find the mean of the given data.
   (a)

x_i	0	1	2	3	4	5	6
f_i	1	4	18	53	37	31	9

   (b)

x_i	10	20	30	40	50	60	70	80
f_i	3	7	18	43	45	28	9	2

   (c)

x_i	0	0.1	0.2	0.3	0.4	0.5	0.6	0.7
f_i	1	2	5	3	8	11	1	3

5. An experiment was repeated 15 times in an attempt to estimate the amount of a particular chemical found in a fixed volume of air. The results are shown in table 6.16. Find the mean weight of chemical.

chemical (mg)	<1	1-2	2-3	3-4	4-5	5-6
frequency	4	5	3	2	0	1

Table 6.16

201

6.	The data set below shows 50 random numbers rounded to three decimal places.

9.730	1.518	6.825	9.700	2.273	3.977	5.560	3.262	5.990	2.673
3.495	3.848	6.336	9.670	1.316	3.502	6.193	7.471	0.036	1.070
1.433	8.220	5.605	8.333	3.249	4.190	4.552	5.532	4.776	6.419
9.905	6.565	3.127	0.902	3.284	0.298	4.359	8.463	5.786	2.637
2.938	7.822	2.333	2.253	7.135	8.022	3.206	3.714	6.640	1.897

(a) Find the mean of the raw data, correct to four decimal places.
(b) Arrange the data into class intervals: $0 - 1, 1 - 2, 2 - 3 \ldots$.
(c) Construct a frequency table with a column for the mid-interval value for each interval and hence find the mean value of the grouped data.
(d) Explain the difference between your answers in (a) and (c).

6.4.2 The Median

The median is the value taken by the middle data point when the data is arranged in ascending order.

Ungrouped Data: For an odd number of data points arranged in ascending order, $\{x_1, x_2, x_3 \ldots x_{2n-1}\}$, the median is the value of x_n. For an even number of data points arranged in ascending order , $\{x_1, x_2, x_3 \ldots x_{2n}\}$, the median is the mean of x_n and x_{n+1}. For example, the median of the data set $\{61, \ 63, \ 63, \ 64, \ \mathbf{66}, \ 69, \ 70, \ 70, \ 71\}$ is the value of x_5 because there are nine data points, so $2n - 1 = 9 \Rightarrow n = 5$. The value of $x_5 = 66$, so the median is 66.

However, the median of the data set $\{61, \ 63, \ 63, \ 64, \ \mathbf{66}, \ \mathbf{69}, \ 70, \ 70, \ 71, \ 72\}$ is the mean of x_5 and x_6, the fifth and sixth data points, because, when there is an even number of data points, there is no middle data point. Therefore the median is $\dfrac{66 + 69}{2} = 67.5$.

It is important to note that the calculation of the mean involves the value of each data point whereas the calculation of the median involves the value of only one, or possibly, two data points. If, for example, the previous set of data had been $\{61, \ 63, \ 63, \ 64, \ 66, \ 69, \ 70, \ 70, \ 71, \ 350\}$, the median would still have been 67.5, whereas the mean would have changed from 66.9 to 94.7. This has important implications when deciding whether the median rather than the mean is an appropriate measure of central tendency.

When data is presented in the form of a frequency table, it is convenient to add a cumulative frequency column and obtain the median from the cumulative frequency column (table 6.17).

There are 49 data points, so the median is the value of the 25th. The cumulative frequency column shows that all the data points from the 13th to the 25th are 4s, so the median is 4.

Grade, x_i	Frequency, f_i	Cumulative Frequency, cf_i
1	0	0
2	2	2
3	10	12
4	13	25
5	10	35
6	8	43
7	6	49

Table 6.17

The following example demonstrates the use of the graphing calculator in determining the median of a set of ungrouped data.

Example 6.3: Table 6.18 shows a frequency table of the number of power cuts which occurred in a given month in 23 cities of a particular country. Find the median number of power cuts.

Number of Power Cuts	0	1	2	3	4	5	6
Frequency	7	4	3	5	2	1	1

Table 6.18

Solution 6.3:

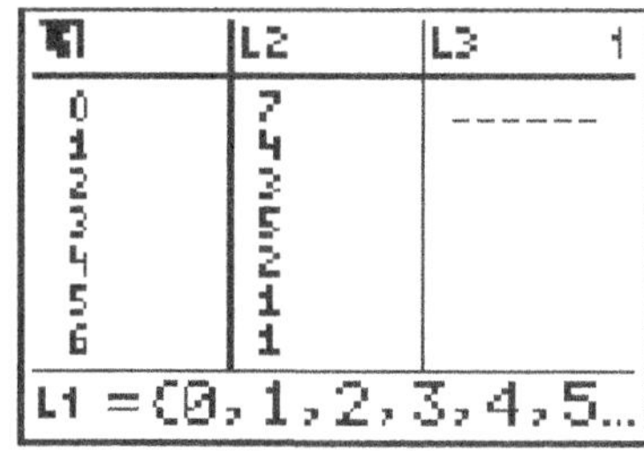

Figure 6.18

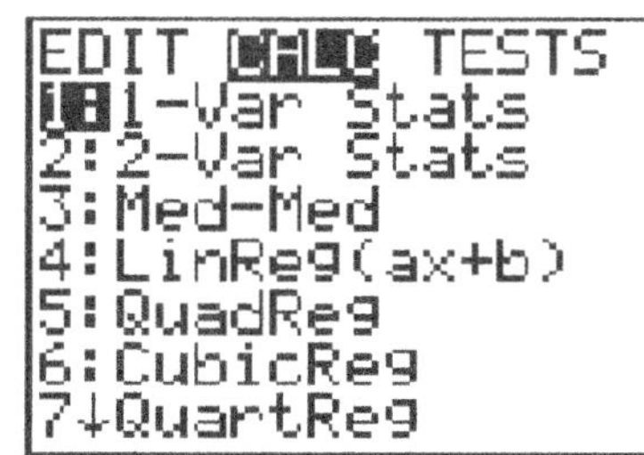

Figure 6.19

Figure 6.20

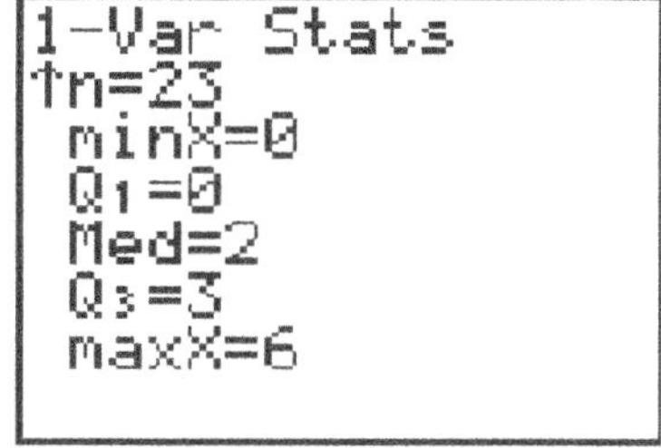

Figure 6.21

Therefore, the median number of power cuts is 2. This is not a difficult calculation and you may feel that, as the median is the value of the 12th data point, it is easier to read the median directly from the frequency table.

Grouped Data: When data is grouped into class intervals, the median is conveniently obtained from a cumulative frequency curve. As the median is the value of the middle data point, it is the 50th percentile and it can be found in a manner similar to that of finding the 90th percentile shown in Example 6.1.

Figure 6.22 shows the cumulative frequency curve of the heights of the hospital employees with the construction lines for estimating the median from the curve. The 50th percentile, from which the median is estimated, corresponds to the 27th data point, because there are 53 data points and so the middle one will be at data point $\left(\dfrac{53+1}{2}\right) = 27$. The construction line starts at 27 on the vertical axis

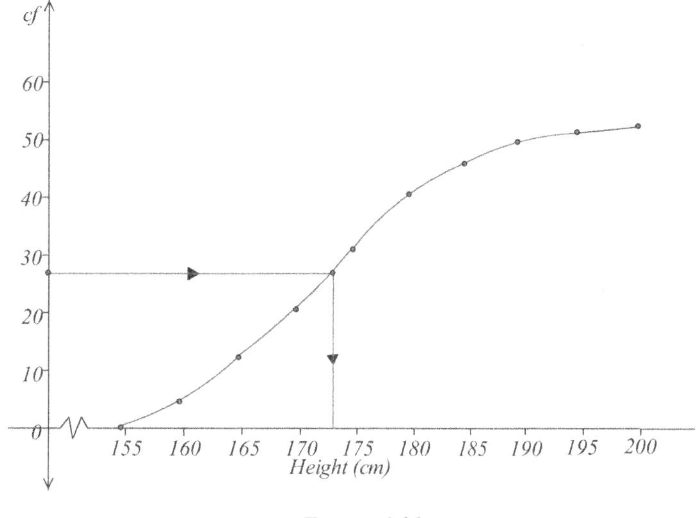

Figure 6.22

and is drawn horizontally until it intersects the curve. Then it is drawn vertically downwards and intersects the horizontal axis at 173. Therefore, the median is 173cm.

The graphing calculator method used to find the median for ungrouped data <u>cannot</u> be used, at least with the TI-84 Plus, to find the median of a set of grouped data because it interprets all data points that lie within a given class interval as taking the mid-interval value of that interval.

However, figures 6.23 to 6.27 demonstrate a method which can be used to estimate the median height of the hospital employees in which the data is grouped.

Use table 6.7 (see section 6.3.1) to enter the upper class boundary values in L_1 and the cumulative frequencies in L_2 .

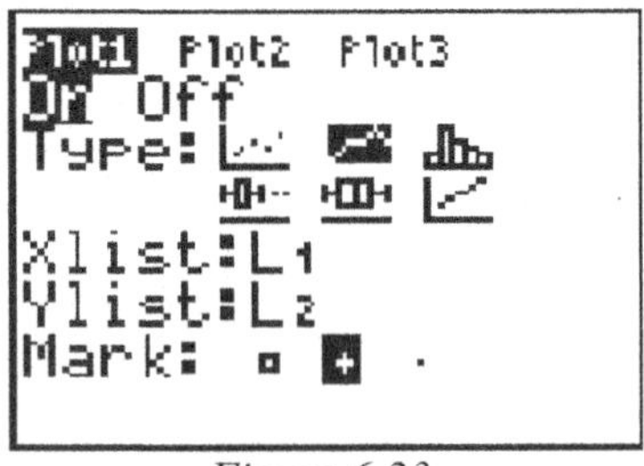

Figure 6.23

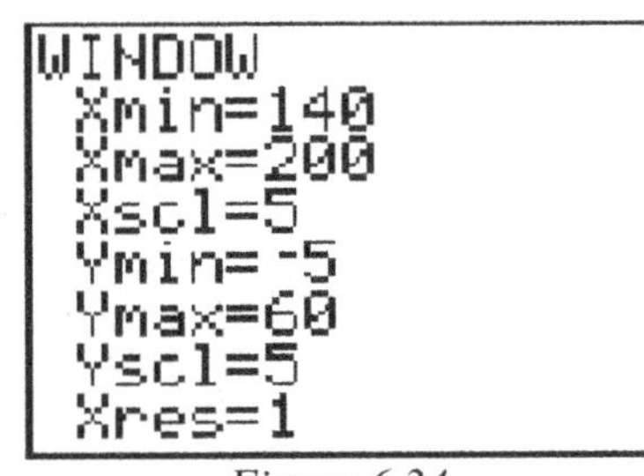

Figure 6.24

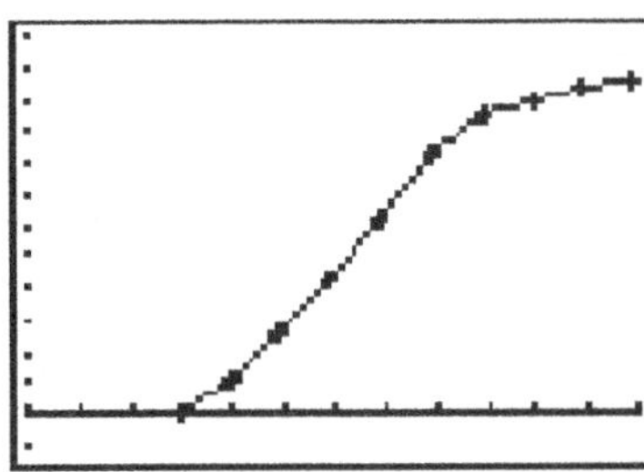

Figure 6.25

Having obtained the cumulative frequency curve, a horizontal line is constructed to the curve from the vertical axis where the cumulative frequency is $\dfrac{53+1}{2} = 27$. Then a line is constructed vertically from there to the horizontal axis.

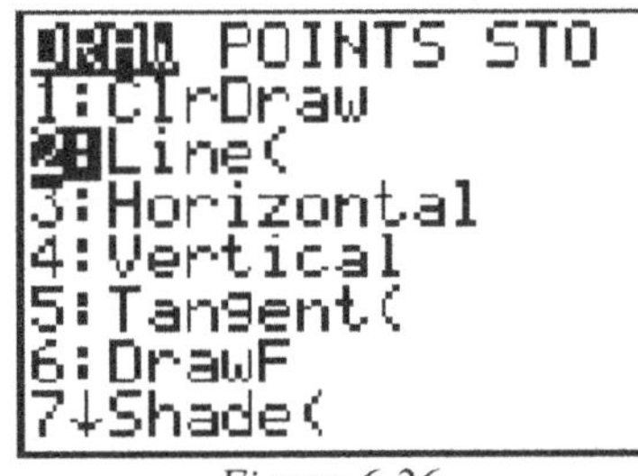

Figure 6.26

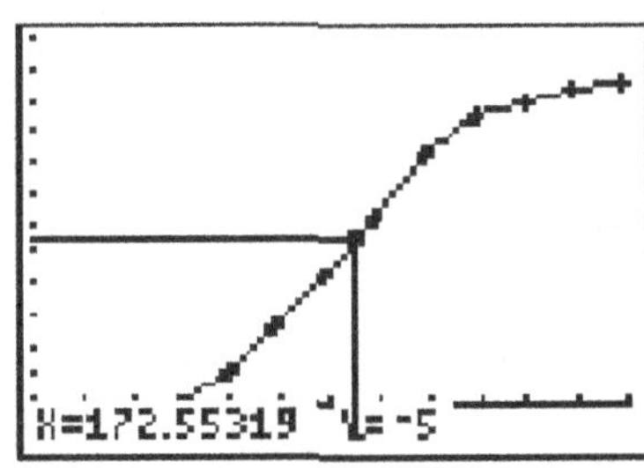

Figure 6.27

In this way, the median is found to be approximately 172.6.

6.4.3 The Mode

The *mode* is the value of the data points which occur most frequently. For example, the mode of the data in table 6.1 on page 191, which shows the IB grades of 49 Mathematical Studies candidates, is 4 because this grade has a higher frequency (13) than any of the other grades. For grouped data, the mode is the value of the class interval which has the highest frequency. It is usually called the *modal class*. For example, the modal class of the data in table 6.2 on page 191 is 60 – 69 because this interval has a higher frequency than any of the other intervals.

Exercise 6.5

1. In each case, arrange the following data in ascending order and hence find the median.
   (a) 42 29 37 50 44 39 32 40 42 45 47
   (b) 1.94 1.86 1.99 2.02 1.88 1.93 1.92 1.95
   (c) 588 595 570 569 589 603 594 577 589 601 585 591 593 583
       577 580 599 602 574 589

2. In each case, find the median of the following data expressed as frequency tables.
   (a)

x_i	1	2	3	4	5	6	7	8	9	10
f_i	0	3	4	9	15	13	14	8	4	1

   (b)

x_i	1	2	3	4	5	6	7	8	9	10
f_i	12	14	9	8	6	3	4	2	0	2

3. Table 6.19 shows, in thousands of dollars, the annual salary distribution of 172 full time employees of a manufacturing company.
   (a) Construct a cumulative frequency table.
   (b) Construct a cumulative frequency curve.
   (c) Use your cumulative frequency curve to find the employees' median salary.

Salary, x_i	Frequency, f_i
$x_i < 20$	28
$20 \le x_i < 30$	59
$30 \le x_i < 40$	37
$40 \le x_i < 50$	24
$50 \le x_i < 60$	11
$60 \le x_i < 70$	13

Table 6.19

4. Use the cumulative frequency curve you drew in answer to question 2 of Exercise 6.3 to find the median mark by drawing appropriate construction lines.

5. Use the cumulative frequency curve you drew in answer to question 3 of Exercise 6.3 to estimate the median area of the farms.

6.5 Measures of Spread

6.5.1 The Inter-Quartile Range

Another important feature of a set of data to consider is the way in which it is spread. Is the data distributed widely or is it clustered closely together?

Data in both (a) and (b), shown below, have the same mean and median, but it is evident that the distribution (i.e., the spread) of the data in (a) is very different from the distribution of the data in (b).

   (a)   31.3 31.3 31.5 31.6 31.7 31.7 32.1
   (b)   9.1 10.3 24.6 31.6 37.8 47.9 54.3

Perhaps the crudest but simplest measure of spread is range. The *range* of a set of data is the difference between the extreme data points. The range of the data in (a) is $32.1 - 31.3 = 0.8$; the range of the data in (b) is $54.3 - 9.1 = 45.2$, which is much greater.

A similar but rather more sophisticated measure of spread is the inter-quartile range in which the range of the middle half of the data is calculated. The lower quartile, Q_L, is the 25th percentile, and the upper quartile, Q_U, is the 75th. The *inter-quartile range* is $Q_U - Q_L$. Figure 6.28 is a reproduction of the cumulative frequency curve drawn in figure 6.08 (section 6.3.2) with the lower and upper quartiles constructed on the figure.

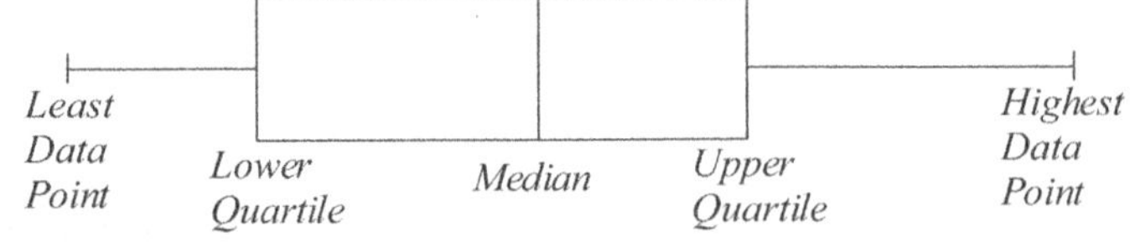

Figure 6.28

The construction lines are drawn from the appropriate positions on the vertical axis. The upper quartile is the height corresponding to the cumulative frequency value of $53 \times 0.75 = 39.75$. The lower quartile is the height corresponding to the cumulative frequency value of $53 \times 0.25 = 13.25$.

Therefore, the value of the upper quartile, Q_U, is 178.5, and the value of the lower quartile, Q_L, is 164.5. The inter-quartile range is $178.5 - 164.5 = 14.0$.

6.5.2 The Box-and-Whisker Plot

A box-and-whisker plot is a diagrammatic way of showing the median, the quartiles and the range of a set of ungrouped data.

An example of a box-and-whisker plot is shown in figure 6.29. The box has a length representing the inter-quartile range, and the whiskers stretch to the extremes of the range.

Figure 6.29

Figures 6.30 – 6.32 show how a TI-84 Plus produces a box and whisker plot for the raw data in data set 9.

17	13	18	11	21	17	14	18
19	15	15	26	19	18	14	14
13	20	18	16	15	19	17	13

Data Set 9

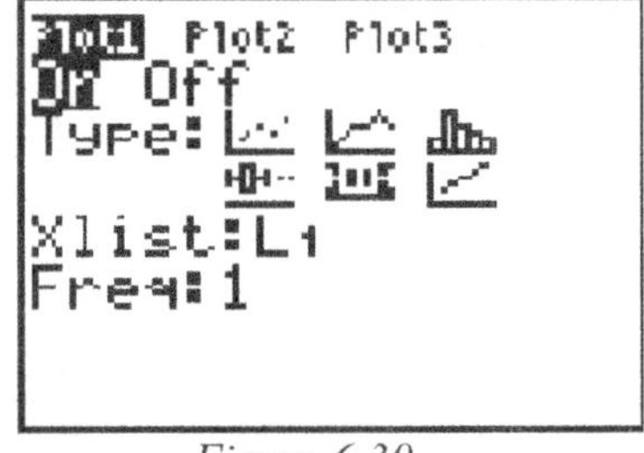

Figure 6.30

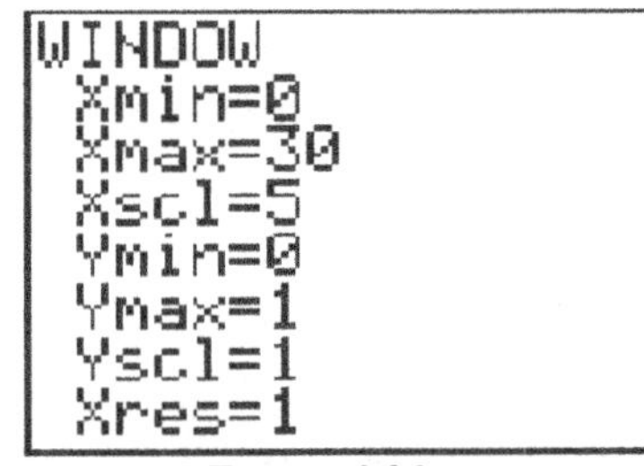

Figure 6.31

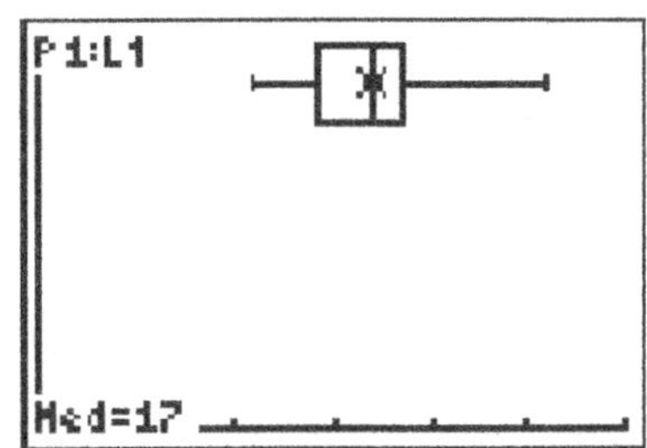

Figure 6.32

The trace enables you to access the median, the quartiles and the extreme values.

6.5.3 The Standard Deviation

The measure of spread which is of most value is called the *standard deviation*. The standard deviation of a set of data which represents an entire population is denoted by σ (sigma). If the data represents a sample from a population, then the standard deviation is denoted by s. It will generally be assumed that the data presented in an example or an exercise is from a sample.

The formula for the standard deviation of a set of ungrouped data, $\{x_i\}$, $i = \mathbb{Z}^+$, is obtained as follows:

- the sum of all the deviations from the mean, $\bar{x}$, would seem to be a measure of the spread of the data. This gives $\displaystyle\sum_{i=1}^{n}(x_i - \bar{x})$.

- However, some of these deviations are positive, others are negative and, when added together, they cancel each other. If the deviations are squared, then they are all positive. When added together, they represent a reasonable measure of the spread of the data. This gives $\displaystyle\sum_{i=1}^{n}(x_i - \bar{x})^2$.

- But, $\displaystyle\sum_{i=1}^{n}(x_i - \bar{x})^2$ depends on n, the number of data points. The more data points there are, the larger the sum of the squares of the deviations will be. This problem can be overcome by dividing the sum by n. This gives $\displaystyle\frac{1}{n}\sum_{i=1}^{n}(x_i - \bar{x})^2$.

- In order to rectify the earlier need to square the deviations, the square root of the expression is taken. This gives the standard deviation formula $s = \sqrt{\dfrac{1}{n}\displaystyle\sum_{i=1}^{n}(x_i - \bar{x})^2}$.

For data grouped into N class intervals, the standard deviation formula is $s = \sqrt{\dfrac{1}{n}\displaystyle\sum_{i=1}^{N}f_i(x_i - \bar{x})^2}$,

where $n = \displaystyle\sum_{i=1}^{N}f_i$. The standard deviation of a population is denoted by σ so that, in the (unlikely) circumstance that all data points of a population are known, $\sigma = \sqrt{\dfrac{1}{n}\displaystyle\sum_{i=1}^{n}(x_i - \mu)^2}$.

However, in general, σ is not known and is estimated from a sample. If the sample is of size n data points, then the best estimate of σ, based on this sample, s_n, is given by $s_n = \sqrt{\dfrac{1}{n-1}\displaystyle\sum_{i=1}^{n}(x_i - \bar{x})^2}$.

The TI-84 Plus uses the notation σ_x to denote the standard deviation of a set of data and s_x to denote an estimate of the standard deviation of a population from a sample. In this course, you will only need to use σ_x.

Example 6.4:   Copy table 6.20 and add to it columns for $f_i x_i$ and $f_i\left(x_i - \bar{x}\right)^2$. Use this extended table to find the standard deviation, s.

x_i	1	2	3	4	5	6	7
f_i	2	4	7	23	20	11	5

Table 6.20

Solution 6.4:   Table 6.21 is a copy of table 6.20 with the required columns added.

x_i	f_i	$f_i x_i$	$f_i\left(x_i - \bar{x}\right)^2$
1	2	2	$2\left(1-4.5\right)^2 = 24.5$
2	4	8	$4\left(2-4.5\right)^2 = 25$
3	7	21	$7\left(3-4.5\right)^2 = 15.75$
4	23	92	$23\left(4-4.5\right)^2 = 5.75$
5	20	100	$20\left(5-4.5\right)^2 = 5$
6	11	66	$11\left(6-4.5\right)^2 = 24.75$
7	5	35	$5\left(7-4.5\right)^2 = 31.25$
	$\displaystyle\sum_{i=1}^{7} f_i = 72$	$\displaystyle\sum_{i=1}^{7} f_i x_i = 324$	$\displaystyle\sum_{i=1}^{7} f_i\left(x_i - \bar{x}\right)^2 = 132$
		$\bar{x} = \dfrac{324}{72} = 4.5$	$s = \sqrt{\dfrac{1}{72} \times 132} = 1.35$

Table 6.21

This is a tedious process and will usually be carried out with the aid of the graphing calculator statistical functions. Figures 6.33 – 6.36 show how to obtain the standard deviation of the data given in table 6.20 using a TI-84 Plus.

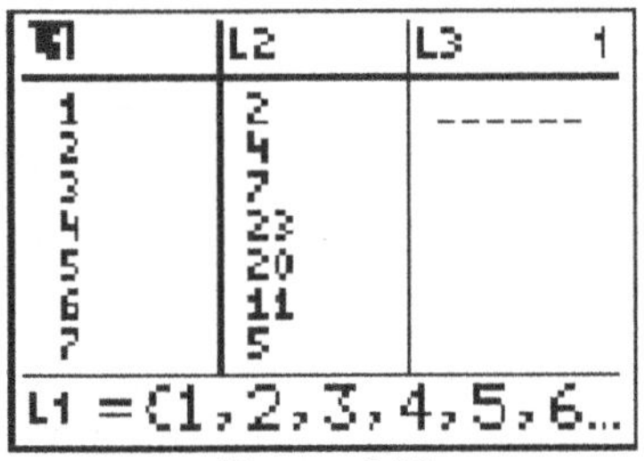

Figure 6.33

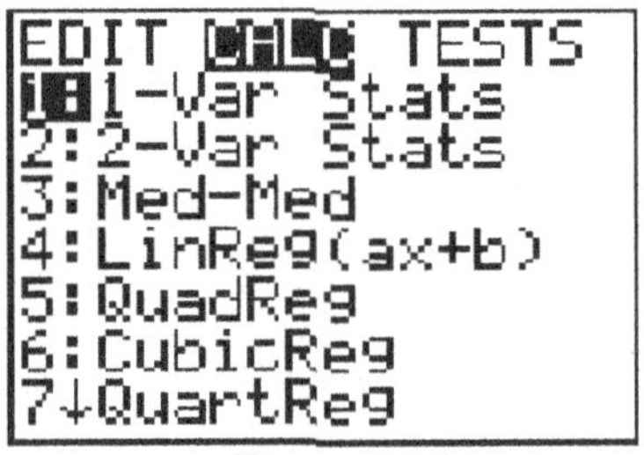

Figure 6.34

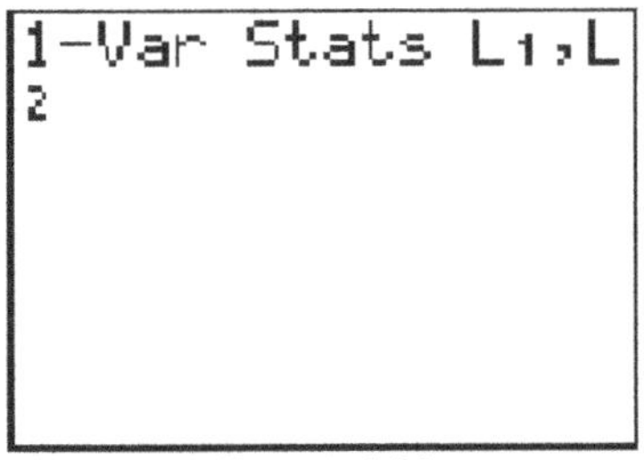

Figure 6.35

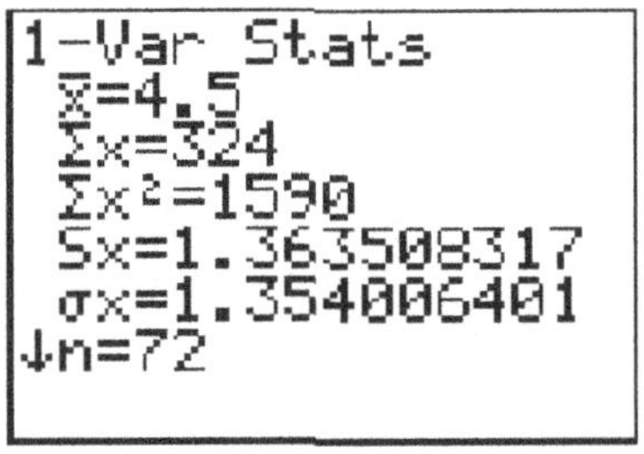

Figure 6.36

Exercise 6.6

1. Use your graphing calculator to make a box-and-whisker plot of the following sets of data. In part (b), the data is shown in the form of a frequency table. In each case, use the plot to obtain the median and inter-quartile range of the data.

   (a)   384, 371, 399, 383, 377, 402, 390, 394, 385

   (b)

x	1	2	3	4	5	6	7
f	2	8	25	47	41	18	8

2. Minimum annual temperatures were recorded at a city in Canada over a period of 100 years. The results are shown table 6.22.

Temperature, T, $^{\circ}C$	$-40^{\circ} \leq T < -35^{\circ}$	$-35^{\circ} \leq T < -30^{\circ}$	$-30^{\circ} \leq T < -25^{\circ}$
Frequency	5	11	26
Temperature, T, $^{\circ}C$	$-25^{\circ} \leq T < -20^{\circ}$	$-20^{\circ} \leq T < -15^{\circ}$	$-15^{\circ} \leq T < -10^{\circ}$
Frequency	37	18	3

   Table 6.22

   (a) Draw a cumulative frequency table and use it to construct a cumulative frequency curve.
   (b) Draw appropriate construction lines to estimate the upper and lower quartiles. Hence estimate the inter-quartile range.

3. In each case, use the formula $s = \sqrt{\dfrac{1}{n}\sum_{i=1}^{n}(x_i - \bar{x})^2}$ to find the standard deviation of the data given.

   (a) 8.5   7.8   8.1   7.9   7.7   8.0
   (b) 123   134   109   115   117   124   130   122   125   111

4. Copy and complete table 6.23 and use it to find the standard deviation, s, of the given data.

The statistical functions on your graphing calculator should be used to obtain answers to the remaining questions in this exercise.

x_i	f_i	$f_i x_i$	$f_i(x_i - \bar{x})^2$
10	2		
20	11		
30	15		
40	9		
50	3		

Table 6.23

5. Find the standard deviation of data set 10
   (a) as ungrouped data.
   (b) grouped into class intervals $0.8 - 0.9$, $0.9 - 1.0$, $1.0 - 1.1$ and $1.1 - 1.2$.
   (c) Explain any differences in your answers to parts (a) and (b).

0.913	1.051	0.876	0.994	0.869	0.904
0.887	1.134	1.002	0.922	0.984	0.970

Data Set 10

6. (a) In each case, find the standard deviation of the following sets of data:
   (i) 33 37 29 31 33 32 35 34   (ii) 1.73 1.91 1.80 1.79 1.83 1.88 1.89
   (b) <u>Write down</u> the standard deviations of the following data:
   (i) 35 39 31 33 35 34 37 36
   (ii) 0.173 0.191 0.180 0.179 0.183 0.188 0.189

7. The distribution of marks in two examinations is shown in table 6.24.
   (a) Find the standard deviation of the marks obtained in the English examination and the Mathematics examination.
   (b) Explain why it might be possible to look at the mark distribution of the English and Mathematics examinations and predict that the standard deviation of the English examination is larger than the standard deviation of the Mathematics examination.

%	English	Math
≤ 9	3	0
10 − 19	17	0
20 − 29	31	4
30 − 39	44	37
40 − 49	71	124
50 − 59	193	298
60 − 69	321	463
70 − 79	215	116
80 − 89	144	16
90 − 99	22	3

Table 6.24

8. Records of rainfall were kept over a period of 80 years in Pluvaria, (A) and in Torrentia (B). Table 6.25 shows a frequency table of this rainfall, measured in centimeters to the nearest centimeter, over an 80 year period.

Find the standard deviation of the rainfall over this 80 year period for both Pluvaria and Torrentia, and describe briefly how the different rainfall trends in the two places are reflected in the two standard deviations.

Rainfall, (cm)	Mid-Interval Value	f_A	f_B
<5	2.5	0	0
5 - 10	7.5	3	0
10 - 15	12.5	15	1
15 - 20	17.5	16	2
20 - 25	22.5	21	23
25 - 30	27.5	12	38
30 - 35	32.5	11	24
35 - 40	37.5	7	2
40 − 45	42.5	4	1
45 - 50	47.5	2	0
>50	-	0	0

Table 6.25

6.6 Regression and Correlation

Regression and correlation provide a tool for investigating whether a relationship exists between two sets of data. Regression is used to estimate a function, the regression function, from a given set of data. Correlation is used to obtain a number which measures the interdependence between two sets of data.

6.6.1 Regression

Consider 10 students, A, B, C, . . ., J who each take a mathematics test and a physics test. Table 6.26 shows, as a percentage, the scores obtained by these students in each of the tests.

Students	A	B	C	D	E	F	G	H	I	J
Mathematics Score, x	53	66	47	48	27	67	49	58	82	85
Physics Score, y	62	71	50	44	42	58	58	65	73	76

Table 6.26

A *scatter diagram* can be plotted in which each student's score is plotted as a point on a grid. For example, the scores of student G would be plotted as a point with coordinates $(49, 58)$ on the grid. Figure 6.37 shows this scatter diagram.

Although the data points do not lie on any clear curve or line they do show a trend so that it might not seem unreasonable to represent the relationship between x and y by a *regression function*, $y = f(x)$. The term regression function is used because it is obtained by working backwards, or regressing, from the given data points to the original function relating x and y.

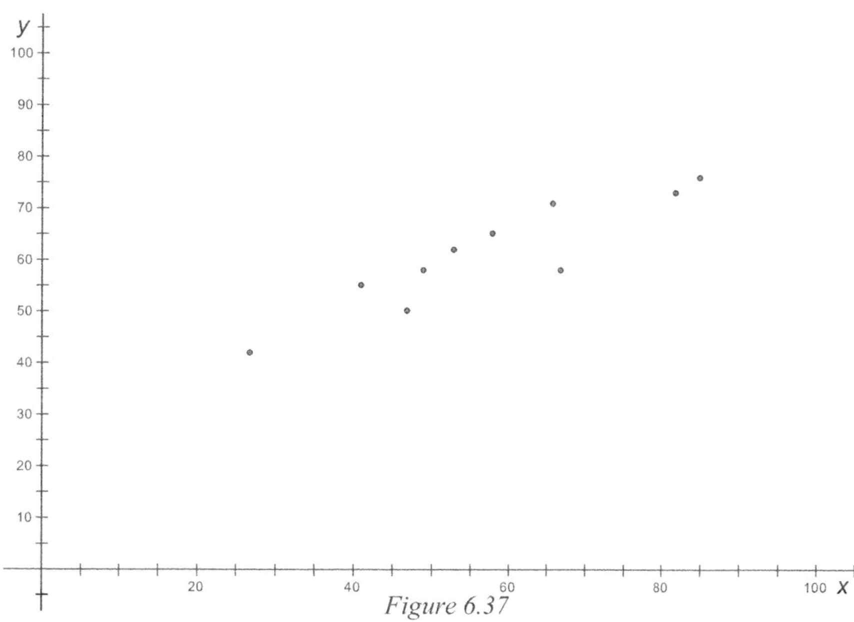

Figure 6.37

Although the regression function can be of any form only linear functions will be considered, in which case, the graph of the regression function is a straight line.

6.6.2 Finding the Regression Line

The simplest method of obtaining the regression line is by drawing a *line of best fit* through the data points on the scatter diagram 'by eye'. However, it is rather subjective and will not be discussed further here. A more objective method is to use *least squares regression of y on x*, described below

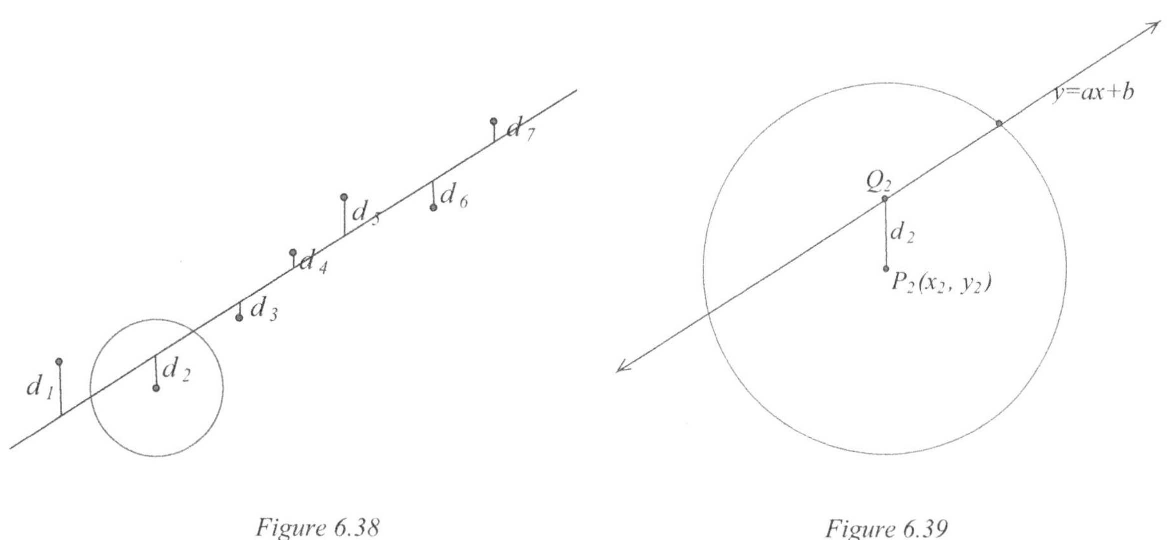

Figure 6.38   *Figure 6.39*

In the method of least squares regression the line of best fit is positioned so that the sums of the squares of the deviations, d_i of the data points from the line of best fit is minimized as shown in figure

6.38. Figure 6.39 is an enlargement of the circle in figure 6.38 and shows a typical data point $P_2(x_2, y_2)$ and a segment of the regression line $y = ax + b$. Q_2 is the corresponding point on the line of best fit and has coordinates $(x_2, ax_2 + b)$. Let $Q_2P_2 = d_2$, where d_2 is the distance parallel to the y-axis of point P_2 from Q_2 so that $d_2 = (ax_2 + b) - y_2$. Then the regression line is obtained by using the condition that the sum of the squares of the distances, d_i for all the data points, $\sum_{i=1}^{n} d_i^2$ is minimized.

For convenience we put $S = \sum_{i=1}^{n} d_i^2$. Therefore, we need to minimize S, where $S = \sum_{i=1}^{n} \left[(ax_i + b) - y_i \right]^2$.

This condition allows a and b to be calculated so that the regression line $y = ax + b$ can be found.

The term 'regression line of y on x' is used because it is assumed that x is accurate and that the data points do not lie on the straight line because of inaccuracies in measuring y. It is these inaccuracies in y which give rise to the deviations, d_i parallel to the y-axis. If it is assumed that y is accurate and that there are inaccuracies in measuring x then we would measure deviations parallel to the x-axis and find the regression line of x on y.

Now we know from Unit 5, section 5.2.2 that minimum values can be found by differentiating the appropriate function and equating the derivative to zero and we can use that fact here by considering both a and b as independent variables. Therefore our criteria for the line of best fit are:

$$\text{(i)} \quad \frac{dS}{da} = 0 \qquad \text{(ii)} \quad \frac{dS}{db} = 0$$

For criterion (i) $\dfrac{dS}{da} = \sum_{i=1}^{n} 2\left((ax_i + b) - y_i\right) \times x_i = 0$ by differentiating $S = \sum_{i=1}^{n} \left[(ax_i + b) - y_i \right]^2$ with respect to a.

$$\Rightarrow 2a\sum_{i=1}^{n} x_i^2 + 2b\sum_{i=1}^{n} x_i - 2\sum_{i=1}^{n} x_i y_i = 0$$

$$\Rightarrow a\sum_{i=1}^{n} x_i^2 + b\sum_{i=1}^{n} x_i - \sum_{i=1}^{n} x_i y_i = 0$$

$$\Rightarrow \sum_{i=1}^{n} x_i y_i = a\sum_{i=1}^{n} x_i^2 + b\sum_{i=1}^{n} x_i \dots\dots\dots\dots\dots\dots(1)$$

For criterion (ii) $\dfrac{dS}{db} = \sum_{i=1}^{n} 2\left((ax_i + b) - y_i\right) \times 1 = 0$ by differentiating $S = \sum_{i=1}^{n} \left[(ax_i + b) - y_i \right]^2$ with respect to b.

$$\Rightarrow 2a\sum_{i=1}^{n} x_i + 2b\sum_{i=1}^{n} 1 - 2\sum_{i=1}^{n} y_i = 0$$

$$\Rightarrow a\sum_{i=1}^{n} x_i + nb - \sum_{i=1}^{n} y_i = 0$$

Now $\dfrac{1}{n}\sum_{i=1}^{n} x_i = \bar{x}$, the mean of x. and $\dfrac{1}{n}\sum_{i=1}^{n} y_i = \bar{y}$, the mean of y.

Therefore, $\dfrac{a}{n}\displaystyle\sum_{i=1}^{n} x_i + \dfrac{nb}{n} - \dfrac{1}{n}\displaystyle\sum_{i=1}^{n} y_i = 0 \;\Rightarrow\; \bar{y} = a\bar{x} + b$ . . . . (2)

From equation (2) we can deduce that the point $(\bar{x},\, \bar{y})$ lies on the regression line.

Figures 6.40 to 6.44 show the procedure on a TI-84 Pus graphing calculator:

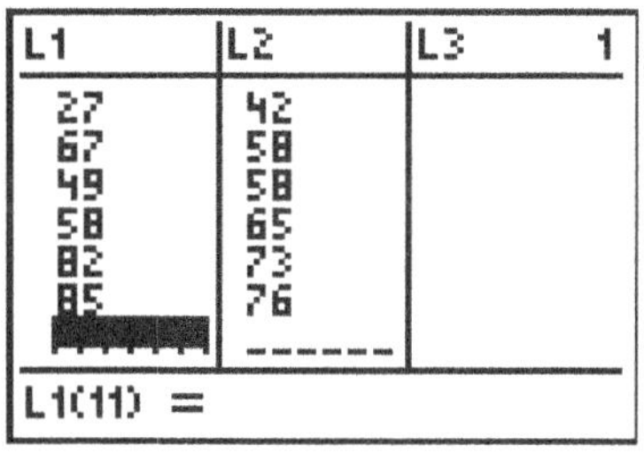

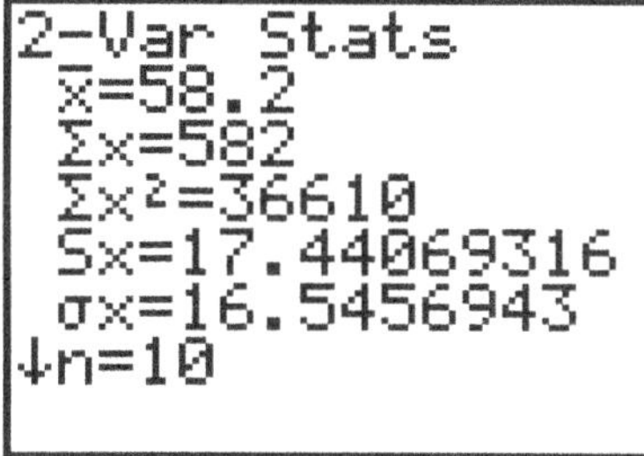

Figure 6.40

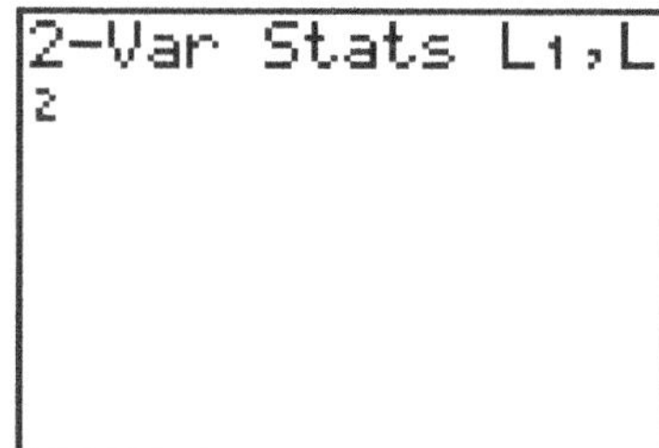

Figure 6.41

Figure 6.42

Figure 6.43

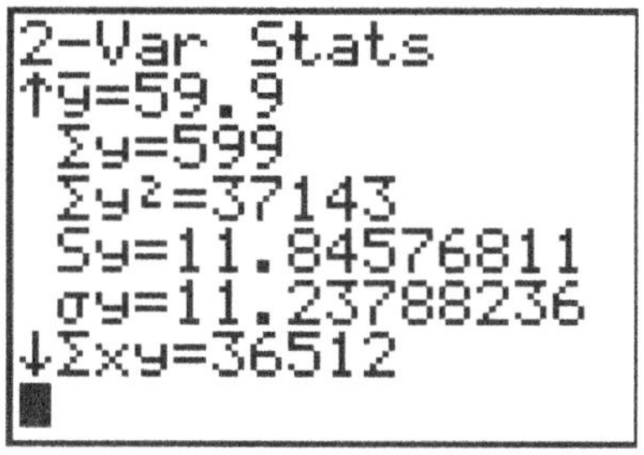

Figure 6.44

So, equation (1) becomes $36512 = 36610a + 582b$ . . . . (3)

And equation (2) becomes $59.9 = 58.2a + b$ . . . . (4)

Equations (3) and (4) are easily solved simultaneously, as shown below:

$$36512 = 36610a + 582\left(59.9 - 58.2a\right)$$

$$1650.2 = 2737.6a \quad \Rightarrow a = 0.60279...$$

$$\Rightarrow b = 59.9 - 58.2 \times 0.60279 = 24.817...$$

So that the equation of the regression line is $y = 0.603x + 24.8$

The TI-84 Plus graphing calculator obtains the regression line without the need to use equations (1) and (2) above, as shown in figures 6.45 – 6.48

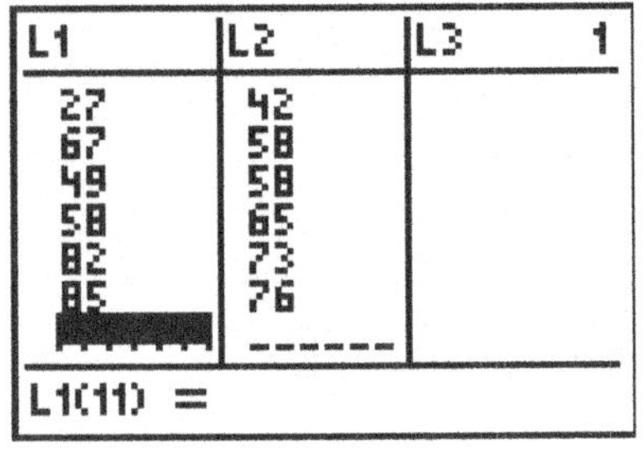

Figure 6.45

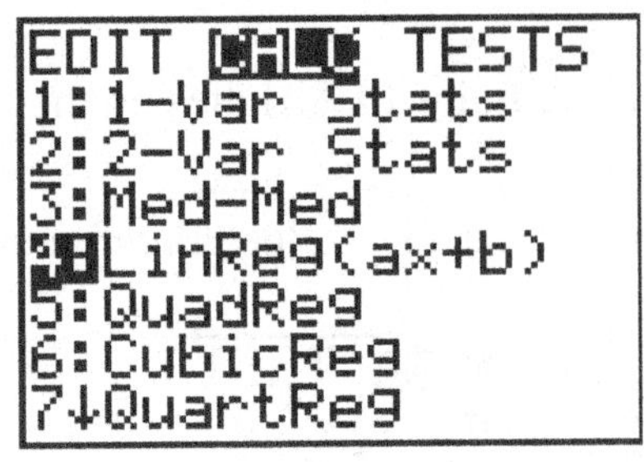

Figure 6.46

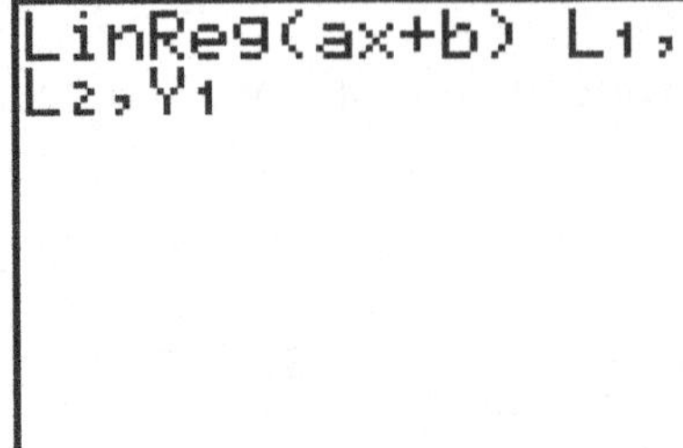

Figure 6.47

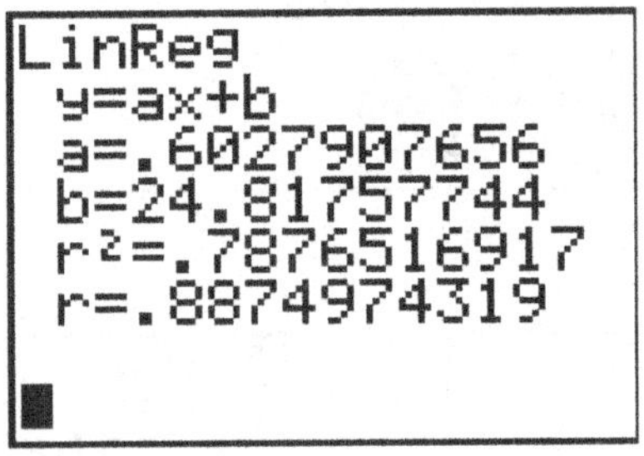

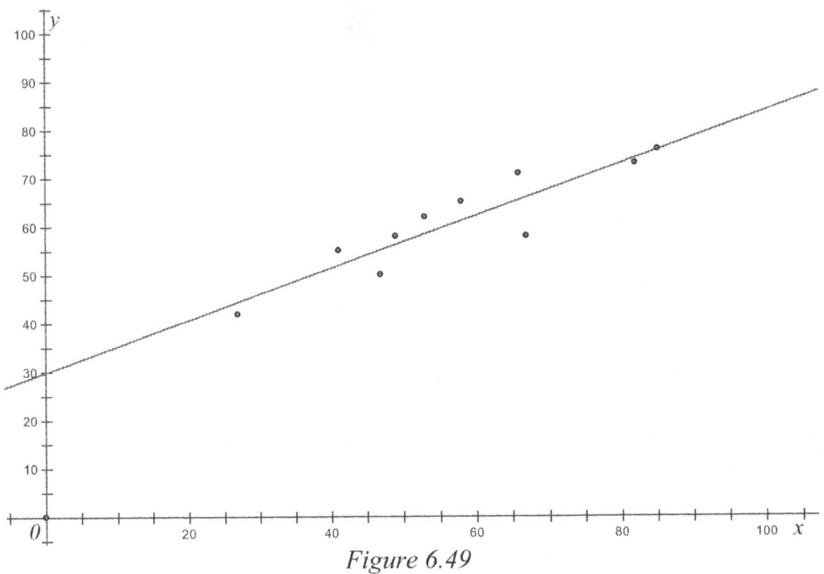

Figure 6.48

The regression line is shown in figure 6.49 superimposed upon the scatter diagram of the original data.

Figure 6.49

The regression line can be used to estimate the value of y for a given value of x.

Example 6.5:

x	3	6	9	12	15	18	21
y	25	31	40	45	49	55	62

Table 6.27

Use the data from table 6.27 to

(a) draw a scatter diagram of the data.

(b) find the equation of the line of best fit in the form $y = ax + b$, using the method of least squares regression.

(c) include the line of best fit on your scatter diagram.

Solution 6.5:

(a)

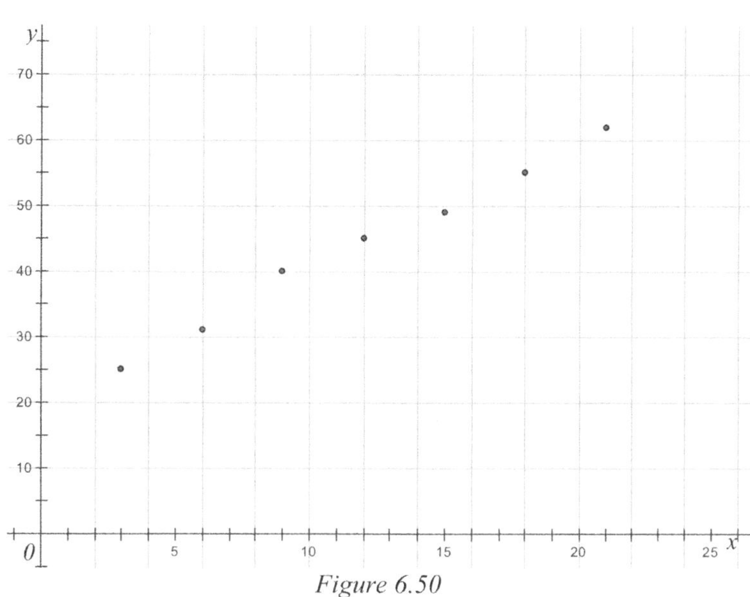

Figure 6.50

(b) The regression line is usually calculated with the aid of the graphing calculator statistical functions without need of written working. The line of best fit is $y = 2x + 19.9$

(c) Figure 6.51 shows the line of best fit superimposed upon the scatter diagram

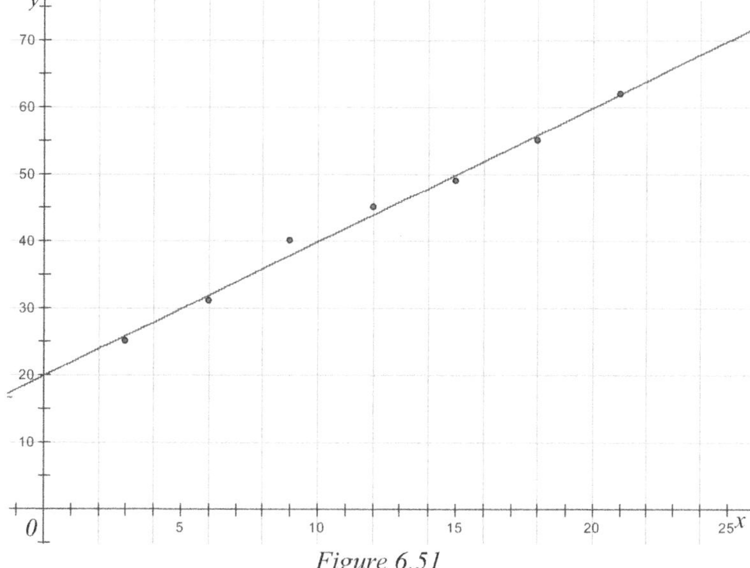

Figure 6.51

215

6.6.3 Correlation

The correlation between two sets of data, $\{x_i\}$ and $\{y_i\}$ provides a measure of the interdependence of two sets of data, $\{x_i\}$ and $\{y_i\}$. It can be measured by Pearson's product-moment correlation coefficient r, where $r = \dfrac{s_{xy}}{s_x s_y}$ (from now on the term *correlation coefficient* will be used).

$s_{xy} = \dfrac{1}{n}\sum_{i=1}^{n}(x-\overline{x})(y-\overline{y})$ is called the *covariance* of the two data sets and s_x is the standard deviation of $\{x_i\}$ and s_y is the standard deviation of $\{y_i\}$.

If these data points are completely interdependent, for example, if $y = ax + b$ then it can be shown that $r = 1$, if $a > 0$ and $r = -1$ if $a < 0$. If, on the other hand, the data sets are independent or only partially interdependent then $s_{xy} < s_x s_y$ and r takes values between -1 and $+1$ so that, in general, $-1 \le r \le 1$, and the correlation coefficient gives us a way of assessing how strongly or weakly the data sets are related or interdependent.

Figure 6.52 shows a scatter diagram of two data sets $\{x_i\}$ and $\{y_i\}$ for which the correlation coefficient is 0.976. Because the data points show a strong linear tendency we say that there exists a strong correlation between x and y. It is clear that if a regression line were drawn through the data it would have a positive gradient, therefore we say that the data points show a strong positive correlation.

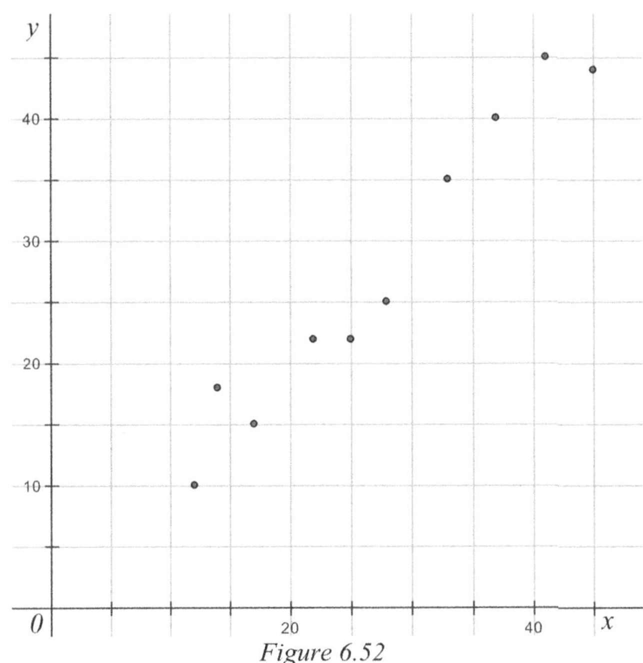

Figure 6.52

Figure 6.53 shows a scatter diagram of data sets $\{x_i\}$ and $\{y_i\}$ for which the correlation coefficient is 0.460. The data points show only a very weak linear tendency and although it would be difficult to draw a meaningful line of best fit through the data it is clear from the positive value of the correlation coefficient that the data shows weak positive correlation between x and y.

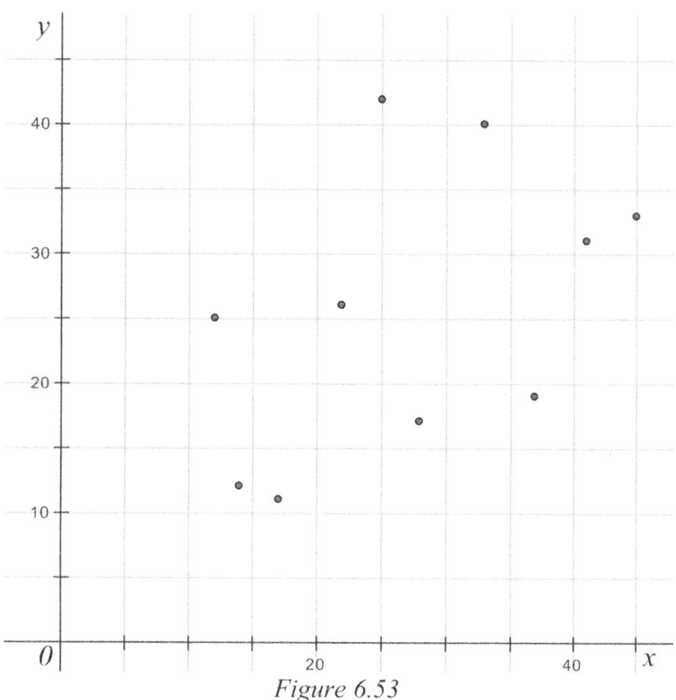

Figure 6.53

Figure 6.54 shows a scatter diagram of data sets $\{x_i\}$ and $\{y_i\}$ for which the correlation coefficient is -0.996. If a line of best fit were drawn through the data points it would have a negative gradient so we say that the data points show a strong negative correlation.

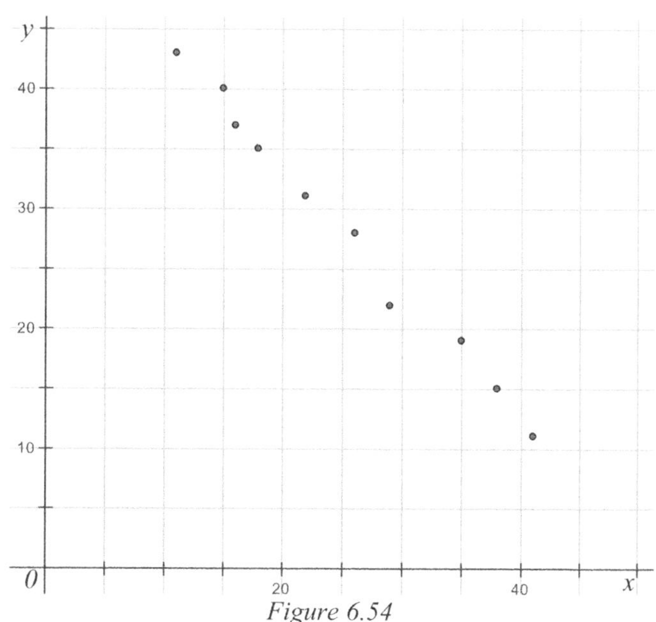

Figure 6.54

Figure 6.55 shows a scatter diagram of data sets $\{x_i\}$ and $\{y_i\}$ for which the correlation coefficient is -0.308 and we say that the data points show a weak negative correlation between x and y.

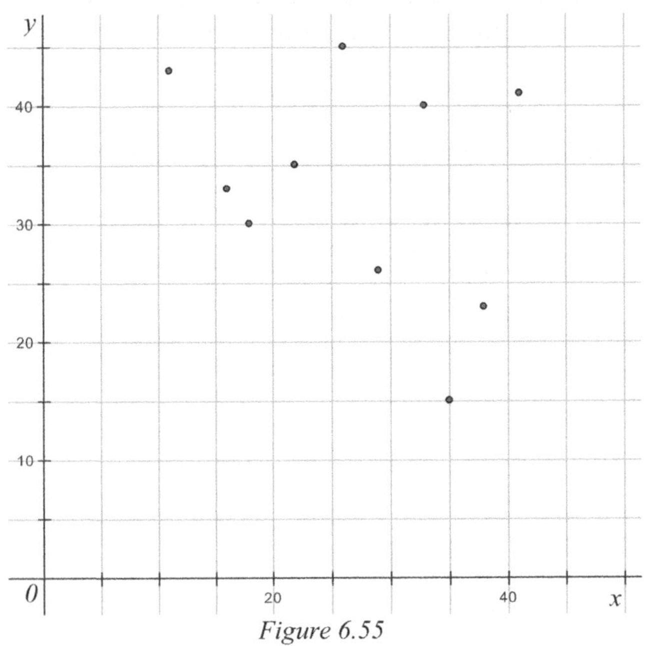

Figure 6.55

Example 6.6: Using the data from table 6.28
    (a) draw a scatter diagram
    (b) calculate the correlation coefficient.

x	13	28	41	72	116	145	183	214
y	38	33	45	59	74	78	85	93

Table 6.28

Solution 6.6:
    (a)

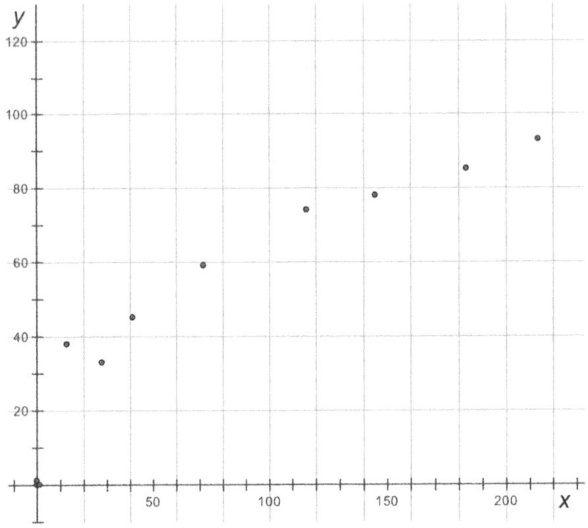

Figure 6.56

(b) Figures 6.57 - 6.60 show how to obtain the correlation coefficient for the data of table 6.28. Figure 6.60 shows that the correlation coefficient, $r = 0.9779798... = 0.978$, correct to three significant figures.

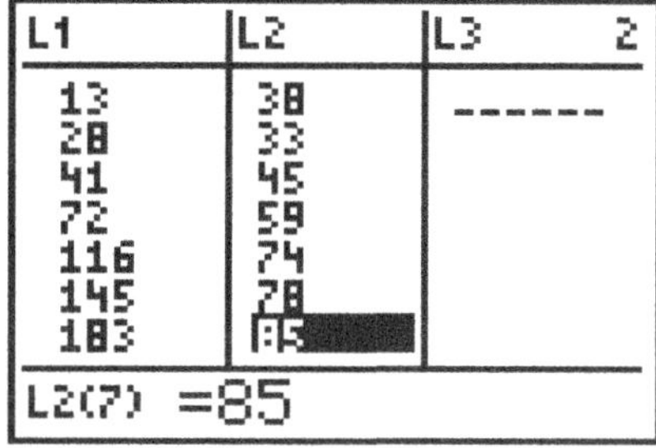

Figure 6.57

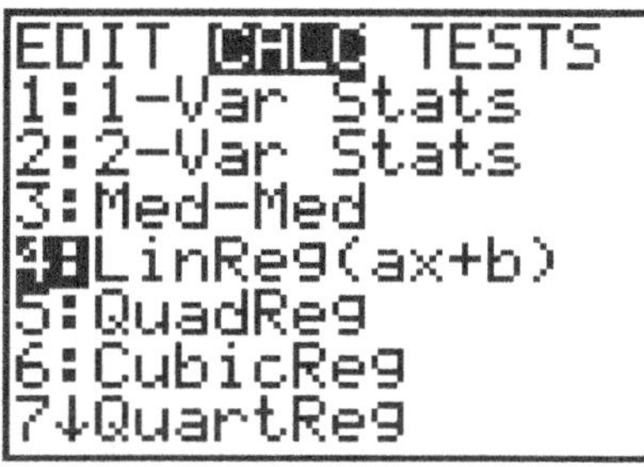

Figure 6.58

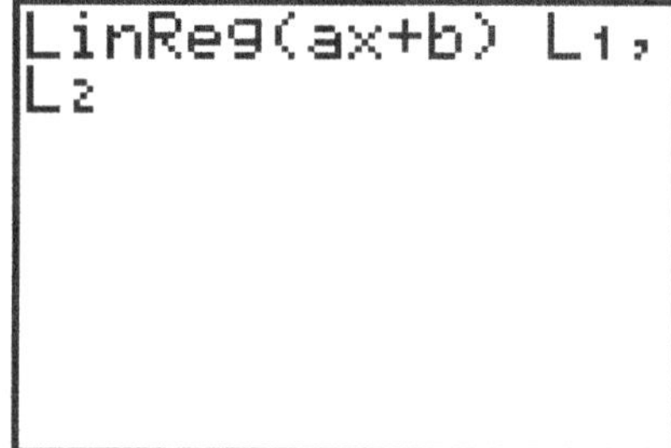

Figure 6.59

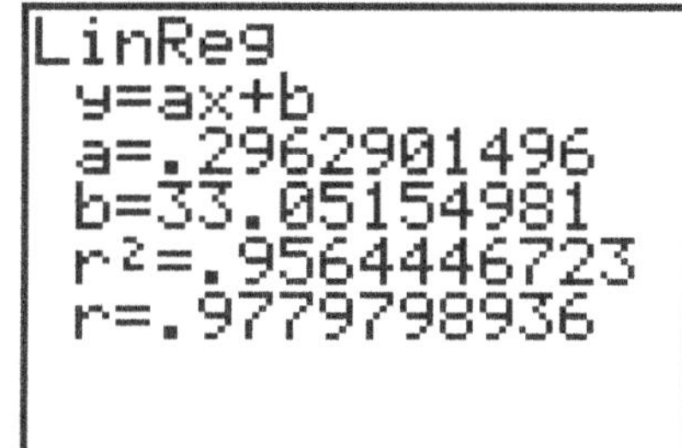

Figure 6.60

NOTE: If r does not appear underneath the value of the constants for the regression line then they can be obtained by going to the catalog of the graphing calculator and changing "diagnostics off" to "diagnostics on" as shown in figures 6.61 - 6.63 below.

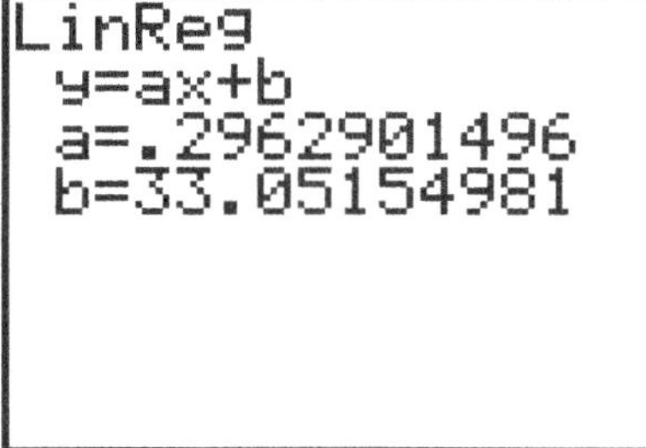

Figure 6.61

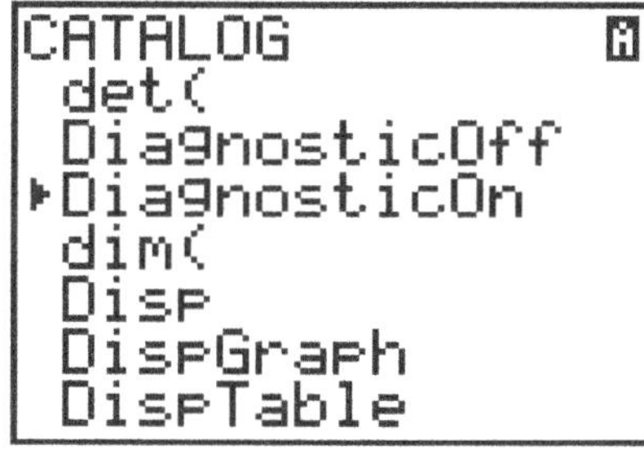

Figure 6.62

a=.2962901496
b=33.05154981

DiagnosticOn
                    Done

Figure 6.63

Exercise 6.7

1. For the following three data sets;
    (i) draw a scatter diagram
    (ii) find the equation of the regression line of y on x.
    (iii) draw the regression line on the scatter diagram.

(a)

x	1	2	3	4	5	6	7	8
y	15	21	28	29	36	42	45	54

(b)

x	10	31	55	70	83	95	102	115	131	146	155
y	77	76	70	65	63	52	47	40	36	33	25

2.   Table 6.29 shows the height, h (in meters) and the weight, w (in kilograms) of 10 children.

h	1.41	1.06	1.53	1.34	1.17	1.01	1.35	1.44	1.62	1.25
w	53.4	40.6	51.9	43.7	44.5	39.7	50.1	46.4	61.3	42.1

Table 6.29

(a) Draw a scatter diagram and mark on it the point $\left(\overline{h},\ \overline{w}\right)$.

(b) Find the equation of the regression line of w on h and draw it on the scatter diagram.

(c) Estimate the weight of a child whose height is 1.50 m.

3.   For the following three sets of data,
(i) draw a scatter diagram   (ii) find the correlation coefficient

(a)

x	10	20	30	40	50	60	70	80	90	100
y	18	27	31	38	47	57	63	70	77	84

(b)

x	0.147	0.231	0.304	0.394	0.422	0.468	0.496	0.513
y	8.46	8.09	8.03	7.94	7.96	7.91	7.61	7.43

(c)

x	1.03	1.08	1.12	1.19	1.23	1.29	1.32	1.37	1.42	1.45	1.49
y	7.31	6.94	6.43	5.98	5.41	5.02	4.66	4.11	3.74	3.21	2.85

4.   Draw scatter diagrams to represent a set of 8 data points for which the correlation coefficient is approximately   (a)   1   (b) -1 (c)   0

5.   Table 6.30 shows the number of games won and the number of goals scored by 20 football teams during a season.

Team	A	B	C	D	E	F	G	H	I	J
Games won (x)	31	30	13	59	25	17	37	51	9	28
Goals scored (y)	64	76	44	93	71	41	50	88	26	43
Children	K	L	M	N	O	P	Q	R	S	T
Games won (x)	34	12	43	36	54	16	8	40	21	39
Goals scored (y)	68	39	85	61	76	31	24	81	48	68

Table 6.30

(a)   (i) Draw a scatter diagram of the data.
       (ii) Comment on any relationship which is suggested by the scatter diagram.
       (iii) Find the correlation coefficient.
(b) Find the equation of the regression line of y on x.
(c) Team U won 30 games during the season. Estimate the number of goals scored by team U during the season.

UNIT 7: INTEGRATION

7.1   Anti-Differentiation

The process of obtaining the original function from its derivative is known as *integration*, sometimes referred to as *anti-differentiation*. For example, if $f(x) = x^2$, then $f'(x) = 2x$, so an anti-derivative of $2x$ is x^2. However, the derivative of $g(x) = x^2 + 1$ is $g'(x) = 2x$, so an anti-derivative of $2x$ is also $x^2 + 1$. In fact, there are a large number of possible anti-derivatives of $2x$, each one differing from the others only by a translation in the direction of the y-axis. Therefore, we write that $f'(x) = 2x \Rightarrow f(x) = x^2 + c$ where $c \in \mathbb{R}$, and c is called the *constant of integration*. Figure 7.01 shows a few examples of the function $y = x^2 + c$ for different values of c.

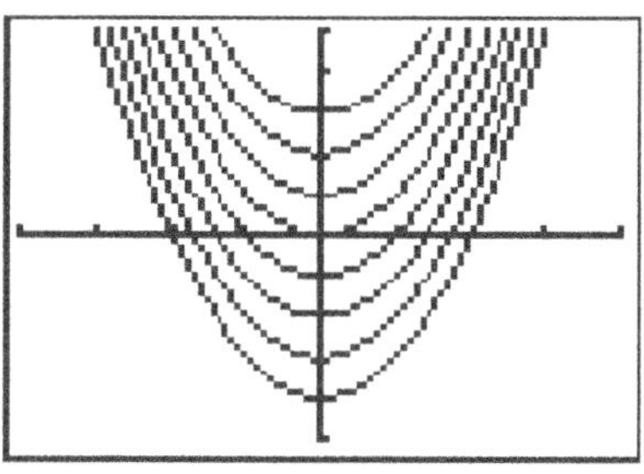

Figure 7.01

To obtain a function from its derivative, one may use inspection (mathematicians' term for 'experienced guesswork') to find that function which, when differentiated, gives the required derivative. For example, the anti-derivative of $f'(x) = 4x$ is $f(x) = 2x^2 + c$, the anti-derivative of $f'(x) = 3x^2 + 6x - 4$ is $f(x) = x^3 + 3x^2 - 4x + c$ and so on. However, it will be helpful to obtain a general formula for integrating a polynomial function:

Let $g(x) = x^m \Rightarrow g'(x) = mx^{m-1}$, and let $m = n+1$ so $g(x) = x^{n+1} \Rightarrow g'(x) = (n+1)x^n$.

If $f(x) = \dfrac{1}{n+1} g(x)$, then $f(x) = \dfrac{1}{n+1} x^{n+1} + c \Leftrightarrow f'(x) = x^n$.

Therefore, $f'(x) = x^n \Leftrightarrow f(x) = \dfrac{1}{n+1} x^{n+1} + c, \; n \neq -1$.

The anti-derivatives of all the basic functions are summarized in table 7.1

$f'(x)$	$f(x)$
x^n	$\dfrac{1}{n+1} x^{n+1} + c, \; n \neq -1$
x^{-1}	$\ln x + c, \; n = -1$
e^x	$e^x + c$
$\sin x$	$-\cos x + c$
$\cos x$	$\sin x + c$

Table 7.1

7.2 Indefinite Integration

A function f may be integrated even if it has not already been differentiated. The result of integrating a function is a family of functions, its *integral*, which is indefinite in the sense that it contains an unknown constant, c, as explained in section 7.1. For that reason, this process is called *indefinite integration*.

New notation is needed to indicate this process of indefinite integration. An instruction to integrate a function, $f(x)$, is written $\int f(x)dx$. '$\int ....dx$' signifies that $f(x)$ is to be integrated. The symbol '$\int$' is a kind of S, indicating summation. Integration is a summation process.

Example 7.1: Find $\int (4x+5)dx$.

Solution 7.1: $\int (4x+5)dx = 4\left(\dfrac{x^2}{2}\right)+5x+c = 2x^2+5x+c$

Exercise 7.1

1. Find the following integrals.

    (a) $\int 3x^2 dx$ (b) $\int x^4 dx$ (c) $\int x^{-2} dx$ (d) $\int (x^2+x+1)dx$

    (e) $\int (2-5x)dx$ (f) $\int x(3x+2)dx$ (g) $\int x^{\frac{1}{3}} dx$ (h) $\int \dfrac{1}{\sqrt{x}}dx$

    (i) $\int x^2(2x-3)dx$ (j) $\int \left(\dfrac{x+1}{\sqrt{x}}\right)dx$

2. Find the following integrals.

    (a) $\int 3\sin x\,dx$ (b) $\int \dfrac{1}{5}\cos x\,dx$ (c) $\int (\cos x+\sin x+1)\,dx$ (d) $\int 4e^x dx$

    (e) $\int (e^x-\cos x)dx$ (f) $\int (2\cos x-\sin x+e^x)dx$

3. Find the following integrals.

    (a) $\int \dfrac{2}{x}\,dx$ (b) $\int \dfrac{3x+2}{x}\,dx$ (c) $\int \dfrac{x^4+3x^2-1}{x}\,dx$ (d) $\int \dfrac{1}{4x}\,dx$

    (a) $\int \dfrac{1+4x-x^2}{3x^2}\,dx$ (f) $\int \left(\dfrac{1}{2x}+2\right)dx$

4. Use the formulae $\sin 2\theta = 2\cos\theta\sin\theta$ and $\cos 2\theta = 1-2\sin^2\theta = 2\cos^2\theta -1$ to show the following integrals. (See Unit 3, section 3.3)

    (a) $\int \sin\dfrac{1}{2}x\cos\dfrac{1}{2}x\,dx = -\dfrac{1}{2}\cos x+c$ (b) $\int \sin^2\dfrac{1}{2}x\,dx = \dfrac{1}{2}x-\dfrac{1}{2}\sin x+c$

    (c) $\int \cos^2\dfrac{1}{2}x\,dx = \dfrac{1}{2}x+\dfrac{1}{2}\sin x+c$

7.3 Integration of Composite Functions

This section limits itself to the integration of composite functions of the form $f(ax+b)$ where f is one of the functions x^n, e^x, $\sin x$ or $\cos x$.

As these functions are easily integrated by inspection, their integrals will be given in table 7.2 as standard forms.

$$\int (ax+b)^n \, dx = \frac{1}{a(n+1)}(ax+b)^{n+1} + c, \ n \neq -1$$

$$\int \frac{1}{ax+b} dx = \frac{1}{a}\ln(ax+b) + c$$

$$\int e^{ax+b} dx = \frac{1}{a}e^{ax+b} + c$$

$$\int \sin(ax+b)dx = -\frac{1}{a}\cos(ax+b) + c$$

$$\int \cos(ax+b)dx = \frac{1}{a}\sin(ax+b) + c$$

Table 7.2

Example 7.2: Find $\int (3x+8)^4 \, dx$.

Solution 7.2: $a = 3$, $b = 8$ and $n = 4$, so that

$$\int (3x+8)^4 \, dx = \frac{1}{3\times 5}(3x+8)^5 + c$$

$$= \frac{1}{15}(3x+8)^5 + c$$

Example 7.3: Find $\int \sin\left(\frac{1}{3}x - \frac{\pi}{4}\right)dx$.

Solution 7.3: $a = \frac{1}{3}$, $b = -\frac{\pi}{4}$, so that $\int \sin\left(\frac{1}{3}x - \frac{\pi}{4}\right)dx = -\frac{1}{\frac{1}{3}}\cos\left(\frac{1}{3}x - \frac{\pi}{4}\right) + c = -3\cos\left(\frac{1}{3}x - \frac{\pi}{4}\right) + c$.

Exercise 7.2

In each case, find the integral.

1. $\int (2x+1)^3 dx$

2. $\int (1-x)^2 \, dx$

3. $\int \sin 2x dx$

4. $\int \frac{1}{2}\cos 2x dx$

5. $\int \frac{1}{3x-2} dx$

6. $\int e^{x+2} dx$

7. $\int e^{3-5x} dx$

8. $\int e^{-0.05x} dx$

9. $\int \frac{4}{10-x} dx$

10. $\int \cos\left(3x - \frac{\pi}{4}\right)dx$

11. $\int \sin\left(\frac{\pi}{3} - 2x\right)dx$

12. $\int \sqrt{4x+3} dx$

13. $\int \frac{3}{x+2} dx$

14. $\int (3-4x)^4 dx$

15. $\int \frac{2}{\sqrt{2x+1}} dx$

16. $\int e^{3x} dx$

17. $\int (5x-2)dx$

18. $\int \left(\cos\frac{1}{5}x - \sin\frac{2}{5}x\right)dx$

19. $\int \cos\frac{\pi x}{3} dx$

20. $\int \sin x^{\circ} dx$

7.4   Integration by Substitution

For more complicated integrals the method of finding the integral used in the previous section becomes impractical. Some of these more complicated integrals can be solved by using a substitution of variables. In this section the method of integration by substitution is shown.

Consider the indefinite integral $\int x\left(x^2+1\right)^3 dx$.

We use the substitution $u = x^2 + 1$ to rewrite the original integral in terms of u. This requires us to first differentiate $u = x^2 + 1$ to get $\dfrac{du}{dx} = 2x$ from which we can write $du = 2x\,dx$ . . . . [1]

Now we use [1] so that the integral can be rewritten entirely in terms of u:

$$\int x\left(x^2+1\right)^3 dx = \int \left(x^2+1\right)^3 \frac{1}{2}\left(2x\,dx\right) = \frac{1}{2}\int u^3\,du$$

$$\Rightarrow \int x\left(x^2+1\right)^3 dx = \frac{1}{2}\int u^3\,du = \frac{1}{2}\left(\frac{1}{4}u^4\right)+c = \frac{1}{8}u^4+c = \frac{1}{8}\left(x^2+1\right)^4+c$$

How do we know which substitution to use? It is not always obvious. No substitution is ever really wrong but it may be unhelpful. For example, suppose the substitution $v = x^2$ were used in the previous example. Then $\dfrac{dv}{dx} = 2x$ so that $dv = 2x\,dx$ and the integral becomes

$$\int x\left(x^2+1\right)^3 dx = \int \left(x^2+1\right)^3 x\,dx = \frac{1}{2}\int \left(v+1\right)^3\left(2x\,dx\right) = \frac{1}{2}\int \left(v+1\right)^3 dv$$

This is not wrong, but the integral in terms of v is not as easy to solve as the one obtained from the substitution $u = x^2 + 1$.

The most helpful substitution is often obvious for many integrals. In this course a helpful substitution will often be provided and certainly, for those integrals where the substitution is not obvious, a helpful substitution will be given.

Example 7.4: Use the substitution $u = e^x + 4$ to find the indefinite integral $\int \dfrac{e^x}{e^x+4}dx$

Solution 7.4: $u = e^x + 4 \quad \Rightarrow \quad \dfrac{du}{dx} = e^x \quad \Rightarrow \quad du = e^x dx$

$$\text{Therefore } \int \frac{1}{e^x+4}e^x dx = \int \frac{1}{u}du = \ln u + c = \ln\left(e^x+4\right)+c$$

Example 7.5: Use the method of substitution to find the indefinite integral $\int x\sqrt{1-x^2}\,dx$

Solution 7.5: Let $u = 1 - x^2 \Rightarrow \dfrac{du}{dx} = -2x \quad du = -2x\,dx$

Therefore, $\int x\sqrt{1-x^2}\,dx = \int\left(1-x^2\right)^{\frac{1}{2}}(x\,dx) = \int -\dfrac{1}{2}\left(1-x^2\right)^{\frac{1}{2}}(-2x\,dx) = -\dfrac{1}{2}\int u^{\frac{1}{2}}\,du$

$$= -\dfrac{1}{2}\dfrac{u^{\frac{3}{2}}}{\frac{3}{2}} + c = -\dfrac{1}{3}u^{\frac{3}{2}} + c = -\dfrac{1}{3}\left(1-x^2\right)^{\frac{3}{2}} + c$$

Example 7.6: Find $\int \cos x \sin^2 x\,dx$

Solution 7.6: Let $u = \sin x \Rightarrow \dfrac{du}{dx} = \cos x \Rightarrow du = \cos x\,dx$

$$\Rightarrow \int \cos x \sin^2 x\,dx = \int (\sin x)^2 (\cos x\,dx) = \int u^2\,du = \dfrac{1}{3}u^3 + c = \dfrac{1}{3}\sin^3 x + c$$

Exercise 7.3
1.   Find the following integrals using the given substitution
   (a) $\int x\left(x^2 + 3\right)dx$ $\hspace{6em}$ $u = x^2 + 3$

   (b) $\int \dfrac{x}{x^2 + 2}\,dx$ $\hspace{6em}$ $u = x^2 + 2$

   (c) $\int e^x\left(3 - e^x\right)^4 dx$ $\hspace{5.5em}$ $u = 3 - e^x$

   (d) $\int \cos x\left(1 + \sin x\right)^2 dx$ $\hspace{4em}$ $u = 1 + \sin x$

   (e) $\int \dfrac{1}{x}\ln x\,dx$ $\hspace{6.5em}$ $u = \ln x$

   (f) $\int \sin x\left(e^{\cos x}\right)dx$ $\hspace{5.5em}$ $u = \cos x$

2.   Use an appropriate substitution to find the given integrals
   (a) $\int \dfrac{1}{3}x^2\left(x^3 + 1\right)^2 dx$ (b) $\int \dfrac{x^2}{x^3 + 1}\,dx$ (c) $\int (2x+1)\sqrt{x^2 + x + 1}\,dx$ (d) $\int \dfrac{2e^{2x}}{e^{2x} + 1}\,dx$

3.   Find $\int \tan x\,dx$ by writing $\tan x = \dfrac{\sin x}{\cos x}$ and using the substitution $u = \cos x$

You may have noticed that the integrals in the previous section all consist of the product $a \times b$, where, $a = (g \circ f)(x) = g(f(x))$ and b is a multiple of $f'(x)$. For example, in the integral $\int \dfrac{x}{x^2 + 2}\,dx$, if $f(x) = x^2 + 2$ and if $g(x) = \dfrac{1}{x}$ then $a = g(f(x))$ and $b = x = \dfrac{1}{2}f'(x)$. This situation serves to simplify the integral obtained when the variable is changed from x to u.

225

In fact, with practice these integrals can be solved quickly by inspection and there is no need to use substitution.

Example 7.7: Find the following integrals

(a) $\int 2x\left(x^2+3\right)^3 dx$

(b) $\int x^2 \cos x^3 dx$

(c) $\int \frac{1}{x}\left(\ln x\right)^2 dx$

Solution 7.7: (a) Since $2x$ is the derivative of x^2+3 the integral is $\frac{1}{4}\left(x^2+3\right)^4+c$

(b) Since $3x^2$ is the derivative of x^3 the integral can be re-written as $\frac{1}{3}\int\left(3x^2\right)\cos x^3 dx$

and then the integral is $\frac{1}{3}\sin x^3+c$

(c) Since $\frac{1}{x}$ is the derivative of $\ln x$ the integral is $\frac{1}{3}\left(\ln x\right)^3+c$

Exercise 7.4   Find the following integrals using inspection.

1. $\int 2xe^{x^2} dx$ 2. $\int \frac{1}{x}\left(\ln x\right)^2 dx$ 3. $\int \frac{3x^2}{x^3+1} dx$ 4. $\int \frac{\cos x}{\sin x} dx$

5. $\int x \sin x^2 dx$ 6. $\int \sin x \cos^2 x\, dx$ 7. $\int 2x\left(x^2+4\right) dx$ 8. $\int x\sqrt{2x^2+3} dx$

9. $\int \frac{\sin x}{3-\cos x} dx$ 10. $\int \frac{2e^x}{e^x+3} dx$ 11. $\int \frac{x^2}{x^3-1} dx$ 12. $\int \cos 2x\left(4+3\sin 2x\right) dx$

13. $\int \sin^3 x\, dx$ (Hint: write $\sin^3 x = \sin x\left(1-\cos^2 x\right)$)

7.5   Evaluation of the Constant of Integration

The constant of integration can be evaluated, provided additional information is given. This information is often called a *boundary condition*.

Example 7.8: If $f'(x)=3x^2+10x-3$ and $f(2)=11$, find $f(x)$.

Solution 7.8: $f(x)=x^3+5x^2-3x+c$, $f(2)=11 \Rightarrow 11=8+20-6+c \Rightarrow c=-11$

Therefore, $f(x)=x^3+5x^2-3x-11$.

Example 7.9:   Given that $\dfrac{dy}{dx} = \sin 2x$ and that $y = 1$ when $x = \dfrac{\pi}{6}$, find y as a function of x.

Solution 7.9: $y = -\dfrac{1}{2}\cos 2x + c$. $1 = -\dfrac{1}{2}\cos\left(2 \times \dfrac{\pi}{6}\right) + c \Rightarrow 1 = -\dfrac{1}{2} \times \dfrac{1}{2} + c \Rightarrow c = \dfrac{5}{4}$.

$$\text{Therefore, } y = \dfrac{5}{4} - \dfrac{1}{2}\cos 2x$$

Exercise 7.5

1.   Find y, given that

(a) $\dfrac{dy}{dx} = 4x + 1$ and that $y = 3$ when $x = 0$ (b) $\dfrac{dy}{dx} = 3 - 2x^2$ and that $y = 15$ when $x = 3$

(c) $\dfrac{dy}{dx} = \sin x$ and that $y = 1$ when $x = \dfrac{\pi}{3}$ (d) $\dfrac{dy}{dx} = 4e^{-2x}$ and that $y = 1$ when $x = 0$

(e) $\dfrac{dy}{dx} = \dfrac{1}{\sqrt{2x+1}}$ and that $y = 4$ when $x = 4$

2.   Given that $y = \displaystyle\int \left(4x^3 - 6x + 5\right) dx$ and $y = 8$ when $x = 3$, find y as a function of x.

3.   If $x = \displaystyle\int \left(3t^2 - \dfrac{2}{t^2}\right) dt$ and $x(1) = 5$, find $x(t)$.

4.   Given that $y = \displaystyle\int e^x dx$ and $y = 2$ when $x = 0$, find y as a function of x.

5.   Find y as a function of x given that $y = \displaystyle\int \dfrac{4}{x} dx$ and $y = 5$ when $x = e^2$.

7.6   Application of Integration to Kinematics

In section 5.6.4, you saw that $v(t) = s'(t)$ and $a(t) = v'(t)$. These ideas, viewed in the context of integration rather than differentiation, give $s(t) = \displaystyle\int v(t) dt$ and $v(t) = \displaystyle\int a(t) dt$. These statements can be used to solve kinematics problems.

Example 7.10: A particle is projected in a straight line relative to a fixed point, O with velocity function $v(t) = 25 - 10t$, $t \geq 0$. If the displacement of the particle from O at time $t = 2$ is 8, find

(a) the displacement function, $s(t)$ of the particle.

(b) the displacement of the particle when $t = 4$.

Solution 7.10: $v(t) = s'(t) = 25 - 10t \Rightarrow s(t) = 25t - 5t^2 + c$. But the boundary condition is $s(2) = 8$ so

that $8 = 25 \times 2 - 5 \times 2^2 + c \Rightarrow 8 = 50 - 20 + c \Rightarrow 8 = 30 + c \Rightarrow c = -22$.

(a) Therefore, the displacement function is $s(t) = 25t - 5t^2 - 22$.

(b) $s(4) = 25 \times 4 - 5 \times 4^2 - 22 = 100 - 80 - 22 = -2$. Therefore, the displacement when $t = 4$ is -2.

Exercise 7.6

1. A ball is thrown vertically upwards with initial velocity 16ms^{-1}. The acceleration of the ball is constant and equal to -10 ms^{-2}. Time, t is measured in seconds.
   (a) Find the velocity, $v(t)$ of the ball.
   (b) Find the velocity of the ball when $t = 2$.
   (c) If the ball is thrown from a height of 2m, show that the ball reaches a maximum height of 14.8m.

2. A particle travels in a straight line relative to a fixed point, O and has acceleration $a(t) = 20 - 3t$, $t \geq 0$. Find
   (a) the acceleration if $t = 5$.
   (b) $v(t)$ if $v(0) = 3$.
   (c) $v(5)$.
   (d) the value of t when $v = 0$.

3. A rocket is projected vertically with velocity $v(t) = t + 2e^{0.5t}$, $0 \leq t \leq 10$, where v is measured in ms^{-1} and time, t is measured in seconds. Find
   (a) the velocity of the rocket after 10 seconds.
   (b) the displacement function, $s(t)$, given that $s(0) = 0$.
   (c) the altitude of the rocket above its point of projection after 10 seconds.

   After 10 seconds, the rocket continues vertically at a constant velocity of $v(10)$.
   (d) Find the height of the rocket after 1 minute from when the rocket was launched.

4. A particle moves in a straight line relative to a fixed point, O so that its acceleration is $a(t) = 4\sin t$, $t \geq 0$, where t is measured in seconds and displacement is measured in meters.
   (a) Find   (i) $v(t)$, given that $v(0) = 8$.
                   (ii) the maximum velocity and the value of t when it *first* occurs.
                   (iii) the displacement function, $s(t)$, given that $s(0) = 0$.
   (b) Draw a large, neat graph of $y = s(t)$ for the interval $0 \leq t \leq 10$.

5. A particle traveling in a straight line relative to a fixed point, O has velocity $v(t) = at^2 + bt + c$ where a, b and c are constants. Displacement, $s(t)$, is measured in meters, and time, t in seconds. The table shows the velocity, in ms^{-1}, of the particle. Use the table to find the values of a, b and c, and show that,

t	0	3	6
$v(t)$	10	8	16

given $s(0) = 0$, after 10 seconds the particle is 168.5m from O. A particle moves in a straight line relative to a fixed point, O, such that $v(t) = 5e^{-0.4t}$, $t \geq 0$.

(a) Find $s(t)$, given that when $t = 0$, $s = 0$. Show that $s(8) = 11.99$, correct to two decimal places.

(b) Find the maximum displacement of the particle from O.

(c) Find the acceleration when $t = 1$.

7.7   Area under a Curve

Figure 7.02 shows a portion of the curve $y = f(x)$. A shaded region, whose area is ΔA, is enclosed between a portion of the curve BF and the straight lines BC, CD and DF. From figure 7.02, it is clear that the area of rectangle $BCDE \leq \Delta A \leq$ the area of rectangle $ACDF$.

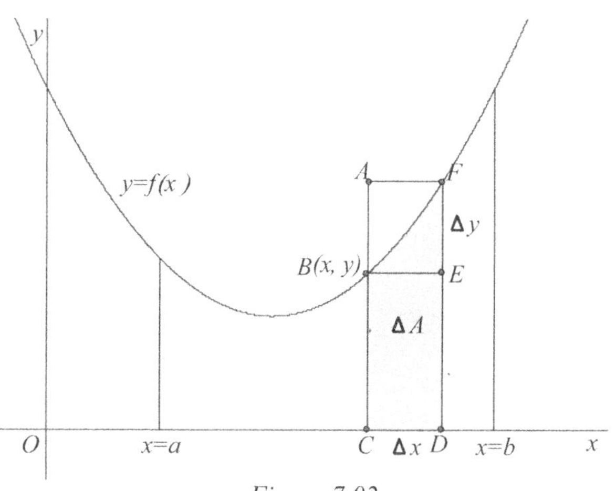

Figure 7.02

Let B have coordinates (x, y). Let $CD = \Delta x$ be a small increase in x. Let $FE = \Delta y$ be a small corresponding increase in y. Then

$$y\Delta x \leq \Delta A \leq (y + \Delta y)\Delta x \implies y \leq \frac{\Delta A}{\Delta x} \leq y + \Delta y.$$

As Δx, Δy and $\Delta A \to 0$, $\dfrac{\Delta A}{\Delta x} \to \dfrac{dA}{dx}$. Therefore, $y \leq \dfrac{dA}{dx} \leq y \implies \dfrac{dA}{dx} = y$, and, if this equation is integrated with respect to x, $\displaystyle\int \frac{dA}{dx}dx = \int y\,dx \implies \int dA = \int y\,dx \implies A(x) = \int y\,dx$.

If we let $\displaystyle\int y\,dx = F(x) + c$, then $A(x) = F(x) + c$.

To find the area A enclosed between the curve $y = f(x)$, the x-axis and the lines $x = a$ and $x = b$, it is helpful to imagine a line L, parallel to the y-axis, moving steadily from its starting position at the line $x = a$ to its finishing position at the line $x = b$.

As it moves in the direction of increasing x, the line L accumulates area between itself and the line $x = a$. Initially, when L is at the line $x = a$, the area accumulated is 0 and $A(a) = F(a) + c = 0$. When L reaches the line $x = b$, the area accumulated is $A(b) = F(b) + c$. Therefore the area,

$$A = A(b) - A(a) = (F(b) + c) - ((F(a) + c) = F(b) - F(a) \text{ and } A = F(b) - F(a).$$

Suppose that $f(x) = 3x^2 + 2x$ and that $a = 2$ and $b = 3$. Therefore, $A = \displaystyle\int (3x^2 + 2x)dx = x^3 + x^2 + c$, so $F(x) = x^3 + x^2$. Then $A = F(3) - F(2)$. $F(3) = 3^3 + 3^2 = 27 + 9 = 36$ and $F(2) = 2^3 + 2^2 = 8 + 4 = 12$. So $A = 36 - 12 = 24$.

7.8 Definite Integration

The notation used in section 7.7 is useful for <u>explaining the process</u> of finding the area under a curve, but a more concise notation is beneficial when <u>actually finding</u> the area under a curve. The area enclosed between $y = f(x)$, the x-axis and the lines $x = a$ and $x = b$ is written $A = \int_a^b f(x)dx$.

Referring to the example of section 7.7, if $f(x) = 3x^2 + 2x$, $a = 2$, $b = 3$,

$$A = \int_2^3 \left(3x^2 + 2x\right)dx = \left[x^3 + x^2\right]_2^3 = (27+9) - (8+4) = 24.$$

This process is called *definite integration*. When carrying out a definite integration, the constant, c is omitted because it always cancels in the enumeration of the integral.

In general, the evaluation of the definite integral $A = \int_a^b f(x)dx$ represents the area enclosed between the graph of $f(x)$, the x-axis and the lines $x = a$ and $x = b$.

Example 7.11: The function f is defined by $f(x) = x^2 - 2x + 15$.

      (a) Draw the graph of $y = f(x)$ for the domain $-5 \le x \le 10$.

      (b) Shade the area enclosed by the graph of $y = f(x)$, the x-axis and the lines $x = 0$ and $x = 5$.

      (c) Evaluate the definite integral $\int_0^5 f(x)dx$ to find the area of the shaded region.

      (d) Check your answer to part (c) using your graphing calculator.

Solution 7.11: (a) and (b) See figure 7.03.

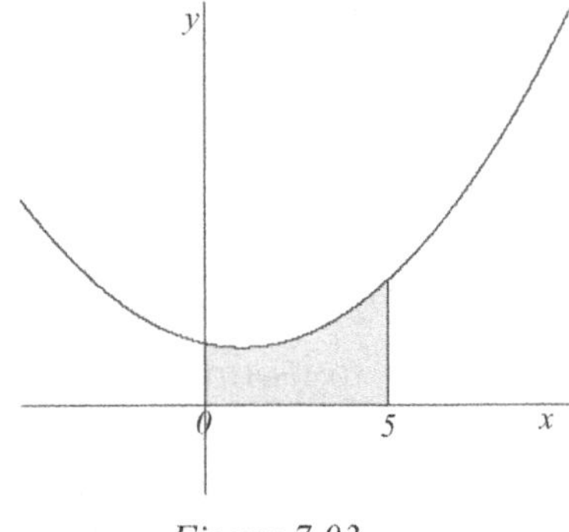

$$\text{(c)} \int_0^5 \left(x^2 - 2x + 15\right)dx = \left[\frac{1}{3}x^3 - x^2 + 15x\right]_0^5$$

$$= \left(\frac{125}{3} - 25 + 75\right) - 0 = 91\tfrac{2}{3}$$

(d) See figure 7.04.

Figure 7.03

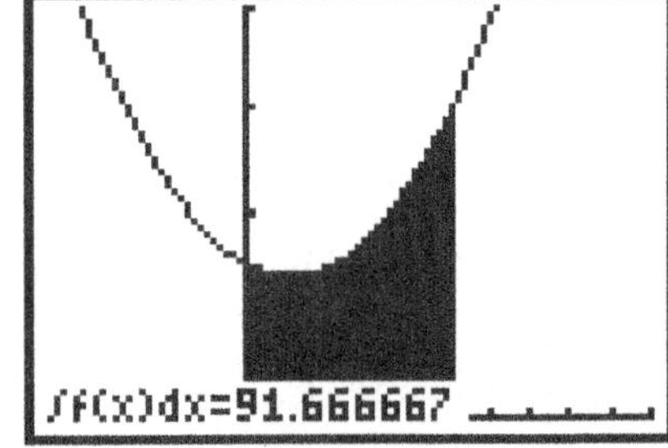

Figure 7.04

Example 7.12: Figure 7.05 shows the graph of the function $y = x^2 - 4x + 3$. Find the area of the region enclosed between the curve and the x-axis.

Solution 7.12: The area, A can be represented by the definite integral, I where

$$I = \int_1^3 \left(x^2 - 4x + 3\right) dx = \left[\frac{1}{3}x^3 - 2x^2 + 3x\right]_1^3$$

$$= (9 - 18 + 9) - \left(\frac{1}{3} - 2 + 3\right) = -\frac{4}{3}.$$

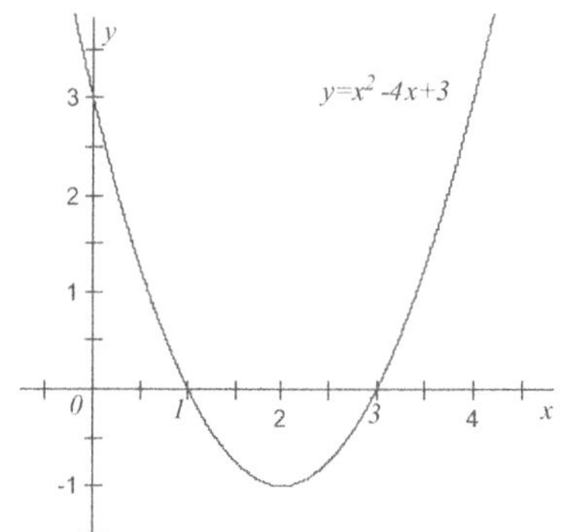

Figure 7.05

The definite integral is negative, indicating that the region whose area is sought is below the x-axis. Therefore the area of the region, A is given by $|I| = \left|-\frac{4}{3}\right| = \frac{4}{3}$.

Example 7.13: Consider the graph of the function $f(x) = x(x-2)(x-3)$.

   (a) Draw the graph of $y = f(x)$ over a suitable domain.

   (b) Evaluate the definite integrals $\int_0^2 f(x)\,dx$ and $\int_2^3 f(x)\,dx$

   (c) Find the area enclosed between the graph of $y = f(x)$ and the x-axis.

Solution 7.13: (a) See figure 7.06.

      (b) The graphing calculator may be used to evaluate these integrals. Figures 7.07 and 7.08 show the graphing calculator solutions.

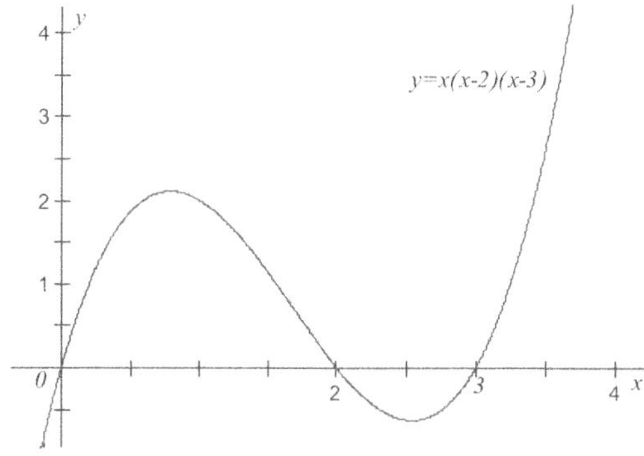

Figure 7.06

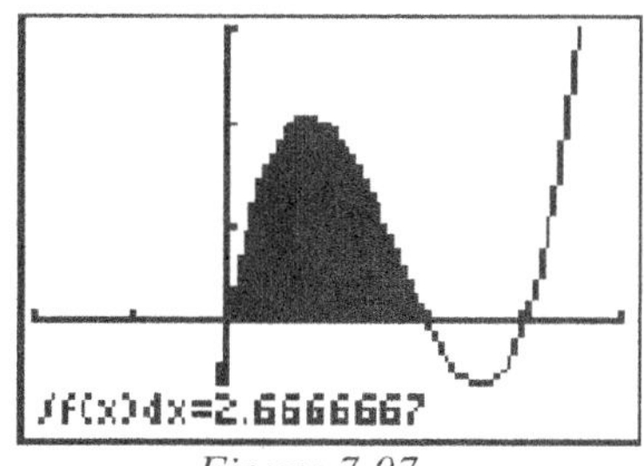

Figure 7.07

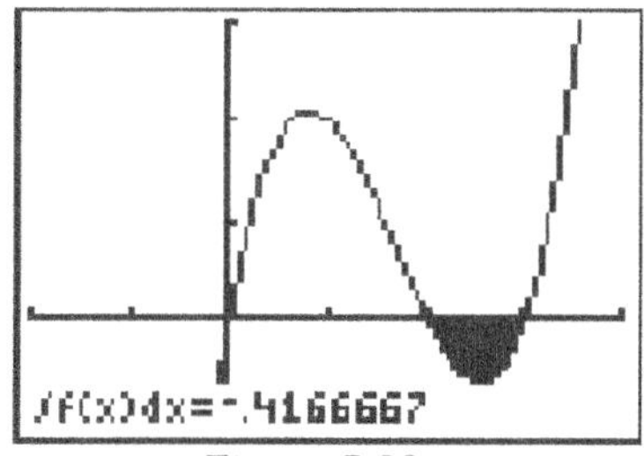

Figure 7.08

Therefore $\int_0^2 x(x-2)(x-3)\,dx = 2.67,\qquad \int_2^3 x(x-2)(x-3)\,dx = -0.417$

(c) The region between $x=0$ and $x=2$ is above the x-axis but the region between $x=2$ and $x=3$ is below the x-axis therefore the area enclosed between the curve and the x-axis is

$$\int_0^2 x(x-2)(x-3)\,dx + \left|\int_2^3 x(x-2)(x-3)\,dx\right| = 2.666666... + |-0.4166666...| = 3.08$$

In the next example the definite integral does not explicitly involve the area under a curve. In this case the definite integral is evaluated without reference to the graph of the function being integrated.

Example 7.14: Show that $\displaystyle\int_0^{\frac{\pi}{2}} (1+\sin 2x)\,dx = 1 + \frac{\pi}{2}$

Solution 7.14: $\displaystyle\int_0^{\frac{\pi}{2}} (1+\sin 2x)\,dx = \left[x - \frac{1}{2}\cos 2x \right]_0^{\frac{\pi}{2}} = \left(\frac{\pi}{2} - \frac{1}{2}\cos\pi \right) - \left(0 - \frac{1}{2}\cos 0 \right)$

$$= \left(\frac{\pi}{2} - \frac{1}{2}(-1) \right) + \frac{1}{2} = \frac{\pi}{2} + 1 = 1 + \frac{\pi}{2}$$

Exercise 7.7

1. Evaluate the following definite integrals.

(a) $\displaystyle\int_1^2 x^3\,dx$ (b) $\displaystyle\int_0^1 x^2(4x+3)\,dx$ (c) $\displaystyle\int_2^3 \frac{1}{x^2}\,dx$ (d) $\displaystyle\int_1^4 \left(\sqrt{x} - \frac{1}{\sqrt{x}}\right)\,dx$

(e) $\displaystyle\int_1^2 \left(\frac{x^3+2}{x^2}\right)\,dx$ (f) $\displaystyle\int_{-2}^1 (x^3 + x^2 + 5)\,dx$

2. In each case, using your graphing calculator if necessary, draw a large neat sketch of the graph of the function to be integrated over a suitable domain and shade the region whose area is represented by each definite integral. Calculate the definite integral in each case, correct to three significant figures, and check your answer using your graphing calculator.

(a) $\displaystyle\int_{-1}^1 e^x\,dx$ (b) $\displaystyle\int_0^2 \frac{1}{2x+3}\,dx$ (c) $\displaystyle\int_0^{\frac{\pi}{3}} \cos x\,dx$

3. Show that

(a) $\displaystyle\int_0^4 e^{-\frac{1}{2}t}\,dt = 2\left(1 - \frac{1}{e^2}\right)$ (b) $\displaystyle\int_1^5 \frac{4}{3x+1}\,dx = \frac{4}{3}\ln 4$ (c) $\displaystyle\int_0^{\frac{\pi}{2}} (1+\sin 2\theta)\,d\theta = 1 + \frac{\pi}{2}$

4. Some functions cannot be integrated using the method of anti-differentiation described in sections 7.1 and 7.3 or by substitution described in section 7.4. This means that, in general, definite integrals involving these functions cannot be given exactly. However, you can use

your graphing calculator to evaluate these definite integrals approximately to any reasonable degree of accuracy.

For each of the following functions $f(x)$,
(i)   use your graphing calculator to draw a large, neat sketch of $y = f(x)$ for the given domain.

(ii) evaluate $\displaystyle\int_a^b f(x)dx$, where a and b are the limits of the domain.

(a) $f(x) = e^{-x^2}$, $-2 \le x \le 2$　　　　　(b) $f(x) = \sqrt{\sin x}$, $0 \le x \le \pi$

(c) $f(x) = \dfrac{2x+9}{\sqrt{x^3+10}}$, $-2 \le x \le 6$

5.　　(a) Draw the graph of $y = \sin x$ for the domain $0 \le x \le 2\pi$.

　　　(b) Evaluate　(i) $\displaystyle\int_0^\pi \sin x\,dx$　　　(ii) $\displaystyle\int_\pi^{2\pi} \sin x\,dx$　　　(iii) $\displaystyle\int_0^{2\pi} \sin x\,dx$

　　　(c) Explain how the answers to parts (i), (ii), and (iii) of (b) are related.

6.　　In each case, draw a graph of the function over a suitable domain and then find the areas enclosed between the graphs of the given function and the x-axis.
　　　(a) $f(x) = x^2 + 5x$　　　　(b) $f(x) = x^3(x-2)$　　　　　(c) $f(x) = x(x+1)(x-2)$
　　　(d) $f(x) = x^3 \ln x$, $x > 0$　(e) $f(x) = x^4 - 3x^3 - x^2 + 3x$　(f) $f(x) = x^5 - 3x^2 + 1$

7.9　Volumes of Revolution

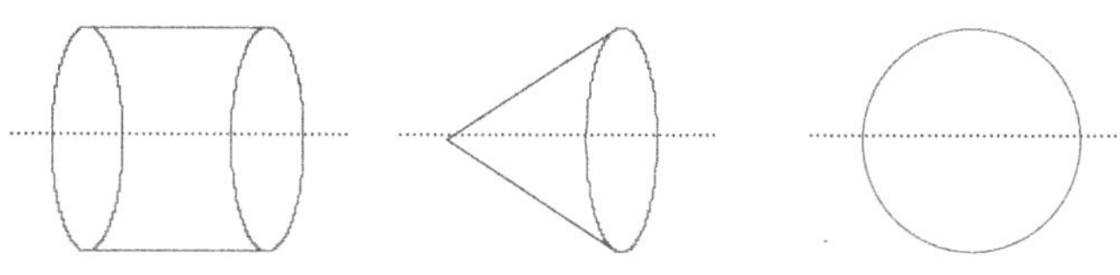

Figure 7.09 shows a cylinder, a cone and a sphere. These are all examples of solids of revolution: shapes formed by completely rotating a line or a curve about a fixed axis. The volumes of these shapes are known as *volumes of revolution*.

Figure 7.09

A cylinder with radius r and height h has a constant circular cross-section; therefore, its volume is simply the area of this cross-section multiplied by the height: $V = \pi r^2 h$.

For all other solids of revolution, although their cross-section is always circular, the radius of the circles varies, so it is more difficult to find their volumes. However, we can use integration to find these volumes.

Suppose we want to find the volume of a cone with height 10cm and radius 2cm. This cone can be generated by completely rotating the line with equation $y = \dfrac{1}{5}x$ about the x-axis, between the origin O and the line $x = 10$ (see figure 7.10).

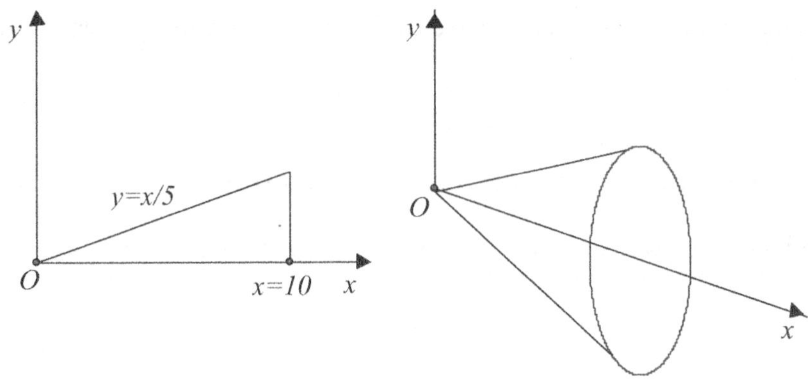

Figure 7.10

We need to think of the cone as being made up of a large number of thin discs and consider a typical disc of volume ΔV with thickness Δx and height varying between y and $y+\Delta y$.

Figure 7.11 shows a side view of such a disc from which it is evident that the volume, ΔV, of this disc lies between the volume of a cylindrical disc with radius y and the volume of a cylindrical disc with radius $y+\Delta y$, therefore,

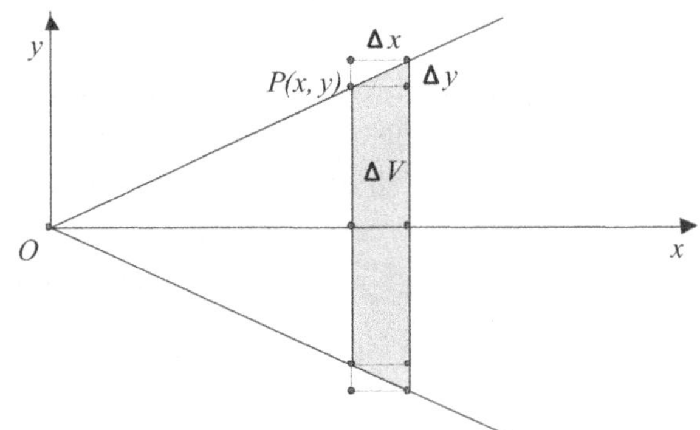

Figure 7.11

$$\pi y^2 \Delta x \leq \Delta V \leq \pi (y+\Delta y)^2 \Delta x$$

$$\Rightarrow \pi y^2 \leq \frac{\Delta V}{\Delta x} \leq \pi (y+\Delta y)^2.$$

As $\Delta x \to 0$, $\Delta y \to 0$ and $\dfrac{\Delta V}{\Delta x} \to \dfrac{dV}{dx}$,

$\dfrac{dV}{dx} = \pi y^2 = \pi \left(\dfrac{x}{5}\right)^2 = \dfrac{1}{25}\pi x^2$. Therefore the volume,

V of the complete cone can be obtained by evaluating the integral between $x=0$ and $x=10$, so that

$$V = \int_0^{10} \frac{1}{25}\pi x^2 \, dx = \frac{\pi}{25}\int_0^{10} x^2 \, dx = \frac{\pi}{25}\left[\frac{x^3}{3}\right]_0^{10} = \frac{\pi}{25}\left[\left(\frac{10^3}{3}\right)-\left(\frac{0^3}{3}\right)\right] = \frac{1000\pi}{75} = \frac{40\pi}{3}$$

This method may be used to find the volume of a solid of revolution formed by completely rotating about the x-axis the region enclosed by the curve $y=f(x)$, the lines $x=a$, $x=b$ and the x-axis.

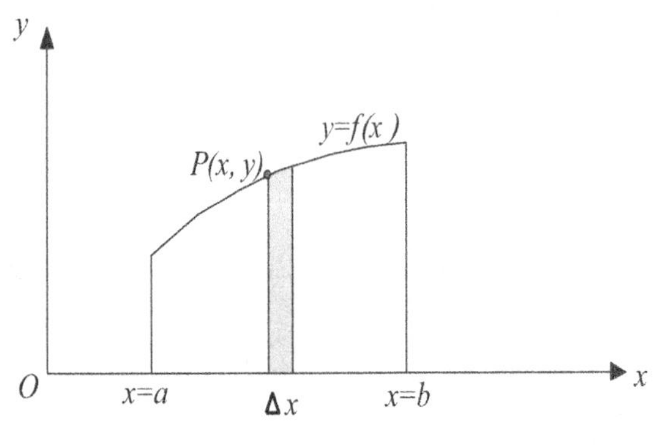

Figure 7.12

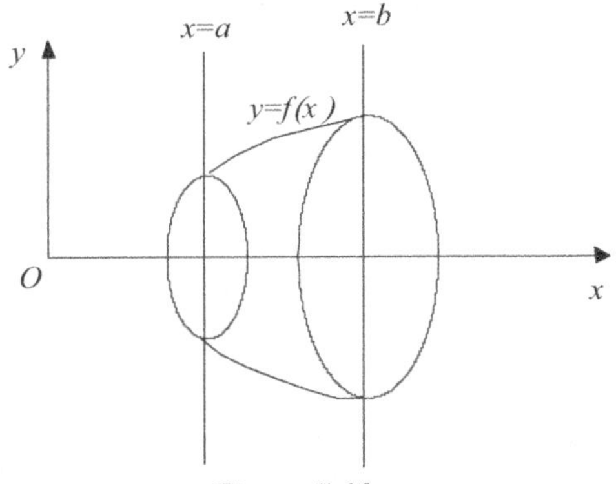

Figure 7.13

Using the same method that was used to find the volume of the cone, we find that the volume, ΔV of the thin disc shown in figure 7.12 is

$$\pi y^2 \Delta x \le \Delta V \le \pi (y + \Delta y)^2 \Delta x$$

$$\Rightarrow \pi y^2 \le \frac{\Delta V}{\Delta x} \le \pi (y + \Delta y)^2$$

and, again, as

$$\Delta x \to 0, \ \Delta y \to 0, \quad \frac{\Delta V}{\Delta x} \to \frac{dV}{dx}, \quad \frac{dV}{dx} = \pi y^2 .$$

The total volume, V of the solid of revolution (see figure 7.13) is obtained by adding the volumes of all the discs between the lines $x = a$ and $x = b$. Therefore $V = \int_a^b \pi y^2 dx$.

Example 7.15: Find the volume of the solid formed when the region enclosed by the curve $y = \sqrt{x + 2}$ is completely rotated about the x-axis between $x = 2$ and $x = 7$.

Solution 7.15: $V = \int_2^7 \pi y^2 dx = \pi \int_2^7 \left(\sqrt{x + 2} \right)^2 dx = \pi \int_2^7 (x + 2) dx$

$$= \pi \left[\frac{1}{2} x^2 + 2x \right]_2^7 = \pi \left[\left(\frac{1}{2} \times 7^2 + 2 \times 7 \right) - \left(\frac{1}{2} \times 2^2 + 2 \times 2 \right) \right]$$

$$= \pi \left[\left(\frac{49}{2} + 14 \right) - (2 + 4) \right] = \frac{65}{2} \pi$$

Example 7.16: A solid of revolution is generated when the region enclosed by the curve $f(x) = x^2$ and the lines $x = 0$ and $x = 1$ is rotated through $360°$ about the x-axis. Show that its volume is $\frac{\pi}{5}$.

Solution 7.16: The required volume, V is given by $V = \int_0^1 \pi y^2 dx = \pi \int_0^1 (f(x))^2 dx = \pi \int_0^1 (x^2)^2 dx$

$$\Rightarrow V = \pi \int_0^1 x^4 dx = \pi \left[\frac{x^5}{5} \right]_0^1 = \pi \left(\frac{1}{5} - \frac{0}{5} \right) = \frac{\pi}{5} .$$

Exercise 7.8

1. In each case, find the volume of the solid generated by rotating, about the x-axis, the region enclosed by the given curves, between the given lines. Give your answers as multiples of π.

   (a) $y = \sqrt{2x + 3}$; $x = 3, \ x = 12$

   (b) $y = 3x + 1$; $x = 0, \ x = 3$

   (c) $y = \sqrt[3]{x}$; $x = 1, \ x = 8$

   (d) $y = \dfrac{1}{\sqrt{4x + 1}}$; $x = 0, \ x = 2$

2. In each case, find the volume of the solid formed by completely rotating about the x-axis, the region enclosed by the given curves between the given lines. Give your answers correct to three significant figures.

(a) $f(x) = 2\cos x$; $x = 0$, $x = \dfrac{\pi}{2}$ (b) $f(x) = x^2 + 3$; $x = 1$, $x = \sqrt{6}$

(c) $f(x) = e^{-x}$; $x = 0$, $x = 1$ (d) $f(x) = \sqrt{x} + 1$; $x = 4$, $x = 9$

3. Let the region, enclosed by the curve $y = x(3 - x)$ and the x-axis, be R. R is rotated through $360°$ about the x-axis, forming a solid of revolution. Show that the volume of this solid is 8.1π.

4. The graph of the function $y = \sqrt{x}$, $0 \le x \le a$ is rotated through $360°$ about the x-axis to form a solid of revolution whose volume is 18π. Find the value of a.

5. Figure 7.14 shows the graphs of $y = x$ and $y = x^2$ for the interval, $0 \le x \le 1$. V is the volume of the solid generated when the shaded region enclosed between the line and the curve is completely rotated about the x-axis.

Show that $V = \dfrac{2\pi}{15}$.

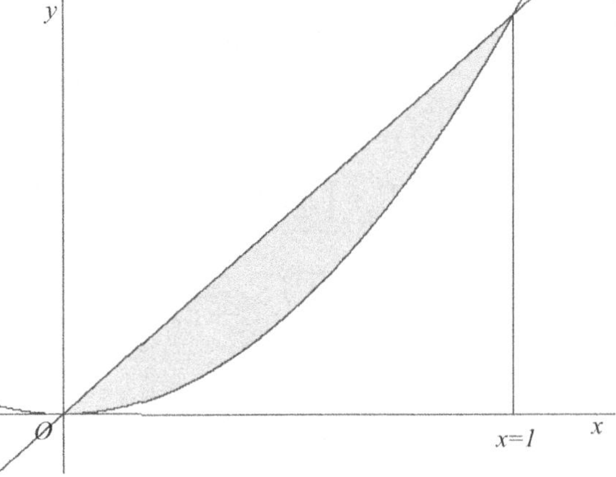

Figure 7.14

6. R is the solid formed when the graph of the function $y = x + 1$ is rotated through $360°$ about the x-axis between $x = 0$ and $x = a$. If the volume of R is 100π find, correct to 3 significant figures, the value of a.

7. The region enclosed by the graph of the function $f(x) = \sqrt{x}\sin x$, between $x = 0$ and $x = \pi$ is rotated through $360°$ about the x-axis to form a solid of revolution.

(a) Show that $\dfrac{d}{dx}\left(\dfrac{1}{4}x^2 - \dfrac{1}{4}x\sin 2x - \dfrac{1}{8}\cos 2x \right) = x\sin^2 x$

(b) Use the result of (a) to show that the volume of this solid of revolution is $\dfrac{\pi^3}{4}$.

Units 5 & 7 Review Exercises

Part 1

No graphing calculators should be used to answer questions in Part 1.

1.   It is given that $\dfrac{dy}{dx} = 3x^2 + 2x - 4$ and that when $x = 1$, $y = 5$. Find y in terms of x.

2.   Figure 7.15 shows part of the graph of $y = 2\sin\left(x - \dfrac{\pi}{3}\right)$

   (a) Write down the exact value of a, the x-intercept of the curve.

   (c) Use your answer in (a) to find the exact value of
   $$\int_0^a 2\sin\left(x - \dfrac{\pi}{3}\right)\,dx\,.$$

   (c) Hence, find the area of the shaded region.

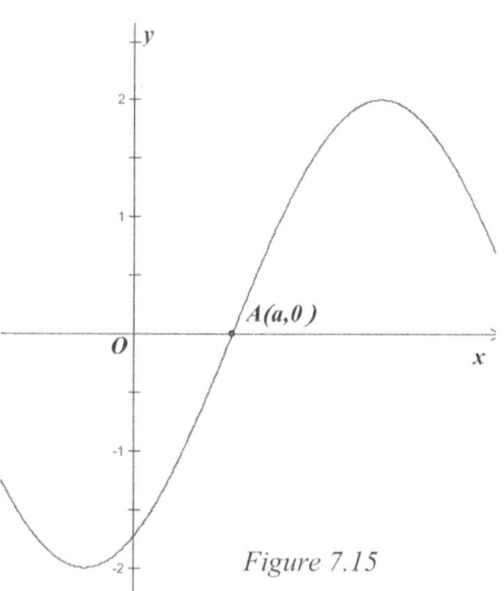

3.   Consider the function $f(x) = 2 - \dfrac{1}{x}$, $x > 0$

   (a) Write down the equations of both asymptotes.

   (b) Solve the equation $f(x) = 0$.

   (c) Sketch the graph of $y = f(x)$ for $0 \le x \le 4$

   The region enclosed by the curve of f, the x-axis and the line $x = 1$ and the line $x = 2$ is rotated through $360°$ about the x-axis. Let V be the volume of the solid formed.

   (d)   (i) Write down an integral for V.

              (ii) Show that $V = \pi\left(\dfrac{9}{2} - \ln 16\right)$.

4.   A particle moves due north so that its displacement, s meters from a fixed point O at time t seconds is given by $s = 9 + 8t - t^2$.

   (a) Find the initial displacement of the particle from O.

   (b) Find the velocity, v of the particle after 6 seconds and state in which direction the particle is moving at that time.

   (c) Find the maximum distance north of O reached by the particle.

5. Figure 7.16 shows part of the graph of the curve
$y = ax^2 + bx + c$, a, b, $c \in \mathbb{Z}$. A is the y-intercept
and B is the vertex.

(a) Write down $\dfrac{dy}{dx}$

A has coordinates $(0,\ 3)$ and B has coordinates
$(3,\ 9)$.

(b) Write down the value of c.

(c) Show that $3a + b = 2$

(d) Show that $a = -\dfrac{2}{3}$ and find the value of b.

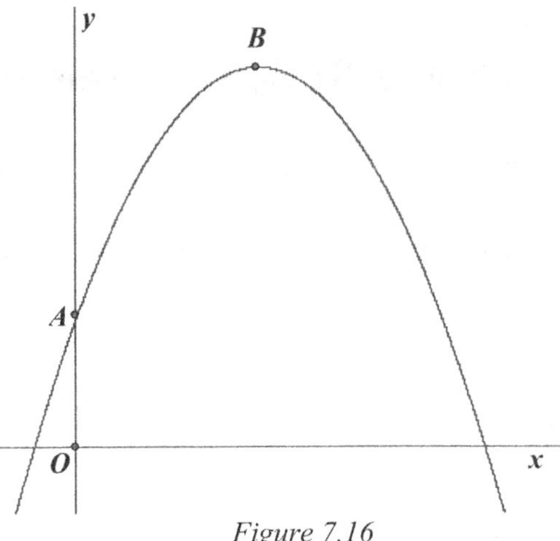

Figure 7.16

The tangent is drawn to the curve at the point A.

(e)   (i) Calculate the gradient of the tangent at A.   (ii) Find the equation of this tangent.

6. Given that $\displaystyle\int_{-1}^{2} f(x)\, dx = 11$, find

(a) $\displaystyle\int_{-1}^{2} \big(f(x)+1\big)\,dx$

(b) $\displaystyle\int_{-1}^{2} 5f(x)\, dx$.

If, in addition, $f(-1) = 1$ and $f(2) = 4$,

(c) find $\displaystyle\int_{-1}^{2} f'(x)\, dx$.

7. The velocity v of a particle at time t is given by $v = 2\sin 3t + 1$, where t is measured in radians.
The displacement at time t is s. If $s = 4$ when $t = 0$, find the an expression for s in terms of t.

8. Figure 7.17 shows the graph of the function
$y = f(x)$, $-4 \le x \le 4$ is shown below.

(a) Write down the value of

   (i) $f'(1)$ (ii) $f'(-1)$

(b) Sketch the graph of $y = f'(x)$, $-4 \le x \le 4$

(c) Sketch the graph of $y = f\left(\dfrac{x}{2}\right)$, $-4 \le x \le 4$.

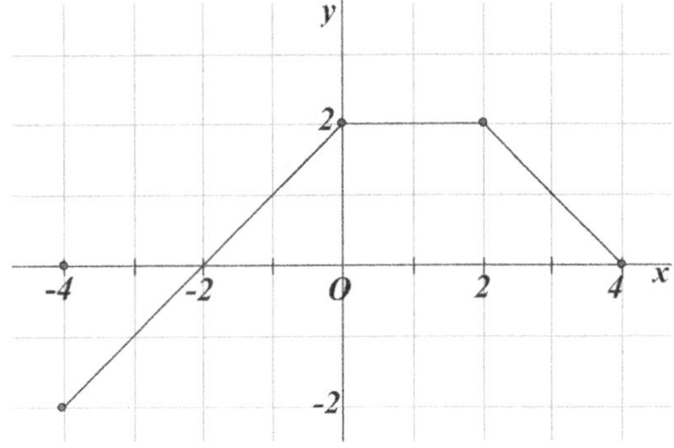

Figure 7.17

9. (a) Find $\displaystyle\int \dfrac{1}{(2x+1)^2}\, dx$

(b) Given that $\displaystyle\int_{1}^{2} \dfrac{1}{(2x+1)^2}\, dx = \dfrac{1}{k}$, $k \in \mathbb{Z}$ find the value of k.

238

10. Figure 7.18 shows the graph of $y = f(x)$

Copy the diagram shown and underneath it, on a separate set of axes sketch the graph of $y = f'(x)$.

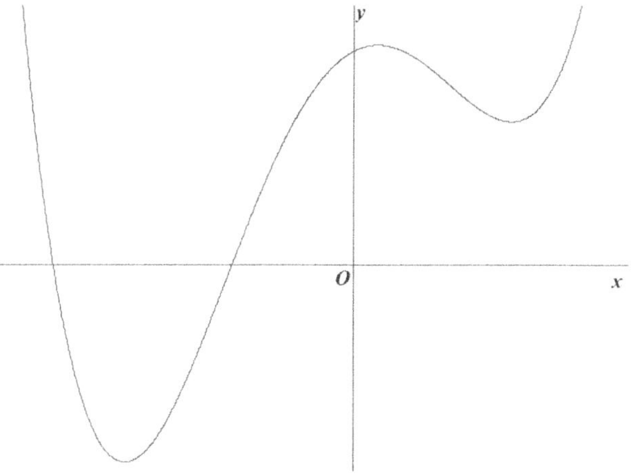

11. Figure 7.19 shows the graph of

$$f(x) = \frac{2}{\sqrt{x+1}}, \quad x > -1.$$ The shaded region

is bounded by the curve, $x = k$, $x = 1$ and the x-axis and is rotated around the x-axis through $360°$.

Figure 7.18

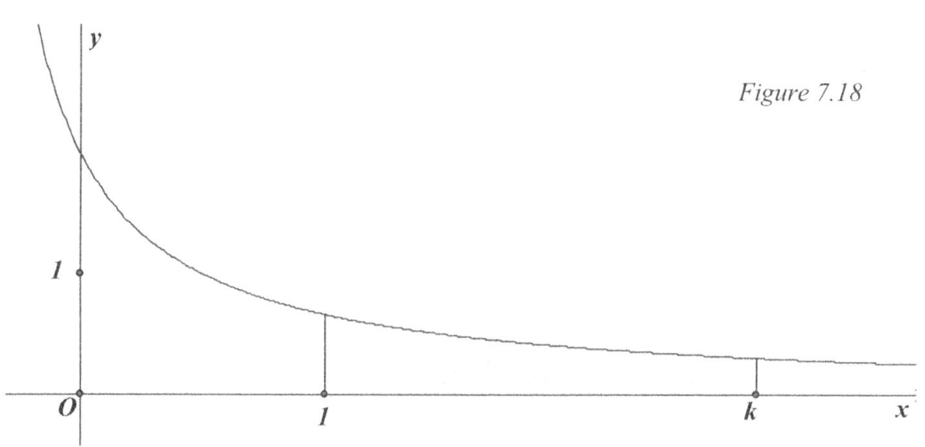

Figure 7.19

The volume of the solid formed is $4\pi \ln 3$. Find the value of k.

12. A rectangle $ABCD$ has side $AB = x$ cm. and area 36 cm^2
(a) Write the side AD in terms of x.
(b) Find an expression for the perimeter, y in terms of x.
(c) Determine $\dfrac{dy}{dx}$.
(d) Hence find the value of x which minimizes the perimeter and show that, for the value of x the perimeter is a minimum rather than a maximum.

Part 2

Graphing calculators will usually be needed to answer questions in Part 2.

13. A particle moves along a straight line so that its acceleration, a ms^{-2} at time t seconds, relative to a fixed point O is given by $a = 2e^{-t} + 4$. When $t = 0$ the particle is situated at O and its velocity is 3 ms^{-1}.
(a) Find an expression for the velocity, v of the particle in terms of t.
(b) Find an expression for the displacement, y of the particle in terms of t.

(c) Hence find the distance of the particle from O at time, $t = 2$.

14. The function f is defined as $f(x) = x^2 e^{-x}$. Part of the curve of f is shown in figure 7.20.

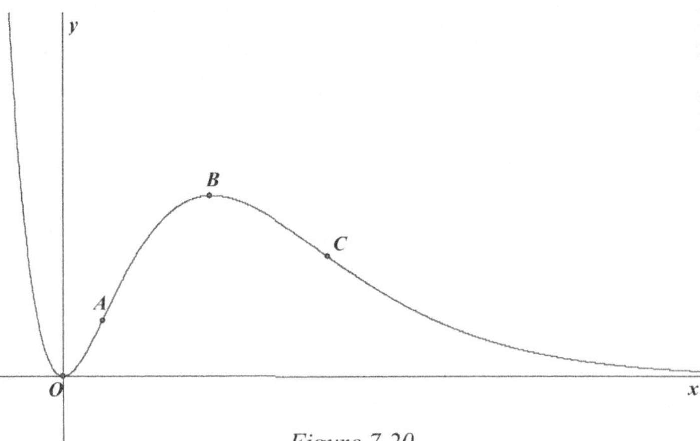

Figure 7.20

There are points of inflection at A and C and a local maximum point at B.

(a) (i) Write down the value of $f'(x)$ at B.

    (ii) Show that
$$f'(x) = e^{-x}\left(2x - x^2\right)$$

    (iii) Hence, find the x-coordinate of B.

(b) Find $f''(x)$

(c) Find the x-coordinates of A and C.

15. Let $f(x) = x\ln\left(x^2 + 5\right)$, $-1 \le x \le 6$

(a) Find $f'(x)$

(b) Sketch the graph of $y = f'(x)$ and find the range of $y = f'(x)$ for the given domain.

16. Let $f(x) = \dfrac{15}{ax + b}$, $x \ne -\dfrac{b}{a}$.

Part of the graph of f is shown in figure 7.21. The graph of f passes through the point with coordinates $(1, 5)$. There is a vertical asymptote at $x = -2$.

(a) Show that $a = 1$ and find the value of b.

(b) (i) Using $a = 1$ and your value of b find $f'(x)$.

    (ii) Show that f is a decreasing function over its entire domain.

(c) Find the volume of the solid formed when the region between $x = 0$ and $x = 4$ is rotated about the x-axis through 360^0.

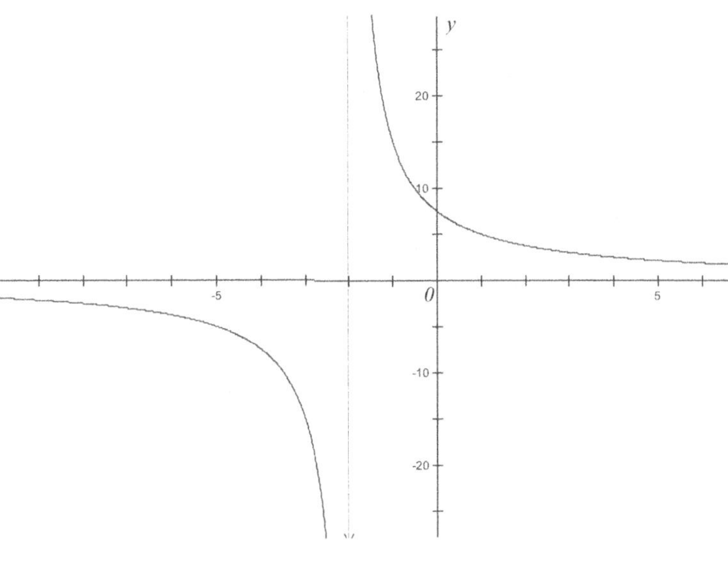

Figure 7.21

17.   The function f is given as
$f(x) = e^{-x}\cos x \quad 0 \le x \le 2\pi$. The graph of f is shown in figure 7.22.

The graph of f cuts the x-axis at A and D and there is a minimum point at B and an inflection point at C.

(a) Write down the coordinates of A and D.
(b) Find $f'(x)$ and hence show that the x-coordinate of B is $\dfrac{3\pi}{4}$.

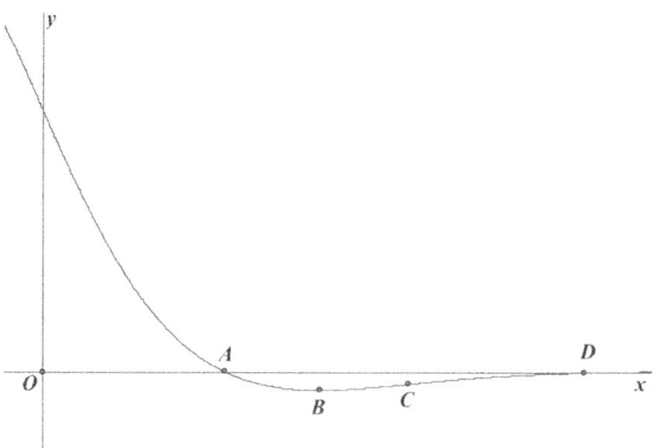
Figure 7.22

(c) Show that $f''(x) = 2e^{-x}\sin x$ and find the coordinates of C.
(d) Find the area enclosed by the graph of f, and the x-axis between A and D.

18.   Let $f(x) = \dfrac{x}{\left(1+e^x\right)}, \quad -1 \le x \le 5$. P is the maximum point and Q is an inflection point on the curve of f, which is shown in figure 7.23.

(a) Write down the equations of any asymptotes.

(b) Show that $f'(x) = \dfrac{1+e^x(1-x)}{\left(1+e^x\right)^2}$.

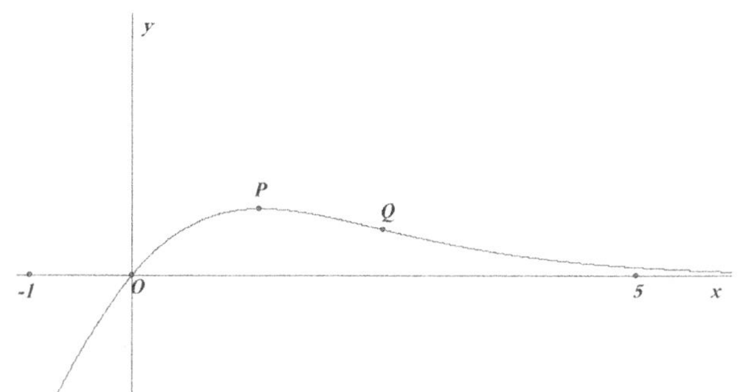
Figure 7.23

(c) Find the coordinates of P.
(d) Without finding $f''(x)$, determine the x-coordinate of Q.

(e) Find the equation of the normal to $y = f(x)$ at the origin.

(f) Let R be the region enclosed by $y = f(x)$ and the line $y = \dfrac{1}{5}x$.

    (i)   Find an expression for the area of R.   (ii)   Determine the area of R.

19.   A hemispherical cup ABC, is used to measure quantities of ingredients. A particular recipe requires half a cup of flour, by volume. The cup is formed by rotating the curve BC, given by the function,
$y = \sqrt{1-x^2}, \quad 0 \le x \le 1$, through $360°$ about the x-axis. Figure 7.24 shows this cup, on its side, related to xy-axes with origin O. The shaded region represents half a cup by volume. Find DB.

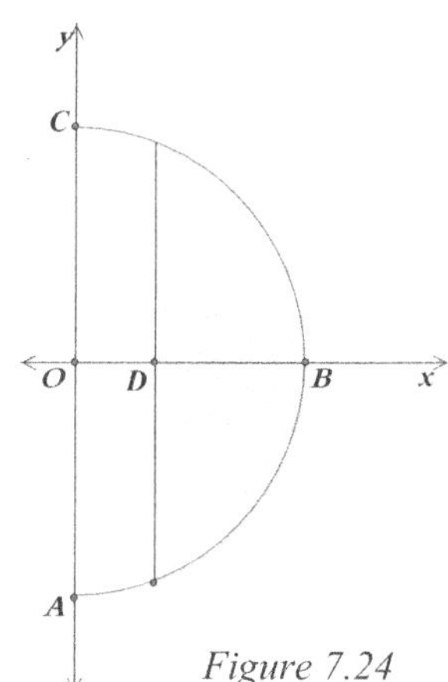
Figure 7.24

UNIT 8: PROBABILITY

8.1 Elementary Probability

Consider the following six examples of the use of probability:

1) A fair die is thrown. The probability of obtaining a 'six' is $\dfrac{1}{6}$.

2) A bag contains 4 black balls and 6 white balls which are otherwise identical in size and texture. A ball is chosen from the bag at random. The probability of choosing a white ball is $\dfrac{3}{5}$.

3) A room contains 30 people: 14 with blue eyes, the other 16 with brown eyes. The probability that a person, selected at random, has brown eyes is $\dfrac{8}{15}$.

4) In a class of 12 girls, 8 study Spanish, 6 study Arabic and 3 study neither of these languages. A girl is selected at random. The probability that she studies both Spanish and Arabic is $\dfrac{5}{12}$.

5) The probability that Maria's favorite soccer team will win its next match is $\dfrac{3}{4}$.

6) A thumb tack is thrown in the air and falls to the ground. The probability that it lands on its back is 0.8.

These six examples of the use of probability have a certain feature in common. They all involve some kind of action: throwing a die, choosing a ball from a bag, selecting a person from a class or a room, playing a soccer match, throwing a thumb tack. We call each of these actions a *trial*.

Each trial involves a number of *outcomes*. The outcomes are:

1) a number from 'one' to 'six'
2) the color of the ten balls
3) the eye colors of the 30 people
4) the languages studied by the 12 girls
5) a win, a draw or a loss for Maria's favorite team
6) the thumb tack lands on its back or on its side

We call the set of outcomes of each trial a *sample space*.

In addition, each example involves an *event*. This is the outcome, or set of outcomes, in whose probability we are interested. The events are:

1) obtaining a 'six'
2) selecting a white ball

3) selecting a person with brown eyes
4) selecting a girl who studies both Spanish and Arabic
5) Maria's team wins
6) the thumb tack lands on its back.

In the first four examples, the outcomes are equally likely provided the die is unbiased and the selections of balls and people are carried out randomly. A selection is *random* if each item to be selected is equally likely to be chosen.

In example 1), there are six equally likely outcomes, only one of which represents the event, so the probability is $\dfrac{1}{6}$.

In example 2), there are ten equally likely outcomes, and six of these represent the event, so the probability is $\dfrac{6}{10} = \dfrac{3}{5}$.

In example 3), there are 30 equally likely outcomes, and 16 of these represent the event, so the probability is $\dfrac{16}{30} = \dfrac{8}{15}$.

In example 4), there are 12 equally likely outcomes, and five of these represent the event, so the probability is $\dfrac{5}{12}$.

In examples 5) and 6), the outcomes are not necessarily equally likely.

In example 5), there are three outcomes, but there is nothing to suggest that they are equally likely.

How might a probability value of $\dfrac{3}{4}$ have arisen? One possibility is that the results of the previous four matches between the two teams had been recorded, and Maria's team had won three of these four. In other words, the data from previous results could have been used as a basis for making an estimate of the probability.

Similarly, in example 6), there are two outcomes: the thumb tack lands either on its back or on its side. However, there is no symmetry about a thumb tack, as opposed to a die, so the two outcomes are not necessarily equally likely. A probability value can be obtained by carrying out a series of trials. If a thumbtack is thrown 100 times and lands on its back 80 of these times, then the relative frequency of $\dfrac{80}{100} = 0.8$ forms an estimate of the probability.

Note that the probabilities of examples 5) and 6) are only estimates. If the results of more matches played between the two soccer teams had been used, we should have expected a different result. Similarly, if the thumb tack had been thrown 1000 times instead of 100, we should have expected a different estimate of the probability.

Figure 8.01 shows the results of a TI-84 Plus simulation of the random selection of a red ball from a bag containing a red, a blue and a yellow ball. The relative frequency is plotted every 20 trials from 20 to 800 trials. In addition, a line is drawn at 1/3, the probability of randomly selecting the red ball. As the number of trials increases, the relative frequency appears to approach this line. The program for the simulation appears at the end of Unit 9 and is called 'Probability Simulation Program'.

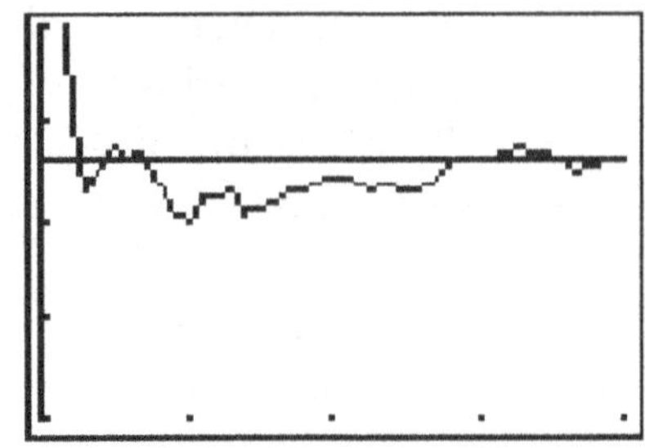

Figure 8.01

As figure 8.01 strongly suggests, the more trials that are carried out, the closer the estimate of the probability approaches the probability of $\dfrac{1}{3}$, obtained from assuming equally likely outcomes.

The simulation suggests why the process of random sampling is so important. Suppose you need to find the probability of selecting at random from a large truck load of potatoes a potato whose mass is less than 100g. It could be done by choosing at random 500 potatoes from the load and by finding the relative frequency of the potatoes whose mass is less than 100g, in the knowledge that this provides a reasonable estimate of the required probability.

Ultimately, of course, the number of trials taken in an attempt to establish a probability value is a compromise between the need for accuracy, which requires a large number of trials, and the need for practicality and economy, which demands a small number of trials.

The following is a summary of the definitions encountered above.

Trial: an action or experiment for which there are a number of outcomes.

Sample Space, U: the set of all possible outcomes.

Event, A : A is a subset of $U\left(A\subset U\right)$ which contains all the outcomes for which the event occurs.

Probability of an Event, A : $P\left(A\right)=\dfrac{n\left(A\right)}{n\left(U\right)}$, where $n\left(U\right)$ is the total number of equally likely outcomes and $n\left(A\right)$ is the number of equally likely outcomes for which the event occurs.

Events which cannot occur together are called *mutually exclusive* events. Looking back at the six examples of the use of probability, in the first three, the events cannot occur together. In example 1), when a die is thrown, the score on the die is either one of six numbers; it cannot be two numbers simultaneously. A similar situation occurs in examples 2) and 3). However, in example 4), the events 'studying Spanish' and 'studying Arabic' are not mutually exclusive because there are five girls who study both of these languages.

For two mutually exclusive events, A and B, $P\left(A\cup B\right)=P\left(A\right)+P\left(B\right)$, or alternatively, $P\left(A\cap B\right)=0$.

Events which include the entire sample space are known as *exhaustive* events. In example 1), the events of 'throwing a 'six'' and 'not throwing a 'six'' are exhaustive. If events A and B are exhaustive, then $U = A \cup B$ and $P(A \cup B) = 1$. A particularly important example of exhaustive events is provided by an event A and its complement, the event 'not A', written A', so that $P(A \cup A') = 1$. Since A and A' are also mutually exclusive $P(A \cup A') = P(A) + P(A') = 1$.

Example 8.1: A regular six-sided die is tossed. Find the probability that the result is a prime number.

Solution 8.1: $U = \{1, 2, 3, 4, 5, 6\}$, $n(U) = 6$, $X = \{2, 3, 5\}$, $n(X) = 3$. Therefore, $P(X) = \dfrac{3}{6} = \dfrac{1}{2}$.

Example 8.2: Two regular six-sided dice are thrown. Find the probability that the difference of the two scores is an odd number.

Solution 8.2: For each outcome of one die (A), there are six equally likely outcomes of the second die (B). Therefore, there are 36 equally likely outcomes altogether.
It is convenient to represent the differences in the scores of the two dice by a table (table 8.1).

		A					
		1	2	3	4	5	6
	1	0	1	2	3	4	5
	2	1	0	1	2	3	4
B	3	2	1	0	1	2	3
	4	3	2	1	0	1	2
	5	4	3	2	1	0	1
	6	5	4	3	2	1	0

Table 8.1

The number of equally likely outcomes of the scores on the two dice is 36, so $n(U) = 36$. If X is the event of obtaining an odd numbered difference, $n(X) = 18$ so that

$P(X) = \dfrac{18}{36} = \dfrac{1}{2}$. As the throw of each die is equally likely to result in an even number or an odd number, the solution could, more quickly, have been found by considering table 8.2, which shows all possibilities of 'odds' and 'evens'. We note that O – O = E, E – E = E, O – E = O and E – O = O.

	O	E
O	E	O
E	O	E

Table 8.2

Then $n(U) = 4$, $n(X) = 2$ and $P(X) = \dfrac{2}{4} = \dfrac{1}{2}$.

Exercise 8.1

1. A fair die is thrown once and the result noted. In each case, find the probability of events A and B and state whether the events A and B are (i) exhaustive (ii) mutually exclusive.
   (a) A : a 'four' is obtained B : a 'five' is obtained
   (b) A : a 'two' is obtained B : an even number is obtained
   (c) A : an even number is obtained B : an odd number is obtained
   (d) A : neither a 'four' nor a 'five' is obtained B : a 'five' is obtained
   (e) A : a prime number is obtained B : a 'one' or an even number is obtained

2.	Two fair dice are thrown and the sum of the scores, S, on the two dice is noted. Draw a table showing all possible values of S. For each of the following pairs of events, A and B, find the probability of events A and B and state whether they are (i) exhaustive (ii) mutually exclusive.

(a) A: exactly one die shows a 'three'　　　　$B: S = 5$
(b) A: at least one die is a 'four'　　　　　　$B: S = 12$
(c) $A: S < 5$　　　　　　　　　　　　　　　　$B: S \geq 5$
(d) A: the score on both dice is 'one'　　　　$B: S = 2$
(e) $A: 2 \leq S \leq 12$　　　　　　　　　　　　$B: S = 7$

3.	Peter applies to a university. The university can make three decisions:

　　　　A: accept him
　　　　R: reject him
　　　　W: put him on a waiting list

Experience suggests that for a student of Peter's ability, $P(A) = 0.2$, $P(R) = 0.5$.
(a) Use the fact that events A, R and W are exhaustive to find $P(W)$.
(b) Find the probability that Peter is either accepted or put on the waiting list.

4.	A bag contains 10 balls: 4 red, 3 blue, 2 green and 1 yellow. A ball is removed at random from the bag. Find the probability of the selection of a
(a) red ball　　　　(b) blue ball　　　　(c) blue or yellow ball　　　　(d) ball that is not red

5.	Table 8.3 shows the number of seeds in a seed pod of a sample of 100 plants of a specific species.

Seeds per Pod	≤ 10	11	12	13	14	15	16	17	≥ 18
Number of Pods	2	8	21	24	19	11	8	4	3

Table 8.3

Find the probability that a seed pod selected at random from the sample contains
(a) 14 seeds　　　　　　(b) more than 14 seeds

6.	A bag contains 2 red balls, 2 green balls and 1 blue ball. A ball is removed at random, its color is noted, and then the ball is replaced. A second ball is removed at random and its color noted. Draw a table to show the number of outcomes in the sample space and use it to find the probability that both balls removed are
(a) red　　　　　　　　(b) the same color

7.	$U = \{1, 2, 3, \ldots, 10\}$, $P = \{2, 5, 9\}$, $Q = \{1, 4, 5, 7\}$. If an integer is chosen at random from U, find the probability that
(a) it belongs to P　　(b) it belongs to P and Q　　(c) it belongs to neither P nor Q

8. Table 8.4 shows the frequency table of the IB results of 80 students. A student from this group is selected at random. Find the probability that the student's grade is

Grade	1	2	3	4	5	6	7
Frequency	11	5	8	21	15	12	8

Table 8.4

(a) a '7'   (b) less than a '4'

8.1.1 Non-Exclusive Events

In question 7 of Exercise 8.1, you may have noticed that events P and Q were not mutually exclusive. In that question,

$U = \{1, 2, 3, \ldots, 10\}$, $P = \{2, 5, 9\}$, $Q = \{1, 4, 5, 7\}$ and integers were selected at random from U. A Venn diagram, shown in figure 8.02, is more appropriate for displaying the outcomes than a table when events are non-exclusive.

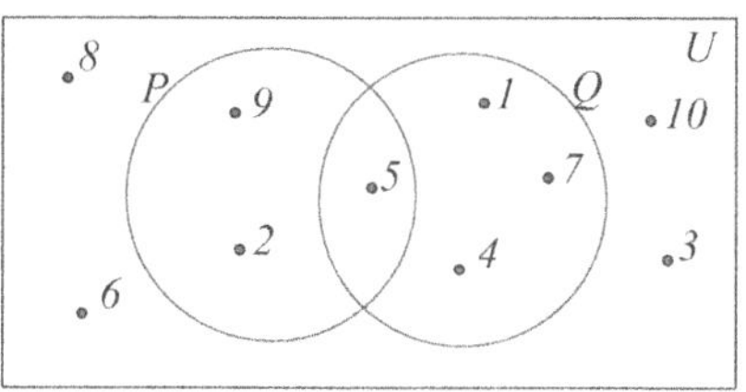

Figure 8.02

Now, $P \cup Q = \{1, 2, 4, 5, 7, 9\}$, $n(P \cup Q) = 6$, $n(P) = 3$ and $n(Q) = 4$ so that $P(P \cup Q) = \dfrac{6}{10}$,

$P(P) = \dfrac{3}{10}$, $P(Q) = \dfrac{4}{10}$ and $P(P \cup Q) \neq P(P) + P(Q)$ because the outcome, 'choosing a 5', occurs in both P and Q.

Consider, in figure 8.03, two sets A and B containing x outcomes in common.

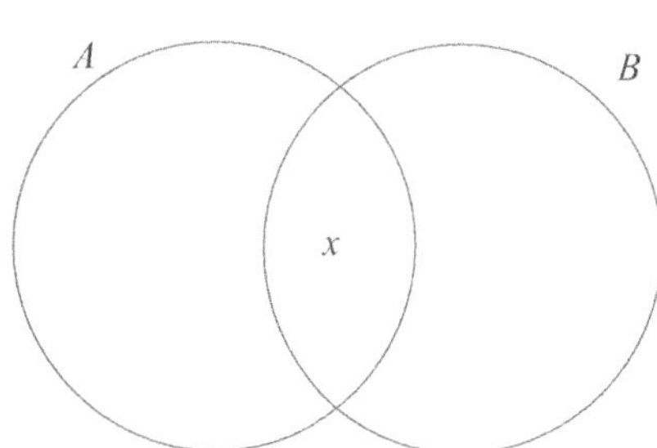

Then $n(A \cup B) = (n(A) - x) + x + (n(B) - x) = n(A) + n(B) - x$

but $x = n(A \cap B) \Rightarrow n(A \cup B) = n(A) + n(B) - n(A \cap B)$

$$\Rightarrow \frac{n(A \cup B)}{n(U)} = \frac{n(A)}{n(U)} + \frac{n(B)}{n(U)} - \frac{n(A \cap B)}{n(U)}$$

$$\Rightarrow P(A \cup B) = P(A) + P(B) - P(A \cap B)$$

Figure 8.03

Example 8.3: A regular six-sided die is thrown. Find the probability of obtaining

(a) an even number   (b) a prime number   (c) an even number or a prime number

Solution 8.3: Let X be the event of obtaining an even number, and let Y be the event of obtaining a prime number. Then $X = \{2, 4, 6\}$, $n(X) = 3$, $Y = \{2, 3, 5\}$, $n(Y) = 3$.

$U = \{1, 2, 3, 4, 5, 6\}$, $n(U) = 6$. Therefore,

(a) $P(X) = \dfrac{n(X)}{n(U)} = \dfrac{3}{6} = \dfrac{1}{2}$

(b) $P(Y) = \dfrac{n(Y)}{n(U)} = \dfrac{3}{6} = \dfrac{1}{2}$

(c) $X \cap Y = \{2\}$, $n(X \cap Y) = 1$. Therefore, $P(X \cap Y) = \dfrac{n(X \cap Y)}{n(U)} = \dfrac{1}{6}$ and,

$$P(X \cup Y) = P(X) + P(Y) - P(X \cap Y) = \frac{1}{2} + \frac{1}{2} - \frac{1}{6} = \frac{5}{6}$$

Example 8.4:   If $P(A \cup B) = \dfrac{5}{6}$, $P(A) = \dfrac{2}{5}$ and $P(B) = \dfrac{3}{4}$, find $P(A \cap B)$.

Solution 8.4: $P(A \cup B) = P(A) + P(B) - P(A \cap B) \Rightarrow \dfrac{5}{6} = \dfrac{2}{5} + \dfrac{3}{4} - P(A \cap B)$

$$\Rightarrow P(A \cap B) = \frac{2}{5} + \frac{3}{4} - \frac{5}{6} = \frac{24 + 45 - 50}{60} = \frac{19}{60}$$

Example 8.5:   If $P(A \cup B') = 0.72$ and $P(B) = 0.43$, find $P(A \cap B)$.

Solution 8.5:   The application of a Venn diagram is often an effective way of dealing with this kind of problem. The shaded region of figure 8.04 represents $A \cup B'$.

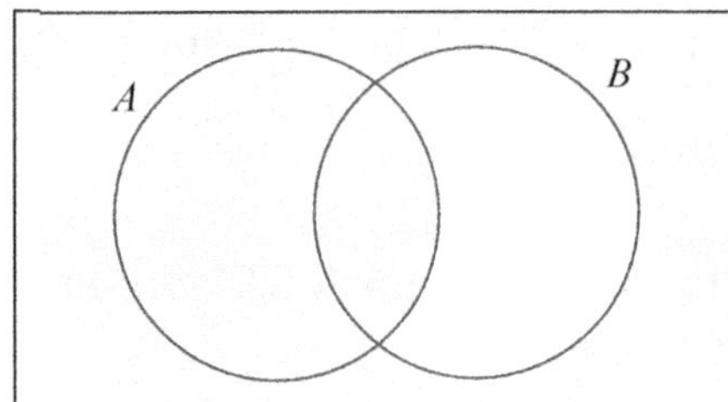

Figure 8.04

From figure 8.04 we see that

$$A \cup B' = U - B + (A \cap B)$$

$$\Rightarrow P(A \cup B') = P(U) - P(B) + P(A \cap B)$$

$$\Rightarrow 0.72 = 1 - 0.43 + P(A \cap B) \Rightarrow P(A \cap B) = 0.15$$

Exercise 8.2

1. $U = \{a,b,c,d,e,f,g,h\}$, $A = \{a,b,c,d,e\}$, $B = \{c,d,e,f\}$. If a letter is chosen at random from U, find (a) $P(A)$ (b) $P(A \cap B)$ (c) $P(A \cup B)$ (d) $P(A' \cap B')$, where A is the event 'the letter is a member of set A', B is the event 'the letter is a member of set B' and so on.

2. $U = \{1,2,3,4,5,6,7,8\}$, $A = \{3,4,5,6\}$, $B = \{2,3,4\}$. If an element is chosen at random from U, find (a) $P(A)$ (b) $P(A \cap B)$ (c) $P(B')$ (d) $P(A \cup B)$, where A is the event 'the number is a member of set A', B is the event 'the number is a member of set B' and so on.

3.   (a) Draw a Venn diagram to represent two events P and Q which are not mutually exclusive within a sample space U.

$n(U)=50$, $n(P'\cap Q')=18$ and $n(P)=n(Q)=20$

(b) Find $n(P\cap Q)$.

(c) If an element is randomly chosen from U, find $P(P\cap Q)$.

4. Figure 8.05 shows a Venn diagram with sample space U and events A and B.

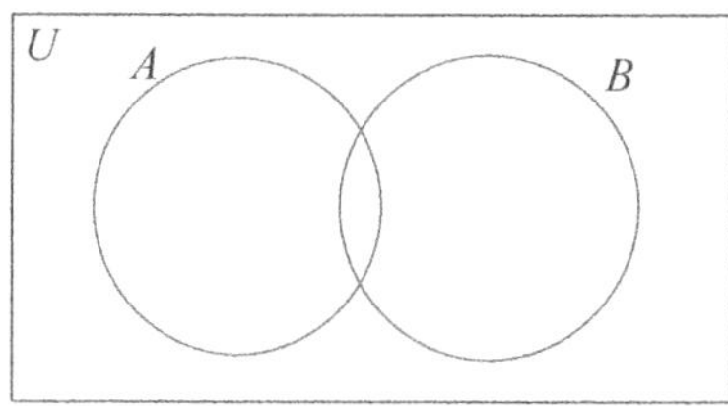

Figure 8.05

$n(U)=80$, $n(A)=40$, $n(B)=25$ and $n(A\cap B)=15$

(a) Copy the Venn diagram shown in figure 8.05 and shade the region $(A'\cap B)$.

(b) Write $n(A'\cap B)$ in terms of $n(B)$ and $n(A\cap B)$.

An element is chosen at random from U.

(c) Use your answer to part (b) to find (i) $n(A'\cap B)$ (ii) $P(A'\cap B)$

(d) Explain why events A and B are not mutually exclusive

5. If $P(A)=0.62$, $P(B)=0.51$ and $P(A\cup B)=0.95$, find (a) $P(A\cap B)$ (b) $P(A\cup B')$.

6. If $P(A)=\dfrac{1}{3}$, $P(B)=\dfrac{3}{4}$ and $P(A\cap B)=\dfrac{1}{5}$, find (a) $P(A\cup B)$ (b) $P(A'\cup B')$.

7. If $P(A)=\dfrac{1}{2}$, $P(A\cap B)=\dfrac{1}{6}$ and $P(A\cup B)=\dfrac{3}{4}$, find (a) $P(B)$ (b) $P(A'\cap B)$.

8. If $P(A')=0.5$, $P(B')=0.6$ and $P(A'\cup B')=0.8$, find $P(A\cup B)$.

9. If $P(A\cap B')=0.23$, $P(A'\cap B)=0.42$ and $P(A\cup B)=0.93$, find $P(A\cap B)$.

10. Of a total of 200 students, 163 study mathematics, 94 study physics and 26 study neither mathematics nor physics. Draw a Venn diagram to represent this data and on it show the number of students studying (a) only mathematics (b) only physics (c) both physics and mathematics. Hence find the probability that a student selected at random studies mathematics but not physics.

11. In a roadworthiness test, cars may fail due to bad lights or due to bad brakes. For a vehicle selected at random, experience shows that the probability of failure due to bad lights is 0.16 and the probability of failure due to bad brakes is 0.28. In addition, the probability of failure due to both bad brakes and bad lights is 0.08. Find the probability that a randomly selected car fails the road test.

8.2 Conditional Probability

The probability value which is given to an event depends on the extent of the information that is available. For example, suppose a bag contains 6 red balls and 4 blue balls. Mr. Patel removes a ball, notes that it is blue, and puts it in his pocket without showing it to his friend Mr. Quail, who knows that Mr. Patel has removed a ball but does not know its color. If Mr. Quail now removes a ball from the bag, what is the probability that it is red?

Mr. Patel would argue that the probability is $\dfrac{2}{3}$, because he knows that there are 6 red balls in the bag which now contains a total of 9 balls. On the other hand, Mr. Quail would state that the probability is $\dfrac{3}{5}$, because he does not know which ball was removed, and so, from his point of view, the bag might as well contain the original 10 balls of which 6 would be red.

The reason that Mr. Patel and Mr. Quail give different probabilities to the same event is that Mr. Patel's sample space is different from Mr. Quail's sample space. In fact, Mr. Patel's sample space is more restricted than Mr. Quail's.

To generalize, consider two events, A and B, which are not mutually exclusive. Suppose that event B has already occurred; that is, B is *given*. Then the probability of event A, given event B, written $P(A\,|\,B)$, has sample space restricted to $n(B)$ because all outcomes not in B may be eliminated because they are no longer relevant.

Figure 8.06 shows set B, the sample space, as a shaded circle. The lined region is the set $A \cap B$, and the probability that A occurs given that B has occurred is given by

$$P(A\,|\,B) = \frac{n(A \cap B)}{n(B)} = \frac{n(A \cap B)\,|\,n(U)}{n(B)\,|\,n(U)} = \frac{P(A \cap B)}{P(B)}$$

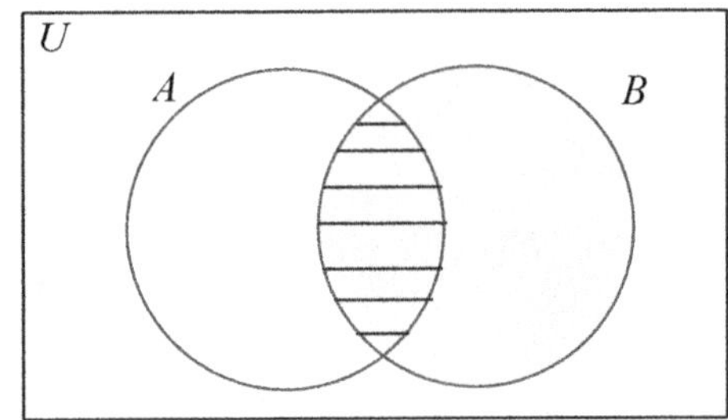

Figure 8.06

In a similar manner the probability that A occurs given that B has not occurred is $P(A\,|\,B') = \dfrac{P(A \cap B')}{P(B')}$.

8.2.1 Contingency Tables

Contingency tables are a useful means of solving conditional probability problems. Table 8.5 is an example of a contingency table and shows an analysis in terms of gender of the International Baccalaureate Mathematics courses in a specific school.

	Higher Level	Mathematics SL	Mathematical Studies	
Girls	7	16	24	47
Boys	19	19	9	47
	26	35	33	94

Table 8.5

The probability that a male student selected at random is taking Mathematical Studies can be read from the row labeled 'Boys' of the contingency table and is $\dfrac{9}{47} = 0.191$. The probability that a randomly selected student taking Higher Level Mathematics is a girl can be read from the column labeled 'Higher Level' and is $\dfrac{7}{26} = 0.269$.

8.2.2 Probability Tree Diagrams

The tree diagram depicts a set of events in which each event corresponds to a branch. It is customary to write the event at the end of its branch and the probability of the event adjacent to the branch.

Successive events are represented by adjoining branches, and the probability of successive events is obtained by taking the <u>product</u> of the probabilities of the successive events.
Alternative events are represented by alternative branches, and the probability of alternative events is obtained by taking the <u>sum</u> of the probabilities of the alternative events.
A probability tree representing events A and B is shown in figure 8.07.
The thick lines show the branches which correspond to event A and can be used to find $P(A)$:

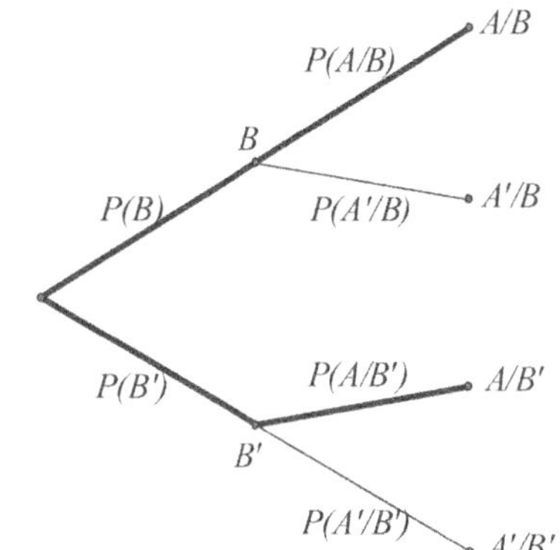

Figure 8.07

$$P(A\,|\,B) = \frac{P(A \cap B)}{P(B)} \Rightarrow P(A \cap B) = P(B) \times P(A\,|\,B)$$

Therefore, by multiplying probabilities along the upper thick line, $P(A \cap B)$ can be determined. In a similar manner, the lower thick line determines $P(B' \cap A)$.

Now, event $A = (A \cap B) \cup (A \cap B')$, as shown in figure 8.08.
The lined region represents $(A \cap B)$, and the shaded region represents $(A \cap B')$.

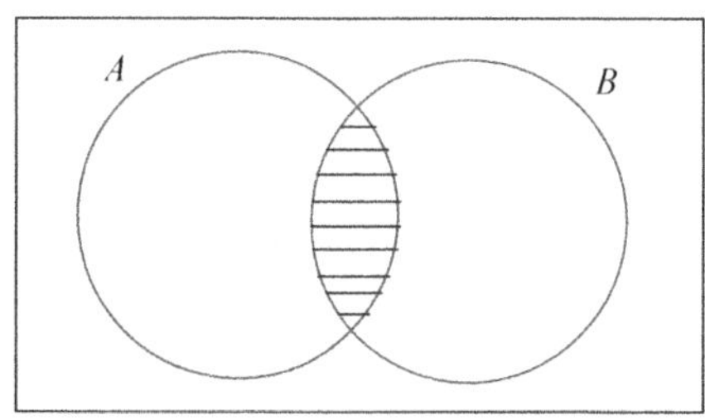

Figure 8.08

Therefore, $P(A) = P(A \cap B) + P(A \cap B')$, and since
$P(A \cap B) = P(A\,|\,B) \times P(B)$ and $P(A \cap B') = P(A\,|\,B') \times P(B')$,
$P(A) = P(A\,|\,B) \times P(B) + P(A\,|\,B') \times P(B')$ so that $P(A)$ can be determined from figure 8.07 by adding the probabilities represented by the two thick lines.

In a similar manner, $P(B) = P(B\,|\,A) \times P(A) + P(B\,|\,A') \times P(A')$.

Example 8.6: A bag contains 3 red balls and 2 blue balls. A ball is
removed at random and not replaced. A second ball is removed at random. Find the
probability that
  (a) the second ball is red, given that the first is blue
  (b) a red ball is taken on the second removal

Solution 8.6: Let R be the event 'the second ball removed is red'
and let B be the event 'the first ball removed is blue'.

Figure 8.09 shows the tree diagram.

(a) $P(R\,|\,B)$ is required. $P(R\,|\,B) = \dfrac{3}{4}$

(b) $P(R) = P(B \cap R) + P(B' \cap R)$

$$= \frac{3}{4} \times \frac{2}{5} + \frac{2}{4} \times \frac{3}{5} = \frac{3}{10} + \frac{3}{10} = \frac{3}{5}$$

The fact that $P(R) < P(R\,|\,B)$ confirms our
intuition that, if we had known that a blue ball
was chosen on the first removal, then it appears to
us more likely that a red ball would be chosen on the
second removal.

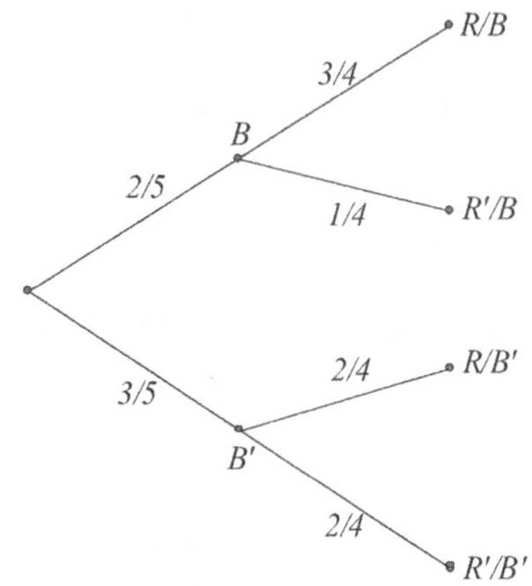

Figure 8.09

Example 8.7: Given that $P(A) = \dfrac{2}{5}$ and $P(A \cap B) = \dfrac{1}{3}$, find $P(B\,|\,A)$. If $P(A\,|\,B) = \dfrac{7}{8}$, find $P(B)$.

Solution 8.7: $P(B\,|\,A) = \dfrac{P(A \cap B)}{P(A)} = \dfrac{1}{3} \div \dfrac{2}{5} = \dfrac{5}{6}$

$P(A\,|\,B) = \dfrac{P(A \cap B)}{P(B)} \Rightarrow P(B) = \dfrac{P(A \cap B)}{P(A\,|\,B)} = \dfrac{1}{3} \div \dfrac{7}{8} = \dfrac{8}{21}$

Exercise 8.3
1.   Table 8.6 shows the results of a survey
      which compares eye color and eyesight for
      125 adults. Find the probability that
      (a) a person chosen at random has bad
          eyesight.
      (b) a blue-eyed person chosen at random has
          good eyesight.
      (c) a randomly chosen person with bad eyesight has brown eyes.

	Blue Eyes	Brown Eyes	
Good Eyesight	31	55	86
Bad Eyesight	26	13	39
	57	68	125

Table 8.6

2.   250 boys and 250 girls are each offered a lunch choice of (A)
      burger and fries, (B) soup and salad or (C) lasagna. Table 8.7
      shows the results of their preferences. Find the probability
      that
          (a) a boy chosen at random prefers A.
          (b) a girl chosen at random prefers C

	A	B	C	
Boys	131	81	38	250
Girls	94	74	82	250
	225	155	120	500

Table 8.7

252

(c) a randomly chosen child who prefers
   lasagna is a girl.

Give your answers, correct to two decimal places.

3. Table 8.8 shows a survey which was carried out on a sample of airline pilots and taxi drivers by a research student investigating left-handedness. The contingency table shows the results.

	Airline Pilots	Taxi Drivers	
Left-handedness	7	11	18
Right-handedness	21	61	82
	28	72	100

Table 8.8

Find the probability that
(a) a taxi driver selected at random from the survey was left-handed.
(b) a randomly selected left-handed member of the survey was an airline pilot.

4. Given that $P(B) = \dfrac{2}{3}$ and $P(A \cap B) = \dfrac{1}{6}$ find (a) $P(A|B)$.

If, in addition, $P(A) = \dfrac{3}{8}$, find (b) $P(A|B')$.

5. If $P(A) = \dfrac{4}{5}$, $P(B) = \dfrac{2}{3}$ and $P(A \cap B) = \dfrac{1}{2}$, find (a) $P(A|B)$ (b) $P(B|A)$.

6. Events P and Q are such that $P(P|Q) = \dfrac{3}{7}$ and $P(Q|P) = \dfrac{4}{5}$. If, in addition, $P(Q) = \dfrac{2}{3}$, find
(a) $P(P \cap Q)$ (b) $P(P)$

7. $P(A) = \dfrac{3}{8}$, $P(B|A') = \dfrac{2}{5}$ and $P(B'|A) = \dfrac{1}{3}$. Copy and complete the tree diagram shown in figure 8.10 and hence find
(a) $P(A \cap B)$ (b) $P(A' \cap B')$

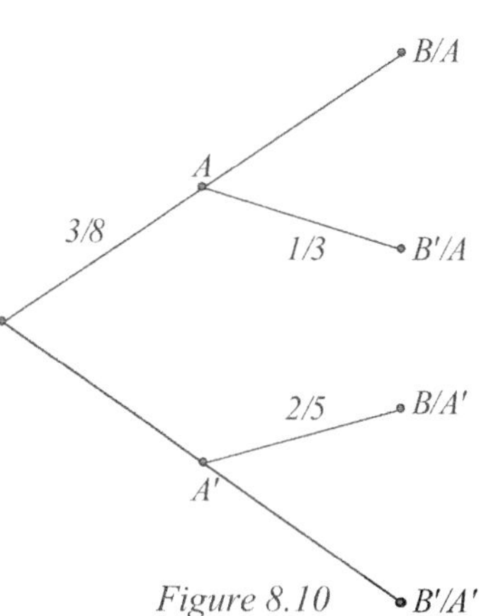

8. If $P(A) = 0.48$, $P(A \cap B) = 0.36$ and $P(A' \cap B') = 0.26$, find
(a) $P(B|A)$ (b) $P(B'|A')$

Draw a tree diagram and hence find
(c) $P(B)$ (d) $P(A|B)$.

9. If $P(B|A) = \dfrac{2}{5}$, $P(A|B) = \dfrac{1}{3}$ and $P(B) = \dfrac{4}{5}$, find $P(A)$.

10. In a country, 30% of the population is vaccinated against a specific disease. Those who are vaccinated have a 5% chance of getting the disease, and those not vaccinated have a 20% chance of getting the disease. Let V be the event 'a member of the population is vaccinated'.

Let D be the event 'a member of the population has the disease'.

(a) Copy and complete the tree diagram shown in figure 8.11 by writing the probability values against each branch.

(b) What percentage of the population can be expected to suffer the disease?

(c) If a sufferer of the disease is selected at random, what is the probability that he or she was vaccinated?

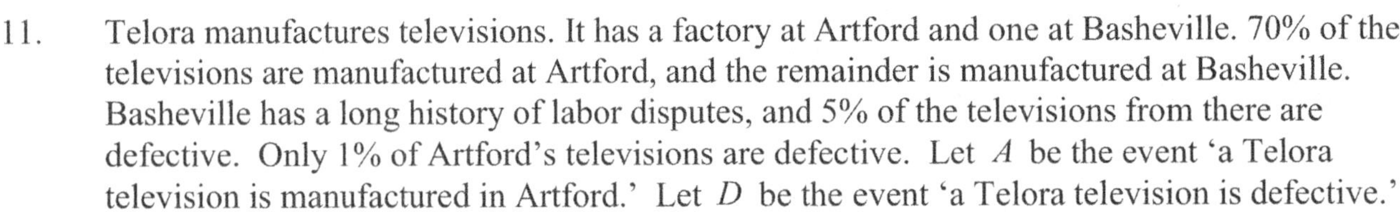

Figure 8.11

11. Telora manufactures televisions. It has a factory at Artford and one at Basheville. 70% of the televisions are manufactured at Artford, and the remainder is manufactured at Basheville. Basheville has a long history of labor disputes, and 5% of the televisions from there are defective. Only 1% of Artford's televisions are defective. Let A be the event 'a Telora television is manufactured in Artford.' Let D be the event 'a Telora television is defective.'

(a) Write down $P(A)$, $P(D\,|\,A)$ and $P(D\,|\,A')$.

(b) A man buys a Telora television which turns out to be defective. Find the probability that it was manufactured in Basheville.

12. A bag contains 100 coins. 99 are standard pennies, and one is similar except that it is double-headed. A coin is removed at random from the bag and tossed 10 times. Let A be the event 'a double-headed penny is removed from the bag'. Let B be the event 'the penny which has been removed from the bag lands 'heads' on each of the 10 occasions it is tossed'.

(a) Write down $P(A)$, $P(B\,|\,A)$ and $P(B\,|\,A')$.

(b) Find the probability that, if the coin lands 'heads' each of the 10 times, then it is the double-headed coin.

8.3 Independent Events

Suppose that a bag contains 3 white balls and 5 black balls. A ball is randomly selected, its color noted, and then the ball is <u>replaced</u>. Let A be the event 'a black ball is selected'.

Now suppose that a second ball is randomly selected and its color noted. Let B be the event 'a black ball is selected on the second selection'. Figure 8.12 shows a tree diagram representing the situation.

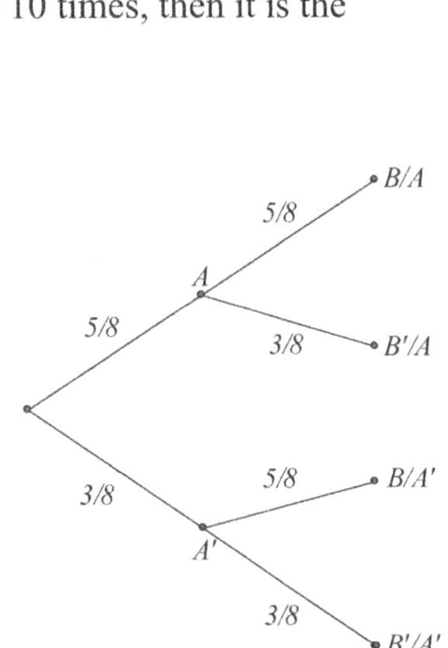

Figure 8.12

Because the first ball removed is replaced before the second ball is removed, the situation just before the second ball is removed is

254

identical to the original situation. Therefore, it does not matter which ball is removed first, and the

probability that the second ball removed is black is $\dfrac{5}{8}$. Hence $P(B\,|\,A)=P(B)$.

We can drop the reference to event A because the probability of event B is *independent* of event A.

Therefore, if $P(B\,|\,A)=P(B)$ and $P(B\,|\,A)=\dfrac{P(A\cap B)}{P(A)}$, then $P(B)=\dfrac{P(A\cap B)}{P(A)}$

$\Rightarrow P(A\cap B)=P(A)\times P(B)$.

Therefore, if events A and B are independent,
- $P(A\,|\,B)=P(A)$
- $P(B\,|\,A)=P(B)$
- $P(A\cap B)=P(A)\times P(B)$

Once you are satisfied that successive events are independent, a tree diagram can be used without the need to use conditional probability notation.

A little thought should convince you that when a coin is tossed twice, the outcome of one coin is unaffected by the outcome of the other. The results when a coin is tossed two or more times, therefore, represent independent events.

Figure 8.13 shows a tree diagram of the result of tossing a coin twice.
H is the event 'a 'head' is obtained', and T is the event 'a 'tail' is obtained'.

The probability of any required events can be read from the diagram.
The probability of

- two 'heads' is $P(H)\times P(H)=\dfrac{1}{2}\times\dfrac{1}{2}=\dfrac{1}{4}$

- a 'head' and a 'tail' is

$$P(H)\times P(T)+P(T)\times P(H)=\dfrac{1}{2}\times\dfrac{1}{2}+\dfrac{1}{2}\times\dfrac{1}{2}=\dfrac{1}{4}+\dfrac{1}{4}=\dfrac{1}{2}$$

- two 'tails' is $P(T)\times P(T)=\dfrac{1}{2}\times\dfrac{1}{2}=\dfrac{1}{4}$

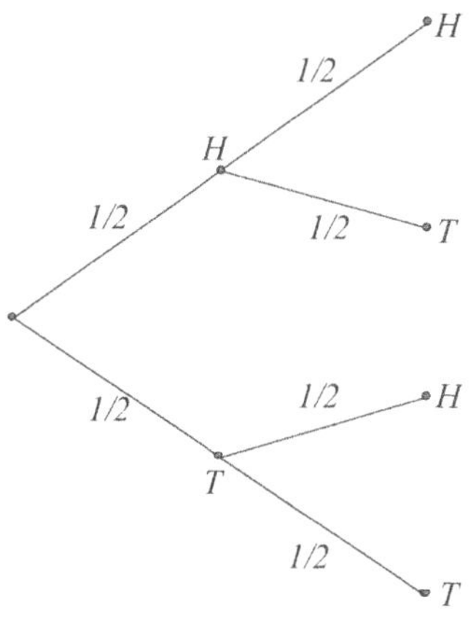

Figure 8.13

Example 8.8:   A coin is tossed and a die is thrown. Find the probability that a 'head' and a 'six' are obtained.

Solution 8.8:   Figure 8.14 shows the relevant tree diagram. The probability of obtaining a 'head' and a 'six' is calculated from the thick line on the diagram and is

$$\frac{1}{2}\times\frac{1}{6}=\frac{1}{12}.$$

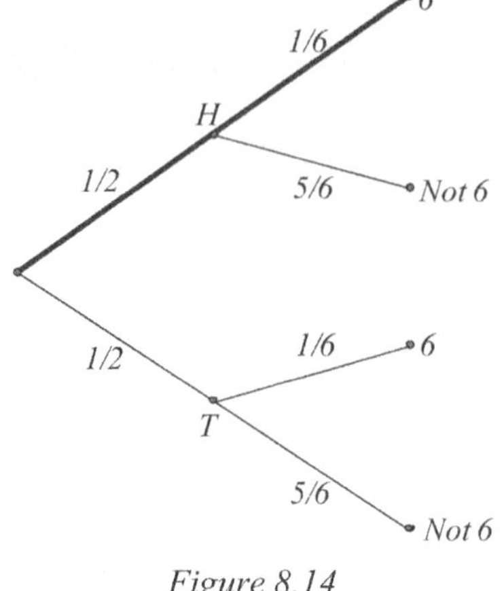

Figure 8.14

Example 8.9:   Find the number of times a die needs to be thrown in order that there is a 90% chance that a 'six' is obtained.

Solution 8.9:   There are many ways in which one or more 'sixes' can be obtained in n throws of a die but there is only one way that <u>no</u> 'sixes' can be obtained.

Suppose that the die is thrown n times; then the probability of no 'sixes' is $\left(\dfrac{5}{6}\right)^{n}$, so

the probability of one or more 'sixes' in n throws of the die is $1-\left(\dfrac{5}{6}\right)^{n}$. Therefore,

because we want this probability to be more than 90%,

$$1-\left(\frac{5}{6}\right)^{n}\geq 0.9 \Rightarrow -\left(\frac{5}{6}\right)^{n}\geq -0.1 \Rightarrow \left(\frac{5}{6}\right)^{n}\leq 0.1 \Rightarrow \log_{10}\left(\frac{5}{6}\right)^{n}\leq -1 \Rightarrow n\log_{10}\left(\frac{5}{6}\right)\leq -1$$

$$\Rightarrow n\geq \frac{-1}{\log_{10}\left(5/6\right)}=12.6 \text{, and, as } n\in\mathbb{Z}^{+},\ n=13.$$ (Note that it is necessary to change

the inequality sign when dividing by $\log_{10}\left(\dfrac{5}{6}\right)$ because $\log_{10}\left(\dfrac{5}{6}\right)<0$). Therefore, it is

necessary to throw a die 13 times to be 90% sure of obtaining a 'six'.

Exercise 8.4

1.   Events A and B are independent and such that $P(A)=\dfrac{1}{8}$ and $P(B)=\dfrac{2}{9}$. Leaving your

answers as fractions in their lowest terms, calculate (a) $P(A\cap B)$ (b) $P(A\cup B)$.

2.   Independent events A and B are such that $P(A\cup B)=\dfrac{7}{8}$ and $P(A)=\dfrac{3}{5}$.

Find (a) $P(B)$ (b) $P(A\cap B)$

3.   If A and B are events such that $P(A)=0.6$, $P(B)=0.5$ and $P(A\cup B)=0.8$, show that A and B are independent events.

4.   If $P(A \cup B) = 0.78$, $P(A) = 0.45$ and $P(A \cap B) = 0.27$, find $P(B)$. Show also that A and B are independent events.

5.   Events P and Q are independent and such that $P(P) = 3P(Q)$. Find $P(Q)$ if

(a) P and Q are such that $P(P \cup Q) = 0.8$ (b) P and Q are exhaustive.

6.   The probability that a man will live for ten more years is $\frac{1}{4}$, and the probability that his wife will live for ten more years is $\frac{1}{2}$. Assuming independence, find the probabilities that

(a) both man and wife will be alive in ten years' time.
(b) neither will be alive in ten years' time.
(c) at least one will be alive in ten years' time.
(d) only the wife will be alive in ten years' time.

7.   A bag contains 3 red, 2 blue and 4 green balls. A ball is selected at random, its color noted, and then the ball is replaced. A second ball is then selected at random and its color noted. Find the probability that
(a) the first ball is green and the second ball is blue.
(b) a green and a blue ball are selected.
(c) both balls selected are red.
(d) both balls are the same color.

8.   One letter is randomly chosen from the word MISSISSIPPI, and one letter is chosen randomly and independently from the word MISSOURI. Show that the probability that the letters chosen are the same is just under 20%.

9.   (a) A coin is tossed until a 'head' is obtained. Find the probability that the coin is tossed four times.
(b) How many times should a coin be tossed in order to be 90% certain that at least one 'head' is obtained?

10.   A basketball player has a probability of 0.5 of scoring on a free throw. How many free throws would the player have to take in order for his/her probability of scoring on at least one throw to be 0.99?

11.   (a) How many times should a fair die be thrown in order to be 95% certain of obtaining a 'six'?
(b) A die is thrown until a 'six' is obtained. Find the probability that the die is thrown at least 10 times.

12.   In arms negotiations between two countries, it is estimated that there is a 50% chance that agreement will be reached on limiting chemical weapons and a 70% chance that agreement will be reached on limiting nuclear weapons. There is a 20% chance that no agreement will be reached on either issue. Find the probability of achieving agreement on limiting both chemical and nuclear weapons. In the negotiations, are the two issues treated independently?

UNIT 9: PROBABILITY DISTRIBUTIONS

9.1   Discrete Probability Distributions

Consider an experiment consisting of n trials, in which a trial consists of throwing two dice and recording the sum of the scores on both dice. Figures 9.01 to 9.04 show four histograms in which $n = 25,\ 50,\ 100$ and 350, respectively.

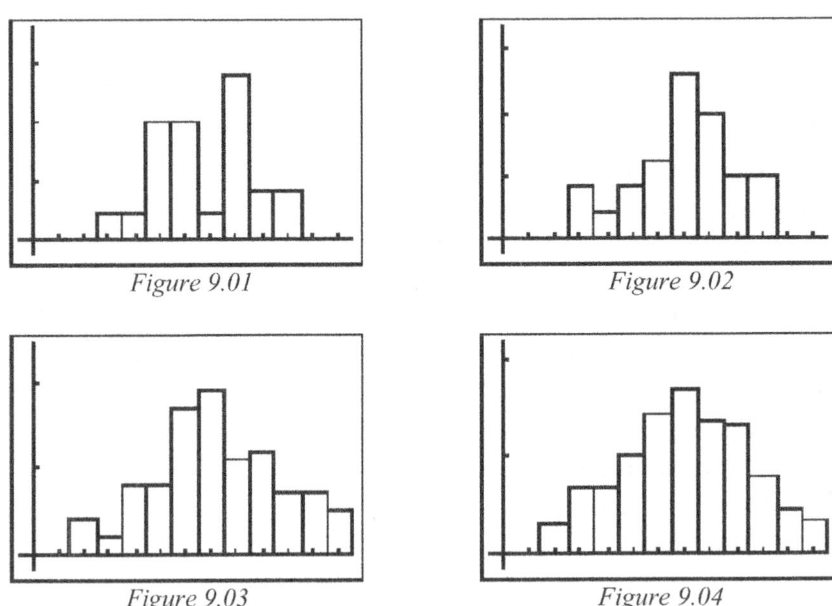

Figure 9.01

Figure 9.02

Figure 9.03

Figure 9.04

If, instead of actually carrying out this experiment, the probability of each of the possible outcomes is found, then a probability distribution is obtained (see section 9.2 for the definition of a probability distribution).

Table 9.1 shows a table of outcomes of the sum of the scores on the two dice, and figure 9.05 shows a diagram of the corresponding probability distribution.

	1	2	3	4	5	6
1	2	3	4	5	6	7
2	3	4	5	6	7	8
3	4	5	6	7	8	9
4	5	6	7	8	9	10
5	6	7	8	9	10	11
6	7	8	9	10	11	12

Table 9.1

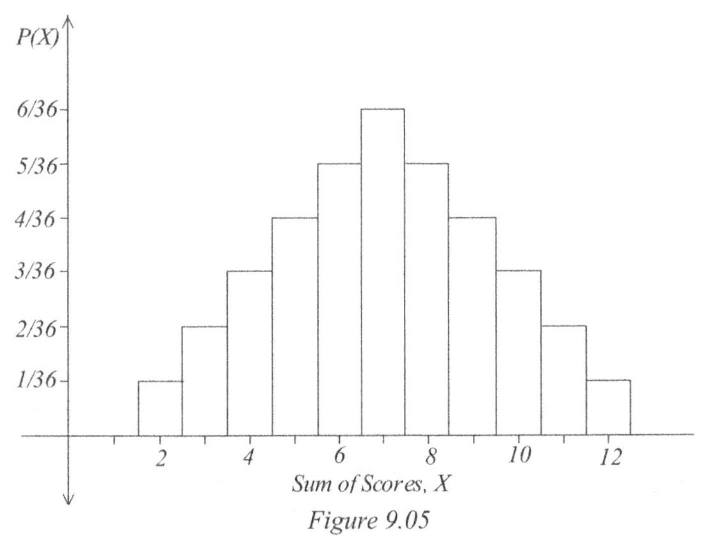

Figure 9.05

The figures 9.01 to 9.04 indicate that as more trials are carried out, that is, as n increases, the shape of the frequency distribution approaches the shape of the probability distribution shown in figure 9.05. The variable that is used for a probability distribution is called a *random variable*.

9.1.1 Discrete Random Variables

X is called a *discrete random variable* if it has the following properties:

- X takes values $x_1, x_2, x_3 \ldots, x_n$. It is not necessary that n be finite.

- Each value of X is associated with a probability, $P(X = x_i)$.

- $\displaystyle\sum_{i=1}^{n} P(X = x_i) = 1$

The values of a random variable are real numbers, usually integers, assigned to the elements of a sample space. For example, when a regular die is thrown, the random variable, X, can take the value 1, 2, 3, 4, 5, 6, so that $X = \{1, 2, 3, 4, 5, 6\}$ and to each of these values is assigned the probability, $\dfrac{1}{6}$.

This assignation of probabilities to each random variable is called the *probability distribution*.

Example 9.1: The random variable, X, takes the probability distribution shown in table 9.2. Find the value of c.

x	0	1	2	3	4
$p(X = x_i)$	1/5	1/6	3/10	1/5	c

Table 9.2

Solution 9.1: As $\displaystyle\sum_{i=1}^{5} P(X = x_i) = 1$ so $\dfrac{1}{5} + \dfrac{1}{6} + \dfrac{3}{10} + \dfrac{1}{5} + c = 1 \Rightarrow \dfrac{26}{30} + c = 1 \Rightarrow c = \dfrac{2}{15}$.

Example 9.2: A bag contains 5 red balls and 3 blue balls which are randomly removed one at a time, without being replaced, until a red ball is removed. Find the probability distribution of X, the number of balls removed.

Solution 9.2: As there are only 3 blue balls, $X \in \{1, 2, 3, 4\}$. The corresponding probabilities are found using a probability tree, as shown in figure 9.06.

$$P(X = 1) = \frac{5}{8}, \quad P(X = 2) = \frac{3}{8} \times \frac{5}{7} = \frac{15}{56}$$

$$P(X = 3) = \frac{3}{8} \times \frac{2}{7} \times \frac{5}{6} = \frac{5}{56}$$

$$P(X = 4) = \frac{3}{8} \times \frac{2}{7} \times \frac{1}{6} \times 1 = \frac{1}{56}$$

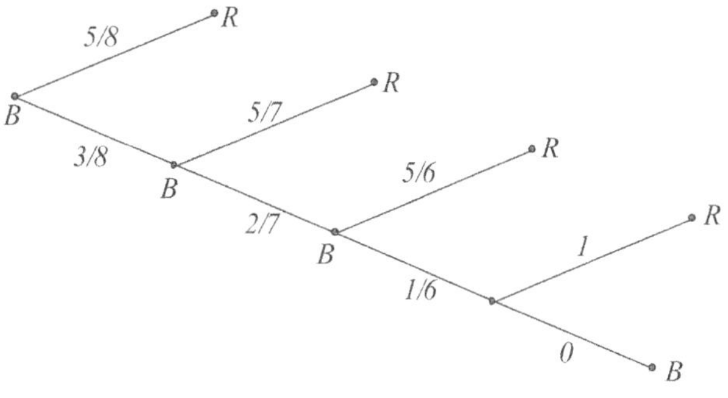

Figure 9.06

259

At this stage, it might be wise to check that the sum of all the probabilities is 1:

x	1	2	3	4
$P(X=x)$	5/8	15/56	5/56	1/56

Table 9.3

$$\frac{5}{8}+\frac{15}{16}+\frac{5}{56}+\frac{1}{56}=\frac{35+15+5+1}{56}=1$$

Table 9.3 shows the probability distribution of X.

9.1.2 The Expectation of X

The expectation of a random variable, X is rather like the mean of a set of data. Without collecting any experimental data, we can calculate the expectation of a random variable by simply using the probabilities of the values assigned to X.

The mean of a population of n data points, grouped into N classes, is $\mu=\frac{1}{n}\sum_{i=1}^{N}f_i x_i=\sum_{i=1}^{N}x_i\left(\frac{f_i}{n}\right)$. If a random variable, X is defined so that $X\in\{x_1,x_2,x_3,\ldots,x_n\}$, then $P(X=x_i)=\frac{f_i}{n}$. Hence, $\mu=\sum_{i=1}^{N}x_i P(X=x_i)$. However, for probability distributions, the mean μ is usually written in terms of the random variable X as $E(X)$, so that $E(X)=\sum_{i=1}^{N}x_i P(X=x_i)$.

Example 9.3: Find $E(X)$ if X is defined as in Example 9.2.

Solution 9.3: $E(X)=\sum_{i=1}^{4}x_i P(X=x_i)=1\times\frac{5}{8}+2\times\frac{15}{56}+3\times\frac{5}{56}+4\times\frac{1}{56}\approx1.5$. Therefore, if we repeat the trial of taking balls from a bag containing 5 red and 3 blue balls until we obtain a red ball, we should expect, on average, to take 1.5 balls per trial.

A sequence of trials such as those defined in Example 9.2 may be simulated on a graphing calculator. A TI-84 Plus program in which these trials are repeated n times and the mean number of times, m that a ball is removed from the bag until a red ball is obtained, is calculated. Table 9.4 shows one example of the results obtained when this program is run. The program is shown at the end of Unit 9. It is called 'Expectation Simulation Program'.

n	10	50	100	500	1000	1500	2000
m	1.6	1.48	1.45	1.438	1.520	1.481	1.4965

Table 9.4

The data in table 9.4 suggest that n gets large so m approaches $E(X)$ or, more concisely $\lim_{n\to\infty}m=E(X)$.

Exercise 9.1

1. The random variable X has the probability distribution shown in the tables below. Find, in each case, the value of c, if c is a constant.

(a)

x	0	1	2	3
$P(X=x)$	c	$2c$	0.4	$3c$

(b)

x	-2	-1	0	1	2	3	4
$P(X=x)$	0.15	0.20	$3c$	$4c$	$3c$	0.05	0.10

(c)

x	0	1	2	3
$P(X=x)$	c	$0.1c$	c^2	0.2

2. Two regular dice are thrown. If X is the random variable representing the difference of the scores obtained, construct
(a) a table showing the value of X for each combination of scores on the two dice.
(b) a table showing the probability distribution of X.

3. A bag contains 5 blue balls and 3 red balls. A ball is removed and not replaced. A second ball is removed. X is the number of blue balls removed.
   (a) Copy and complete the probability tree diagram shown in figure 9.07.
    (b) Find the probability distribution of X.

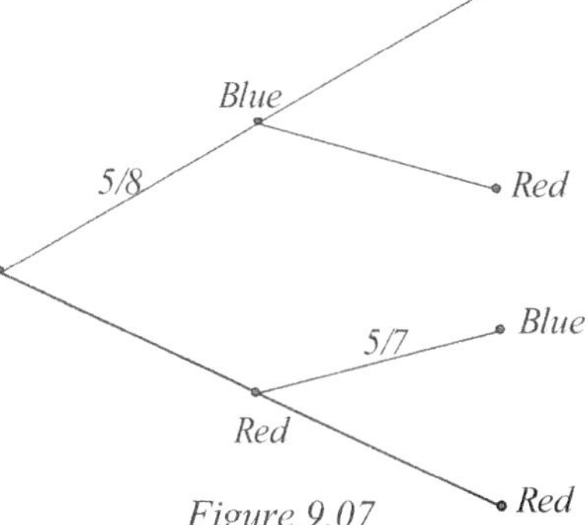

Figure 9.07

4. A bag contains 3 red balls, 4 white balls and 3 green balls.
   Three balls are removed at random from the bag, one at a time, and are not replaced.
   (a) Copy and complete the probability tree diagram shown in figure 9.08.
   (b) Find the probability distribution for the number of white balls removed.
   (c) Find the expectation of the number of white balls removed.

Figure 9.08

5. A game is played in which a fair coin is tossed three times. Three points are scored if a 'head' is obtained on the first toss, two points are scored for a 'head' obtained on the second toss and one point is scored for a 'head' obtained on the third toss. No points are scored for a 'tail'. Let X be the random variable representing the total score.
   (a) By considering the outcome TTH, where
      H represents obtaining a 'head' and T represents obtaining a 'tail', find $P(X=1)$.

(b) By considering appropriate outcomes, show that $P(X=3)=\dfrac{1}{4}$.

(c) Copy and complete table 9.5 showing the probability distribution of X.

(d) Find $E(X)$.

X	0	1	2	3	4	5	6
$P(X=x)$	1/8			1/4			

Table 9.5

6. A game is played in which two fair coins are tossed. If two heads are obtained the player scores 5 points. If two tails are obtained the player scores 3 points and if a head and a tail are obtained the player scores 1 point. Find the player's expected score.

7. A slot machine has three rotating discs. When a dollar bill is fed into the machine, each disc displays, in a window, either an orange, a cherry, a plum or a star. Table 9.6 shows the probability, for each disc, that the symbols appear in the window.

orange	cherry	plum	star
0.5	0.3	0.15	0.05

Table 9.6

The discs operate independently. A player wins if the three discs each display the same symbol. The prizes are shown in table 9.7.

3 oranges	3 cherries	3 plums	3 stars
$2	$10	$50	$1000

Table 9.7

Let X be the random variable representing the amount, in dollars, that a player wins each time the machine is 'played'.

(a) Explain why $X \in \{-1,\ 1,\ 9,\ 49,\ 999\}$.

(b) Find the probability distribution of X.

(c) Find $E(X)$ and hence estimate the amount of money that the owner of the machine makes when the machine is played 1000 times.

8. An aircraft has three lavatories, each of which is occupied 40% of the time during a particular flight. Assuming that the lavatories are occupied independently,
(a) draw a table for the probability distribution of the number of lavatories which are occupied.
(b) find the expected number of occupied lavatories.

9. Sarah has a cash card but has forgotten the three digit PIN number except that exactly one of the digits is a 1, the sum of the digits is 7 and the PIN does not start with a 0. She attempts to draw cash from a machine. Assume that the machine permits her sufficient attempts and that she works by a process of elimination.
(a) Draw a table of the probability distribution of the number of guesses needed before the correct PIN number is found.
(b) Find the mean number of guesses that would be needed in order to find the correct PIN number.

9.2   The Binomial Distribution

You have already met, in the repeated tossing of a coin, the repeated throwing of a die or in the selection, <u>with replacement</u>, of balls from a bag, a situation which may be modeled by a random variable having the following characteristics:

    a)   each trial may be viewed as either 'a success' or 'a failure'
    b)   the outcome of each trial is independent of the previous trial
    c)   each trial has a constant probability of success

Consider a bag containing three white balls and one black ball, and suppose that a trial is defined to be the random removal of a single ball. When the color of the ball has been noted, it is replaced in the bag.

If we consider choosing a black ball as 'a success' and choosing a white ball as 'a failure', property a) is satisfied because the ball removed is either black or white. Property b) is satisfied provided we have an effective procedure for random selection. Property c) is satisfied because the probability of selecting a black ball is $\dfrac{1}{4}$.

Suppose that the trial is repeated n times, and the random variable, X , is defined as the number of times that the black ball is removed. If n is small, it is possible to analyze the distribution of X using probability trees as shown in figures 9.09, 9.10 and 9.11.

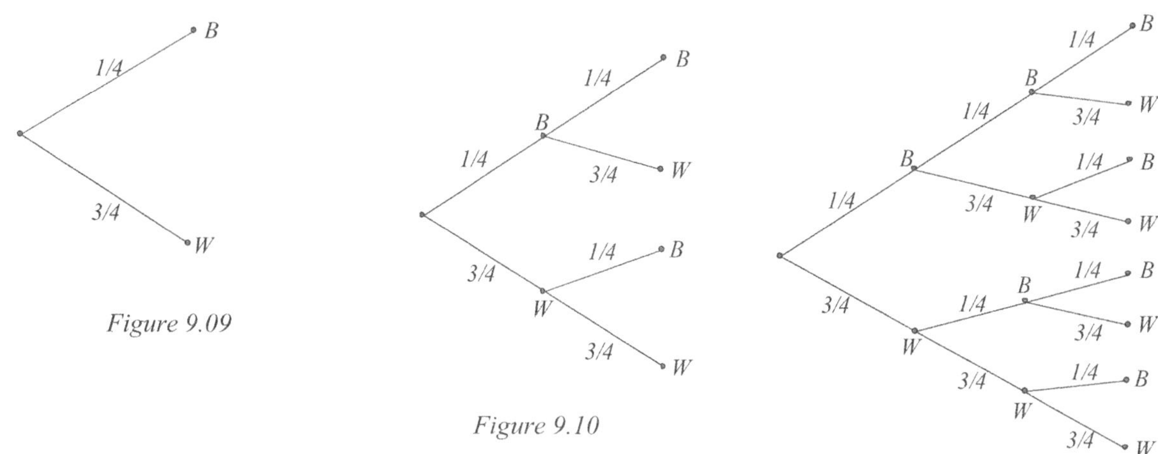

Figure 9.09

Figure 9.10

Figure 9.11

Referring to figure 9.09, for $n = 1$,

$$P(X = 0) = \frac{3}{4}, \quad P(X = 1) = \frac{1}{4}$$

Referring to figure 9.10, for $n = 2$,

$$P(X = 0) = \left(\frac{3}{4}\right)^2 , \quad P(X = 1) = 2\left(\frac{1}{4}\right)\left(\frac{3}{4}\right), \quad P(X = 2) = \left(\frac{1}{4}\right)^2$$

Referring to figure 9.11, for $n = 3$,

$$P(X=0)=\left(\frac{3}{4}\right)^3, \; P(X=1)=3\left(\frac{3}{4}\right)^2\left(\frac{1}{4}\right), \; P(X=2)=3\left(\frac{3}{4}\right)\left(\frac{1}{4}\right)^2, \; P(X=3)=\left(\frac{1}{4}\right)^3$$

Summarizing these results in a triangular form should suggest to you why this distribution is called *binomial* (Unit1):

$$n=1 \qquad 1\left(\tfrac{3}{4}\right)\; 1\left(\tfrac{1}{4}\right)$$

$$n=2 \qquad 1\left(\tfrac{3}{4}\right)^2\; 2\left(\tfrac{3}{4}\right)\left(\tfrac{1}{4}\right)\; 1\left(\tfrac{1}{4}\right)^2$$

$$n=3 \qquad 1\left(\tfrac{3}{4}\right)^3\; 3\left(\tfrac{3}{4}\right)^2\left(\tfrac{1}{4}\right)\; 3\left(\tfrac{3}{4}\right)\left(\tfrac{1}{4}\right)^2\; 1\left(\tfrac{1}{4}\right)^3$$

For $n = 4$, a probability tree diagram becomes too cumbersome. Fortunately, it is not necessary to draw one because the coefficients in the probability expressions correspond to the numbers in Pascal's triangle:

$$
\begin{array}{ccccccccc}
 & & & & 1 & & & & \\
 & & & 1 & & 1 & & & \\
 & & 1 & & 2 & & 1 & & \\
 & 1 & & 3 & & 3 & & 1 & \\
n=4\rightarrow 1 & & 4 & & 6 & & 4 & & 1 \\
\end{array}
$$

$$x=0\uparrow\; x=1\uparrow\; x=2\uparrow$$

Therefore, you can use Pascal's triangle to find the appropriate coefficient. For example, if you need $P(X=2)$, then you know that this probability is of the form $k\left(\frac{3}{4}\right)^2\left(\frac{1}{4}\right)^2$. The value of k is obtained by going to the $n=4$ row (actually row 5) and counting along the row, starting at $x=0$, until you reach the coefficient corresponding to $x=2$. In this case, $k=6$ and the required probability is

$$6\left(\frac{3}{4}\right)^2\left(\frac{1}{2}\right)^2 = \frac{27}{32}.$$

Example 9.4: A die is thrown 5 times. Find the probability of getting 2 'sixes'.

Solution 9.4: The probability of obtaining a 'six' on a single throw is $\frac{1}{6}$. The probability of not obtaining a 'six' is $\frac{5}{6}$. For 2 'sixes' from 5 throws, there will, necessarily, be 3 'non-sixes'. Therefore, $P(X=2)=k\left(\frac{5}{6}\right)^3\left(\frac{1}{6}\right)^2$, and k is the number of branches of the corresponding probability tree that yields 2 'sixes' and 3 'non-sixes'. The following portion of Pascal's triangle shows that $k = 10$.

$$
\begin{array}{c}
1 \\
1 \quad\quad 1 \\
1 \quad\quad 2 \quad\quad 1 \\
1 \quad\quad 3 \quad\quad 3 \quad\quad 1 \\
1 \quad\quad 4 \quad\quad 6 \quad\quad 4 \quad\quad 1 \\
n=5 \quad 1 \quad 5 \quad 10 \quad 10 \quad 5 \quad 1
\end{array}
$$

$$x=0\uparrow \quad x=1\uparrow \quad x=2\uparrow$$

$$\text{So, } P(X=2)=10\left(\frac{5}{6}\right)^3\left(\frac{1}{6}\right)^2=10\left(\frac{125}{216}\right)\left(\frac{1}{36}\right)=0.1607510\ldots=0.161 .$$

Exercise 9.2

In each question, use Pascal's triangle to find the binomial coefficients.

1.   A fair coin is tossed 4 times. Find the probability of getting

      (a) exactly three 'heads'   (b) more than 2 'heads'

2.   A fair coin is tossed six times. Find the probability of obtaining

      (a) no 'heads'   (b) exactly three 'heads'   (c) exactly 5 'heads'

3.   When a fair die is thrown 8 times, find the probability of obtaining

      (a) exactly 1 'three'   (b) exactly 2 'threes'   (c) more than 2 'threes'

4.   A computer outputs integers from 0 to 9 randomly. Find the probability that a string of 7 integers contains

      (a) no 'twos'   (b) exactly 2 'threes'   (c) exactly 3 'nines'

Pascal's triangle has limitations. Suppose you want to know the probability of obtaining exactly 4 'sixes' when a die is thrown 20 times. You know that in 20 throws you need 4 'sixes' and 16 'non-sixes', so, $P(X=4)=k\left(\dfrac{5}{6}\right)^{16}\left(\dfrac{1}{6}\right)^{4}$. The binomial coefficient k might be obtained from the 5th number along the 21st row of Pascal's triangle, but you can obtain k more conveniently by using your graphing calculator. k is the number of ways of choosing 4 items from 20 distinct items, written $\begin{pmatrix}20\\4\end{pmatrix}$, or, on most graphing calculators, written as $20C4$.

Figures 9.12 and 9.13 show you how to obtain the value of $\begin{pmatrix} n \\ r \end{pmatrix}$ (or nCr) for the particular case where $n = 20$ and $r = 4$ on a TI-84 Plus. First, enter 20 on the home screen and then go to the MATH menu.

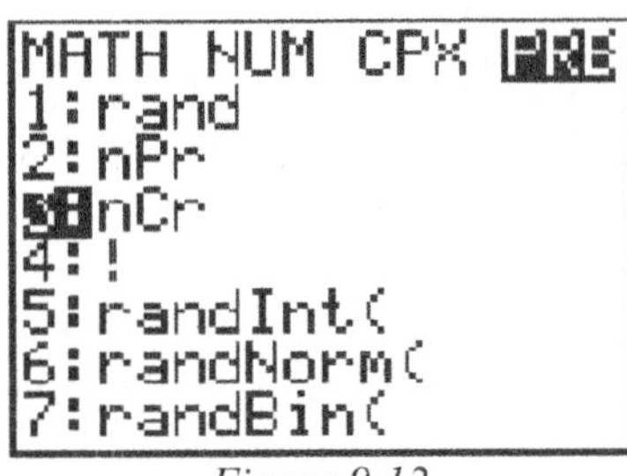

Figure 9.12

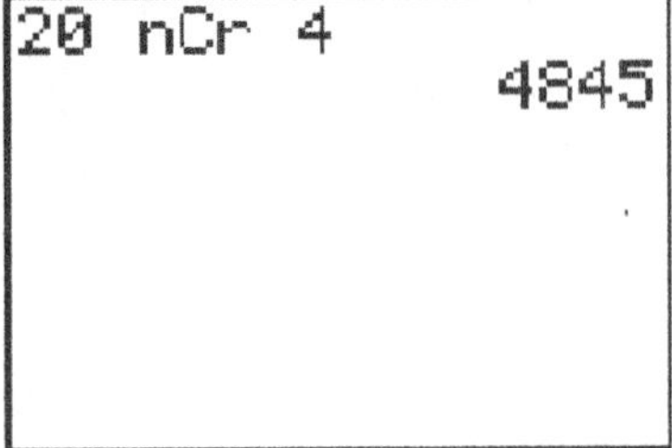

Figure 9.13

Therefore, $P(X=4) = \begin{pmatrix} 20 \\ 4 \end{pmatrix}\left(\frac{5}{6}\right)^{16}\left(\frac{1}{6}\right)^{4} = 4845\left(\frac{5}{6}\right)^{16}\left(\frac{1}{6}\right)^{4} = 0.2022035... = 0.202.$

More generally, if a random variable X is binomially distributed with n trials and the probability of 'success' is p for each trial, then this information can be abbreviated by the expression $X \sim B(n,p)$. Let $1-p = q$. Then, if $X \sim B(n,p)$, the probability distribution of X is given by the terms in the expansion of the binomial expression $(q+p)^{n}$.

Example 9.5:   Find the probability distribution of X, where $X \sim B\left(4, \frac{2}{5}\right)$, by expanding the appropriate binomial expression.

Solution 9.5: The probabilities $P(X=r)$ for $r = 0,1,2,3,4$ are obtained by expanding

$$(q+p)^{n} = \left(\frac{3}{5}+\frac{2}{5}\right)^{4} = \left(\frac{3}{5}\right)^{4} + 4\left(\frac{3}{5}\right)^{3}\left(\frac{2}{5}\right) + 6\left(\frac{3}{5}\right)^{2}\left(\frac{2}{5}\right)^{2} + 4\left(\frac{3}{5}\right)\left(\frac{2}{5}\right)^{3} + \left(\frac{2}{5}\right)^{4}$$

$$= \frac{81}{625} + \frac{216}{625} + \frac{216}{625} + \frac{96}{625} + \frac{16}{625}$$

Table 9.8 shows the probability distribution in tabular form.

r	0	1	2	3	4
$P(X=r)$	81/625	216/625	216/625	96/625	16/625

Table 9.8

Example 9.6: Chocolates are manufactured by a machine which, on average, produces 0.4% defective chocolates. The chocolates are put at random into boxes containing 24 chocolates.

   (i) Find the probability that a box selected at random has
      (a) no defective chocolates
      (b) exactly one defective chocolate
      (c) more than one defective chocolate

   (ii) If a store buys 20 boxes of chocolates, find the probability that
      (a) none of these boxes contains defective chocolates
      (b) exactly one box contains defective chocolates
      (c) more than one of these boxes contain defective chocolates

Solution 9.6: Let X be the random variable representing the number of defective chocolates.

(i) $X \sim B(24,\ 0.004)$

  (a) $P(X=0)=\left((1-0.004)^{24}=0.9082891...=0.908\right.$

  (b) $P(X=1)=\binom{24}{1}0.996^{23}0.004=0.08754594...=0.0875$

  (c) $P(X>1)=1-\big(P(X=0)+P(X=1)\big)$

$$=1-(0.908289+0.087546)=0.004164917...=0.00416$$

(ii) Now you are concerned not with the defectiveness of individual chocolates but with <u>boxes</u> which contain one or more defective chocolates. Let Y be the random variable representing the number of boxes containing one or more defective chocolates. The probability that a box contains one or more defective chocolates is $1-0.9082891=0.0917109$. Therefore, $Y \sim B(20,\ 0.0917109)$.

  (a) $P(Y=0)=0.9082891^{20}=0.1460434...=0.146$

  (b) $P(Y=1)=\binom{20}{1}\times 0.9082891^{19}\times 0.0917109=0.2949231...=0.295$

  (c) $P(Y>1)=1-(0.1460434+0.2949231)=0.5590335...=0.559$

Some graphing calculators include the binomial probability function (pdf) and the cumulative binomial distribution function (cdf) so that calculations of the kind just carried out can be done more conveniently.

Suppose X is a binomial random variable such that $X \sim B(10, \, 0.3)$. Figures 9.14 and 9.15 show how to find $P(X = 5)$ by starting at the DISTR menu.

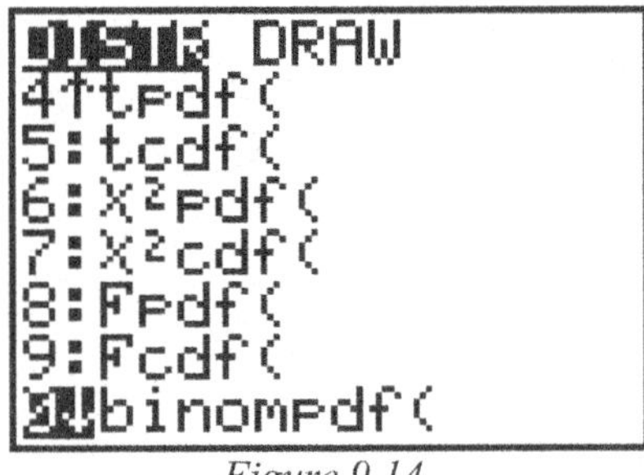

Figure 9.14

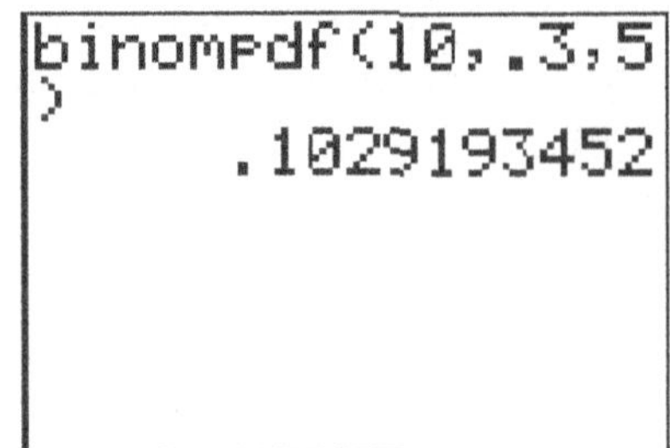

Figure 9.15

The cumulative binomial distribution function maps the random variable, X to $P(X \le x)$. Figures 9.16 and 9.17 show how to find $P(X \le 5)$.

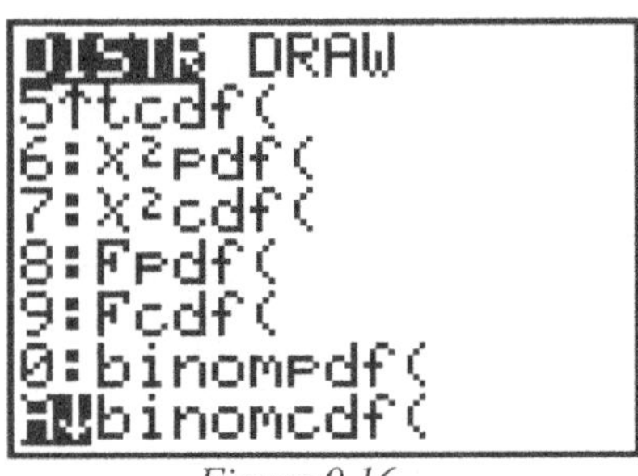

Figure 9.16

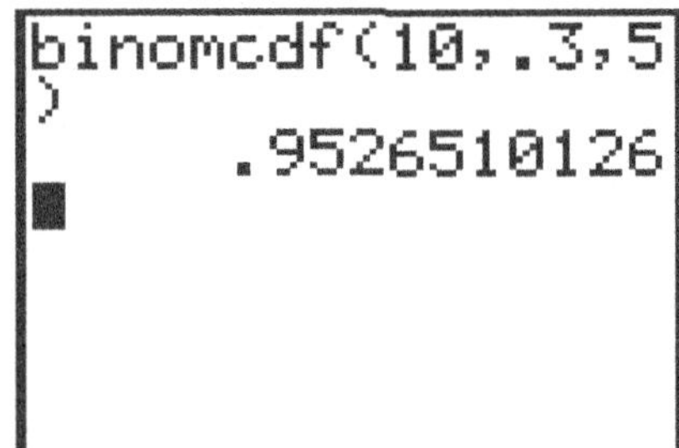

Figure 9.17

Exercise 9.3

1. A bag contains 15 balls, numbered from 1 to 15 inclusive. Five selections, with replacement, are made from the bag. Find the probability that
    (a) all balls selected have numbers greater than 10
    (b) exactly three balls selected have numbers greater than 10

2. Assuming that a woman is equally likely to give birth to a girl or a boy, find the probability that if she has a family of 6 children, 3 of them are girls and 3 are boys.

3. From a regular pack of 52 playing cards, a card is drawn, at random, and replaced. This trial is carried out 8 times. Find the probability of drawing an ace on at least 2 occasions.

4. Over a long period, the proportion of defective items in a manufacturing process is 5%. In a random sample of 25, find the probability of
    (a) no defectives   (b) exactly two defectives   (c) more than two defectives

5. Find the probabilities of obtaining
    (a) at least one 'six' in 6 throws of a die   (b) at least two 'sixes' in 12 throws
    (c) at least three sixes' in 18 throws of a die   (d) at least 4 'sixes' in 24 throws of a die

6.   A bus holds 48 passengers. On average, one passenger in every 16 who books a seat fails to
       turn up. Stating carefully any assumptions made, find the probability that, if the booking office
       accepts 50 bookings, there will not be enough seats.

7.   A bag contains a large number of balls of which 25% are blue. The rest are yellow. If 7 balls
       are taken from the bag, find the probability that
       (a) they are all yellow   (b) exactly 2 are blue

8.   A company makes 1.5 volt batteries, and its quality control department estimates that 0.7% of
       the batteries are defective. If a sample of 20 batteries is taken, find the probability that
       (a) exactly one is defective   (b) more than one are defective   (c) less than two are defective

       The batteries are sold in packets of 8. Find the probability that
       (d) a pack contains no defective batteries
       (e) a consignment of 10 packs contains less than 2 packs with defective batteries

9.   Experience has shown that at a busy highway intersection, the probability on any given day that
       an accident occurs is 0.008.
       (a) Show that the probability of at least one accident in a randomly selected week is
            approximately 5%
       (b) Find the probability that, in a randomly selected four-week period,
                (i) there are no accidents   (ii) 3 of the weeks are accident free
                (iii) 2 of the weeks are accident free

9.2.1 The Expectation of a Binomial Distribution

Suppose that 3 dice are thrown simultaneously, and X is the random variable representing the number

of 'sixes' obtained. Then, using the result obtained in section 9.1.2, $E(X) = \sum_{i=0}^{3} x_i P(X = x_i)$.

As $X \sim B\left(3, \dfrac{1}{6}\right)$, the terms of the expansion of $\left(\dfrac{5}{6} + \dfrac{1}{6}\right)^3$ will give $P(X = x_i)$ for $i = 0, 1, 2, 3$:

$$\left(\frac{5}{6} + \frac{1}{6}\right)^3 = \left(\frac{5}{6}\right)^3 + 3\left(\frac{5}{6}\right)^2\left(\frac{1}{6}\right) + 3\left(\frac{5}{6}\right)\left(\frac{1}{6}\right)^2 + \left(\frac{1}{6}\right)^3$$

So, $E(X) = 0 \times \left(\dfrac{5}{6}\right)^3 + 1 \times 3\left(\dfrac{5}{6}\right)^2\left(\dfrac{1}{6}\right) + 2 \times 3\left(\dfrac{5}{6}\right)\left(\dfrac{1}{6}\right)^2 + 3 \times \left(\dfrac{1}{6}\right)^3 = \dfrac{1}{216}(0 + 75 + 30 + 3) = \dfrac{108}{216} = \dfrac{1}{2}$.

This is an increasingly long and tedious calculation as n increases. As it turns out, it is unnecessary,
because, if X is binomially distributed, i.e. $X \sim B(n, p)$, then $E(X) = np$. The proof is as follows:

Let $X \sim B(n, p)$ and $q = 1 - p$. Then $E(X) = \sum_{r=0}^{n} x_r P(X = x_r)$ and $x_r \in \{0, 1, 2, \ldots, n\}$.

Therefore, $E(X) = \sum_{r=0}^{n} r \binom{n}{r} q^{n-r} p^r = 0 \times q^n + 1 \times \binom{n}{1} q^{n-1} p + 2 \times \binom{n}{2} q^{n-2} p^2 + \ldots + n \times p^n$

$$= nq^{n-1} p + \frac{2n(n-1)}{2!} q^{n-2} p^2 + \frac{3n(n-1)(n-2)}{3!} q^{n-3} p^3 \ldots + np^n$$

$$= np \left\{ q^{n-1} + (n-1) q^{n-2} p + \frac{(n-1)(n-2)}{2!} q^{n-3} p^2 + \ldots + p^{n-1} \right\}$$

$$= np(q+p)^{n-1} \text{ but } (q+p) = 1 \Rightarrow (q+p)^{n-1} = 1. \text{ Therefore, } E(X) = np.$$

Example 9.7: If X is a random variable such that $X \sim B(18, p)$ and $E(X) = 7.2$, find the value of p.

Solution 9.7: As $E(X) = np$, $7.2 = 18p \Rightarrow p = \frac{7.2}{18} = 0.4$.

Exercise 9.4

1. (a) A fair coin is tossed 50 times. Find the expected number of 'heads'.
    (b) An unbiased six-sided die is rolled 60 times. Find the expected number of times a 'six' is rolled.

2. (i) A regular die is rolled 6 times.
        (a) Write down the expected number of 'sixes' obtained.
        (b) Find the probability that this number of 'sixes' is obtained.
    (ii) A regular die is rolled 60 times.
        (a) Write down the expected number of 'sixes' obtained.
        (b) Find the probability that this number of 'sixes' is obtained.

3. (a) A machine produces electronic components, 6% of which are defective. If a batch of 120 components is tested, find the expected number of defective components in the batch.
    (b) As a result of a major disaster, a city is faced with having to care for a large number of refugees. A survey reveals that 35% of families are willing to host a refugee. The city has 25 000 families. Estimate the number of refugees that can be expected to obtain care from a family.

4. (a) If $X \sim B(n, p)$, $E(X) = 15$ and $p = 0.3$, find the value of n.
    (b) If $X \sim B(8, p)$ and $E(X) = 5$, find p.

5. If $X \sim B(10, p)$, $Y \sim B(30, p^2)$ and $E(X) = E(Y)$, find the non-zero value of p.

6. X is a random variable such that $X \sim B(3, p)$ and $P(X = 2) = 0.2$.
    (a) Show that $15p^3 - 15p^2 + 1 = 0$.
    (b) Draw a graph of $f(p) = 15p^3 - 15p^2 + 1$ and use it to find three solutions of the equation of part (a).
    (c) Explain why only two of these solutions are valid.

7. X is a random variable with a binomial distribution such that $X \sim B(n, p)$. The table shows the distribution of X.

x	0	1	2	3
$P(X = x)$	$64/125$	$48/125$	$12/125$	$1/125$

(a) Write down n.   (b) Find $E(X)$.   (c) Hence find p.

8. Tulip bulbs produce plants whose flowers are either red or yellow. The bulbs are sold in packets of six, and these packets are tested in order to monitor the relative proportion of red flowers to yellow flowers. A sample of 120 packets of bulbs were planted and grown, and the results obtained are shown in the table.

Number of red flowers, x	0	1	2	3	4	5	6
Number of samples with x red flowers	16	33	31	18	14	7	1

(a) Find the mean number of red flowers per packet.
(b) Use your answer to part (a) to show that if the number of red flowers per packet is binomially distributed, then the probability that a bulb gives a red flower is 0.342.
(c) Find the frequency distribution of red flowers, assuming that it is binomial with probability 0.342.

9.3 The Normal Distribution

9.3.1 A Continuous Random Variable

Consider the discrete random variable X which represents the number of 'heads' obtained when a fair coin is tossed n times. Figures 9.18 to 9.21 show histograms for the distribution of X for $n = 4$, 8, 16 and 32. The height of the rectangle representing x 'heads' is equal to $P(X = x)$.

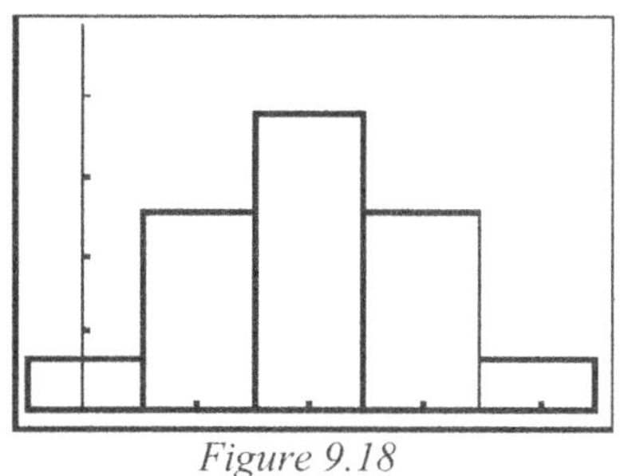

Figure 9.18

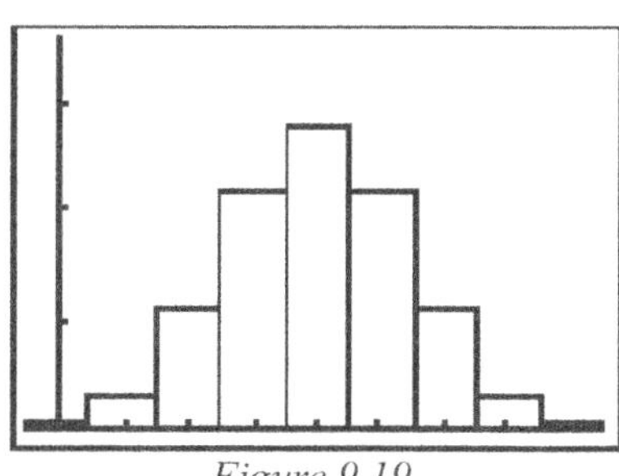

Figure 9.19

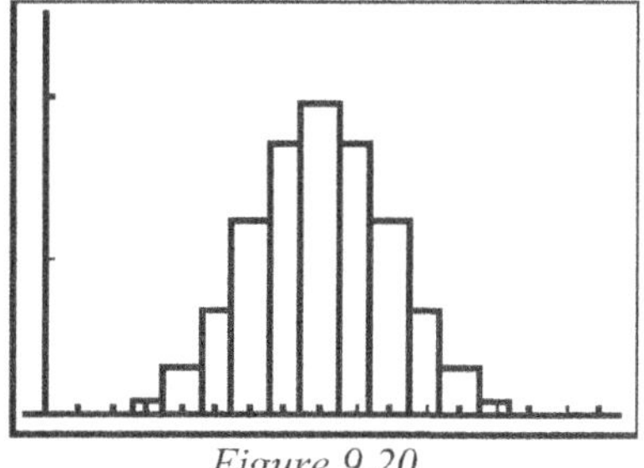

Figure 9.20

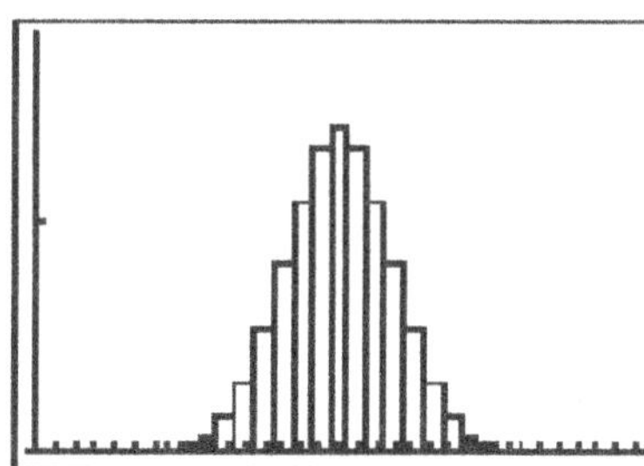

Figure 9.21

As n increases, so the outline of the distribution becomes smoother and takes on a bell shape. In addition, as n increases, the probability that X takes on any specific value gets less and less. In the limit, as n increases towards infinity, the distribution of X becomes a smooth continuous curve called the *normal distribution* curve.

When the sample space is infinite and X becomes a *continuous random variable*, the probability that X takes any specific value becomes zero. It is only meaningful to consider the probability that X lies within a certain interval, and that probability can be found by measuring the area under the normal distribution curve between the limits of that interval.

Figure 9.22 shows the normal distribution curve for a continuous normal random variable, X whose mean is 50 and whose standard deviation is 6. Since the sum of all the probabilities taken by a continuous random variable is 1, the area under the normal distribution curve is 1. The probability that X lies in the interval 55 to 60 is shown by the shaded region.

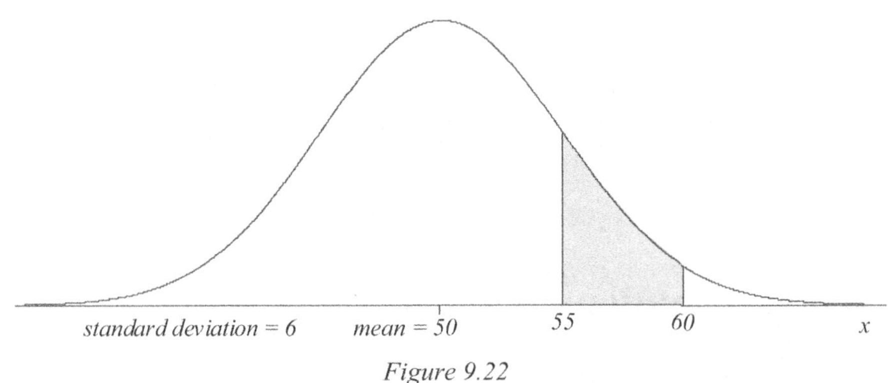

Figure 9.22

The area of the shaded region is approximately 0.155 and we write that $P(55 < X < 60) = 0.155$.

The probability that a continuous random variable takes any specific value is 0. Because of this, we could equally have written $P(55 \le X \le 60) = 0.155$.

The normal distribution is a central feature of statistics and is important as a model for the distribution of a wide variety of <u>continuous</u> variables.

The following is a list of examples which may be usefully modeled by a normal distribution:

- the height of a cow in a herd of cattle
- the mass of a tomato in a load of tomatoes
- the time from conception to birth of a human fetus
- the error in a set of measurements
- the mean of a sample taken from a normally distributed population

The normal distribution is represented by a family of functions. All have graphs with a bell shape but differ in their position on the horizontal axis and the steepness of the bell. The curve is symmetric about the line $x = \mu$, and the curve never meets the x-axis, which it approaches asymptotically.

Despite the fact that the graph never touches the x-axis, the area between the curve and the x-axis is always finite. In fact, this area is always 1. The shape of a normal curve is completely defined if its mean, μ and standard deviation, σ are known.

It is convenient, at this stage, to define the variance of a population, σ^2, and it is customary to use the variance, as opposed to the standard deviation, σ, in certain situations. One such situation is in defining a normally distributed variable, X in which case we write $X \sim N\left(\mu,\ \sigma^2\right)$. This simply means that X is a normally distributed random variable whose mean is μ and whose standard deviation is σ.

9.3.2 The Standardized Normal Distribution Function

The simplest and most important normal curve is that possessed by the standardized normal variable, Z which has $\mu = 0$ and $\sigma^2 = 1$ so that $Z \sim N\left(0,\ 1\right)$. The standardized normal distribution is described by the function $f\left(z\right) = \dfrac{1}{\sqrt{2\pi}} e^{-\frac{z^2}{2}}$. Figure 9.23 shows the graph of this function.

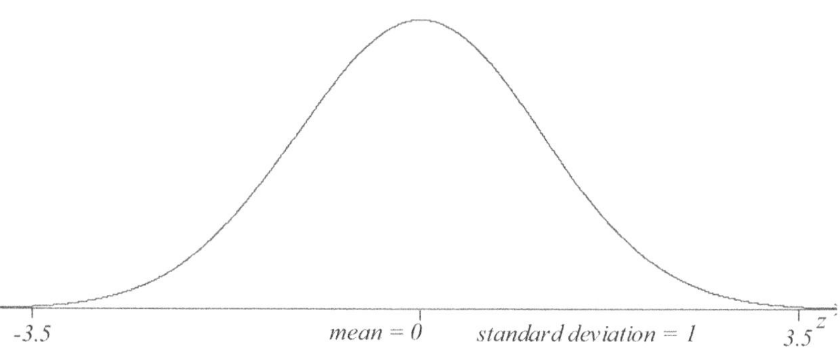

Figure 9.23

It should be noted that if the z-axis extends from -3.5 to 3.5, 99.95% of the area enclosed by the graph and the z-axis is included.

In order to find the probability that Z lies in a given interval, it is necessary to measure the area enclosed between the curve, the z-axis and the extremes f the interval. Although areas under curves are found by integration, (see Unit 7 section 7.7) this is not possible for the normal curve because it cannot be integrated analytically. However, the function which measures the area between the curve and the z-axis to the left of z (a specific value of Z) is available in most graphing calculators. This area, denoted by $P\left(Z < z\right)$, is shown in figure 9.24.

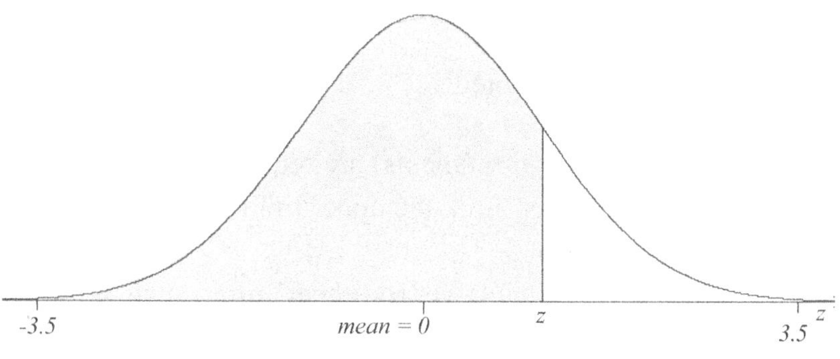

Figure 9.24

Use of the graphing calculator to find areas under the standardized normal curve:
The following example shows how to find areas under the normal curve for a TI-84 Plus.

Example 9.8: Find (i) $P(Z > 0.72)$ (ii) $P(Z < -1.65)$ (iii) $P(-0.4 < Z < 1.1)$

Solution 9.8: Figures 5.100 to 5.103 indicate the method.

(i) $P(Z > 0.72) = 0.236$

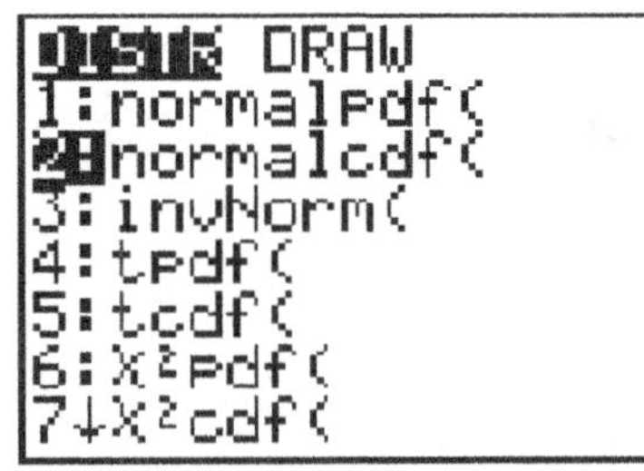

Figure 9.25

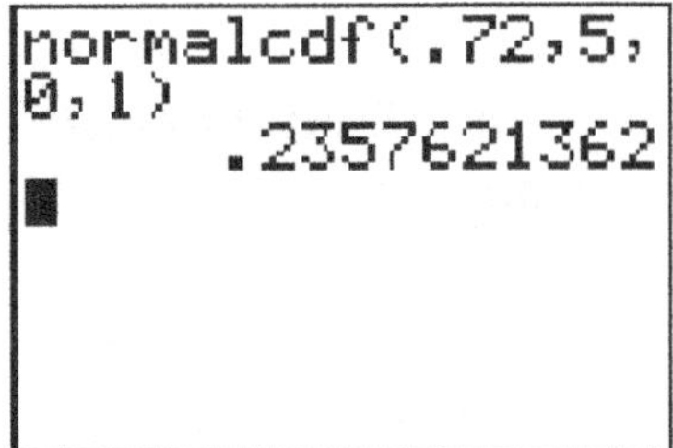

Figure 9.26

(ii) $P(Z < -1.65) = 0.0495$ (iii) $P(-0.4 < Z < 1.1) = 0.520$

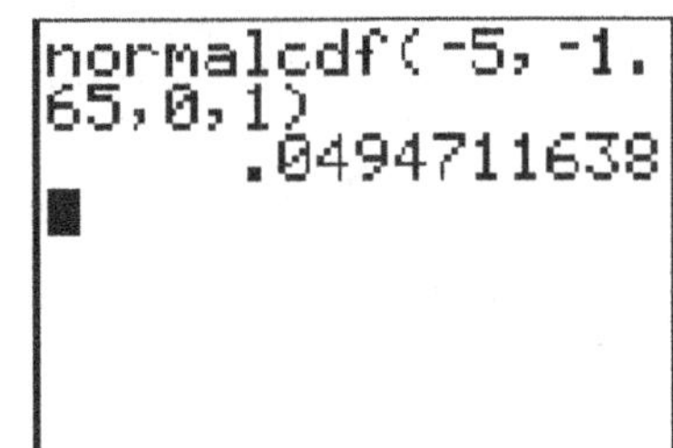

Figure 9.27

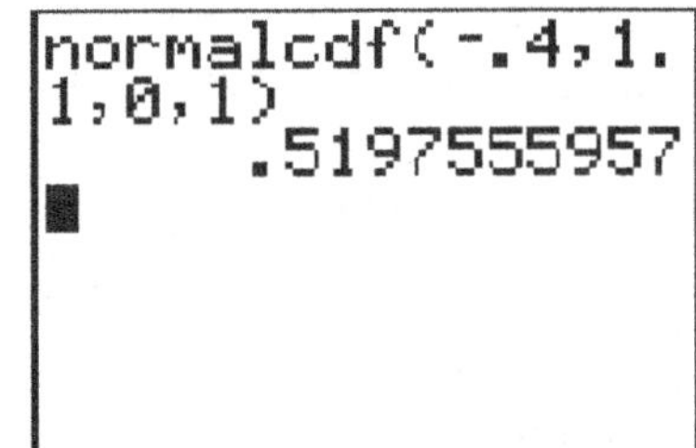

Figure 9.28

Two points should be noted:

- Four numbers (arguments) are required to find the area under the curve when *normalcdf* is used: the lower limit, the upper limit, the mean, and the standard deviation, written in that order.
- When there is no lower (or upper) limit, such as $P(Z < -1.65)$, one has to be inserted. In the example above, -5 was used as a suitable lower limit, since the area under the standardized normal curve below -5 is negligible. Clearly, any value less than -5 could have been used. If -4 had been used, then $P(Z < -1.65) = 0.04943976...$, and there would have been an error in the fourth decimal place.

Exercise 9.5

1. In each case find

(a) $P(Z < 2.3)$ (b) $P(Z < 0.35)$ (c) $P(Z < 1.424)$ (d) $P(Z < -0.88)$

(e) $P(Z < -2.71)$ (f) $P(Z < -0.473)$ (g) $P(Z > 0.53)$ (h) $P(Z > -1.81)$

(i) $P(Z > 2.065)$ (j) $P(-1 < Z < 2)$ (k) $P(-0.31 < Z < 1.69)$ (l) $P(1.3 < Z < 2.1)$

2. Note that $|Z| < a \Rightarrow -a < Z < a$. In each case find

(a) $P(|Z| < 1)$ (b) $P(|Z| < 2.5)$ (c) $P(|Z| > 3)$

(d) $P(|Z - 1| < 0.1)$ (e) $P(|Z - 0.5| > 1.5)$

9.3.3 The Normal Distribution

A random variable, X which is normally distributed with mean μ and variance σ^2 can be transformed into the standardized variable, Z. In this way, it is possible to find areas under the curve of any normal distribution. $X \sim N(\mu,\ \sigma^2)$ is transformed to $Z \sim N(0,\ 1)$ by

- a translation
- a stretch parallel to the x-axis

The translation is carried out by subtracting μ from X, which has the effect of changing the mean of X to zero without altering the variance.

The stretch is carried out by dividing $X - \mu$ by σ. This has the effect of changing the variance to 1 without altering the mean.

Therefore, the transformation $Z = \dfrac{X - \mu}{\sigma}$ changes $X \sim N(\mu,\ \sigma^2)$ to the standardized variable $Z \sim N(0,\ 1)$, and areas under the standardized normal curve are equal to corresponding areas under the equivalent normal curve with mean, μ and variance, σ^2.

Suppose that X is normally distributed with mean 8 and variance 4, that is $X \sim N(8,\ 4)$ and you require $P(X < 0.95)$. Then, using the transformation, $Z = \dfrac{X - \mu}{\sigma}$, $P(X < 9.5) = P\left(\dfrac{X - \mu}{\sigma} < \dfrac{9.5 - 8}{2} \right)$

$= P(Z < 0.75) = 0.7733727 \ldots$

The TI-84 Plus will evaluate $P(X < 9.5)$ in a manner similar to that used to find areas under the standardized normal curve. The mean, 8, and the standard deviation, 2, replace 0 and 1 respectively, therefore $P(X < 9.5) = normalcdf(-100, 9.5, 8, 2) = 0.7733727 \ldots$

Because most graphing calculators can evaluate probabilities without the need for standardization it is not immediately obvious why the standardization process is necessary. However, in section 9.3.4 you will see that standardization is essential in solving certain problems.

Note, once again, that a lower limit is necessary. Any value will work provided that it is negative and large. One method that may be used is to remember that, for a standardized normal variable, -5 is a suitable lower limit, so that, by rearranging the transformation formula $z = \dfrac{x - \mu}{\sigma}$ to $x = \sigma z + \mu$, a suitable lower limit can be found. In this case it would be $-5 \times 2 + 8 = -2$.

Example 9.9: If $X \sim N(35,\ 9)$, find $P(X > 40)$.

Solution 9.9: $P(X > 40) = normalcdf\,(40, 50, 35, 3) = 0.04779033... = 0.478$. An upper limit greater than about 50 is suitable.

Example 9.10: Given that X is normally distributed with mean μ and standard deviation σ, find $P\big(|X - \mu| < 2\sigma\big)$.

Solution 9.10: $P\big(|X - \mu| < 2\sigma\big) = P\left(\left|\dfrac{X - \mu}{\sigma}\right| < 2\right) = P\big(|Z| < 2\big) = P(-2 < Z < 2)$.

$P(-2 < Z < 2) = normalcdf\,(-2, 2, 0, 1) = 0.9544998... \approx 0.954$

Example 9.11: The masses of potatoes from a consignment of potatoes are normally distributed with mean 237g and standard deviation 31g. Potatoes whose mass is greater than 300g are used for making fries. If 1000 potatoes are randomly selected, estimate the number of potatoes which could be used for making fries.

Solution 9.11: Let M be a continuous random variable representing the mass of a randomly selected potato. Then $M \sim N(237,\ 31^2)$ and we require $P(M > 300)$.

$P(M > 300) = normalcdf\,(300, 400, 237, 31) = 0.02106362... = 0.0211$. Therefore, the estimated number of potatoes would be about 20.

Exercise 9.6:
1.   (i) For $X \sim N(5,\ 1)$, find

        (a) $P(X < 6.7)$ (b) $P(X < 4.2)$ (c) $P(3.7 < X < 5.7)$

    (ii) For $X \sim N(12,\ 1.5)$, (remember that if $\sigma^2 = 1.5$ then $\sigma = \sqrt{1.5}$) find

        (a) $P(X < 14)$ (b) $P(X > 9.6)$ (c) $P(9 < X < 10)$

(iii) Given that X is normally distributed with mean 0.2 and standard deviation 0.01, find

      (a) $P(X>0.205)$ (b) $P(X>0.198)$ (c) $P(0.199<X<0.201)$

(iv) Given that X is normally distributed with mean 31.8 and standard deviation 2.07, find

      (a) $P(X>34.6)$ (b) $P(X<30)$ (c) $P(|X-31.8|<3)$

2.   (i) If X is normally distributed with mean 10 and variance 4, find

      (a) $P(|X-10|<1)$ (b) $P(|X-9|<3)$

  (ii) If X is normally distributed with mean μ and standard deviation σ, find

      (a) $P(|X-\mu|<\sigma)$ (b) $P(|X-\mu|<3\sigma)$

3.   For a population of a specific breed of cat, the weight of individual cats is normally distributed with mean 4.55kg and standard deviation 0.3kg. Find the percentage of cats with weights

(a) less than 4kg   (b) between 4.2kg and 4.4kg   (c) between 4.5kg and 5kg

4.   A bus is scheduled to arrive at a specific bus stop at 08:00. A survey shows that the actual arrival time is normally distribute. The mean time of arrival is 08:04, and the standard deviation of its arrival time is 2 minutes. Find

(a) the probability that, on a randomly chosen day, it arrives within 2 minutes of 08:04.

(b) the percentage of times when the bus arrives after 08:05.

A man walks to the bus stop each morning and gets there at 08:00.

(c) Estimate how many times in a 3 month period (90 days), he misses the bus.

5.   A sample of size n is taken from a population whose distribution is normal with mean μ and

standard deviation σ. The mean of the sample, $\bar{X}$, is such that $\bar{X} \sim N\left(\mu, \dfrac{\sigma^2}{n}\right)$. A random

sample of 10 tomatoes is selected from a batch whose mean mass is known to be normally distributed with mean 45g and standard deviation 3g. Find the probability that the mean of the sample is less than 43g.

6.   The weights of oranges are normally distributed with a mean of 96g and standard deviation 25g. For selling purposes, oranges are divided into 3 grades: A, B and C. Grade A oranges weigh more than 126g, grade C oranges weigh less than 60g. All remaining oranges are classified as grade B.

(a) Calculate the percentage of oranges in each grade.

(b) Grade A, B and C oranges sell for 30¢, 25¢ and 15¢ respectively. Calculate the expected receipts from the sale of 80 000 oranges.

9.3.4 The Inverse of the Standardized Normal Distribution Function

The inverse standardized normal distribution function is the method by which you can find z from $P(Z < z) = k$ for a given value of k. The inverse function maps $P(Z < z) \to z$, where $P(Z < z)$ is the area under the standardized normal curve to the left of z (figure 9.29).

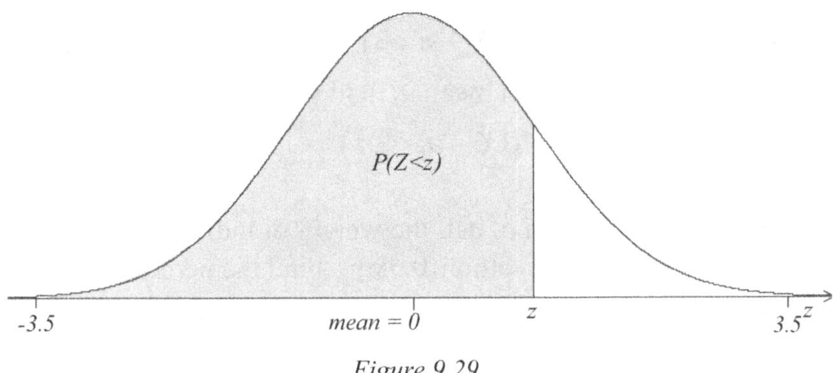

Figure 9.29

Use of the graphing calculator to find z for a given value of k: Many graphing calculators have an inverse cumulative normal distribution function which enables z to be found for a given value of $P(Z < z)$.

Suppose you want to find the value of z for which $P(Z < z) = 0.7$. Figures 9.30 and 9.31 show how to find z using a TI-84 Plus from which we obtain $P(Z < z) = 0.7 \Rightarrow z = 0.5244005... = 0.524$.

Figure 9.30

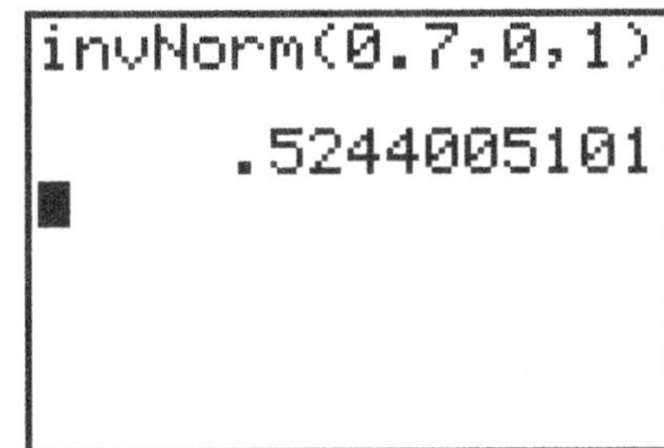

Figure 9.31

A similar procedure can be used for the inverse of a general normal cumulative distribution function. Suppose you need to find the value of x such that $P(X < x) = 0.87$, where X is a normal variable whose mean is 33 and standard deviation is 6. Figure 9.32 shows how this is done with a TI-84 Plus, giving a result of

$$P(X < x) = 0.87 \Rightarrow x = 39.75834... = 39.8$$

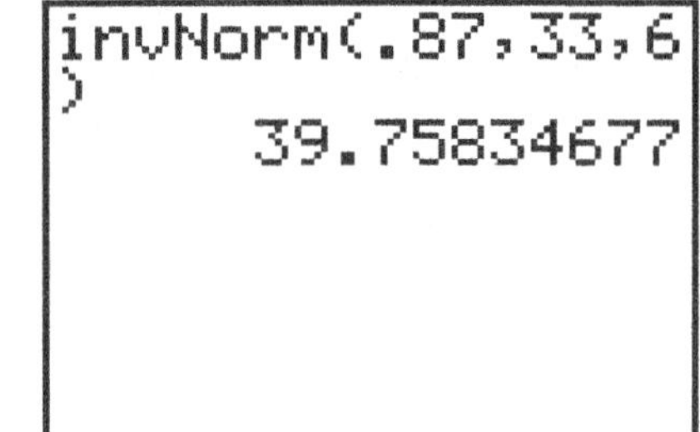

Figure 9.32

A sketch of the normal curve with appropriate shading can serve as a way of showing work when solving problems using the graphing calculator (figure 9.33).

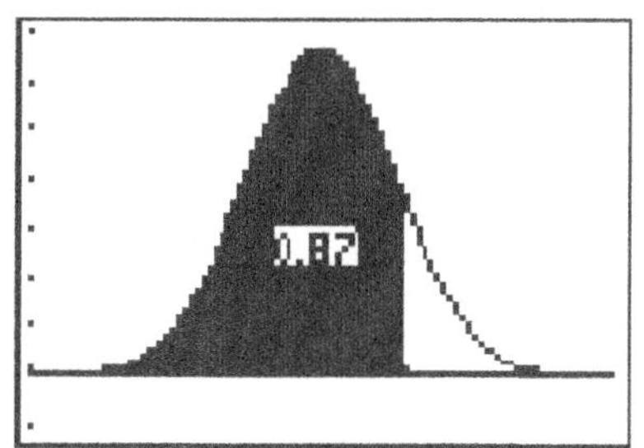

Figure 9.33

Example 9.12: In each case, find the value of k.

    (a) $P(Z < k) = 0.637$ (b) $P(Z > k) = 0.637$

    (c) $P(|Z| < k) = 0.637$

Solution 9.12: (a) The shaded area of figure 9.34 has an upper limit of k. $P(Z < k) = 0.637$

$$\Rightarrow k = invNorm(0.637, 0, 1) = 0.3504513... = 0.350$$

(b) The shaded area of figure 9.35 has a lower limit of k. By symmetry, $k = -0.350$. (Alternatively

$$P(Z > k) = 0.637 \Rightarrow P(Z < k) = 1 - 0.637 = 0.363$$

and so $k = invNorm(0.363, 0, 1) = -0.350$)

(c) The shaded area of figure 9.36 has a lower limit of $-k$ and an upper limit of k. $P(|Z| < k) = 0.637$

$$\Rightarrow P(-k < Z < k) = 0.637 \Rightarrow P(0 < Z < k) = 0.3185$$

$$\Rightarrow P(Z < k) = 0.50 + 0.3185 = 0.8185 \Rightarrow$$

$$k = invNorm(0.8185, 0, 1) = 0.9096634... = 0.910$$

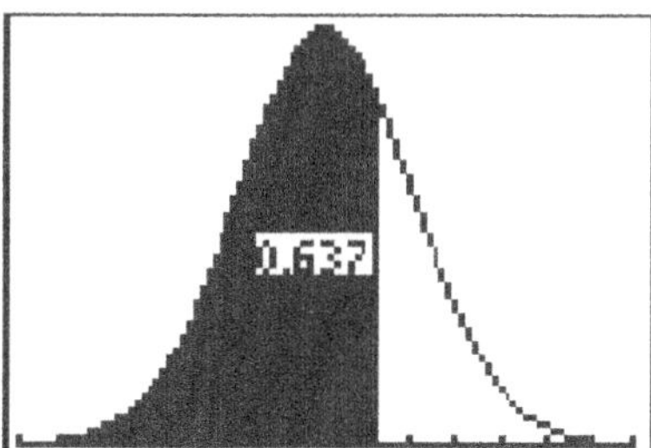

Figure 9.34

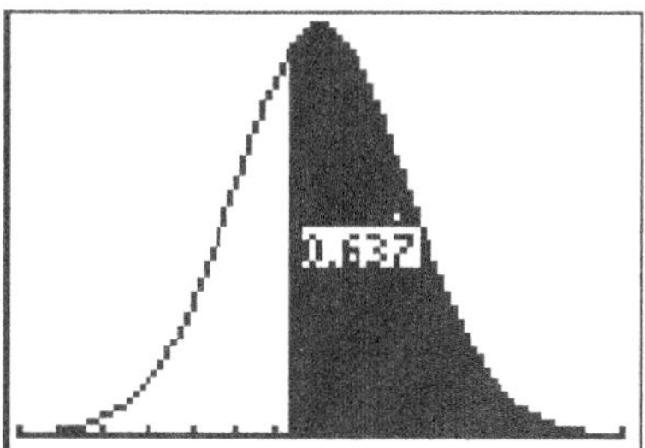

Figure 9.35

Example 9.13: X is a normal random variable whose mean is 34.2. It is found that 75% of X are less than 35.7. Find the standard deviation of X.

Solution 9.13: Although it is known that $P(X < 35.7) = 0.75$, you cannot work with X because the standard deviation is unknown. The problem is resolved by transforming X into the standardized variable, Z. The shaded area of figure 9.37 has an upper limit of z_1.

$$P(Z < z_1) = 0.75 \Rightarrow z_1 = invNorm(0.75, 0, 1)$$

$$= 0.6744897..., \text{ but } z_1 = \frac{x_1 - \mu}{\sigma}, \text{ so}$$

$$0.6744897 = \frac{35.7 - 34.2}{\sigma} = \frac{1.5}{\sigma}$$

$$\Rightarrow \sigma = \frac{1.5}{0.6744897} = 2.223903.... \text{ Therefore, the}$$

standard deviation is 2.22

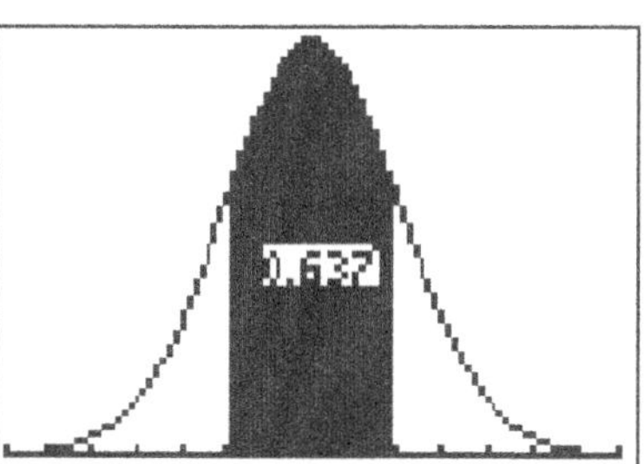

Figure 9.36

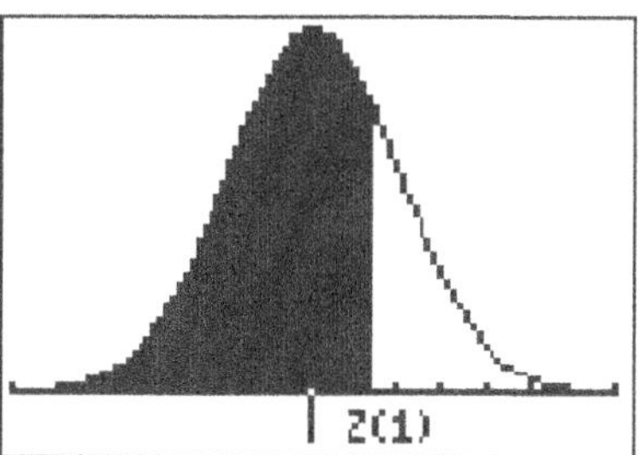

Figure 9.37

Example 9.14: The lifetime of Superglow light bulbs is normally distributed. It is found that 12% fail within 2000 hours and 5% continue to operate after 2500 hours. Find the mean and standard deviation of the lifetime of Superglow light bulbs.

Solution 9.14: The un-shaded area of figure 9.38 has a lower limit of 2000 and an upper limit of 2500.

$$P(Z > z) = 0.05 \Rightarrow P(Z < z) = 0.95 \Rightarrow z = 1.645$$

$$\Rightarrow \frac{2500 - \mu}{\sigma} = 1.645 \Rightarrow 2500 - \mu = 1.645\sigma \ \ldots\ldots\ldots\ldots (1)$$

$P(Z > z) = 0.12 \Rightarrow P(Z < z) = 0.88 \Rightarrow z = 1.175$ and so, by symmetry, $P(Z < 0.12) = -1.175$.

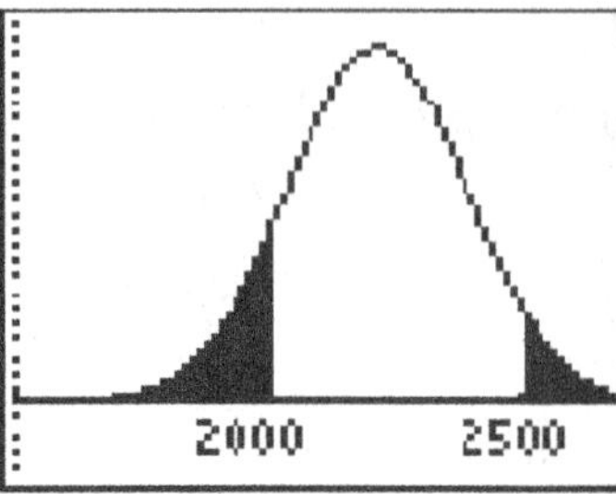

Figure 9.38

Then $\dfrac{2000 - \mu}{\sigma} = -1.175 \Rightarrow 2000 - \mu = -1.175\sigma \ \ldots\ldots (2)$

μ and σ can now be found by solving (1) and (2) simultaneously:

$$
\begin{aligned}
2500 - \mu &= 1.645\sigma \\
2000 - \mu &= -1.175\sigma \\
\hline
500 &= 2.820\sigma \quad \Rightarrow \sigma = 177.3
\end{aligned}
$$

By replacing $\sigma = 177.3$ in (1), we obtain $2500 - 1.645 \times 177.3 = 2208$. Therefore, the mean life time is 2210 hours and the standard deviation is 177 hours.

Exercise 9.7

1. In each case, find z

(a) $P(Z < z) = 0.995$ (b) $P(Z < z) = 0.216$ (c) $P(Z > z) = 0.658$

(d) $P(Z > z) = 0.115$ (e) $P(|Z| < z) = 0.5$ (f) $P(|Z| > z) = 0.1$

2. If $X \sim N(15,\ 3^2)$ (remember that if $\sigma^2 = 3^2$ then $\sigma = 3$) find x if

(a) $P(X < x) = 0.85$ (b) $P(X > x) = 0.99$ (c) $P(X < x) = 0.5965$

(d) $P(X < x) = 0.10$ (e) $P(|X - 15| < x) = 0.25$

3. A random variable X is normally distributed with standard deviation 0.56. It is known that the probability of choosing, at random, a value of X less than 4.04 is 0.772. Find the mean of the distribution.

4. In a study of the time taken to complete an online survey 50% of the study group completed the survey in 8 minutes and 15% took longer than 10 minutes. Assuming that the time taken to complete the survey is normally distributed, find the standard deviation of the survey time.

5. A random variable X, whose mean is 20, is normally distributed such that, $P(|X| > 5) = 0.4$.

(a) Draw a diagram to illustrate this information.   (b) Find the standard deviation of X.

6. A random variable X is normally distributed and is such that there is a probability of 8% that X is greater than 53 and a probability of 15% that it is less than 41. Find the mean and standard deviation of the distribution.

7. The results of an examination, given as a percentage, are normally distributed with a standard deviation of 14%. The pass mark is 45%. If 16% of the candidates fail, find the mean percentage mark.

8. Records kept over a period of 134 years show that the quantity of rainfall during November in a certain city is normally distributed. Rainfall only exceeded 30cm in five of those years, and rainfall was only less than 10cm in 12 of those years. Use the data to estimate the mean November rainfall in the city.

9. The weights of tomatoes are normally distributed and have a mean mass of 85g with standard deviation of 15g. They are sold in three grades: A (large), B (medium) and C (small). All tomatoes with a mass less than the mean are graded C. If the rest are graded so that there are equal numbers of grade A and B, find the minimum mass of a grade A tomato.

10. A machine produces cylindrical washers which have a mean diameter of 3.751cm and a standard deviation of 0.003cm. The machine is upgraded so that it now produces washers whose diameters have a standard deviation of 0.002cm. It is required that the proportion of washers whose diameter exceeds 3.758cm remains equal to the proportion produced by the machine before it was upgraded. Find what the new mean diameter of the washers should be in order to achieve this. Assume that the diameter of the washers is normally distributed, both before and after the upgrade.

Units 6, 8 & 9 Review Exercises

Part 1

No graphing calculators should be used to answer questions in Part 1.

1. A and B are events in which $P(A) = \dfrac{2}{3}$, $P(A \cap B) = \dfrac{1}{4}$, $P(A \cup B) = \dfrac{5}{6}$.

(a) Calculate $P(B)$.

(b) Explain why events A and B are not independent.

(c) Calculate $P(B \mid A)$

2.   The random variable X is normal with mean 32. $P(X > 30) = 0.65$.

(a) Draw this information on a diagram.

(b) Find $P(X < 34)$

(c) Find $P(|X - 32| < 2)$

3.   (a) For the data given in table 9.9 find

     (i) $\sum xy$ (ii) $\sum x^2$ (iii) $\sum x$ (iv) $\bar{x}$

     (v) $\sum y$ (vi) $\bar{y}$

x	1	2	3	4	5
y	0	5	10	15	20

Table 9.9

(b) Use your answers from (a) to find the equation of the regression line.

(c) Show that the regression line passes through all five data points and hence write down the correlation coefficient for these data.

4.   Table 9.10 shows the probability distribution of a discrete random variable X.

Find (a) the value of k

    (b) $E(X)$

x	0	1	2	3	4
$P(X = x)$	k^2	$\dfrac{3}{10}$	$\dfrac{3}{5}k$	$\dfrac{1}{10}$	$\dfrac{1}{5}$

Table 9.10

5.   The probability distribution of a discrete random variable is

$$P(X = x) = \frac{1}{50}(16 - x^2), \quad x = 0,\ 1,\ 2,\ 3.\ \text{Find}$$

(a) $P(X = 2)$ (b) $P(X \le 1)$ (c) $E(X)$

6.   A bag contains four blue balls and six red balls. Three balls are chosen at random from the bag and each one taken has its color noted and is then replaced before the next one is taken. Calculate the probability of obtaining

(a) exactly 3 blue balls.

(b) less than 2 blue balls.

7.   Figure 9.39 shows a Venn diagram. The universal set contains 100 elements, so $n(U) = 100$.

Using similar notation, $n(A) = 71$, $n(B) = 55$, $n(C) = 23$. $B \subset A$, $C \subset A$ and $B \cap C \ne \varnothing$.

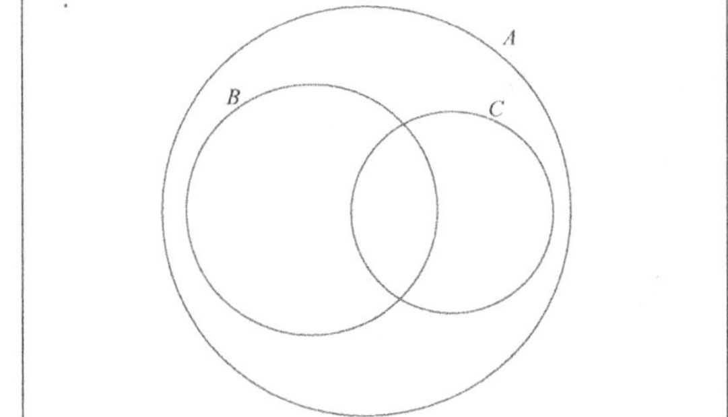

Figure 9.39

The probability that, for example, an element from A is selected is $P(A) = \dfrac{n(A)}{n(U)} = \dfrac{71}{100}$.

(a)   (i) Write down $P(A')$

      (ii) Find the maximum possible value of $P(B \cap C)$

      (iii) Find the minimum possible value of $P(B \cap C)$

Let $n(B \cap C) = 18$

(b)   Find $P(B \cap C')$ (c)   Find $P(C' | A)$.

8.   A bag contains 6 red balls and 2 green balls. Hans and Maria play a game in which they take turns in removing, at random, a ball from the bag. The ball is not replaced. The one who first obtains a green ball is the winner. Hans removes the first ball.
(a) Find the probability that Hans selects a green ball at his first go.
(b) Find the probability that Maria selects a green ball at her first go.
(c) Find the probability that Maria wins the game.

9.   Betty and Dora play a game in which they each write down on a separate piece of paper a random integer between 1 and 3 inclusive. The numbers are compared. If the numbers are equal Betty wins \$3. If the numbers differ by one number Dora wins \$9. If the numbers differ by two Betty wins \$9. Let X be the random variable for the difference between the two numbers.
(a) Find the probability distribution of X.

Let Y be the random variable for the amount won by Betty and Z be the random variable for the amount won by Dora.
(b) (i) Show that $E(Y) = 3$ (ii) Find $E(Z)$

Now the amounts won are changed so that, if the numbers written down are equal Betty wins \$6. If the numbers differ by one Dora wins the same amount as Betty wins if the numbers differ by two.
(c) Find how much this amount should be if the expected amounts won by Betty and Dora are equal.

Part 2

Graphing Calculators will usually be needed to answer questions in Part 2.

10.   A continuous random variable X is normally distributed with mean 5 and standard deviation 2. Calculate

(a) $P(X > 10)$

(b) $P(3 < X < 6)$

(c) $P(X < a) = 0.75$

(d)   (i) Shade on a sketch of the normal curve representing the probability distribution of X the region where $|X - 5| < 3$.

(ii) Calculate $P(|X - 5| < 3)$

11. A continuous random variable, X is normally distributed with mean 5.4 and standard deviation 1.6. In addition, it is known that $P(X < a) = 0.4$.

(a) Sketch a graph of a normal distribution function which shows this information
(b) Find the value of a.

Y is a normally distributed random variable with mean μ and standard deviation 1.6 for which

$P(Y < 1) = 0.05$

(c) Find the mean μ of the random variable Y.

12. A six-sided die is thrown 12 times.
(a) Write down the expected number of 'sixes'

Calculate the probability of obtaining
(b) (i) exactly 2 sixes (ii) more than 3 sixes (iii) the most likely number of sixes.

13. In an ice-skating competition, two judges were asked to score the performance of competitors on a scale of 0 to 10. Table 9.11 shows the two judges' scores of eight competitors, A - H.

Competitor	A	B	C	D	E	F	G	H
Judge X	9	5	7	7	8	4	5	9
Judge Y	8	5	8	7	9	6	6	7

Table 9.11

Find a correlation coefficient for the two sets of data.

14. The temperature of patients visiting the emergency room of a hospital is always taken and is found to be normally distributed. The mean temperature of the patients is $37.3°C$ with standard deviation $0.8°C$.
(a) A patient is chosen at random from this population. Find the probability that this patient has a temperature higher than $38°C$.
(b) 70% of patients are found to have temperatures lower than $t°C$. Find the value of t.

Another hospital does not know the mean temperature nor standard deviation temperature of the patients who visit its emergency room. However, this hospital does know that
- 30% of patients have temperatures in excess of $37.5°C$
- 40% of patients have temperatures less than $37.0°$
(c) Calculate the mean and standard deviation of the temperature of these patients.

15. Figure 9.40 shows a cumulative
frequency graph of the time taken
by 60 runners to finish a race.
(a) Use the graph to estimate
    (i)   the median time of
        the race.
    (ii)   the interquartile
        range
(b) Copy and complete the table
    9.12 shown below.

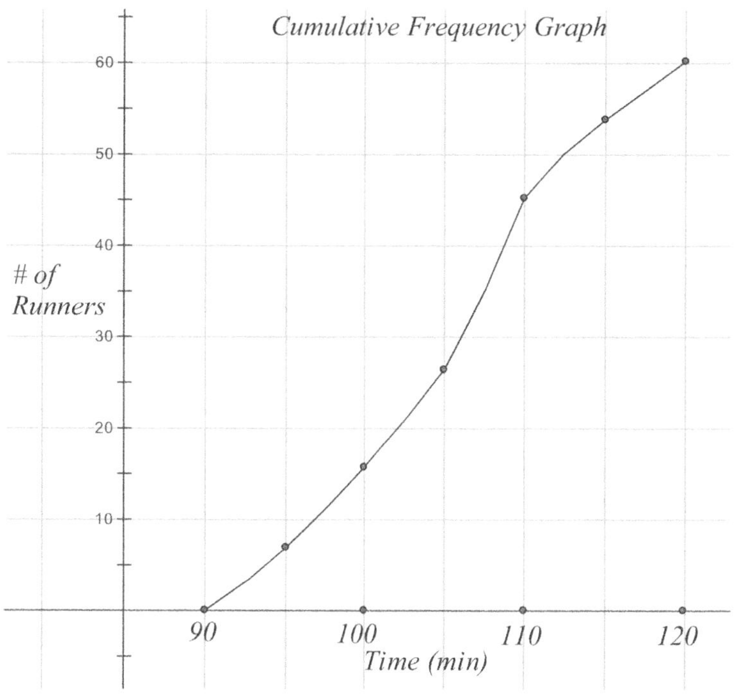

Figure 9.40

Time, t (min)	Frequency, f
<90	0
$90 \leq t < 95$	7
$95 \leq t < 100$	
$100 \leq t < 105$	
$105 \leq t < 110$	18
$110 \leq t < 115$	
$115 \leq t < 120$	

Table 9.12

(c) Use the table you obtained in (b) to find the mean time of the race.

16. The following set of data points are arranged in ascending order:

$$3, \ 3, \ 5, \ 6, \ 6, \ 6, \ 7, \ 7, \ 7, \ 8, \ 9$$

(a) Draw a box and whisker diagram and mark against it the extreme values, the lower and upper quartiles and the median.
(b) Calculate the mean and standard deviation of the data.

17. Consider the following data: 35, 45, x, y, z, 51, 74. This data is subject to the following conditions:

    (i) $x, y, z \in \mathbb{Z}^+$ (ii) $45 < x \leq y < z < 51$

Write down four possible values of x.
(b) Given that the mode is equal to the median and that the mean is 50, find possible values for x, y and z.

18. A juggler estimates that he has a 3% probability of dropping his juggling clubs when he performs a particular routine. In a practice session he carries out the routine 40 times.
(a) Estimate the mean number of times the juggler drops his clubs when practicing the routine 40 times.

Find the probability that, in practicing the routine 40 times, he drops the clubs
(b) exactly 5 times.
(c) more than 3 times.
(d) p times where $2 \leq p \leq 6$.

19. Figure 9.41 shows a frequency histogram
of 20 data points, $\{x_i\}$ $i = 1, 2, \ldots, 20$

(a) Write down the modal value of the
data

(b) For the data shown in figure 9.41 the

(i) mean

(ii) the standard deviation.

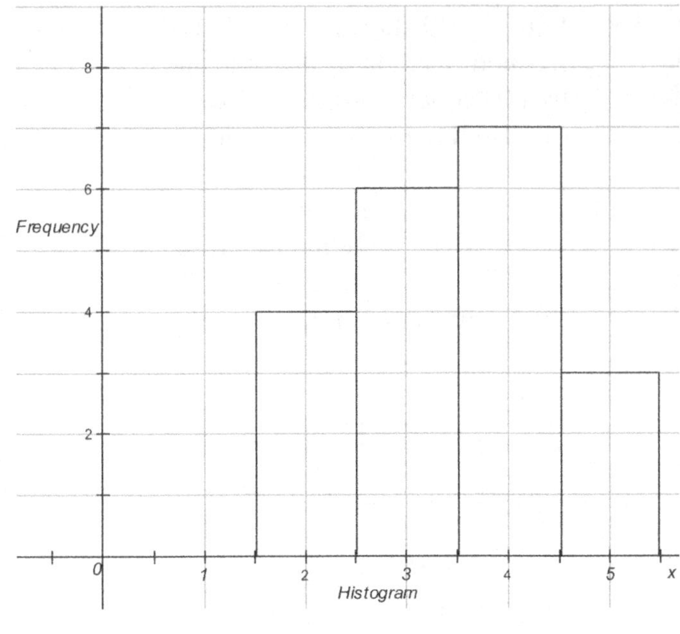

Figure 9.41

20. Random integers from 1 to 10 inclusive
are generated by a calculator until the
integer 10 is obtained or until 4 integers
have been generated. Let X be the random
variable for the number of integers
generated.

(a) Find $P(X = 2)$

(b) Copy and complete table 9.13 shown below

x	1	2	3	4
$P(X = x)$			0.081	

Table 9.13

(c) Calculate $E(X)$

TI-84Plus Simulation Programs

```
Probability Simulation Program

40→dim(L₁)
40→dim(L₂)
1→A
1→B
1→K
0→N
0→R
Lbl 1
iPart(rand*3+1)→X
N+1→N
If X=3
R+1→R
If N=20K
Then
K→L₁(A)
K+1→K
A+1→A
R/N→L₂(B)
B+1→B
End
If K=41
Goto 2
Goto 1
Lbl 2
0→Xmin
40→Xmax
10→Xscl
.2→Ymin
.4→Ymax
0.05→Yscl
Plot1(xyLine,L₁,L₂,·)
PlotsOff
PlotsOn 1
Horizontal 1/3
DispGraph
```

```
Expectation Simulation Program

Disp "NUMBER OF TRIALS"
Input N
0→Y
1→R
Lbl 1
1→K
Lbl 4
1+iPart((rand*(9-K))→X
If X≤5
Goto 2
K+1→K
Goto 4
Lbl 2
R+1→R
Y+K→Y
If R=N+1
Goto 3
Goto 1
Lbl 3
Disp Y/N
```

Solutions to Unit 1 Exercises

Exercise 1.1

1. (a) $\dfrac{1}{32}, \dfrac{1}{64}, \dfrac{1}{128}$ (b) $\dfrac{1}{6}, \dfrac{1}{7}, \dfrac{1}{8}$ (c) 13, 16, 19 (d) $-3, 3, -3$

(e) 32, 64, 128 (f) 216, 343, 512 (g) $-14, -17, -20$ (h) 16, $-17, 21$

(i) $\dfrac{11}{16}, \dfrac{13}{19}, \dfrac{15}{22}$ (j) 120, 720, 5040

2. (a) 2, 5, 8, 11 (b) $-3, -1, 3, 11$ (c) 1, 5, 11, 19 (d) $\dfrac{1}{2}, \dfrac{1}{3}, \dfrac{1}{4}, \dfrac{1}{5}$

(e) 0, 2, 0, 2

3. (a) $1+2+3+4+5$ (b) $\dfrac{1}{2}+\dfrac{2}{3}+\dfrac{3}{4}+\dfrac{4}{5}+\dfrac{5}{6}$ (c) $0+\dfrac{1}{2}+\dfrac{2}{4}+\dfrac{3}{8}+\dfrac{4}{16}$ (d) $u_1+u_3+u_5+u_7$

(e) $u_1^2+u_2^2+u_3^2+\ldots+u_n^2$ (f) $3+9+27+81+243$ (g) $1+1+1+1+1+1$

(h) $1-2+3-4$

4. (a) $\displaystyle\sum_{r=1}^{5} r$ (b) $\displaystyle\sum_{r=1}^{5} \dfrac{1}{r}$ (c) $\displaystyle\sum_{r=1}^{5} \dfrac{1}{2^{r-1}}$ (d) $\displaystyle\sum_{r=1}^{5} r^2$ (e) $\displaystyle\sum_{r=1}^{5} r^3$ (f) $\displaystyle\sum_{r=1}^{n} u_r$

Exercise 1.2

1. (a) $u_1=1,\ d=2,\ u_5=1+4\times2=9,\ u_{10}=19,\ u_n=2n-1$

(b) $u_1=5,\ d=7,\ u_5=5+4\times7=33,\ u_{10}=68,\ u_n=7n-2$

(c) $u_1=12,\ d=-2,\ u_5=4,\ u_{10}=-6,\ u_n=-2n+14$

(d) $u_1=4,\ d=-6,\ u_5=4+4\times-6=-20,\ u_{10}=-50,\ u_n=10-6n$

(e) $u_1=\dfrac{2}{3},\ d=\dfrac{5}{3},\ u_5=\dfrac{22}{3},\ u_{10}=\dfrac{47}{3},\ u_n=\dfrac{5n}{3}-1$

2. The graphing calculator method described in Example 1.3 may be used to find n in each part of this question as an alternative to the method shown in part (a).

(a) $u_n=2n-1=99 \Rightarrow n=50,\ S_{50}=25(2+49\times2)=2500$ (b) $n=27,\ S_{27}=2592$

(c) $n=42,\ S_{42}=-1218$ (d) $n=32,\ S_{32}=-2848$ (e) $n=60,\ S_{71}=2990$

3. $u_1=18,\ u_{23}=656 \Rightarrow u_1+22d=656 \Rightarrow 22d+18=656 \Rightarrow d=29$

4. $u_1=14,\ S_{13}=\dfrac{13}{2}(2\times14+12d)=572 \Rightarrow 13(14+6d)=572 \Rightarrow d\approx5$

5. $u_{18}=u_1+17d=469,\ u_{33}=u_1+32d=784$. Solving simultaneously gives $u_1=112$ and $d=21$.

6. $u_{10} = 25$, $u_{41} = 304 \Rightarrow u_1 + 9d = 25$, $u_1 + 40d = 304$. Solving simultaneously gives $u_1 = -56$ and $d = 9$. $u_{75} = -56 + 74 \times 9 = 610$

7. (a) $u_n = 21 + (n-1)28 = 1197 \Rightarrow n = 43$ (b) $S_{43} = \dfrac{43}{2}(42 + 42 \times 28) = 26187$

8. $u_1 = 110$, $d = -7$. We require the least value of n such that $u_1 + (n-1)d < 0$. So, $110 + (n-1)(-7) < 0 \Rightarrow -7n + 117 < 0 \Rightarrow n > 16\frac{5}{7}$. Therefore the least value of n is 17, and the first negative term is $u_{17} = 110 + 16 \times -7 = -2$. The graphing calculator method described in Example 1.3 will provide a much quicker solution.

9. (a) $u_n > 2500$, $u_n = 1 + (n-1)14 = 14n - 13$ so that $14n - 13 > 2500 \Rightarrow n > 179.5$. Therefore the least value of $n = 180$.

(b) $S_n = \dfrac{n}{2}(2 + 14(n-1)) > 2500 \Rightarrow 7n^2 - 6n - 2500 > 0 \Rightarrow n^2 - \dfrac{6}{7}n - \dfrac{2500}{7} > 0$

$\Rightarrow \left(n - \dfrac{3}{7}\right)^2 - \dfrac{9 + 17500}{49} > 0 \Rightarrow n - \dfrac{3}{7} > 18.90308... \Rightarrow n = 20$.

10. (a) $\displaystyle\sum_{r=1}^{n} r = 1 + 2 + \ldots + n$ so that $u_1 = 1$ and $d = 1$. Therefore letting $\displaystyle\sum_{r=1}^{n} r = S_n$,

$S_n = \frac{n}{2}(2u_1 + (n-1)d) = \frac{n}{2}(2 + n - 1) = \frac{1}{2}n(n+1)$.

(b) $S_n = 3n^2 + 5n$, $S_{n-1} = 3(n-1)^2 + 5(n-1) = 3n^2 - n - 2$. Now $S_{n-1} + u_n = S_n$

$\Rightarrow u_n = S_n - S_{n-1} = (3n^2 + 5n) - (3n^2 - n - 2) = 6n + 2$.

Therefore, $u_1 = 6 \times 1 + 2 = 8$, $u_2 = 6 \times 2 + 2 = 14$ so that $d = 6$.

11. (a) $d_u = 7$, $d_v = -5$

(b) $u_n = u_1 + (n-1)d_u = 3 + (n-1)7 = 7n - 4$, $v_n = v_1 + (-1)d_v = 115 + (n-1)(-5) = 120 - 5n$

(c) If $u_n > v_n$ then $7n - 4 > 120 - 5n \Rightarrow n > 10\frac{1}{3}$, so the least value of n is 11.

12. $u_k = 594 + (k-1)(-4) = 598 - 4k$ and $S_k = \dfrac{k}{2}(1188 + (k-1)(-4)) = 596k - 2k^2$.

As $u_k = S_k$, $598 - 4k = 596k - 2k^2 \Rightarrow k^2 - 300k + 299 = 0 \Rightarrow (k - 299)(k - 1) = 0$; therefore, $k = 1$, $k = 299$ but $k \neq 1$, so $k = 299$.

Exercise 1.3

1. (a) $u_1 = 10\ 000$, $d = 1\ 200$. Note that u_2 is the value of the investment after 1 year. Therefore u_6 is the value of the investment after 5 years.

(b) (i) $u_6 = u_1 + 5d = 10\ 000 + 6\ 000 = 16\ 000$ (ii) $u_{11} = 22\ 000$ (iii) $u_{26} = 40\ 000$

2.   (a) $u_1 = 8$, $d = -3$. Note that the 3rd term of the sequence represents the velocity after 2 seconds and the 4th term of the sequence represents the velocity after 3 seconds and so on.

       (b) (i) $u_4 = u_1 + 3d = 8 - 9 = -1$, so velocity after 3 seconds is -1ms^{-1}.

          (ii) $u_{21} = u_1 + 20d = 8 - 60 = -52$, so velocity after 20 seconds is -52ms^{-1}.

3.   (a) (i) 784   (ii) 768   (iii) 752   (iv) $800 - 16n$

       (b) $u_1 = 800$, $d = -16$. When $u_n = 0$ cars pass each other, $u_n = 800 - 16n = 0$, so $n = 50$ (or, $u_1 = 784$, $d = -16 \Rightarrow 784 - (n-1)16 = 0$). Therefore cars pass each other after 50 seconds.

       (c) As the cars are approaching each other at 16ms^{-1} and need to travel 800m before passing each other, we can use the formula relating distance, constant velocity and time : time $=$ distance/speed. Therefore time $= 800/16 = 50$ seconds.

4.   (a) The nested carts can be modeled by an arithmetic sequence in which $u_1 = 0.95$, $d = 1.13 - 0.95 = 0.18$ and the nth term is the length of n nested carts. So the length of 12 nested carts is $u_{12} = 0.95 + 11 \times 0.18 = 2.93\text{m}$.

       (b) Assume that a nested row of carts is stored lengthwise in the storage space. Then, since the width of a cart is 0.48m and the width of the storage space is 3m, there is space for 6 rows of nested carts. The maximum length of each row is 5m so we require n such that $u_n \leq 5$

$$\Rightarrow 0.95 + (n-1)0.18 \leq 5 \Rightarrow 0.18(n-1) \leq 4.05 \Rightarrow n \leq 23.5 \text{ but } n \in \mathbb{Z}^+ \text{ so } n = 23.$$ Therefore the total number of carts which can be stored is $6 \times 23 = 138$.

          If the carts are stored widthwise in the storage space there is room for 10 rows. The maximum length of each row is 3m so we require n such that $u_n \leq 3$

$$\Rightarrow 0.95 + (n-1)0.18 \leq 3 \Rightarrow n \leq 12.4 \text{ but } n \in \mathbb{Z}^+ \text{ so } n = 12.$$ Therefore the total number of carts which can be stored is $12 \times 10 = 120$.

          Therefore it is more efficient to store the carts lengthwise and the total number of carts which can be stored is 138.

5.   (a) $u_1 = 20$, $d = -2$

       (b) $\$2 = 8 \times \$0.25$ so the time that eight 25¢ coins will buy is the sum of the first 8 terms of the series which is $S_8 = \frac{8}{2}(2 \times 20 + 7 \times -2) = 4(40 - 14) = 104$, 104 minutes is 1 hour, 44 minutes.

       (c) $S_{10} = \frac{10}{2}(2 \times 20 + 9 \times -2) = 5(40 - 18) = 110$. So the maximum parking time is 1 hour 50 minutes.

6.   (a) $u_1 = 300$, $d = 300 \times 0.04 = 12$. The value of his investment after 5 years is represented by the sixth term of the arithmetic sequence and $u_6 = u_1 + 5d = 360$. Therefore the value of his investment after 5 years is \$360.

(b) The bonds purchased in January 2000 have value \$360. The bonds purchased in January 2001 have value \$348 and so on. Therefore the value of each years' bonds forms an arithmetic sequence whose first term is 360 and whose common difference is -12. Thus, the total value of the investment after 5 years is, in dollars,

$$S_5 = \frac{5}{2}\left(2 \times 360 + (5-1)(-12)\right) = 1680.$$

7. (a) $u_1 = 10$, $d = 1$ and $u_{50} = u_1 + 49d = 10 + 49 = 59$ $\Rightarrow$ length of longest rod is 59cm.

(b) $\displaystyle\sum_{r=31}^{50} u_r = u_{31} + u_{32} + \ldots + u_{50} = 40 + 41 + \ldots + 59$. This is an arithmetic series with $u_1 = 40$,

$d = 1$. Therefore, $S_{20} = 10\left(2 \times 40 + 19 \times 1\right) = 990$ and the maximum length of 20 rods is

990cm.

(c) $u_{15} = 43 = u_1 + 14 \times 1 \Rightarrow u_1 = 29$. Therefore $S_{15} = \frac{15}{2}\left(2 \times 29 + 14 \times 1\right) = 540$ and so the length of

the rods is 540cm.

(d) For a minimum number of rods we need to use the longest rods, so, as $u_n = 59$ and $d = 1$,

we have $u_1 + (n-1)1 = 59 \Rightarrow u_1 = 60 - n$. $S_n = \frac{n}{2}\left(2u_1 + (n-1)1\right) = 495$ so

$S_n = \frac{n}{2}\left(2(60-n) + n - 1\right) = \frac{n}{2}(119 - n) = 495 \Rightarrow n^2 - 119n + 990 = 0 \Rightarrow (n-9)(n-110) = 0$

$\Rightarrow n = 9, 110$ but $n < 50$ so $n = 9$ and the minimum number of rods is 9.

8. (a) $u_1 = 0$, $d = 8$ (b) $u_r = 8(r-1)$ seconds.

(c) Times measured in seconds.

Machine 1: 60 (to load) + 8 (to go up) + 60 (to unload) + 8 (to go down) = 136

Machine 2: $60 + 8 \times 2 + 60 + 8 \times 2 = 152$

Machine 3: $60 + 8 \times 3 + 60 + 8 \times 3 = 168$

These times form an arithmetic sequence in which $a = 136$, $d = 16$. Since there are 24

machines, one for each floor 2nd and above, $n = 24$.

$S_{24} = 12\left(2 \times 136 + 23 \times 16\right) = 7680$ seconds = 128 minutes = 2 hours 8 minutes.

Exercise 1.4

1. (a) $1, 3, 9, 27, 81$ (b) $32, 8, 2, \dfrac{1}{2}, \dfrac{1}{8}$ (c) $3, -2, \dfrac{4}{3}, -\dfrac{8}{9}, \dfrac{16}{27}$ (d) $\dfrac{4}{9}, \dfrac{2}{3}, 1, \dfrac{3}{2}, \dfrac{9}{4}$

2. (a) $\dfrac{1}{125}, \dfrac{1}{625}, \dfrac{1}{3125}$; $u_n = \left(\dfrac{1}{5}\right)^{n-1}$ (b) $54, 162, 486$; $u_n = 2 \times 3^{n-1}$

(c) $-\dfrac{8}{81}, \dfrac{16}{243}, -\dfrac{32}{729}$; $u_n = \dfrac{1}{3}\left(-\dfrac{2}{3}\right)^{n-1}$ (d) $1.331, 1.4641, 1.61051$; $u_n = 1.1^{n-1}$

(e) $8k, 16k, 32k$; $u_n = k \times 2^{n-1}$ (f) $-\dfrac{1}{8}k^4, \dfrac{1}{16}k^5, -\dfrac{1}{32}k^6$; $u_n = k\left(-\dfrac{1}{2}k\right)^{n-1}$

3. (a) $1 \to X$, $2X \to X$; $n = 18$ (b) $3 \to X$, $\dfrac{11}{3}X \to X$; $n = 11$

(c) $1 \to X$, $1.06X \to X$; $n = 29$ (d) $\dfrac{1}{2} \to X$, $\dfrac{4}{3}X \to X$; $n = 12$

(e) $1 \to X$, $1.08X \to X$; $n = 40$

4. (a) $u_1 = 1$, $r = 2$, $S_{10} = \dfrac{1\left(2^{10} - 1\right)}{2 - 1} = 1023$ (b) $u_1 = 2$, $r = \dfrac{1}{2}$, $S_5 = \dfrac{2\left(1 - 0.5^5\right)}{1 - 0.5} = \dfrac{31}{8}$

(c) $u_1 = 1$, $r = 1.06$, $S_{25} = 54.86451\ldots = 54.86$, correct to 2 decimal places

(d) $u_1 = \dfrac{1}{2}$, $r = \dfrac{4}{3}$, $S_6 = \dfrac{3367}{486} = 6.927983\ldots = 6.93$

(e) $u_1 = 1$, $r = \dfrac{1}{2}$, $S_{20} = 1.999998\ldots = 2.00$, correct to 2 decimal places

5. (a) $u_1 = 18$, $r = \dfrac{2}{3}$ so $S_\infty = \dfrac{18}{1 - 2/3} = 54$ (b) $\dfrac{3}{4}$ (c) $\dfrac{25}{3}$

6. (a) $1.01010101\ldots = 1 + 0.01 + 0.0001 + \ldots = 1 + \dfrac{1}{100} + \dfrac{1}{10000} + \ldots$. Therefore, $u_1 = 1$,

$r = \dfrac{1}{100}$ and $S_\infty = 1.01010101\ldots = \dfrac{1}{99/100} = \dfrac{100}{99}$.

(b) (i) $0.44444\ldots = \dfrac{4}{10} + \dfrac{4}{100} + \dfrac{4}{1000} + \ldots$ so $u_1 = \dfrac{4}{10}$, $r = \dfrac{1}{10}$, $S_\infty = \dfrac{4/10}{1 - 1/10} = \dfrac{4}{9}$.

(ii) $0.91919191\ldots = \dfrac{91}{100} + \dfrac{91}{10000} + \ldots$ so $u_1 = \dfrac{91}{100}$, $r = \dfrac{1}{100}$, $S_\infty = \dfrac{91}{99}$.

(iii) $47.5555\ldots = 47.5 + 0.05 + 0.005 + \ldots = 47 + 0.5(1 + 0.1 + 0.01 + \ldots)$

Therefore, $u_1 = 1$, $r = \dfrac{1}{10}$, $S_\infty = 47 + 0.5\left(\dfrac{1}{1 - 0.1}\right) = 47 + \dfrac{0.5}{0.9} = 47\dfrac{5}{9}$.

7. (a) $u_5 = 64$, $u_8 = 8 \Rightarrow u_1 r^4 = 64$, $u_1 r^7 = 8 \Rightarrow r^3 = \tfrac{1}{8} \Rightarrow r = \tfrac{1}{2}$. So first term is $u_1 = 64 \times 16 = 1024$

and common ratio is $\dfrac{1}{2}$.

(b) $u_1 = 20$, $u_{37} = u_1 r^{36} = 30 \Rightarrow r^{36} = 1.5 \Rightarrow r = 1.011326\ldots = 1.0113$, correct to 4 decimal places.

8. $S_n = \dfrac{1}{2}\left(3^n - 1\right)$. Therefore, $S_1 = u_1 = \dfrac{1}{2}\left(3^1 - 1\right) = 1$, $S_2 = u_1 + u_1 r = \dfrac{1}{2} \times 8 = 4 \Rightarrow 1 + r = 4 \Rightarrow r = 3$

and $u_5 = 1 \times 3^4 = 81$.

9.	(a) $u_1 = 13.5$, $r = \dfrac{1}{3}$, $u_n = \dfrac{1}{162} \Rightarrow 13.5\left(\dfrac{1}{3}\right)^{n-1} = \dfrac{1}{162} \Rightarrow \left(\dfrac{1}{3}\right)^{n-1} = \dfrac{1}{13.5 \times 162} = \dfrac{1}{2187}$. Therefore,

$3^{n-1} = 2187 \Rightarrow 3^{n-1} = 3^7 \Rightarrow n = 8$.

(b) $u_1 = 10$, $r = -2$, $u_n = u_1 r^{n-1} = -5120 \Rightarrow 10(-2)^{n-1} = -5120 \Rightarrow (-2)^{n-1} = -512$

$\Rightarrow (-2)^{n-1} = -2^9 \Rightarrow n = 10$.

10.	(a) $S_6 = \dfrac{u_1\left(r^6 - 1\right)}{r - 1} = 3$, $S_{10} = \dfrac{u_1\left(r^{10} - 1\right)}{r - 1} = 18 \Rightarrow u_1\left(r^6 - 1\right) = 3r - 3$ .... (1) and

$$u_1\left(r^{10} - 1\right) = 18r - 18 \ldots \ldots (2)$$

Equations (1) and (2) can be solved simultaneously by division:

$$\frac{u_1\left(r^{10} - 1\right)}{u_1\left(r^6 - 1\right)} = \frac{18(r - 1)}{3(r - 1)} \Rightarrow r^{10} - 1 = 6\left(r^6 - 1\right) \Rightarrow r^{10} - 6r^6 + 5 = 0$$

(b) Use the graphing calculator to solve $r^{10} - 6r^6 + 5 = 0$. Therefore,
$r = 1$, $r = -1$, $r = 1.540051...$, $r = -1.540051...$. The trivial solutions are $r = 1$ and $r = -1$
because they give rise to geometric series whose terms are all equal $(r = 1)$ or whose terms
alternate with equal magnitude $(r = -1)$. So, the non-trivial solutions are $r = -1.54,\ 1.54$.

Exercise 1.5

1.	Working in millions, the situation can be modeled by a geometric sequence with $u_1 = 280$,

$r = 1 + \dfrac{1.1}{100} = 1.011$. Therefore,

(i) in 2 years time, $u_3 = 280\left(1.011^2\right) = 286.1938...$, so the population will be about 286 million.

(ii) in 10 years time, population will be $u_{11} = 280(1.011)^{10} = 312.3701... = 312$ million.

(iii) in 20 years time, population will be $u_{21} = 280(1.011)^{20} = 348.4826... = 348$ million.

2.	Vladimir's investment can be modeled by a geometric sequence with $u_1 = 50\ 000$ and

$r = 1 + \dfrac{5.6}{100} = 1.056$. After 5 years, $u_6 = 50000(1.056)^5 = 65658.29... = 65658.29$ and so
Vladimir's investment is worth \$65658.29. Galina's investment can be modeled by a

geometric sequence with $u_1 = 15000$ and $r = 1 + \dfrac{4.7}{100} = 1.047$. After 20 years,

$u_{21} = 15000(1.047)^{20} = 37585.89...$, and so her investment is worth \$37585.89.

3.	(a) $14000(1.045) = 14630$, so 630 antelope need to be culled each year.

(b) $34000(1.04)^6 = 43020.84...$, so that 6 years later the population is about 43 000.

4. (i) $P(1.07)^n = 2P \Rightarrow 1.07^n = 2$. Using the method of Example 1.1: $1 \rightarrow X$, $1.07X \rightarrow X$
        which gives $X = 1.967151...$ when $n = 10$ and $X = 2.104851...$ when $n = 11$, so that it will
        take almost 11 years for the population to double.

    (ii) The time period between the beginning of 2001 and the beginning of 2006 is 5 years.
        Therefore the value of the stamp is given by $u_6 = 65000(1.12)^5 = 11455.22...$ which is
        about \$11 500.

    (iii) In January 2004, 10 units $\equiv$ \$1. One year later, $10 \equiv \$1 \times 0.65$ so that 6 years later
        10 units $\equiv \$1 \times 0.65^6$. Therefore, at this time \$1 is worth $\dfrac{10}{1 \times 0.65^6} = 132.5927... 132.59$
        $= 132.59$ units of the currency.

5. This situation is modeled by a geometric sequence with $u_1 = 38000$ and $r = 1 - \dfrac{23}{100} = 0.77$.

    Peter sells the vehicle after 5 years, so we require $u_6 = 38000(0.77)^5 = 10285.78....$ Therefore
    the Gladiator's value is \$10300.

6. (a) (i) $\dfrac{4.8}{12} = 0.4\%$ per month.   (ii) $\dfrac{4.8}{365} = 0.01315068... = 0.0132\%$ per day.

    (b) (i) The situation is modeled by a geometric sequence with $u_1 = 1000$, $r = 1 + \dfrac{.4}{100} = 1.004$.

        Five years is 60 months, and therefore you require $u_{61} = 1000(1.004)^{60} = 1270.640...$
        and the value of the investment is \$1270.64.

        (ii) The geometric sequence has $u_1 = 1000$, $r = 1 + \dfrac{0.0131507}{100} = 1.000131507$, and 5 years is

        1825 days (neglecting Leap Years) so you require $u_{1826} = 1000(1.000131507)^{1825} =$
        1271.23. Note that it is important to give the common ratio, r, very accurately, in order
        to be able to give a final answer correct to the nearest cent. If you had used
        $r = 1.000132$ you would have obtained a final answer of \$1272.37. \$1.14 over 5 years
        might not seem very much, but financial institutions need to work to the nearest cent.

7. This can be modeled by a geometric <u>series</u> with $u_1 = 18$, working in billions, and

    $r = 1 + \dfrac{8}{100} = 1.08$. You need to find n so that $S_n = 1000$. $S_n = \dfrac{18(1.08^n - 1)}{1.08 - 1} = 1000$

    $\Rightarrow 1.08^n - 1 = \dfrac{40}{9} \Rightarrow 1.08^n = 5.444444....$ Using the method of Example 1.1,

    $1 \rightarrow X$, $1.08X \rightarrow X$ gives $1.08^{22} = 5.436540...$ and $1.08^{23} = 5.871463...$ so that n is very close
    to 22 and the mineral resource will become exhausted after 22 years.

8. (a) The situation is modeled by a geometric sequence with $u_1 = 1000$, $r = 1 - \dfrac{10}{100} = 0.9$.

        The value of the first machine, in dollars, at the time of bankruptcy is given by
        $u_6 = 1000(0.9)^5 = 590.49$.

(b) The machine bought at the start of the second year will have value at bankruptcy given by u_5. The value of the machine bought at the start of the third year will have value at bankruptcy given by u_4 and so on. Therefore, the total value of all 5 machines at bankruptcy will be $u_5 + u_4 + u_3 + u_2 + u_1 = S_5$. Now the first term is $u_1 = 1000 \times 0.9 = 900$ so $u_1 = 900$ and $r = 0.9$.

(c) $S_5 = \dfrac{900\left(1 - 0.9^5\right)}{1 - 0.9} = 3685.59$. Therefore, the value of all five machines, to the nearest $100, at the time of bankruptcy is $3700.

9. The population of rats is given by the terms of a geometric sequence with $u_1 = 160$ (working in thousands) and $r = 1 + \dfrac{1.7}{100} = 1.017$. The population of mice is given by the terms of a geometric sequence with $u_1 = 70$ and $r = 1 + \dfrac{3.8}{100} = 1.038$. You need to find n such that

$$160(1.017)^n = 70(1.038)^n \Rightarrow \left(\frac{1.038}{1.017}\right)^n = \frac{16}{7} \Rightarrow 1.02065^n = 2.2857 \Rightarrow n = 40.4.$$

Using the method of Example 1.1, $1 \to X$, $1.02065X \to X$ gives $1.02065^{40} = 2.265028\ldots$ and $1.02065^{41} = 2.311800\ldots$. Therefore, it will take about 40 years for the mice population to exceed the rat population.

10. (a) Interest is 1% per month. Therefore, after 6 months, she owes $1000(1.01)^6$, but then she repays an amount P and so she owes $1000(1.01)^6 - P$.

(b) At the start of the second six month period, she owes $1000(1.01)^6 - P$. Therefore, by the end of the year she owes $\left(1000(1.01)^6 - P\right)(1.01)^6$, at which time she makes the second payment, P. Therefore, as the loan is paid off at the end of the year:

$$\left(1000(1.01)^6 - P\right)(1.01)^6 - P = 0 \Rightarrow 1000(1.01)^{12} = P\left(1 + (1.01)^6\right)$$

$$\Rightarrow P = \frac{1000(1.01)^{12}}{1 + (1.01)^6} = 546.5990\ldots \Rightarrow P = 546.60.$$

Exercise 1.6

1. (i) $(ab)^2 = a^2b^2$. Therefore, $\left(abc^3\right) \times (ab)^2 = a^1b^1c^3a^2b^2 = a^3b^3c^3 = (abc)^3$

(ii) $\sqrt{\dfrac{a^5b^3c}{abc^3}} = \sqrt{\dfrac{a^4b^2}{c^2}} = \left(\dfrac{a^4b^2}{c^2}\right)^{\frac{1}{2}} = \dfrac{a^2b}{c}$

2. (i) $2a^3b^2$ (ii) a^3b^2c (iii) $2a^2b^{-1}c \left(\text{or } \dfrac{2a^2c}{b}\right)$ (iv) $a^2bc^{-2} \left(\text{or } \dfrac{a^2b}{c^2}\right)$

   (v) $a^{-5}b^{-2} \left(\text{or } \dfrac{1}{a^5b^2}\right)$ (vi) $a^{-7}b^{-8} \left(\text{or } \dfrac{1}{a^7b^8}\right)$ (vii) $4^{-3}ab \left(\text{or } \dfrac{ab}{64}\right)$

3. (a) 8 (b) 32 (c) $\dfrac{4}{9}$ (d) 2 (e) 9 (f) $\dfrac{2}{5}$

   (g) $\dfrac{625}{256}$ (h) 16 (i) $\dfrac{8}{27}$ (j) $\dfrac{27}{100}$ (k) $\dfrac{24}{5}$

4. (a) 2 (b) 1 (c) $\dfrac{3}{7}$ (d) $-\dfrac{1}{6}$ (e) $\dfrac{13}{8}$ (f) $-\dfrac{1}{2}$

   (g) $\dfrac{2}{3}$ (h) 2

Exercise 1.7

1. (a) $x = \log_2 3$ (b) $x = \log_3 2$ (c) $x = \log_6 5$ (d) $x = \log_{10} 7 - 1$

   (e) $x = \log_5 4$ (f) $x = \dfrac{1}{2}\log_3\left(\dfrac{7}{4}\right)$ (g) $x = \log_2 9 + 3$ (h) $x = \log_3\left(\dfrac{5}{2}\right)$

   (i) $x = \dfrac{1}{\log_3 11}$ (j) $x = \dfrac{1}{3}\left(2 - \log_4 12\right)$ (k) $x = \pm\sqrt{\log_6 3}$

2. (a) 9 (b) 64

   (c) $x = 9^{\frac{3}{2}} = \left(9^{\frac{1}{2}}\right)^3 = 27$ (d) $2x = 6^3 = 216 \Rightarrow x = 108$

   (e) $x + 1 = 32 \Rightarrow x = 31$ (f) $\dfrac{1}{x} = 5^3 \Rightarrow x = \dfrac{1}{125}$

   (g) $\dfrac{x-1}{2} = 2^7 = 128 \Rightarrow x - 1 = 256 \Rightarrow x = 257$ (h) $\sqrt{x} = 4^2 = 16 \Rightarrow x = 256$

3. (a) $u + v$ (b) $u - v$ (c) $2u + 3v$ (d) $v - 2u$ (e) $u^2 + 2v$ (f) $3u + 2v$
   (g) $4u - v$ (h) $2u - v$

4. (a) $x = 2.69$ (b) $x = 1.26$ (c) $x = 0.112$ (d) $x = \log_{10} 3.1 + 6 = 6.491361... = 6.49$
   (e) $x - 1 = \log_{10} 2097 = 3.321598... \Rightarrow x = 4.32$
   (f) $10^x = 4 - 2.916 = 1.084 \Rightarrow x = 0.03502928... = 0.0350$
   (g) $100^x = 10^{2x} = 78.5 \Rightarrow 2x = \log_{10} 78.5 \Rightarrow x = \dfrac{1}{2}\log_{10} 78.5 = 0.9474348... = 0.947$
   (h) $0.1^x = 10^{-x} = 19 \Rightarrow -x = \log_{10} 19 \Rightarrow x = -1.278753... = -1.28$

5.   (a) $\log 6$ (b) $\log 5$ (c) $\log 6$ (d) $\log \dfrac{2}{3}$ (e) $\log 3$ (f) $2\log 5 \ (\text{or } \log 25)$

      (g) $\log 6$ (h) $\log 72$ (i) $\log 2$ (j) $\log \dfrac{3}{8}$ (k) $\log 8a^2$

6.   (a) 1 (b) 4 (c) $\dfrac{1}{2}$ (d) 0 (e) -3 (f) -1 (g) 4 (h) -8 (i) $\dfrac{5}{2}$

7. $\log_2\left(x^3 y^2\right) = 28 \Rightarrow 3\log_2 x + 2\log_2 y = 28$, $\log_2\left(\dfrac{x^4}{y^3}\right) = -8 \Rightarrow 4\log_2 x - 3\log_2 y = -8$

      $\Rightarrow 3u + 2v = 28, \ 4u - 3v = -8 \Rightarrow u = 4, \ v = 8 \Rightarrow \log_2 x = 4, \ \log_2 y = 8$. Therefore,

      $x = 2^4 = 16, \ y = 2^8 = 256$.

Exercise 1.8

1.   (a) $2^x = 5 \Rightarrow x = \log_2 5 = \dfrac{\log_{10} 5}{\log_{10} 2} = 2.321928... = 2.32$

      (b) $3^x = 88 \Rightarrow x = \log_3 88 = \dfrac{\log_{10} 88}{\log_{10} 3} = 4.075447... = 4.08$

      (c) $4^{x+1} = 15 \Rightarrow x + 1 = \log_4 15 = \dfrac{\log_{10} 15}{\log_{10} 4} = 1.953445... \Rightarrow x = 0.953$

      (d) $2^{3-x} = 22 \Rightarrow 3 - x = \log_2 22 \Rightarrow x = 3 - \dfrac{\log_{10} 22}{\log_{10} 2} = -1.459431... = -1.46$

      (e) $5^{x-2} = 338 \Rightarrow x - 2 = \log_5 338 = \dfrac{\log_{10} 338}{\log_{10} 5} = 3.618061... \Rightarrow x = 5.62$

      (f) $3^{-x^2} = \dfrac{1}{2} \Rightarrow -x^2 = \log_3 \dfrac{1}{2} \Rightarrow x^2 = \log_3 2 \Rightarrow x = \pm\sqrt{\dfrac{\log_{10} 2}{\log_{10} 3}} \Rightarrow x = \pm 0.7943108... = \pm 0.794$

2.   (a) $2t$ (b) $1+t$ (c) $3t-1$ (d) $\dfrac{t}{2}$ (e) $\dfrac{1}{t}$ (f) $\dfrac{t}{1+t}$

3.   (a) (i) $\log_2 4 = \log_2 2^2 = 2\log_2 2 = 2 \times 1 = 2$ **or** let $a = \log_2 4 \Rightarrow 2^a = 4 \Rightarrow a = 2$

       (ii) $\log_4 25 = \dfrac{\log_2 25}{\log_2 4} = \dfrac{1}{2}\log_2 25$ so $\log_2 x = \dfrac{\log_2 25}{2} = \log_2 \sqrt{25} = \log_2 5 \Rightarrow x = 5$

      (b) $\log_3 x = \dfrac{\log_3(7x-6)}{\log_3 9} = \dfrac{1}{2}\log_3(7x-6) \Rightarrow x = \sqrt{7x-6} \Rightarrow x^2 = 7x - 6 \Rightarrow x^2 - 7x + 6 = 0$

       $\Rightarrow (x-6)(x-1) = 0 \Rightarrow x = 1, \ 6$

      (c) $\log_x 81 = \log_3 x \Rightarrow \dfrac{\log_3 81}{\log_3 x} = \log_3 x \Rightarrow 4 = \left(\log_3 x\right)^2 \Rightarrow \log_3 x = \pm 2 \Rightarrow x = 3^{\pm 2} \Rightarrow x = 3^2 = 9$

       and $x = 3^{-2} = \dfrac{1}{9}$.

4.   (a) $\dfrac{\log_{10} x}{\log_{10} 2} = \dfrac{\log_{10} 3}{\log_{10} 5} \Rightarrow \log_{10} x = \dfrac{\log_{10} 3 \times \log_{10} 2}{\log_{10} 5} \Rightarrow x = 1.605036... = 1.61$

       (b) $\dfrac{\log_{10} x}{\log_{10} 7} = \dfrac{\log_{10} 9}{\log_{10} 2} \Rightarrow \log_{10} x = \dfrac{\log_{10} 9 \times \log_{10} 7}{\log_{10} 2} \Rightarrow x = 477.416... = 477$

       (c) $\dfrac{\log_{10} x}{\log_{10} 2} + \dfrac{\log_{10} x}{\log_{10} 4} = \dfrac{\log_{10} 5}{\log_{10} 2} \Rightarrow \dfrac{\log_{10} x}{\log_{10} 2} + \dfrac{\log_{10} x}{2\log_{10} 2} = \dfrac{\log_{10} 5}{\log_{10} 2} \Rightarrow \dfrac{3}{2}\log_{10} x = \log_{10} 5$

           $\Rightarrow x^{\frac{3}{2}} = 5 \Rightarrow x = 5^{\frac{2}{3}} = 2.924017... = 2.92$

       (d) $\dfrac{\log_{10} x}{\log_{10} 3} = \log_{10} 32 \Rightarrow \log_{10} x = \log_{10} 32 \times \log_{10} 3 = 0.7181390... \Rightarrow x = 5.225634... = 5.23$

5.   (a) $x > \dfrac{\log_{10} 5000}{\log_{10} 2} = 12.28771... = 12.3$ (b) $x > \dfrac{\log_{10} 697}{\log_{10} 3} = 5.959140... = 5.96$

       (c) $x > \dfrac{1}{\log_{10} 1.01} = 231.4078... = 231$ (d) $x + 1 > \dfrac{\log_{10} 33}{\log_{10} 1.15} = 25.01759... \Rightarrow x > 24.0$

       (e) $x \log_{10} 0.99 < \log_{10} 0.5 \Rightarrow x > \dfrac{\log_{10} 0.5}{\log_{10} 0.99} = 68.96756... = 69.0$.

       (Note that $\log_{10} 0.99 < 0$ so that the inequality sign must change when the inequality is divided by $\log_{10} 0.99$.)

       (f) $x \log_{10} 0.5 < \log_{10} 0.001 \Rightarrow x > \dfrac{\log_{10} 0.001}{\log_{10} 0.5} = 9.965784... = 9.97$. (Note that $\log_{10} 0.5 < 0$ so that the inequality sign must change when the inequality is divided by $\log_{10} 0.5$.)

Exercise 1.9

1.   (a) $(x+y)^4 = x^4 + 4x^3 y + 6x^2 y^2 + 4xy^3 + y^4$ (b) $(1+y)^3 = 1 + 3y + 3y^2 + y^3$

       (c) $(1+x)^7 = 1 + 7x + 21x^2 + 35x^3 + 35x^4 + 21x^5 + 7x^6 + x^7$

       (d) $(1-x)^6 = 1 - 6x + 15x^2 - 20x^3 + 15x^4 - 6x^5 + x^6$

       (e) $(k-1)^8 = k^8 - 8k^7 + 28k^6 - 56k^5 + 70k^4 - 56k^3 + 28k^2 - 8k + 1$

2.   (a) $(9x^2)(16x^4) = 144x^6$ (b) $(8a^3)(25b^2) = 200a^3 b^2$ (c) -1152

       (d) $\left(\dfrac{x^3}{8}\right)(4y^2) = \dfrac{1}{2}x^3 y^2$ (e) $8x^3$ (f) $\left(\dfrac{1}{81}a^4\right)\left(-\dfrac{27}{8}b^3\right) = -\dfrac{1}{24}a^4 b^3$

3.   (a) $x^4 + 4x^3(2y) + 6x^2(2y)^2 + 4x(2y)^3 + (2y)^4 = x^4 + 8x^3 y + 24x^2 y^2 + 32xy^3 + 16y^4$

       (b) $1 + 5(2x) + 10(2x)^2 + 10(2x)^3 + 5(2x)^4 + (2x)^5 = 1 + 10x + 40x^2 + 80x^3 + 80x^4 + 32x^5$

       (c) $2^3 + 3(2^2)\left(-\dfrac{k}{2}\right) + 3(2)\left(-\dfrac{k}{2}\right)^2 + \left(-\dfrac{k}{2}\right)^3 = 8 - 6k + \dfrac{3k^2}{2} - \dfrac{k^3}{8}$

(d) $3^3 + 3 \times 3^2 \times 2x + 3 \times 3 \times (2x)^2 + (2x)^3 = 27 + 54x + 36x^2 + 8x^3$

(e) $1 + 4\left(\dfrac{2}{3}a\right) + 6\left(\dfrac{2}{3}a\right)^2 + 4\left(\dfrac{2}{3}a\right)^3 + \left(\dfrac{2}{3}a\right)^4 = 1 + \dfrac{8}{3}a + \dfrac{8}{3}a^2 + \dfrac{32}{27}a^3 + \dfrac{16}{81}a^4$

(f) $\left(\dfrac{1}{3}x\right)^3 + 3\left(\dfrac{1}{3}x\right)^2\left(\dfrac{3}{2}y\right) + 3\left(\dfrac{1}{3}x\right)\left(\dfrac{3}{2}y\right)^2 + \left(\dfrac{3}{2}y\right)^3 = \dfrac{x^3}{27} + \dfrac{x^2 y}{2} + \dfrac{9xy^2}{4} + \dfrac{27y^3}{8}$

(g) $\left(x^2\right)^4 + 4\left(x^2\right)^3\left(-\dfrac{1}{x}\right) + 6\left(x^2\right)^2\left(-\dfrac{1}{x}\right)^2 + 4\left(x^2\right)\left(-\dfrac{1}{x}\right)^3 + \left(-\dfrac{1}{x}\right)^4 = x^8 - 4x^5 + 6x^2 - \dfrac{4}{x} + \dfrac{1}{x^4}$

(h) $\left(\dfrac{x}{2}\right)^6 + 6\left(\dfrac{x}{2}\right)^5\left(\dfrac{2}{x}\right) + 15\left(\dfrac{x}{2}\right)^4\left(\dfrac{2}{x}\right)^2 + 20\left(\dfrac{x}{2}\right)^3\left(\dfrac{2}{x}\right)^3 + 15\left(\dfrac{x}{2}\right)^2\left(\dfrac{2}{x}\right)^4 + 6\left(\dfrac{x}{2}\right)\left(\dfrac{2}{x}\right)^5 + \left(\dfrac{2}{x}\right)^6$

$= \dfrac{x^6}{64} + \dfrac{3x^4}{8} + \dfrac{15x^2}{4} + 20 + \dfrac{60}{x^2} + \dfrac{96}{x^4} + \dfrac{64}{x^6}$

4.   (a) 36   (b) 364   (c) 924   (d) 190   (e) 6435

5.   (a) $\dbinom{10}{3} = 120$ (b) $\dbinom{18}{11} = 31824$ (c) $\dbinom{14}{9}(-1)^9 = -2002$ (d) $\dbinom{21}{6} = 54264$

   (e) $\dbinom{15}{10}(2^5) = 3003 \times 32 = 96096$ (f) $\dbinom{9}{5}2^5 = 126 \times 32 = 4032$

   (g) $\dbinom{16}{6}2^{10}\left(-\dfrac{1}{2}\right)^6 = 8008 \times 2^4 = 128128$ (h) $\dbinom{11}{4}3^7 \times 2^4 = 330 \times 2187 \times 16 = 11547360$

6.   (a) $\dbinom{9}{3}(3x)^3 = 84 \times 27x^3 = 2268x^3$ (b) $\dbinom{13}{1}2^{12}(-x) = -13 \times 4096x = -53248x$

   (c) $\dbinom{25}{2}x^2 = 300x^2$ (d) $\dbinom{19}{16}(-x)^{16} = 969x^{16}$

   (e) $\dbinom{5}{2}a^3\left(\dfrac{b}{3}\right)^2 = \dfrac{10a^3 b^2}{9}$ (f) $\dbinom{11}{4}3^7\left(\dfrac{1}{2}x\right)^4 = \dfrac{330 \times 2187}{16}x^4 = \dfrac{360855}{8}x^4$

   (g) $\dbinom{7}{2}(3x)^5\left(-\dfrac{1}{3}\right)^2 = 567x^5$ (h) $\dbinom{6}{4}(2x)^2\left(\dfrac{y}{2}\right)^4 = \dfrac{15}{4}x^2 y^4$

7.   (a) $(1+x)^4 = 1 + 4x + 6x^2 + 4x^3 + x^4 = 1 + 0.4 + 0.06 + 0.004 + 0.0001 = 1.4641$

   (b) $(1+x)^7 = 1 + 7x + 21x^2 + 35x^3 + 35x^4 + 21x^5 + 7x^6 + x^7$

$= 1 + 7 \times 10^{-2} + 21 \times 10^{-4} + 35 \times 10^{-6} + 35 \times 10^{-8} + 21 \times 10^{-10} + 7 \times 10^{-12} + 10^{-14}$

$= 1.07213535210701$

   (c) $1.01^8 = 1.0828567056280801$, the digits being the appropriate row of Pascal's triangle.

8.   (a) $\left(x+\dfrac{1}{x}\right)^7 = x^7 + \binom{7}{1}x^6\left(\dfrac{1}{x}\right)^1 + \ldots = x^7 + 7x^5 + \ldots$ so term in x^5 is $7x^5$

(b) $\left(x^2+\dfrac{2}{x}\right)^5 = x^{10} + \binom{5}{1}(x^2)^4\left(\dfrac{2}{x}\right)^1 + \binom{5}{2}(x^2)^3\left(\dfrac{2}{x}\right)^2 + \ldots = x^{10} + 10x^7 + 40x^4 + \ldots$

so term in x^4 is $40x^4$

(c) $\left(2x-\dfrac{3}{x^2}\right)^4 = (2x)^4 + 4(2x)^3\left(-\dfrac{3}{x^2}\right)^1 + 6(2x)^2\left(-\dfrac{3}{x^2}\right)^2 + \ldots$

$\qquad = 16x^4 - 96x + 216x^{-2} + \ldots$ so term in x^{-2} is $216x^{-2}$

9.   (a) $\binom{5}{4}\dfrac{1}{k}(kx)^4 = 625x^4 \Rightarrow k^3 = 125 \Rightarrow k = 5$

(b) $\binom{7}{3}k^4(-x)^3 = -560x^3 \Rightarrow -35k^4 = -560 \Rightarrow k^4 = 16 \Rightarrow k = \pm 2$ but $k>0$. Therefore, $k=2$.

(c) $\binom{12}{7}(kx)^7 = \dfrac{11264}{243}x^7 \Rightarrow 792k^7 = \dfrac{11264}{243} \Rightarrow k^7 = \dfrac{128}{2187} \Rightarrow k = \dfrac{2}{3}$

10.   (a) $(1+y)^4 = 1 + 4y + 6y^2 + 4y^3 + y^4$

(b) $y = x + x^2 \Rightarrow y^2 = (x+x^2)^2 = x^2 + 2x^3 + x^4$, $\qquad y^3 = (x+x^2)^3 = x^3 + 3x^4 + 3x^5 + x^6$

$y^4 = (x+x^2)^4 = x^4 + 4x^5 + 6x^6 + 4x^7 + x^8$. Therefore, $(1+x+x^2)^4 = 1 + 4(x+x^2) +$

$6(x+x^2)^2 + 4(x+x^2)^3 + (x+x^2)^4 = 1 + 4x + 10x^2 + 16x^3 + 19x^4 + 16x^5 + 10x^6 + 4x^7 + x^8$.

Solutions to Unit 1 Review Exercise

Part 1

1.   (a) $u_{51} = 3 + (51-1)\times 2 = 103$

(b) $u_n = 3 + (n-1)2 = 87 \Rightarrow 2(n-1) = 84 \Rightarrow n-1 = 42 \Rightarrow n = 43$

2.   (a) $u_4 = u_1 r^3 = 9r^3 = \dfrac{8}{3} \Rightarrow r^3 = \dfrac{8}{27} = \left(\dfrac{2}{3}\right)^3 \Rightarrow r = \dfrac{2}{3}$

(b) $S_\infty = \dfrac{u_1}{1-r} = \dfrac{9}{1-\frac{2}{3}} = \dfrac{9}{\frac{1}{3}} = 27$

3.   (a) (i) $S_\infty = \dfrac{u_1}{1-r} = \dfrac{1}{1-\frac{1}{2}} = 2$ (ii) $S_\infty = \dfrac{u_1}{1-r} = \dfrac{1}{1-\left(-\frac{1}{2}\right)} = \dfrac{1}{\frac{3}{2}} = \dfrac{2}{3}$

(b) (i) $S_\infty = \dfrac{a}{1-r} = \dfrac{1}{1-x}$, $\ -1 < x < 1$ (ii) $S_\infty = \dfrac{1}{1-(-x^2)} = \dfrac{1}{1+x^2}$, $-1 \le x < 1$

4.   (a) $4^{x-1}=4^3 \Rightarrow x-1=3 \Rightarrow x=4$ (b) $3^{2-x}=3^4 \Rightarrow 2-x=4 \Rightarrow x=-2$

    (c) $4^x=2^{2x}=2^{-5} \Rightarrow 2x=-5 \Rightarrow x=-\dfrac{5}{2}$ (d) $a^{3x-2}=a^{-1} \Rightarrow 3x-2=-1 \Rightarrow x=\dfrac{1}{3}$

5.   (a) $2x+1=3^2=9 \Rightarrow 2x=8 \Rightarrow x=4$

    (b) $4x^2=2^1=2 \Rightarrow x^2=\pm\dfrac{1}{2} \Rightarrow x=\pm\dfrac{1}{\sqrt{2}}$

    (c) $3x-2=x^2 \Rightarrow x^2-3x+2=0 \Rightarrow (x-2)(x-1)=0 \Rightarrow x=1,\ x=2$

    (d) $5x=a^3 \Rightarrow x=\dfrac{a^3}{5}$

6.   (a) $\log_2(xy)=\log_2 x+\log_2 y=a+b$

    (b) $\log_2\left(\dfrac{x^2}{y}\right)=\log_2 x^2-\log_2 y=2\log_2 x-\log_2 y=2a-b$

    (c) $\log_2\left(2\sqrt{xy}\right)=\log_2 2+\log_2\sqrt{xy}=1+\dfrac{1}{2}\log(xy)=1+\dfrac{1}{2}(a+b)=1+\dfrac{1}{2}a+\dfrac{1}{2}b$

    (d) $\log_4(xy)=\dfrac{\log_2(xy)}{\log_2 4}=\dfrac{a+b}{\log_2 2^2}=\dfrac{a+b}{2\log_2 2}=\dfrac{1}{2}(a+b)=\dfrac{1}{2}a+\dfrac{1}{2}b$

    (e) $\log_x(xy)=\dfrac{\log_2(xy)}{\log_2 x}=\dfrac{\log_2 x+\log_2 y}{\log_2 x}=1+\dfrac{\log_2 y}{\log_2 x}=1+\dfrac{b}{a}$

7. $(x+y)^4=x^4+4x^3y+6x^2y^2+4xy^3+y^4,\quad (x-y)^4=x^4-4x^3y+6x^2y^2-4xy^3+y^4$

    Therefore, $(x+y)^4-(x-y)^4=\left(4x^3y+4xy^3\right)-\left(-4x^3y-4xy^3\right)$, other terms cancelling,

    so that $(x+y)^4-(x-y)^4=8x^3y+8xy^3=8xy\left(x^2+y^2\right)$.

8. $(x-ky)^5=x^5-5kx^4y+10x^3\left(-ky\right)^2+10x^2\left(-ky\right)^3+5x\left(-ky\right)^4+ky^5$

    Simplifying: $(x-ky)^5=x^5-5kx^4y+\boldsymbol{10k^2x^3y^2-10k^3x^2y^3+5k^4xy^4}-ky^5$.

9.   (a) $(x-2)^3=x^3+3x^2\left(-2\right)+3x\left(-2\right)^2+\left(-2\right)^3=x^3-6x^2+12x-8$

    (b) $\left(x^3-6x^2+12x-8\right)(3x+1)=\ldots\ x^3\times 1+\left(-6x^2\right)\times 3x=x^3-18x^3=-17x^3$

Part 2

10. (a) 8, 11, 14, 17 (b) $S_{15} = \dfrac{15}{2}\left(2\times8+(15-1)3\right) = \dfrac{15}{2}(16+42) = 15\times29 = 435$

(c) $S_n = \dfrac{n}{2}\left(16+(n-1)3\right) = \dfrac{n}{2}(13+3n) = \dfrac{3}{2}n^2 + \dfrac{13}{2}n$

11. (a) $S_6 = 3\times6^2 + 7\times6 = 150$ (b) $S_7 = 3\times7^2 + 7\times7 = 196$ (c) $u_7 = S_7 - S_6 = 196-150 = 46$

12. (a) $u_{31} = 8+(31-1)\times5 = 158$ (b) $S_{20} = \dfrac{20}{2}\left(2\times8+19\times5\right) = 10(16+95) = 1110$

13. (a) $u_{18} = -4+(18-1)\times3 = -4+51 = 47$

(b) $u_n = -4+(n-1)3 = 3n-7 > 300 \Rightarrow 3n > 307 \Rightarrow n > 102.3333... $ and since $n \in \mathbb{Z}^+$, $n = 103$.

14. (a) $u_{10} = 8\times\left(-\dfrac{2}{3}\right)^9 = -0.2080983... = -0.208$ (b) $S_{10} = 8\left(\dfrac{1-\left(-\frac{2}{3}\right)^{10}}{1-\left(-\frac{2}{3}\right)}\right) = 4.716760... = 4.72$

(c) $S_\infty = \dfrac{8}{1-\left(-\frac{2}{3}\right)} = \dfrac{24}{5} = 4.8$

15. (a) Interest rate per month is $5.4/12 = 0.45\,\%$ per month.

(b) (i) $8000(1.054)^5 = 10406.22...$ so the value of the investment is $10\,406.22

(ii) $8000(1.0045)^{12\times5} = 8000(1.0045)^{60} = 10473.37...$ so the value of the investment is $10\,473.37.

16. (a) The population of rats after 10 years is $500\times1.03^{10} = 671.9581... = 672$

(b) Let N_t be the number of rats at the end of t years.

Then $N_0 = 500$

$N_1 = 500\times1.03 - 100$

$N_2 = (500\times1.03-100)1.03-100 = 500\times1.03^2 - 100(1+1.03)$

$N_3 = ((500\times1.03-100)1.03-100)1.03-100 = 500\times1.03^3 - 100(1+1.03+1.03^2)$

Continuing in this way we get

$N_t = 500\times1.03^t - 100(1+1.03+1.03^2+...1.03^{t-1})$

And so there will be no more rats when

$500\times1.03^t - 100(1+1.03+1.03^2+...1.03^{t-1}) = 0$

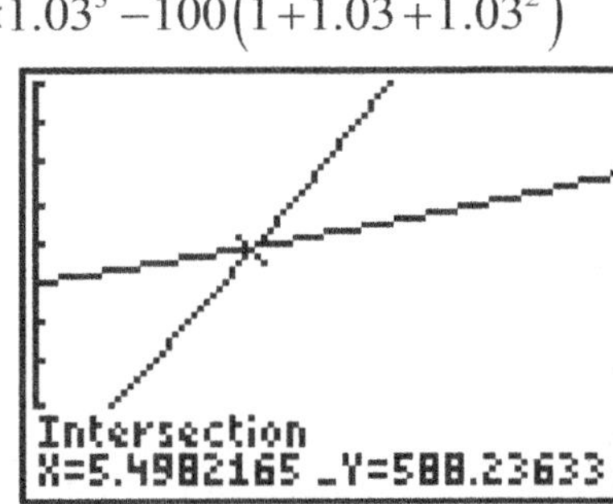

$$500 \times 1.03^t = 100\left(\frac{1.03^t - 1}{1.03 - 1}\right) = 0 \Rightarrow 500 \times 1.03^t = 3333.3\left(1.03^t - 1\right).$$

So there will be no more rats after $5.498216... = 6$ years.

17.   (a) There are 13 terms in the expansion.

    (b) $(2x)^{12} + \binom{12}{1}(2x)^{11} y + \binom{12}{2}(2x)^{10} y^2 = 4096x^{12} + 12 \times 2^{11} x^{11} y + 66 \times 2^{10} y^2$

$$= 4096x^{12} + 24576x^{11}y + 67584x^{10}y^2$$

    Therefore, $a = 24576$, $b = 67584$

    (c) $k = \binom{12}{6} \times 2^6 = 924 \times 64 = 59136$

18.   (a) $\left(x + \dfrac{1}{y}\right)^9 = \ldots + \binom{9}{4}x^5\left(\dfrac{1}{y}\right)^4 + \ldots = \ldots + 126\dfrac{x^5}{y^4} + \ldots$

    (b) $\left(x + \dfrac{1}{x}\right)^9 = \ldots + \binom{9}{3}x^6\left(\dfrac{1}{x}\right)^3 + \ldots$ so the term in x^3 is $\binom{9}{3}x^3 = 84x^3$.

19.   (a) $8000(1.0623)^{10} = 14640.69...$ so his investment is worth $14\,640.70 = 14\,641$ pesos

    (b) $8000(1.0623)^n = 50000 \Rightarrow 1.0623^n = 6.25 \Rightarrow n = \dfrac{\log 6.25}{\log 1.0623} = 30.32249...$

    Therefore, 31 complete years are needed.

    (c) $8000k = 40000 \Rightarrow k = 5$, where k is ratio by which the investment needs to increase over 10 years. $k^{10} = 5 \Rightarrow k = 5^{0.1} = 1.174618...$. So, the interest needs to be $17.46189... = 17.5\%$.

Solutions to Unit 2 Exercises

Exercise 2.1

1.   (a) $x = 1$ has no image because $\dfrac{2}{0}$ is undefined.

    (b) x does not have a unique image because each value of x is assigned to two values in the range.

    (c) x has no image for all values of $x > \dfrac{2}{3}$, $x \in \mathbb{R}$.

    (d) $n = 1$ and $n = 2$ have no image.

    (e) n is mapped to more than one image. For example, $10 \to 1, 2, 5, 10$.

    (f) $x = 0.5$ is mapped to 0 and 1.

2.   (a) 1   (b) -2 (c) -14

3.   (a) 9   (b) 0   (c) $-\dfrac{336}{25}$ (d) $4\sqrt{10}-2=10.64911...=10.6$

4.   (a) $-\dfrac{2}{9}$ (b) $\dfrac{3}{4}$ (c) $\dfrac{11}{100}$

5.   (a) $-7 \le f(x) \le 13$ (b) $-5 < f(x) < 5$ (c) $0 \le f(x) \le 100$ (d) $-1 < f(x) \le 1$

6.   (a) 0.00366 (b) $f(1.732050...) = 1.73$

      (c) $f(0.6989700...) = 0.699$ (d) $f(0.8027415...) = 0.803$

7.   (a) 7   (b) 3   (c) 0   (d) 6   (e) -3

Exercise 2.2

1. (a)   (b)   (c)

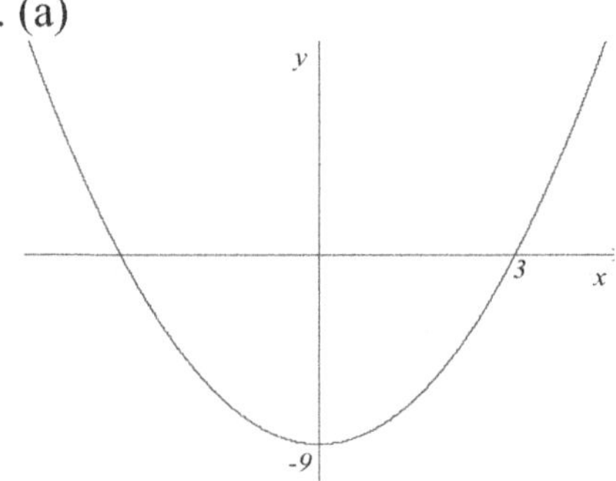 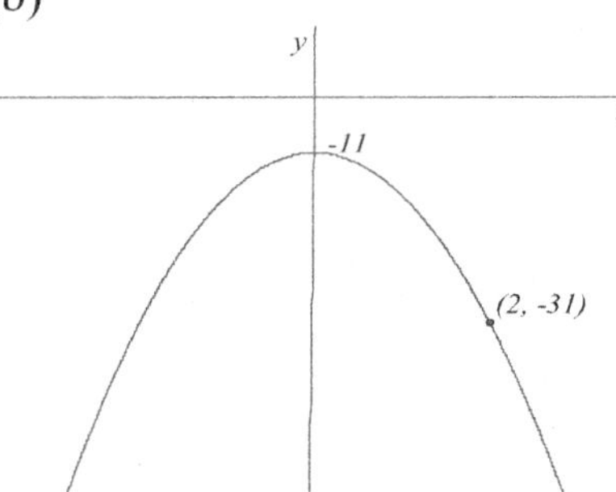 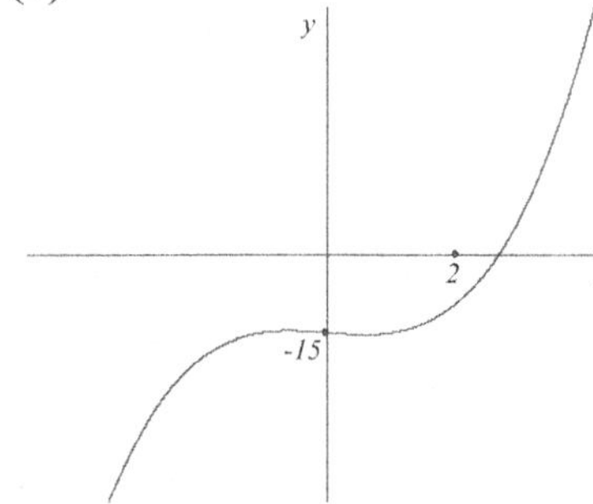

(d)   (e)   (f)

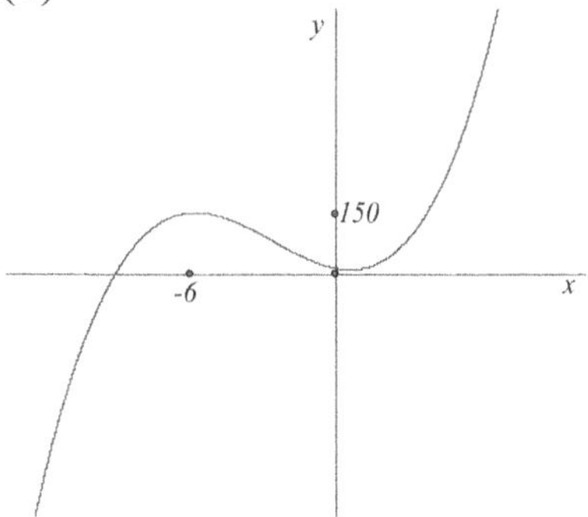 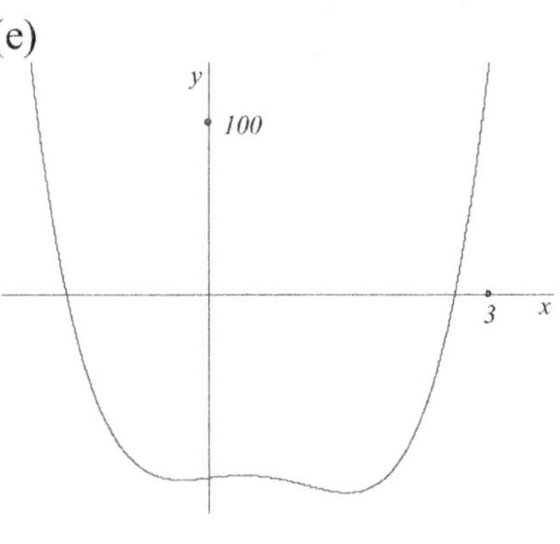 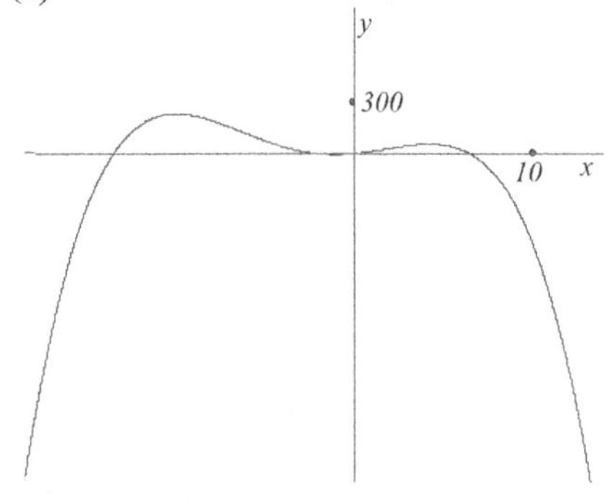

2. (a)   (b)   (c)

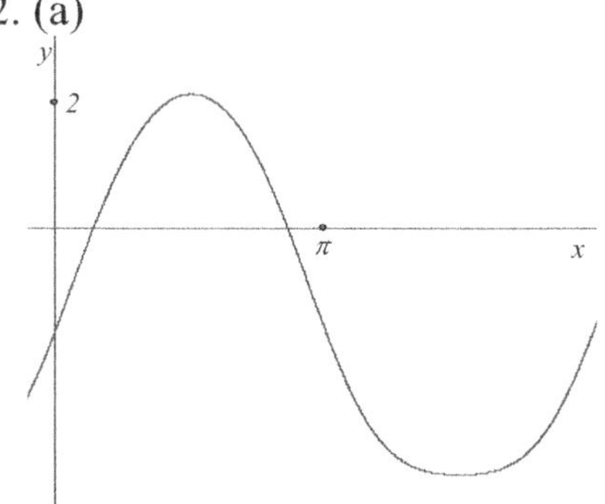 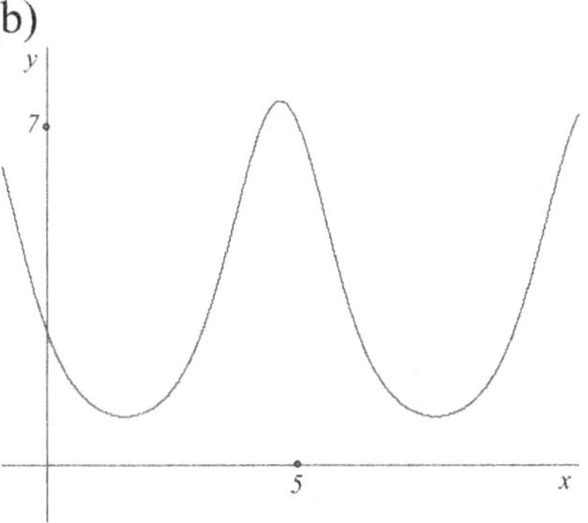 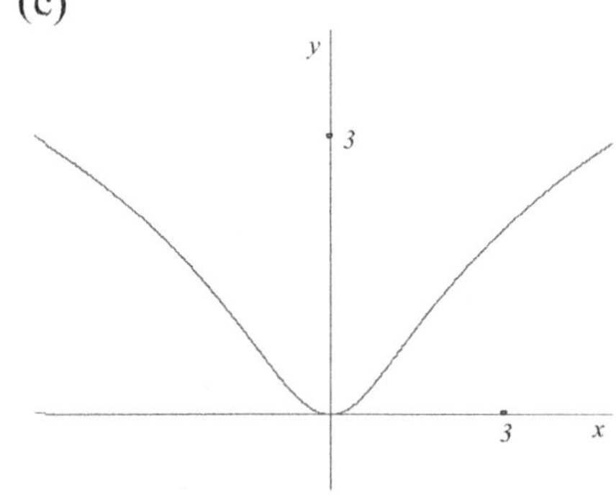

(d)
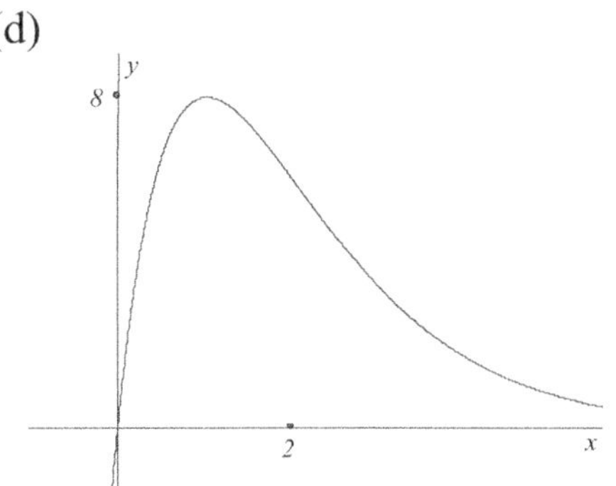

(e)
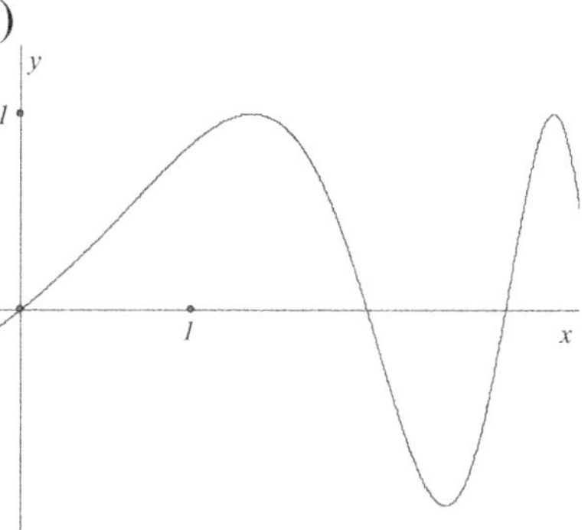

(f)
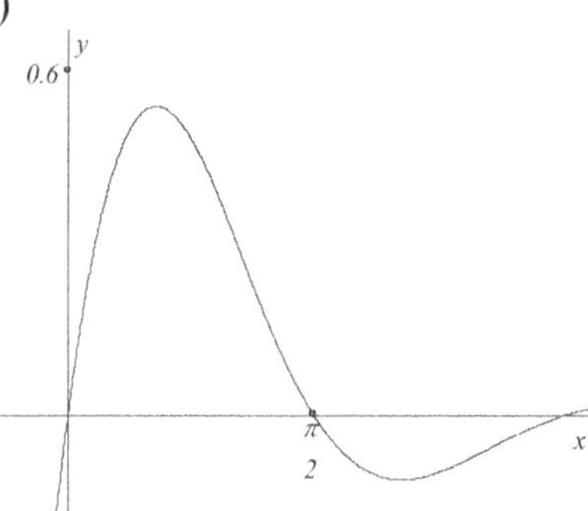

(g)
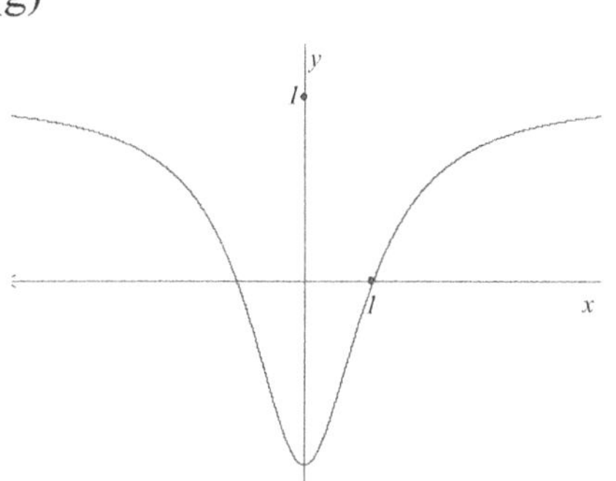

(h)
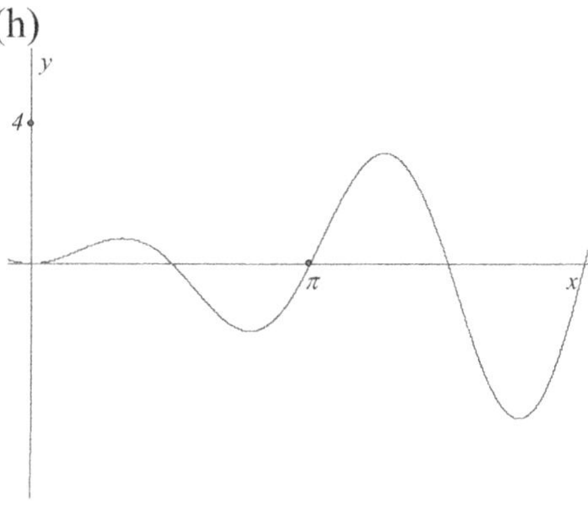

3. (a)
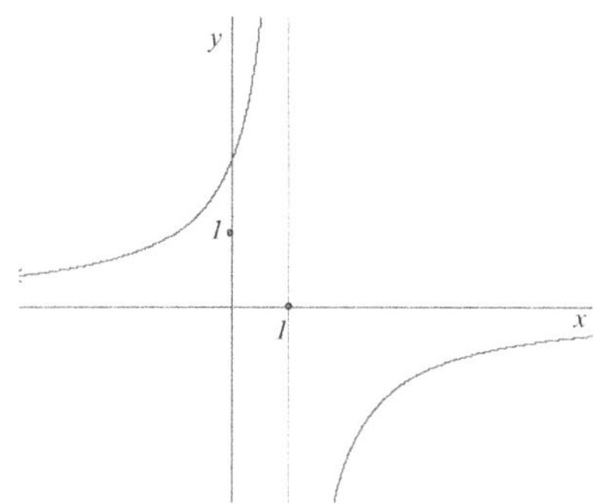

(i) $x = 1$ (ii) $y = 0$

(b)
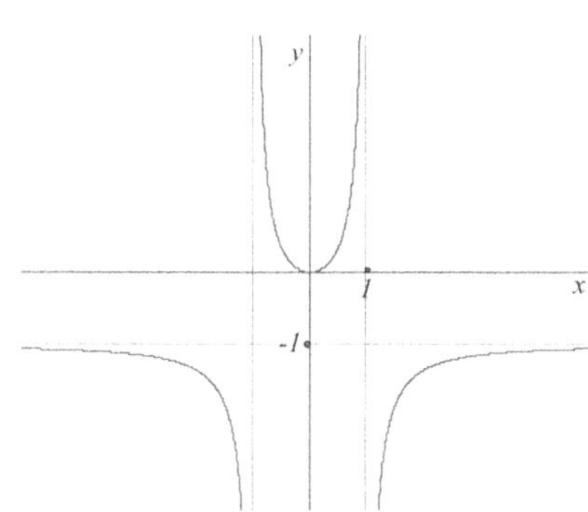

(i) $x = \pm 1$ (ii) $y = -1$

(c)
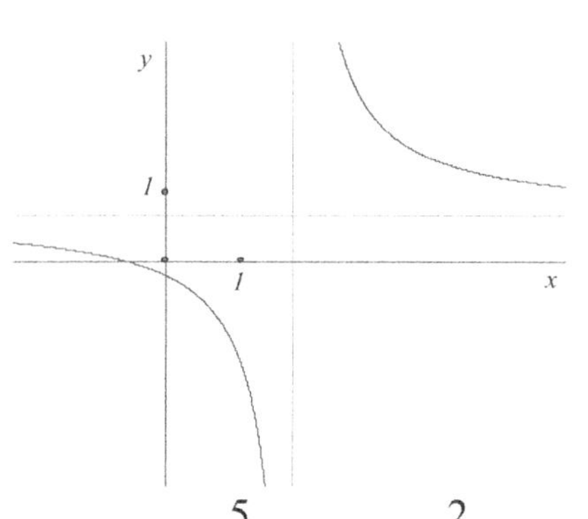

(i) $x = \dfrac{5}{3}$ (ii) $y = \dfrac{2}{3}$

(d)
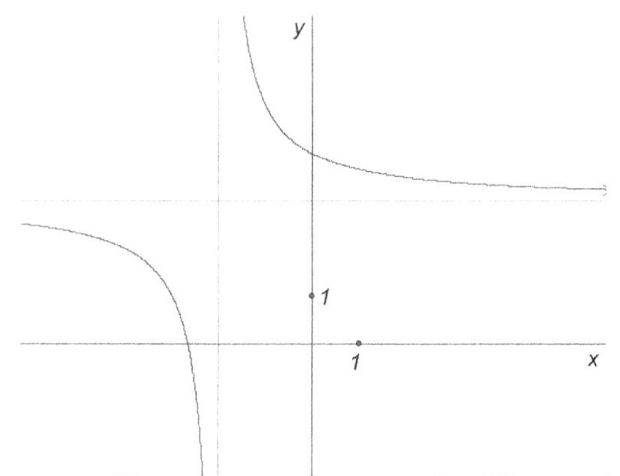

(i) $x = -2$ (ii) $y = 3$

(e) 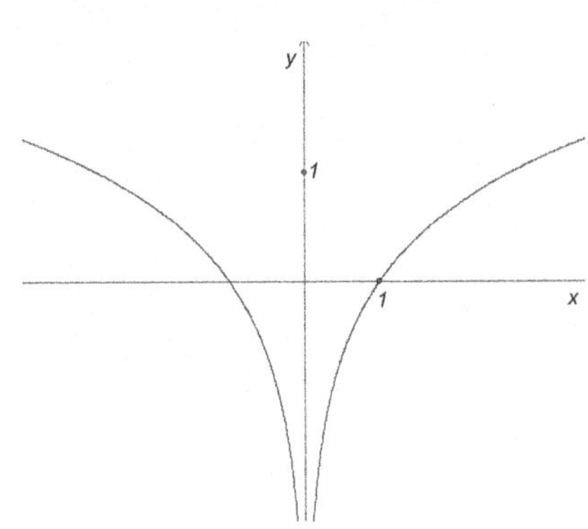

(f) 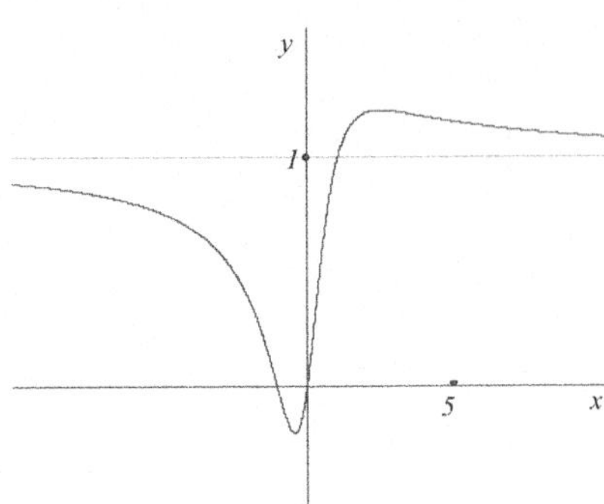

            (i) $x=0$ (ii) None   (i) None (ii) $y=1$

4.   (a) $k=1$ (b) $b=2$ (c) y-intercept is $\left(0,-\dfrac{2}{3}\right)$

5.   (a)   (i) $x=-b$ (ii) $y=a$.
      (b) $a=2$

6.   (a) A vertical asymptote occurs when $x^2+k=0$, but since $k\geq 0$, x^2+k is positive for all values of x and so x^2+k is never equal to zero.
      (b) $y=2$
      (c) The range of f is $-1.75\leq f(x)<2$
      (d) 2 solutions

7.   (a) $f(x)=\dfrac{x^2+2k}{x^2-2kx+k^2}=\dfrac{1+2/k}{1-2k/x+k^2/x^2}=1$ as $x\to\infty$. Therefore the equation of the horizontal asymptote is $y=1$
      (b) $k=2$
      (c)

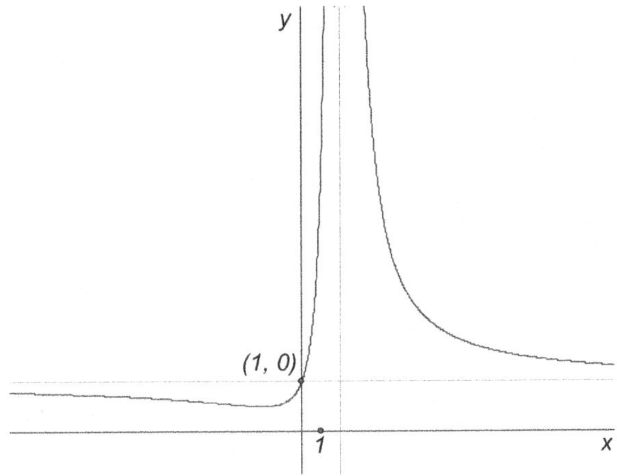

Exercise 2.3

1. A translation of $\begin{pmatrix} 2 \\ 4 \end{pmatrix}$ (translation of 2 units in the positive x direction and 4 units in the positive y direction).

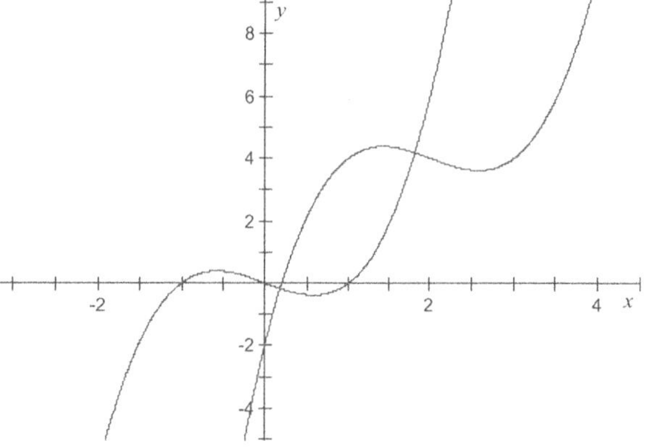

2. (a) A reflection in $y = 0$ (x-axis).

   (b) A translation of $\begin{pmatrix} 2 \\ 3 \end{pmatrix}$ (translation of 2 units in the positive x direction and 3 units in the positive y direction).

   (c) A stretch of factor 2 parallel to the x-axis or stretch of factor $\dfrac{1}{4}$ parallel to the y-axis.

   (d) A translation of $\begin{pmatrix} -4 \\ 0 \end{pmatrix}$ (translation of 4 units in the negative x direction).

3. (a) A translation of $\begin{pmatrix} -3 \\ 1 \end{pmatrix}$.

   (b) A stretch of factor $\dfrac{1}{3}$ parallel to the y-axis or a stretch of factor $\sqrt{3}$ parallel to the x-axis.

   (c) A translation of $\begin{pmatrix} \frac{\pi}{6} \\ 1 \end{pmatrix}$.

   (d) A translation of $\begin{pmatrix} 4 \\ 5 \end{pmatrix}$.

   (e) A reflection in the x-axis followed by a translation of $\begin{pmatrix} 0 \\ 4 \end{pmatrix}$.

   (f) A translation of $\begin{pmatrix} -1 \\ -5 \end{pmatrix}$.

4. (a) A translation of $\begin{pmatrix} 0 \\ -1 \end{pmatrix}$.

   (b) A stretch of factor $\dfrac{1}{3}$ parallel to the x-axis and a translation of $\begin{pmatrix} 0 \\ 1 \end{pmatrix}$.

   (c) A translation of $\begin{pmatrix} -4 \\ -4 \end{pmatrix}$.

5.

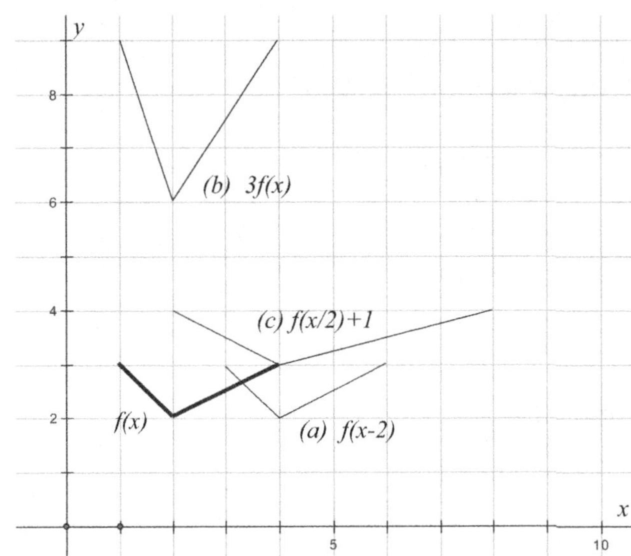

6.

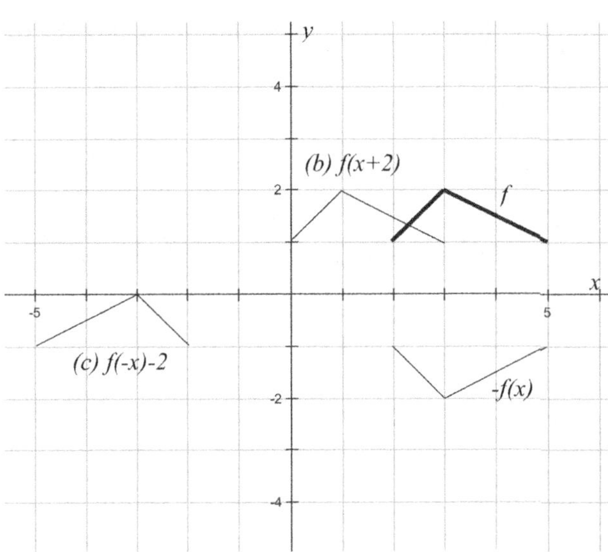

7.　(a)　$a = -5,\quad b = -4$

(b)

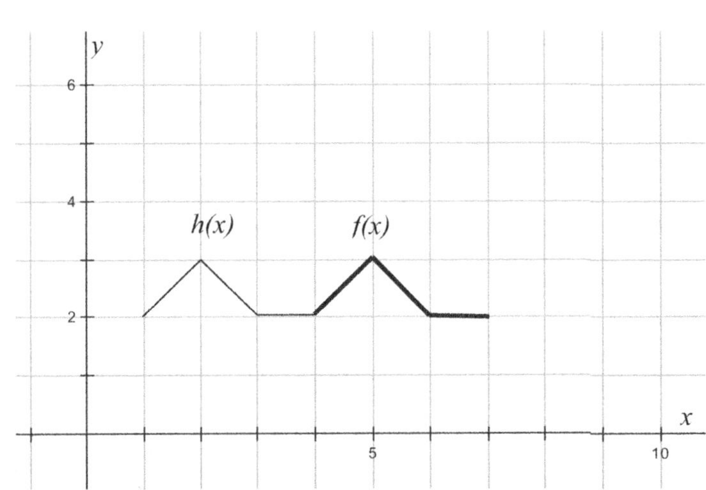

(c)　$c = 2,\quad d = 4$

Exercise 2.4

1.　(a) $\left(x^2+4x+4\right)-4-6=(x+2)^2-10$　　(b) $\left(x^2+12x+36\right)-36+15=(x+6)^2-21$

　　(c) $\left(x^2-2x+1\right)-1+9=(x-1)^2+8$　　(d) $\left(x^2+6x+9\right)-9-11=(x+3)^2-20$

　　(e) $\left(x^2-8x+16\right)-16-1=(x-4)^2-17$　　(f) $\left(x^2-4x+4\right)-4+19=(x-2)^2+15$

　　(g) $\left(x^2+30x+225\right)-225+205=(x+15)^2-20$

(a) Check

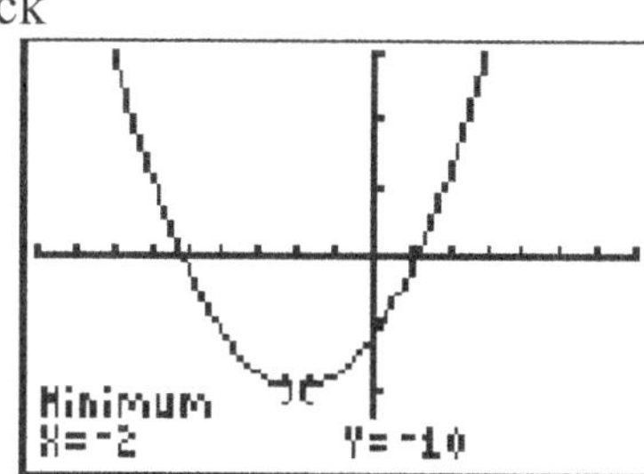

(b) Check

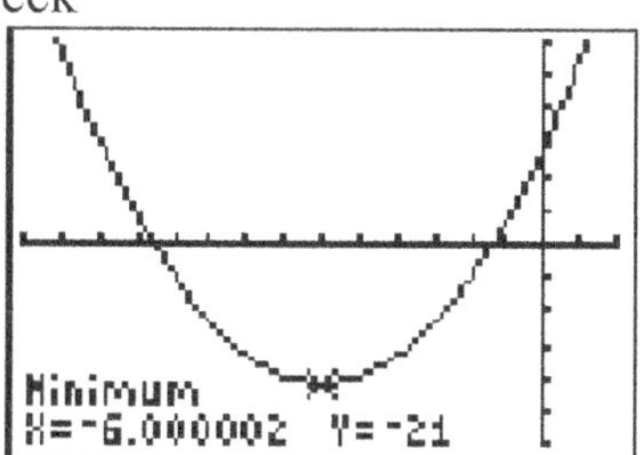

So $x^2+4x-6=\left(x-(-2)\right)^2-10$

$\qquad\qquad\qquad =(x+2)^2-10$

So $x^2+12x+15=\left(x-(-6)\right)^2-21$

$\qquad\qquad\qquad\quad =(x+6)^2-21$

Checks of further parts of this question are omitted.

2.　(a) $2-\left(x^2-4x\right)=2-\left(x^2-4x+4\right)+4=6-(x-2)^2$

　　(b) $9-\left(x^2-6x\right)=9-\left(x^2-6x+9\right)+9=18-(x-3)^2$

　　(c) $6-\left(x^2+18x\right)=6-\left(x^2+18x+81\right)+81=87-(x+9)^2$

　　(d) $5-\left(x^2-16x\right)=5-\left(x^2-16x+64\right)+64=69-(x-8)^2$

　　(e) $-7-\left(x^2-4x\right)=-7-\left(x^2-4x+4\right)+4=-3-(x-2)^2$

　　(f) $17-\left(x^2+8x\right)=17-\left(x^2+8x+16\right)+16=33-(x+4)^2$

　　(g) $-10-\left(x^2+2x\right)=-10-\left(x^2+2x+1\right)+1=-9-(x+1)^2$

(a) Check

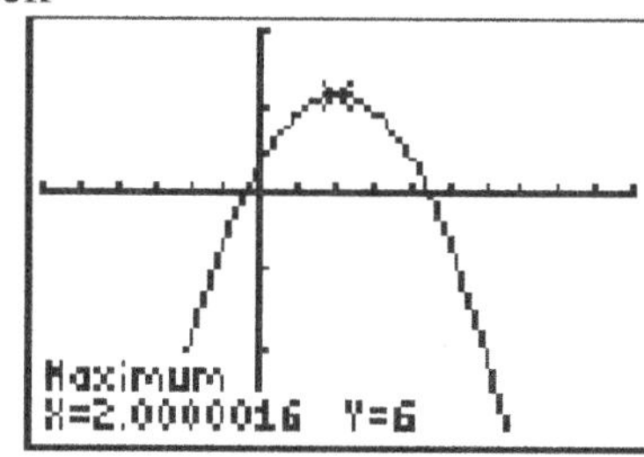

(b) Check

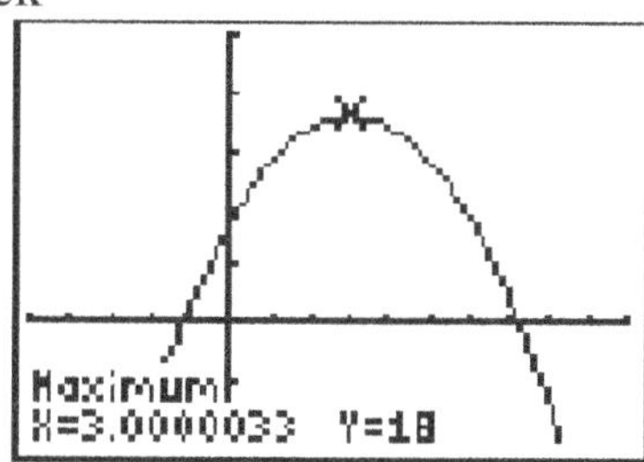

So $2+4x-x^2=6-(x-2)^2$

So $9+6x-x^2=18-(x-3)^2$

Checks of further parts of this question are omitted.

3. (a) $\left(x^2+3x+\dfrac{9}{4}\right)-\dfrac{9}{4}-5=\left(x+\dfrac{3}{2}\right)^2-\dfrac{29}{4}$ (b) $\left(x^2+5x+\dfrac{25}{4}\right)-\dfrac{25}{4}+2=\left(x+\dfrac{5}{2}\right)^2-\dfrac{17}{4}$

(c) $\left(x^2-x+\dfrac{1}{4}\right)-\dfrac{1}{4}-11=\left(x-\dfrac{1}{2}\right)^2-\dfrac{45}{4}$ (d) $\left(x^2-7x+\dfrac{49}{4}\right)-\dfrac{49}{4}+17=\left(x-\dfrac{7}{2}\right)^2+\dfrac{19}{4}$

(e) $\left(x^2+11x+\dfrac{121}{4}\right)-\dfrac{121}{4}-5=\left(x+\dfrac{11}{2}\right)^2-\dfrac{141}{4}$

(f) $\left(x^2-15x+\dfrac{225}{4}\right)-\dfrac{225}{4}+32=\left(x-\dfrac{15}{2}\right)^2-\dfrac{97}{4}$

(g) $\left(x^2+17x+\dfrac{289}{4}\right)-\dfrac{289}{4}+73=\left(x+\dfrac{17}{2}\right)^2+\dfrac{3}{4}$

Checks may be carried out as for questions 1 & 2 and are omitted here.

4. (a) $7-\left(x^2-3x\right)=7-\left(x^2-3x+\dfrac{9}{4}\right)+\dfrac{9}{4}=\dfrac{37}{4}-\left(x-\dfrac{3}{2}\right)^2$

(b) $1-\left(x^2+x\right)=1-\left(x^2+x+\dfrac{1}{4}\right)+\dfrac{1}{4}=\dfrac{5}{4}-\left(x+\dfrac{1}{2}\right)^2$

(c) $5-\left(x^2-5x\right)=5-\left(x^2-5x+\dfrac{25}{4}\right)+\dfrac{25}{4}=\dfrac{45}{4}-\left(x-\dfrac{5}{2}\right)^2$

(d) $1-\left(x^2+3x\right)=1-\left(x^2+3x+\dfrac{9}{4}\right)+\dfrac{9}{4}=\dfrac{13}{4}-\left(x+\dfrac{3}{2}\right)^2$

(e) $-10-\left(x^2-x\right)=-10-\left(x^2-x+\dfrac{1}{4}\right)+\dfrac{1}{4}=-\dfrac{39}{4}-\left(x-\dfrac{1}{2}\right)^2$

(f) $-8-\left(x^2-7x\right)=-8-\left(x^2-7x+\dfrac{49}{4}\right)+\dfrac{49}{4}=\dfrac{17}{4}-\left(x-\dfrac{7}{2}\right)^2$

(g) $21-\left(x^2+7x\right)=21-\left(x^2+7x+\dfrac{49}{4}\right)+\dfrac{49}{4}=\dfrac{133}{4}-\left(x+\dfrac{7}{2}\right)^2$

Checks may be carried out as for questions 1 & 2 and are omitted here.

5. $y=x^2-10x+12=\left(x^2-10x+25\right)-25+12=\left(x-5\right)^2-13$. Therefore, the vertex has coordinates $\left(5,\ -13\right)$. As the axis of symmetry passes through the vertex, it has equation $x=5$. Alternatively, the equation of the axis of symmetry is $x=-\dfrac{b}{2a}=-\dfrac{-10}{2\times 1}=5$.

6. Let $f\left(x\right)=ax^2+bx+c$. As the parabola passes through $\left(0,\ 6\right)$, $f\left(0\right)=6\Rightarrow c=6$. Since the axis of symmetry is $x=\dfrac{5}{2}$, $-\dfrac{b}{2a}=\dfrac{5}{2}\Rightarrow 5a=-b$ and, as the parabola passes through $\left(3,\ 0\right)$, $f\left(3\right)=0\Rightarrow 9a+3b+6=0$. Therefore, $3a+b+2=0$ and $b=-5a\Rightarrow a=1$. So $b=-5$ and $f\left(x\right)=x^2-5x+6$. Alternatively, if the parabola passes through $\left(3,\ 0\right)$ and has axis of

symmetry $x = \dfrac{5}{2}$, then, by symmetry, the parabola also passes through $=(2,\,0)$. Therefore,

$$f(x) = (x-3)(x-2) = x^2 - 5x + 6.$$

Exercise 2.5

1. $\quad 2\left(x^2 - 2x\right) + 1 = 2\left(x^2 - 2x + 1\right) - 1 \times 2 + 1 = 2(x-1)^2 - 1$

2. $\quad 4\left(x^2 + \dfrac{3}{2}x\right) - 9 = 4\left(x^2 + \dfrac{3}{2}x + \dfrac{9}{16}\right) - \dfrac{9}{16} \times 4 - 9 = 4\left(x + \dfrac{3}{4}\right)^2 - \dfrac{45}{4}$

3. $\quad 11 - 2\left(x^2 + 2x\right) = 11 - 2\left(x^2 + 2x + 1\right) + 1 \times 2 = 11 - 2(x+1)^2 + 2 = 13 - 2(x+1)^2$

4. $\quad 3\left(x^2 - \dfrac{10}{3}x\right) + 15 = 3\left(x^2 - \dfrac{10}{3}x + \dfrac{25}{9}\right) - \dfrac{25}{9} \times 3 + 15 = 3\left(x - \dfrac{5}{3}\right)^2 + \dfrac{20}{3}$

5. $\quad 1 - 4\left(x^2 + \dfrac{1}{2}x\right) = 1 - 4\left(x^2 + \dfrac{1}{2}x + \dfrac{1}{16}\right) - \dfrac{1}{16} \times 4 = \dfrac{5}{4} - 4\left(x + \dfrac{1}{4}\right)^2$

6. $\quad -5 - 2\left(x^2 - 4x\right) = -5 - 2\left(x^2 - 4x + 4\right) + 8 = 3 - 2(x-2)^2$

Exercise 2.6

1.   (a)

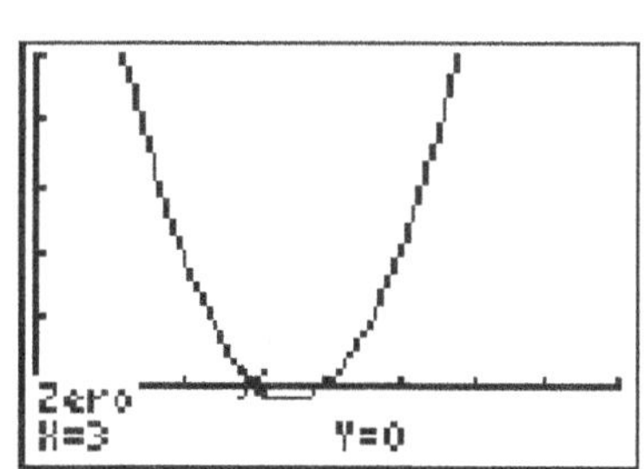

Intercepts are 3 and 4 therefore $x^2 - 7x + 12 = (x-3)(x-4)$

A similar method may be used for the other parts of the question. Only the factored solutions are shown.

(b) $(x+3)(x+2)$ (c) $(x+9)(x-4)$ (d) $(x-10)(x+9)$ (e) $(x-1)^2$

(f) $(5+x)(3-x)$ (g) $(3+x)(2-x)$ (h) $(7-x)(1+x)$ (i) $(x-5)(x+2)$

(j) $(5+x)(1-x)$ (k) $(x+11)(x-3)$ (l) $(x+7)(x+4)$

2. (a) $x^2 - 3x - 4x + 12 = x(x-3) - 4(x-3) = (x-3)(x-4)$

(b) $x^2 + 3x + 2x + 6 = x(x+3) + 2(x+3) = (x+3)(x+2)$

(c) $x^2 + 9x - 4x - 36 = x(x+9) - 4(x+9) = (x+9)(x-4)$

(d) $x^2 - 10x + 9x - 90 = x(x-10) + 9(x-10) = (x-10)(x+9)$

(e) $x^2 - x - x + 1 = x(x-1) - 1(x-1) = (x-1)(x-1) = (x-1)^2$

(f) $15 - 5x + 3x - x^2 = 5(3-x) + x(3-x) = (3-x)(5+x)$

(g) $6 - 3x + 2x - x^2 = 3(2-x) + x(2-x) = (2-x)(3+x)$

(h) $7 + 7x - x - x^2 = 7(1+x) - x(1+x) = (1+x)(7-x)$

(i) $x^2 - 5x + 2x - 10 = x(x-5) + 2(x-5) = (x-5)(x+2)$

(j) $5 - 5x + x - x^2 = 5(1-x) + x(1-x) = (1-x)(5+x)$

(k) $x^2 + 11x - 3x - 33 = x(x+11) - 3(x+11) = (x+11)(x-3)$

(l) $x^2 + 4x + 7x + 28 = x(x+4) + 7(x+4) = (x+4)(x+7)$

Exercise 2.7

1. (a) $x^2 + 10x - 2x - 20 = 0 \Rightarrow x(x+10) - 2(x+10) = 0 \Rightarrow (x+10)(x-2) = 0 \Rightarrow x = -10,\ x = 2$

(b) $x\left(x - \dfrac{72}{x}\right) = x \Rightarrow x^2 - 72 = x \Rightarrow x^2 - x - 72 = 0$

$\Rightarrow x^2 - 9x + 8x - 72 = 0 \Rightarrow x(x-9) + 8(x-9) = 0 \Rightarrow (x-9)(x+8) = 0 \Rightarrow x = 9,\ x = -8$

(c) $3x - x^2 = 9x + 9 \Rightarrow x^2 + 6x + 9 = 0 \Rightarrow x^2 + 3x + 3x + 9$

$\Rightarrow x(x+3) + 3(x+3) = 0 \Rightarrow (x+3)(x+3) = (x+3)^2 = 0 \Rightarrow x = -3$

(d) $x^2 - 8x + 16 = 8x - 47 \Rightarrow x^2 - 16x + 63 = 0 \Rightarrow x^2 - 9x - 7x + 63 = 0$

$\Rightarrow x(x-9) - 7(x-9) = 0 \Rightarrow (x-9)(x-7) = 0 \Rightarrow x = 9,\ x = 7$

2. (a) $x = \dfrac{-8 \pm \sqrt{64+12}}{6} = -\dfrac{4}{3} \pm \dfrac{\sqrt{76}}{6} = -\dfrac{4}{3} \pm \dfrac{\sqrt{19}}{3}$, so $p = -\dfrac{4}{3},\ k = \pm\dfrac{1}{3},\ q = 19$.

(b) $x = \dfrac{-7 \pm \sqrt{49-44}}{2} = -\dfrac{7}{2} \pm \dfrac{1}{2}\sqrt{5}$, so $p = -\dfrac{7}{2},\ k = \pm\dfrac{1}{2},\ q = 5$.

(c) $x = \dfrac{3 \pm \sqrt{9+20}}{10} = \dfrac{3}{10} \pm \dfrac{1}{10}\sqrt{29}$, so $p = \dfrac{3}{10},\ k = \pm\dfrac{1}{10},\ q = 29$.

(d) $x = \dfrac{17 \pm \sqrt{289-272}}{4} = \dfrac{17}{4} \pm \dfrac{1}{4}\sqrt{17}$, so $p = \dfrac{17}{4},\ k = \pm\dfrac{1}{4},\ q = 17$.

3.   (a)

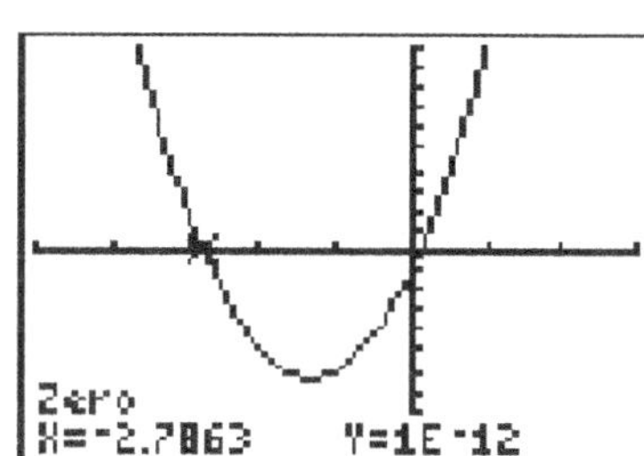

Therefore, $x = -2.79$, $x = 0.120$. As $x = -\dfrac{4}{3} \pm \dfrac{1}{3}\sqrt{19} = -2.786299...,\ 0.1196329...,$

the calculator solutions agree with the algebraic solutions up to at least three significant figures . The other parts may be checked in a similar way.

4.   (a) $\left(x^2 - 4x + 4\right) - 4 - 11 = 0 \Rightarrow (x-2)^2 = 15$; therefore, $a = 2$, $b = 15$ and $x - 2 = \pm\sqrt{15}$

$\Rightarrow x = 2 + \sqrt{15}$, $x = 2 - \sqrt{15}$.

(b) $\left(x^2 - 10x + 25\right) - 25 + 23 = 0 \Rightarrow (x-5)^2 - 2 = 0 \Rightarrow x - 5 = \pm\sqrt{2}$. So $x = 5 - \sqrt{2}$, $x = 5 + \sqrt{2}$

Exercise 2.8

1.   (a) $\Delta = 8$ (b) $\Delta = 0$, equal solutions.   (c) $\Delta = -4$, no real solutions.
     (d) $\Delta = -84$, no real solutions.   (e) $\Delta = 13$ (f) $\Delta = 49$
     (g) $\Delta = 29$ (h) $\Delta = 37$ (i) $\Delta = 0$, equal solutions.

2.   For no real solutions, $\Delta < 0$ and as $\Delta = (-1)^2 - 4k(-1) = 1 + 4k$, $1 + 4k < 0 \Rightarrow k < -\dfrac{1}{4}$.

3.   For equal solutions $\Delta = 0$ and as $\Delta = 4^2 - 4k(k+3) = 16 - 12k - 4k^2$, $16 - 12k - 4k^2 = 0$

$\Rightarrow k^2 + 3k - 4 = 0 \Rightarrow k^2 + 4k - k - 4 = 0 \Rightarrow k(k+4) - 1(k+4) = 0 \Rightarrow (k+4)(k-1) = 0$

$\Rightarrow k = -4,\ 1$.

4.   If $y = kx - k$ and $y = x^2 + 5x + 10$ touch then $kx - k = x^2 + 5x + 10$ has equal solutions.

Therefore $x^2 + (5-k)x + (10+k) = 0$ has $\Delta = 0$. But $\Delta = (5-k)^2 - 4(10+k)$

$= 25 - 10k + k^2 - 40 - 4k = k^2 - 14k - 15$ so that $k^2 - 14k - 15 = 0$

$\Rightarrow k(k-15) + 1(k-15) = 0 \Rightarrow (k-15)(k+1) = 0 \Rightarrow k = 15,\ k = -1$.

5.   (a) If the curves touch, the solution of

$$x^2 + \frac{1}{8} = 3x - x^2 - 1 \Rightarrow 2x^2 - 3x + \frac{9}{8} = 0 \Rightarrow 16x^2 - 24x + 9 = 0$$

will have two equal solutions. Therefore, $\Delta = 0$ but $\Delta = 24^2 - 4 \times 19 \times 9 = 576 - 576 = 0$,

and therefore the curves do touch each other. $16x^2 - 24x + 9 = 0 \Rightarrow (4x - 3)^2 = 0$

$\Rightarrow x = \dfrac{3}{4} \Rightarrow y = \dfrac{11}{16}$. Therefore the coordinates of the point of contact are $\left(\dfrac{3}{4},\ \dfrac{11}{16}\right)$.

(b) If the curves touch, there will be two equal solutions of $x^2 + kx + k = -2 - 2x - x^2$
$\Rightarrow 2x^2 + (2+k)x + (k+2) = 0$. This equation will have equal solutions if $\Delta = 0$ but

$$\Delta = (2+k)^2 - 8(k+2) = (k+2)((k+2)-8) = (k+2)(k-6).$$

Therefore, $(k+2)(k-6) = 0 \Rightarrow k = -2,\ 6$.

For $k = -2$, $2x^2 = 0 \Rightarrow x = 0 \Rightarrow y = -2$ and point of contact is $(0,\ -2)$.

For $k = 6$, $2x^2 + 8x + 8 = 0 \Rightarrow x^2 + 4x + 4 = 0 \Rightarrow (x+2)^2 = 0 \Rightarrow x = -2 \Rightarrow y = -2$ and point of contact is $(-2,\ -2)$.

Exercise 2.9

1.   (a) $(f \circ g)(1) = f(g(1)) = f(3) = 16$ (b) $(g \circ f)(4) = g(f(4)) = g(19) = -15$

     (c) $(f \circ f)(2) = f(f(2)) = f(13) = 46$ (d) $(g \circ g)(-1) = g(g(-1)) = g(5) = -1$

2.   (a) $(g \circ f)(x) = g(f(x)) = g(x+1) = 3(x+1) = 3x+3$

     (b) $(f \circ g)(x) = f(g(x)) = f(3x) = 3x+1$

        $f(x) \times g(x) = (x+1) \times (3x) = 3x^2 + 3x \neq 3x+1 = (f \circ g)(x)$

     (c) $(f \circ f)(x) = f(f(x)) = f(x+1) = (x+1)+1 = x+2$

3.   (a) $(g \circ f)(x) = \dfrac{5}{2}x + \dfrac{7}{2}$ (b) $(f \circ g)(x) = \dfrac{5}{2}x + 19$

     (c) $(g \circ h)(x) = \dfrac{9}{2} - x$ (d) $(f \circ h)(x) = 4 - 10x$

4.   (a) $f \circ g : x \mapsto 11 - 3x^2$ (b) $g \circ h : x \mapsto \dfrac{1}{x^2} - 3,\ x \neq 0$ (c) $h \circ f : x \mapsto \dfrac{1}{2-3x},\ x \neq \dfrac{2}{3}$

     (d) $h \circ h = h^2 : x \mapsto x$ (or $h^2 = i$)   (e) $g \circ g = g^2 : x \mapsto (x^2 - 3)^2 - 3 = x^4 - 6x^2 + 6$

5. $(g \circ h)(x) = (1-x) + 5 = 6 - x \Rightarrow (f \circ (g \circ h)) = f(6-x) = 2(6-x)$

     $(f \circ g)(x) = f(x+5) = 2(x+5) \Rightarrow ((f \circ g) \circ h)(x) = (f \circ g)h(x) = 2(h(x)+5)$

     $= 2((1-x)+5) = 2(6-x)$. Therefore, $f \circ (g \circ h) = (f \circ g) \circ h$.

6.   (a) $(g \circ f)(x) = g(f(x)) = g(2x-1) = \dfrac{1}{2}(2x-1) + \dfrac{1}{2} = x - \dfrac{1}{2} + \dfrac{1}{2} = x$.

       Therefore, $(g \circ f)(x) = x = i(x) \Rightarrow g \circ f = i$.

     (b) $f^2(x) = (f \circ f)(x) = f(f(x)) = f(2-x) = 2 - (2-x) = x$. So $f^2(x) = x = i(x) \Rightarrow f^2 = i$.

7.   (a) $f^{-1}(x) = \dfrac{x-1}{3}$ (b) $g^{-1}(x) = 3-x$ (c) $h^{-1}(x) = \dfrac{1-x}{4}$

      (d) $f^{-1}(x) = \dfrac{3}{x},\ x \neq 0$ (e) $g^{-1}(x) = \dfrac{3x-1}{2}$ (f) $h^{-1}(x) = \dfrac{2(x+2)}{x-1},\ x \neq 1$

8.   (a) $f^{-1}(4) = -1$ (b) $g^{-1}(-1) = -\dfrac{1}{4}$ (c) $h^{-1}(2) = 0$

      (d) $(f \circ g)(x) = 4x+5 \Rightarrow (f \circ g)^{-1}(x) = \dfrac{x-5}{4} \Rightarrow (f \circ g)^{-1}(3) = -\dfrac{1}{2}$

      (e) $\left(h^{-1} \circ g^{-1}\right)(x) = h^{-1}\left(\dfrac{x}{4}\right) = 2 - \dfrac{x}{4} \Rightarrow \left(h^{-1} \circ g^{-1}\right)(-1) = \dfrac{9}{4}$

9. $(f \circ g)(x) = 5(1-2x) = 5-10x \Rightarrow x = 5-10y \Rightarrow 10y = 5-x$. Therefore, $(f \circ g)^{-1}(x) = \dfrac{5-x}{10}$.

      $f^{-1}(x) = \dfrac{x}{5},\ g^{-1}(x) = \dfrac{1-x}{2} \Rightarrow \left(g^{-1} \circ f^{-1}\right)(x) = g^{-1}\left(\dfrac{x}{5}\right) = \dfrac{1 - \frac{x}{5}}{2} = \dfrac{5-x}{10}$.

      Therefore, $(f \circ g)^{-1} = g^{-1} \circ f^{-1}$.

Exercise 2.10

1. (a)   (b)   (c)

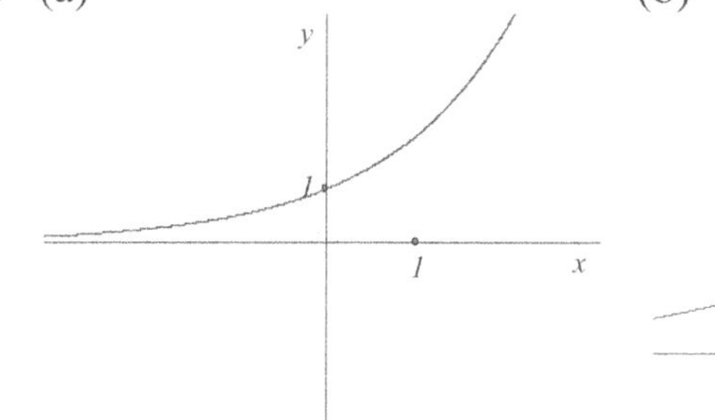 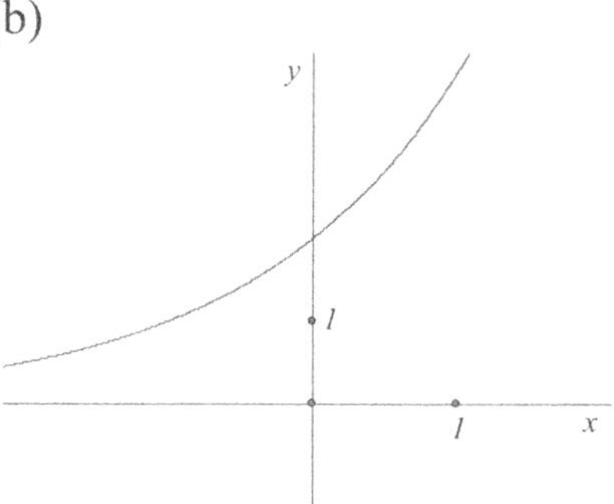 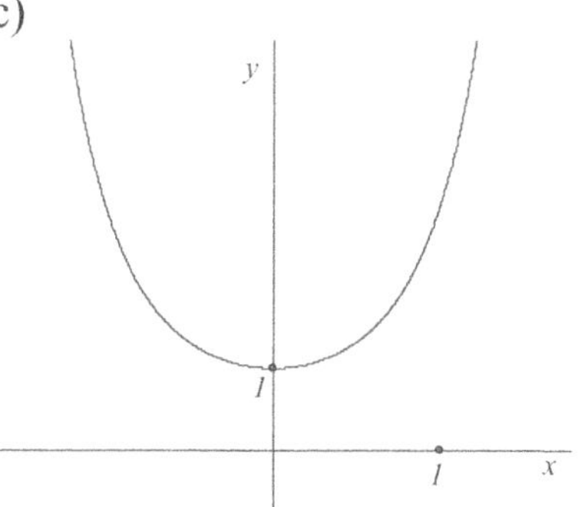

       (d)   (e)

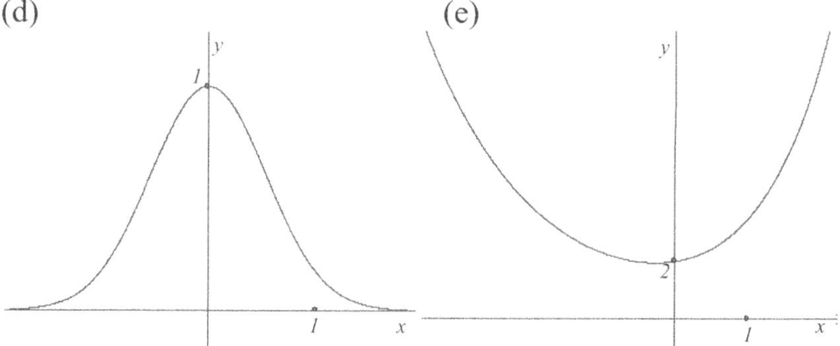

2.   The graph of $y = \log_2 x$ is shown by the thick curve.

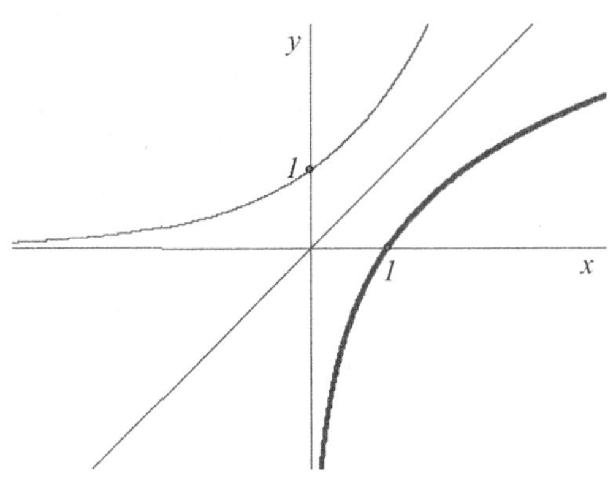

3.   The graph of $y = 2(1 - \log_3 x)$ is shown by the thick line.

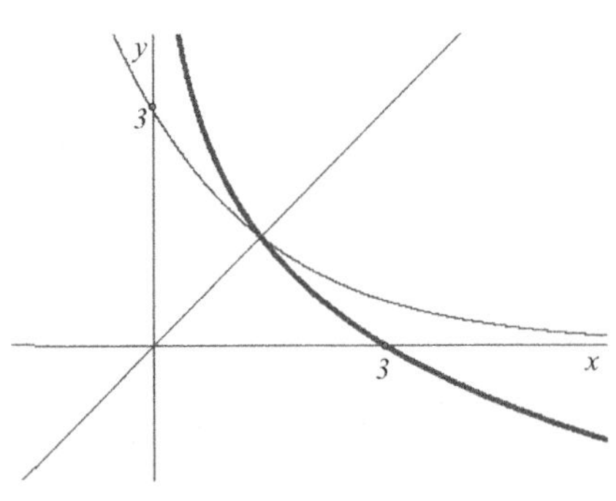

4.   (a) 5

  (b) $a^{3\log_a 2} = a^{\log_a 2^3} = a^{\log_a 8} = 8$

  (c) $a^{-2\log_a 3} = a^{\log_a 3^{-2}} = a^{\log_a \frac{1}{9}} = \dfrac{1}{9}$

  (d) $a^{\frac{1}{2}\log_a 16} = a^{\log_a 16^{\frac{1}{2}}} = a^{\log_a 4} = 4$

  (e) $a^{\log_{a^2} 3} = a^{\frac{\log_a 3}{\log_a a^2}} = a^{\frac{1}{2}\log_a 3} = a^{\log_a \sqrt{3}} = \sqrt{3}$

5.   (a) $\log_3 a = \dfrac{\log_a a}{\log_a 3} = \dfrac{1}{4.931} = 0.2027986... = 0.203$

  (b) $\log_a 27 = 3\log_a 3 = 3 \times 4.931 = 14.8$

  (c) $\log_{a} a^2 = \dfrac{\log_a a^2}{\log_a 3a} = \dfrac{2}{\log_a 3 + \log_a a} = \dfrac{2}{4.931 + 1} = 0.3372112... = 0.337$

6.   (a) $x = \log_2 5 = \dfrac{\log_{10} 5}{\log_{10} 2} = 2.321928... = 2.32$

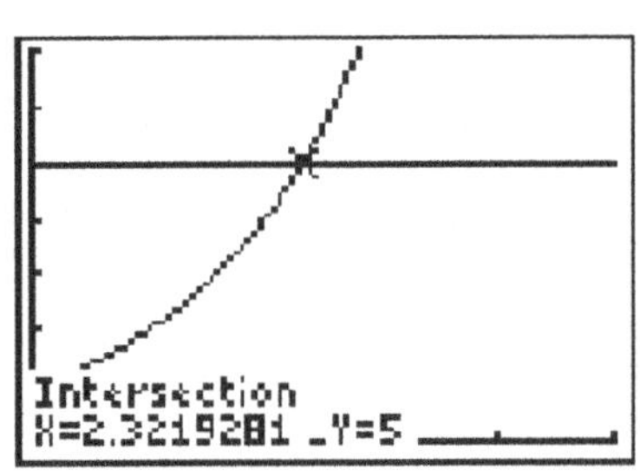

  (b) $2x - 3 = \log_5 39 \Rightarrow x = \dfrac{1}{2}(\log_5 39 + 3)$

$$= \dfrac{1}{2}\left(\dfrac{\log_{10} 39}{\log_{10} 5} + 3\right) = 2.638149... = 2.64$$

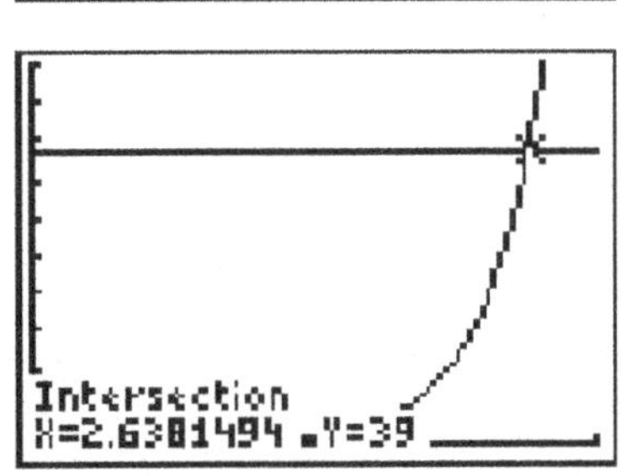

(c) $3^x = 243 = 3^5 \Rightarrow x = 5$

(d) $1 - x = \log_3 4 = \dfrac{\log_{10} 4}{\log_{10} 3}$

$\Rightarrow x = 1 - \dfrac{\log_{10} 4}{\log_{10} 3} = -0.2618595... = -0.262$

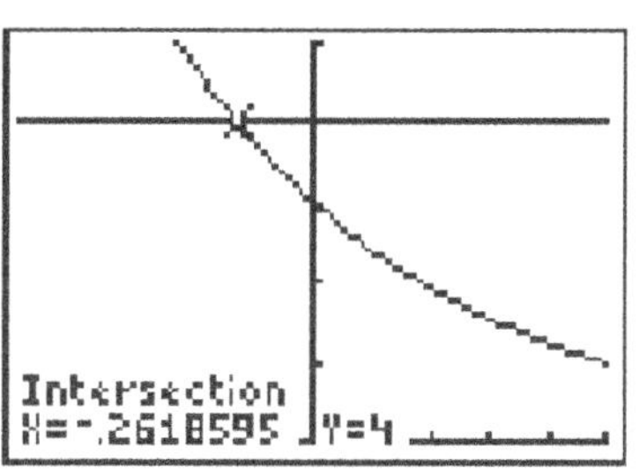

(e) $2^{3-x} = \dfrac{1}{16} = 2^{-4} \Rightarrow 3 - x = -4 \Rightarrow x = 7$

Exercise 2.11

1. $y = e^{3x}$. To find $f^{-1}(x)$ exchange x and y, so

$x = e^{3y} \Rightarrow \ln x = 3y \Rightarrow y = \dfrac{1}{3}\ln x$ and $f^{-1}(x) = \dfrac{1}{3}\ln x,\ x > 0$.

The transformation is a reflection in the line $y = x$.

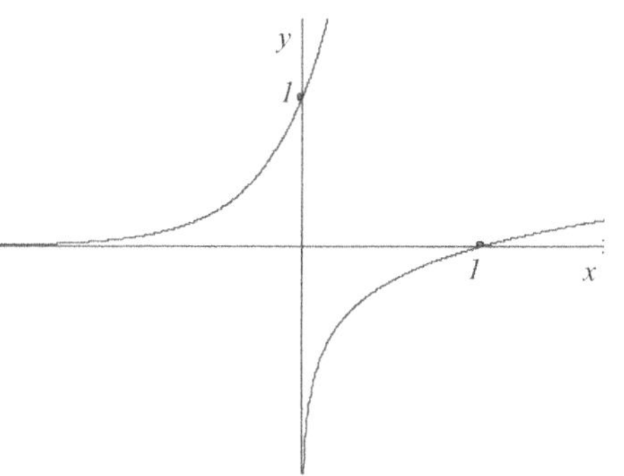

2. (a) $y = e^{\frac{1}{x}}$. To find $g^{-1}(x)$ exchange x and y so $x = e^{\frac{1}{y}} \Rightarrow \ln x = \dfrac{1}{y} \Rightarrow y = \dfrac{1}{\ln x}$.

   Therefore, $g^{-1}(x) = \dfrac{1}{\ln x},\ x > 0$.

   (b) $g(x)$ is shown with a thin line and $g^{-1}(x)$ is shown
   with a thick line. The transformation is a reflection
   in the line $y = x$.

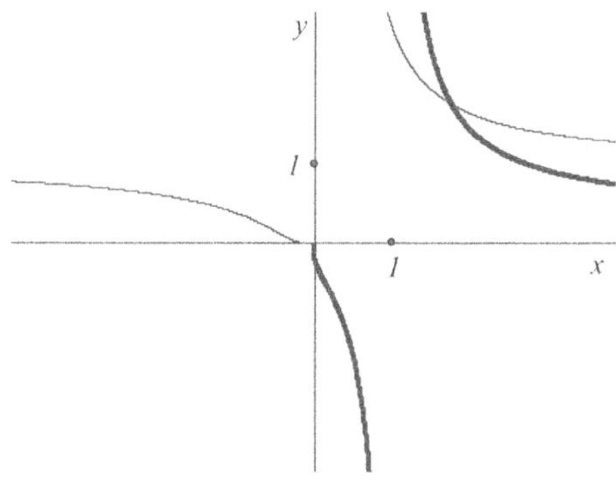

3. (a) $y = 3e^{-x} \Rightarrow x = 3e^{-y} \Rightarrow -y = \ln\dfrac{x}{3} \Rightarrow y = \ln\dfrac{3}{x}$. Therefore, $f^{-1} : x \mapsto \ln\dfrac{3}{x},\ x > 0$.

   (b) $y = 2\ln(x+1) \Rightarrow x = 2\ln(y+1) \Rightarrow y + 1 = e^{\frac{x}{2}} \Rightarrow y = e^{\frac{x}{2}} - 1$. Therefore, $f^{-1} : x \mapsto e^{\frac{x}{2}} - 1$.

   (c) $y = e^{-x^2} \Rightarrow x = e^{-y^2} \Rightarrow -y^2 = \ln x \Rightarrow y = \sqrt{-\ln x} \Rightarrow y = \sqrt{\ln\dfrac{1}{x}}$. So, $f^{-1} : x \mapsto \sqrt{\ln\dfrac{1}{x}},\ x > 0$.

   (d) $y = \ln(3x - 4) \Rightarrow x = \ln(3y - 4) \Rightarrow 3y - 4 = e^x \Rightarrow 3y = e^x + 4 \Rightarrow y = \dfrac{1}{3}(e^x + 4)$.

   Therefore, $f^{-1} : x \mapsto \dfrac{1}{3}(e^x + 4)$.

4.	(a) $3^x = e^{\ln 3^x} = e^{x \ln 3} = e^{1.10x}$ 	(b) $4^{-2x} = e^{\ln 4^{-2x}} = e^{-2x \ln 4} = e^{-2.77x}$

(c) $10^{0.1x} = e^{\ln 10^{0.1x}} = e^{0.1x \ln 10} = e^{0.230x}$ 	(d) $7^{0.5x} = e^{\ln 7^{0.5x}} = e^{0.5x \ln 7} = e^{0.973x}$

(e) $0.4^x = e^{\ln 0.4^x} = e^{x \ln 0.4} = e^{-0.916x}$

5.	(a) $\ln 2^{x+1} = \ln 5 \Rightarrow (x+1)\ln 2 = \ln 5 \Rightarrow x+1 = \dfrac{\ln 5}{\ln 2} \Rightarrow x = \dfrac{\ln 5}{\ln 2} - 1 = 1.321928... = 1.32$

(b) $\ln 3^{2x} = \ln 22 \Rightarrow 2x \ln 3 = \ln 22 \Rightarrow 2x = \dfrac{\ln 22}{\ln 3} \Rightarrow x = \dfrac{\ln 22}{2\ln 3} = 1.406794... = 1.41$

(c) $\ln 1.3^{-2x} = \ln 0.165 \Rightarrow -2x \ln 1.3 = \ln 0.165 \Rightarrow x = -\dfrac{\ln 0.165}{2\ln 1.3} = 3.433794... = 3.43$

(d) $4^{2x-3} = 2.4 \Rightarrow 2x - 3 = \dfrac{\ln 2.4}{\ln 4} \Rightarrow x = \dfrac{1}{2}\left(\dfrac{\ln 2.4}{\ln 4} + 3\right) = 1.815758... = 1.82$

Exercise 2.12

1.	(a) $f(3) = 100e^{0.5 \times 3} = 100e^{1.5} = 448.1689...$, so the population is 448 organisms.

(b) $f(7) = 100e^{0.5 \times 7} = 100e^{3.5} = 3311.545...$, so the population is 3310 organisms.

(c) $f(28) = 100e^{0.5 \times 28} = 100e^{14} = 1.202604... \times 10^8$, so the population is about 120 million organisms.

2.	(a) $n(0) = 250$

(b) Number of birds in January 2005 is $n(3) = 250e^{-0.24} = 196.6569... = 197$. Number of birds in January 2006 is $n(4) = 250e^{-0.32} = 181.5372... = 182$. Therefore, the decline during 2005 is $197 - 182 = 15$ birds.

(c) You need to find t such that $250e^{-0.08t} = 100 \Rightarrow e^{-0.08t} = 0.4 \Rightarrow -0.08t = \ln 0.4$
	$\Rightarrow t = -\dfrac{\ln 0.4}{0.08} \Rightarrow t = 11.45363....$ Therefore, the population of birds falls below 100 after about 11.5 years, that is during the year 2013.

3.	Jim's investment is worth, in dollars, $10000e^{0.043t}$ after t years. So after 5 years, it is worth $10000e^{0.043 \times 5} = 10000e^{0.215} = 12398.62...$, and the interest earned during this time is \$2398.62.

(a) If interest is compounded yearly, Jim's investment is worth, in dollars,
	$10000(1.043)^5 - 10000 = 2343.02$. So Jim overestimates his interest by \$55.60.

(b) If interest is compounded monthly, Jim's investment is worth, in dollars,
	$10000\left(1 + \dfrac{4.3}{1200}\right)^{60} - 10000 = 10000(1.0035833)^{60} - 10000 = 2393.86$. So Jim overestimates his interest by \$4.76.

4.	(a) $n_A(0) = 5000, \ n_B(0) = 3500$

(b) $n_A(10) = 5000e^{0.02 \times 10} = 5000e^{0.2} = 6117.013... = 6110$

(c) $n_B(t) = 7000 \Rightarrow 3500e^{0.025t} = 7000 \Rightarrow e^{0.025t} = 2 \Rightarrow 0.025t = \ln 2 \Rightarrow t = 40\ln 2 = 27.72588....$

Therefore it will take 28 years for the number of B seals to double.

(d) $n_A(t) = n_B(t) \Rightarrow 5000e^{0.02t} = 3500e^{.025t} \Rightarrow \dfrac{e^{0.02t}}{e^{0.025t}} = \dfrac{35}{50} = 0.7 \Rightarrow e^{(0.02-0.025)t} = 0.7$

$\Rightarrow e^{-0.005t} = 0.7 \Rightarrow -0.005t = \ln 0.7 \Rightarrow t = 71.33498....$ Therefore, the number of seals will be equal after about 70 years.

5.   After 3 weeks (21 days), the support for candidate A is estimated at $28e^{0.004\times21} = 28e^{0.084}$ $= 30.45360... = 30.5\%$. Candidate B' support is estimated at $40e^{-0.012\times21} = 31.08978.... = 31.1\%$ Therefore, on the basis of the survey, candidate B is (just) likely to win.

6.   (a) $f(0) = 30 \Rightarrow n_0\left(1 - 0.9e^{-0.15\times0}\right) = n_0(1-0.9) = 0.1n_0 \Rightarrow n_0 = \dfrac{30}{0.1} = 300$

(b)

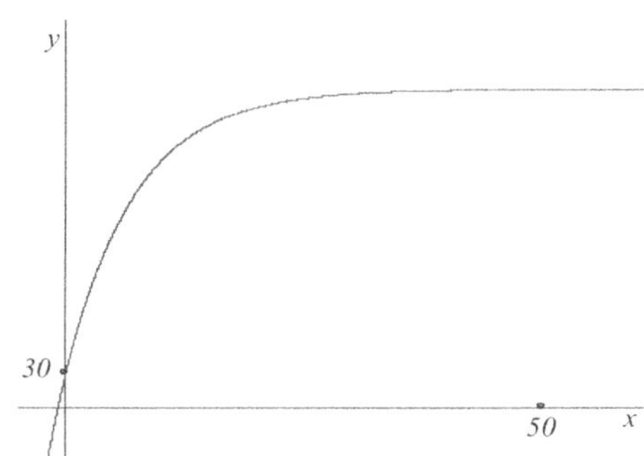

(c) $f(6) = 300\left(1-0.9e^{-0.15\times6}\right) = 300\left(1-0.9e^{-0.9}\right) = 190.2261... = 190$

(d) The graph has a horizontal asymptote at $n = 300$, indicating that after a long time (that is for large values of t) the pond will have 300 trout.

Alternatively: $f(20) = 300\left(1-0.9e^{-3}\right) = 286.5574...$,

$$f(50) = 300\left(1-0.9e^{-7.5}\right) = 299.8506...$$

$$f(100) = 300\left(1-0.9e^{-15}\right) = 299.9999...$$

from which it is clear that, as t increases, n gets closer and closer to 300.

Solutions to Unit 2 Review Exercise

Part 1

1.   (a) $f(x) = \left(x^2 - 4x + 4\right) - 4 + 13 = (x-2)^2 + 9$

Therefore $a = 2,\ b = 9$

(b) $g(x) = \left[(x-2)-2\right]^2 + 9 - 5 = (x-4)^2 + 4 = x^2 - 8x + 20$

Therefore $p = -8,\ q = 20$

(c) The range of f: $f(x) \geq 9$

2.   (a) $a=2, \ c=4$ (b) $5=2+\dfrac{3b}{9-4} \Rightarrow 3 = \dfrac{3b}{5} \Rightarrow b = 5$

3.   (a) $(f \circ g)(x) = f(g(x)) = f\left(1-\dfrac{3}{x}\right) = 2-3\left(1-\dfrac{3}{x}\right) = \dfrac{9-x}{x}$. Therefore $k=9$

       (b) $(g \circ f)(x) = g(2-3x) = 1 - \dfrac{3}{2-3x} = \dfrac{1+3x}{3x-2} = \dfrac{3x+1}{3x-2}$

       (c) $\dfrac{9-x}{x} = 3 \Rightarrow 9-x = 3x \Rightarrow 4x = 9 \Rightarrow x = \dfrac{9}{4}$.

4.   (a) $x=3$ (b) $b=3$

       (c) $ab^2 + c = 24 \Rightarrow 9a + c = 24$. Also $a(2-3)^2 + c = 0 \Rightarrow a + c = 0 \Rightarrow a = 3, \ c = -3$

5.   (a) $f(x) = -(x-2)^2 = -(x^2 - 4x + 4) = -4 + 4x - x^2$, therefore $k = 4$.

       (b)

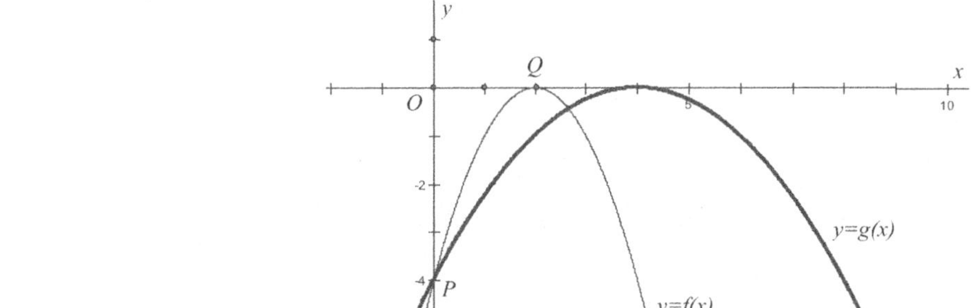

       (c) $g(x) = f\left(\dfrac{x}{2}\right) = -4 + 4\left(\dfrac{x}{2}\right) - \left(\dfrac{x}{2}\right)^2 = -4 + 2x - \dfrac{x^2}{4}$.

6.   (a) $f(x) = x^2 - 6x + 8 = (x-2)(x-4)$ (b) $(2, \ 0), \ (4, \ 0)$

       (c) The coordinates of the vertex of f are $(3, \ 1)$ and the coordinates of the vertex of the graph of g are $(3, \ 0)$. Therefore, $k = 1$.

7.   (a) $f^{-1}(x) = x - 4$

       (b)   (i) $\left(f^{-1} \circ f\right)(x) = i(x) = x$

           (ii) $\left(g^{-1}\right)^{-1}(x) = g(x) = 2x + 1$

       (c) $(g \circ f)(x) = g(f(x)) = g(x+4) = 2(x+4) + 1 = 2x + 9$

320

8.   We require that $\Delta = 0 \Rightarrow k^2 - 4 \times 1 \times (-k+3) = 0 \Rightarrow k^2 + 4k - 12 = 0$.

$k^2 + 6k - 2k - 12 = 0 \Rightarrow (k-2)(k+6) = 0 \Rightarrow k = -6,\ 2$

9.   (a) $g(x) = \dfrac{1}{x+2} + 1 = \dfrac{x-2+1}{x-2} = \dfrac{x-1}{x-2}$ and therefore $a = 1,\ b = 2$.

(b)

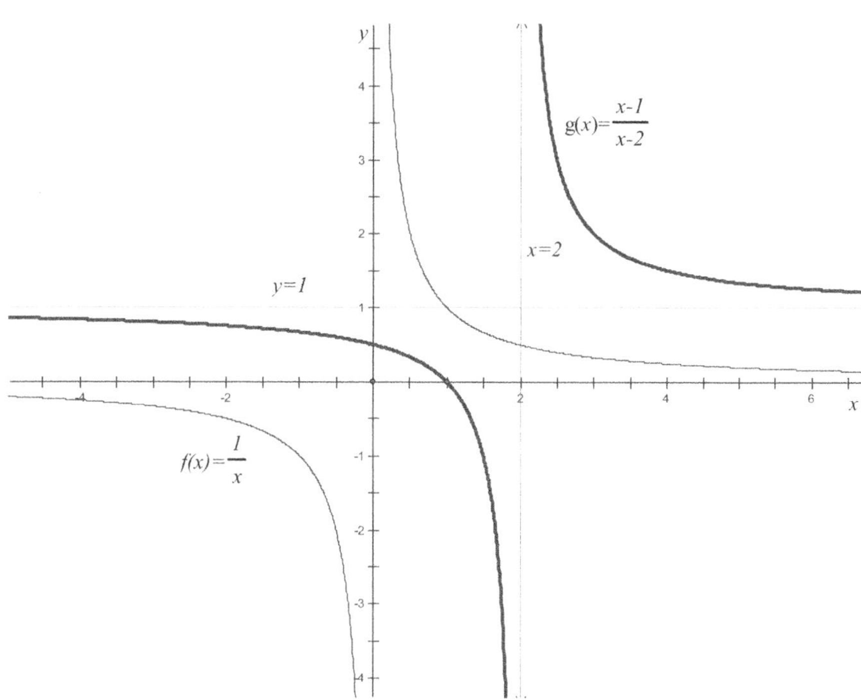

10.   (a) $f(x) = 2x^2 - 4x - 16 = 2(x^2 - 2x) - 16 = 2(x^2 - 2x + 1) - 2 - 16 = 2(x-1)^2 - 18$

(b)   (i)   so the x-intercepts are $(-2,\ 0)$ and $(4,\ 0)$

(ii) $f(0) = 2(-1)^2 - 18 = -16$ so the y-intercept is $(0,\ -16)$

(iii) vertex has coordinates $(1, -18)$

(c)

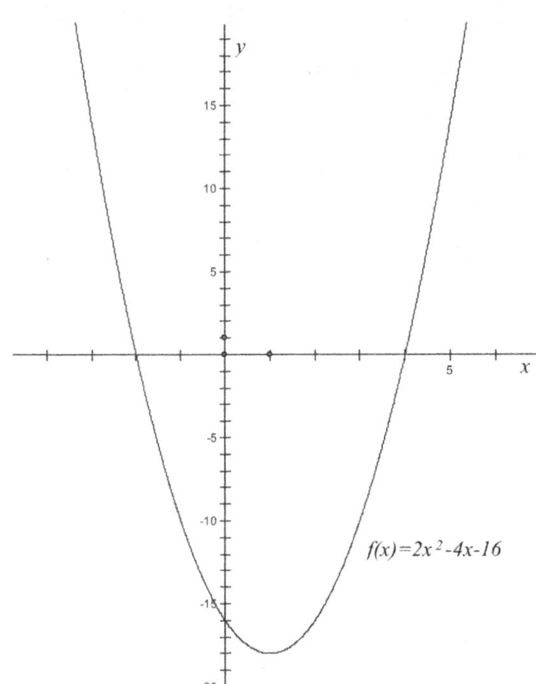

(d) $c = -4$ or $c = 2$

Part 2

11. (a)

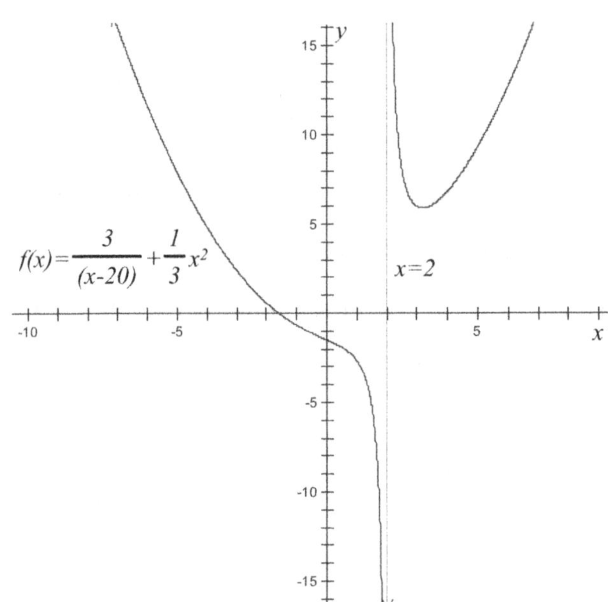

(b) $x = 2$
(c) $(-1.58,\ 0),\ (0,\ -1.5)$

12   (a)

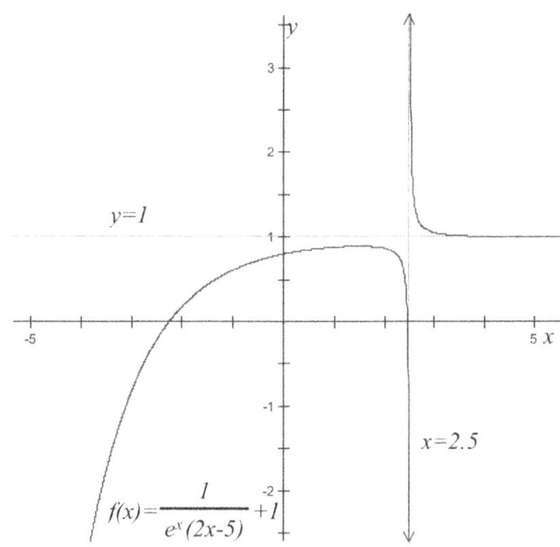

(b) $x = 2.5$

(c) (i) $f(3) = 1.049787...$ (ii) $f(5) = 1.001348...$ (iii) $f(10) = 1.000003...$ (d) $y = 1$

13.   (a) $f < 2$

(b)

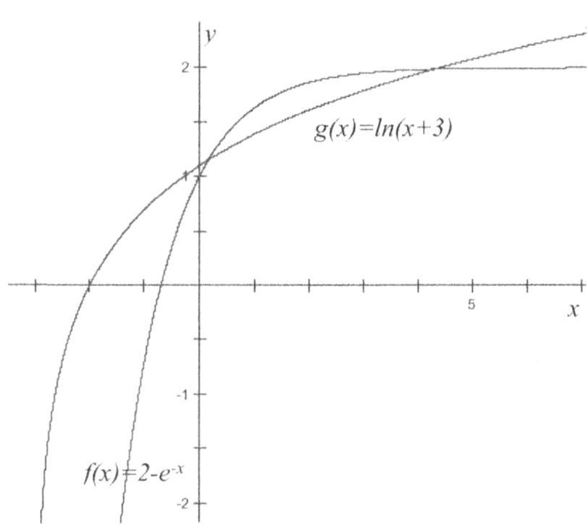

(c) $x = 0.1650879... = 0.165, \quad x = 4.288311... = 4.29$

14.   (a) $m(10) = 30000e^{-10k} = 12000 \Rightarrow e^{-10k} = \dfrac{12}{30} = \dfrac{2}{5} \Rightarrow -10k = \ln\left(\dfrac{2}{5}\right)$

$\Rightarrow k = -\dfrac{1}{10}\ln 0.4 = 0.09162907... = 0.0916$

(b) $m(7) = 30000e^{-0.0916 \times 7} = 15799.80... = 15800$ so, correct to 3 significant figures, the mass of the iceberg is 15800kg.

(c) $m(t) = 30000e^{-0.0916t} \leq 1000 \Rightarrow e^{-0.0916t} \leq \dfrac{1}{30}$

$\Rightarrow -0.0916t \leq \ln\left(\dfrac{1}{30}\right) \Rightarrow t \geq \dfrac{\ln 30}{0.0916} = 37.13097...$

So mass of iceberg is less than 1000kg after 38 complete days.

15.   (a)

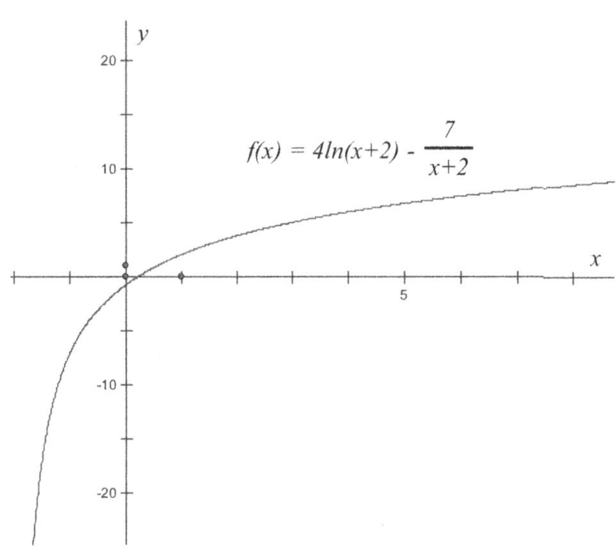

(b) Vertical asymptote which has equation $x = -2$

(c) $f(0) = 4\ln 2 - \dfrac{7}{2} = -0.7274112...$ so the coordinates of the y-intercept are $\left(0, \; -0.727\right)$

(d) $10 - x = 4\ln(x+2) - \dfrac{7}{x+2} \Rightarrow x = 3.999800... = 4.00$

16.   (a) Substituting the coordinates of B into the given function we have $a \times 0^2 + b \times 0 + c = 20$ so that $c = 20$

(b) From $A(4, \; 0)$, $16a + 4b + 20 = 0 \Rightarrow 4a + b = -5$

From $C(2, \; -10)$, $4a + 2b + 20 = -10 \Rightarrow 2a + b = -15$

(c) So that $a = 5$, and $b = -25$ therefore $f(x) = 5x^2 - 25x + 20$

17.   (a)

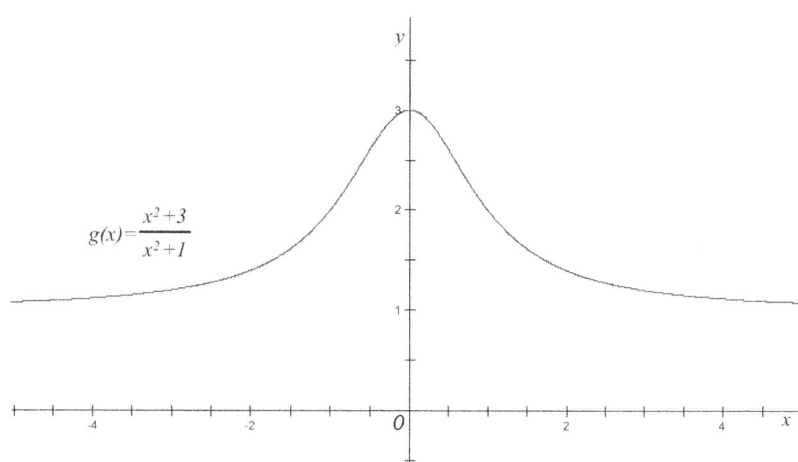

(b) $y = 1$ (c) $1 < g(x) \le 3$

(d) $\dfrac{x^2+3}{x^2+1} = x^2 \Rightarrow x^2+3 = x^4+x^2 \Rightarrow x^4 = 3 \Rightarrow x^2 = \sqrt{3} \Rightarrow x = \pm 3^{1/4} = \pm 1.316074... = \pm 1.32$

18.   (a) $g(x) = (x-5)^2$ (b) $h(x) = 3(x-5)^2$

     (c) $j(x) = 3(x-5)^2 + 2$ (d) $k(x) = 3(x-5)^2 + 6$

19   (a)

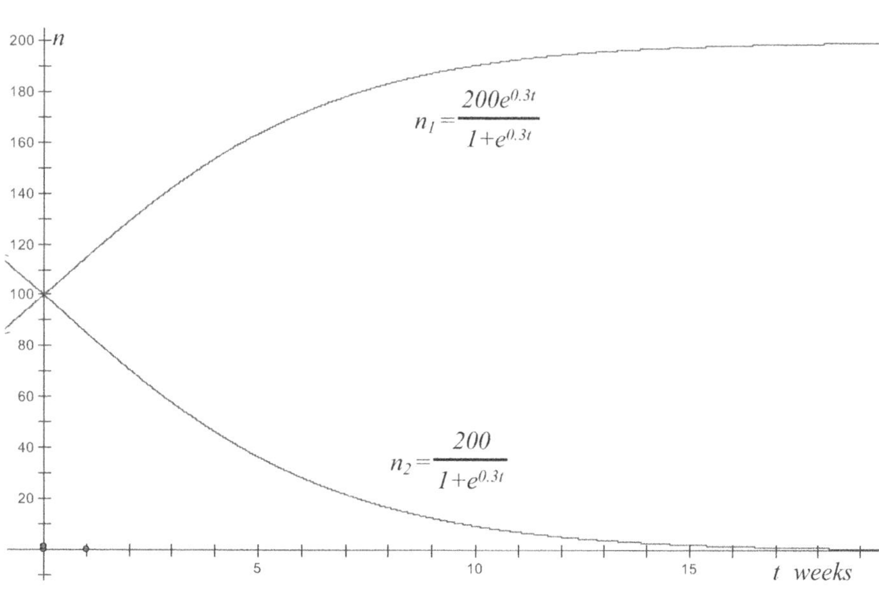

(b) There are 164 green frogs and 36 yellow frogs.

(c) The number of green frogs stabilizes at 200 after about 20 weeks and the yellow frogs disappear after about the same period.

(d) $n_1 + n_2 = 200\left(\dfrac{e^{0.3t}}{1+e^{0.3t}} + \dfrac{1}{1+e^{0.3t}} \right) = 200\left(\dfrac{1+e^{0.3t}}{1+e^{0.3t}} \right) = 200$

325

Solutions to Unit 3 Exercises

Exercise 3.1

1. (a) $120°$ (b) $72°$ (c) $150°$ (d) $80°$ (e) $540°$ (f) $54°$ (g) $15°$ (h) $70°$ (i) $142.5°$

2. (a) $\dfrac{2\pi}{3}$ (b) $\dfrac{5\pi}{9}$ (c) $\dfrac{\pi}{10}$ (d) $\dfrac{\pi}{180}$ (e) $\dfrac{34\pi}{15}$ (f) $\dfrac{7\pi}{36}$ (g) $\dfrac{7\pi}{60}$ (h) $\dfrac{37\pi}{120}$ (i) $\dfrac{7\pi}{6}$

3. (a) 10cm (b) 1.75cm (c) 3.49cm (d) 12.0cm

4. (a) $0.0633 = 3.63°$ (b) $0.949 = 54.4°$ (c) $0.224 = 12.8°$ (d) $0.499 = 28.6°$

5. (a) 28.3cm^2 (b) 18.8cm^2 (c) 23.6cm^2 (d) 23.2cm^2 (e) 5.65cm^2 (f) 22.9cm^2

6. (a) 1.39 (b) 1.04 (c) 0.871

7. Diameter of the moon is approximately equal to the length of the arc of the sector. Arc length is 382100θ, where θ is the angle of the sector in radians. So $\theta = \dfrac{0.5167\pi}{180}$ and the arc length is $382100 \times \dfrac{0.5167\pi}{180} = 3445.822\ldots$. Therefore, the diameter of the Moon is 3450 km.

8. A_1 is the starting position of A. When the triangle is rotated about B, A_1 is rotated about B through $\dfrac{2\pi}{3}$ to A_2. The length of the arc through which A moves is $6 \times \dfrac{2\pi}{3} = 4\pi$ cm. When the

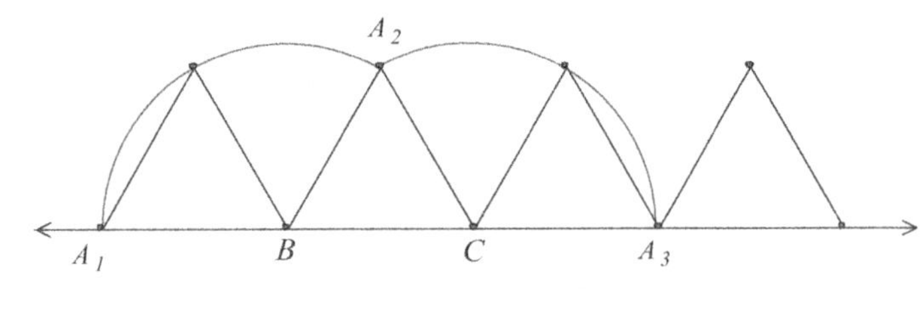

triangle is then rotated about C, A_2 is rotated through an angle of $\dfrac{2\pi}{3}$ about C to A_3. This arc length is also 4π cm. Finally, the triangle is rotated about A, so obviously A does not move. Therefore, the path traveled by A is 8π cm $= 25.13274\ldots = 25.1$ cm.

9. (a) $\dfrac{\theta}{2\pi} = \dfrac{1}{50} \Rightarrow \theta = \dfrac{2\pi}{50} = 0.1256637\ldots$. Therefore, the angle between sun's rays and stick is 0.126 radians.

(b) $r\theta = 850 \Rightarrow r = \dfrac{850}{\theta} = 6764.085\ldots$. Therefore radius of Earth is 6760km.

Exercise 3.2

1. (a) $\sin\left(\pi - \dfrac{3\pi}{4}\right) = \sin\dfrac{\pi}{4}$ (b) $\sin\left(\pi - \dfrac{3\pi}{8}\right) = \sin\dfrac{3\pi}{8}$ (c) $\sin\left(\pi + \dfrac{3\pi}{10}\right) = -\sin\dfrac{3\pi}{10}$

    (d) $\sin\left(2\pi - \dfrac{\pi}{4}\right) = -\sin\dfrac{\pi}{4}$ (e) $\sin\left(360° - 60°\right) = -\sin 60°$ (f) $\sin\left(180° - 45°\right) = \sin 45°$

    (g) $\sin\left(180° + 56°\right) = -\sin 56°$ (h) $\sin\left(360° - 23°\right) = -\sin 23°$

2. (a) $\cos\left(2\pi - \dfrac{\pi}{2}\right) = \cos\dfrac{\pi}{2} \;(= 0)$ (b) $\cos\left(\pi - \dfrac{\pi}{5}\right) = -\cos\dfrac{\pi}{5}$

    (c) $\cos\left(\pi + \dfrac{\pi}{6}\right) = -\cos\dfrac{\pi}{6}$ (d) $\cos\left(\pi - \dfrac{\pi}{6}\right) = -\cos\dfrac{\pi}{6}$

    (e) $\cos\left(180° - 70°\right) = -\cos 70°$ (f) $\cos\left(180° + 30°\right) = -\cos 30°$

    (g) $\cos\left(360° - 50°\right) = \cos 50°$ (h) $\cos\left(180° - 47°\right) = -\cos 47°$

3. (a) $\tan\left(\pi + \dfrac{\pi}{4}\right) = \tan\dfrac{\pi}{4}$ (b) $\tan\left(\pi + \dfrac{\pi}{3}\right) = \tan\dfrac{\pi}{3}$ (c) $\tan\left(\pi - \dfrac{\pi}{8}\right) = -\tan\dfrac{\pi}{8}$

    (d) $\tan\left(\pi + \dfrac{\pi}{6}\right) = \tan\dfrac{\pi}{6}$ (e) $\tan\left(180° - 50°\right) = -\tan 50°$ (f) $\tan\left(360° - 40°\right) = -\tan 40°$

    (g) $\tan\left(180° + 80°\right) = \tan 80°$ (h) $\tan\left(360° - 10°\right) = -\tan 10°$

4. (a) $\dfrac{1}{\sqrt{2}}$ (b) $\dfrac{1}{2}$ (c) $-\dfrac{\sqrt{3}}{2}$ (d) $-\dfrac{1}{\sqrt{2}}$ (e) $-\dfrac{1}{\sqrt{2}}$ (f) $-\dfrac{1}{2}$ (g) $\dfrac{\sqrt{3}}{2}$ (h) $\dfrac{1}{2}$

    Parts (i) to (l) can be done by evaluating the sine and cosine values of the angle and then using

    $\tan\theta = \dfrac{\sin\theta}{\cos\theta}$ (i) -1 (j) $-\sqrt{3}$ (k) $-\dfrac{1}{\sqrt{3}}$ (l) $\dfrac{1}{\sqrt{3}}$

5. The rotational symmetry (half-turn about the origin) of the graph of $y = \sin\theta$ shows that for all values of θ, $\sin(-\theta) = -\sin\theta$.

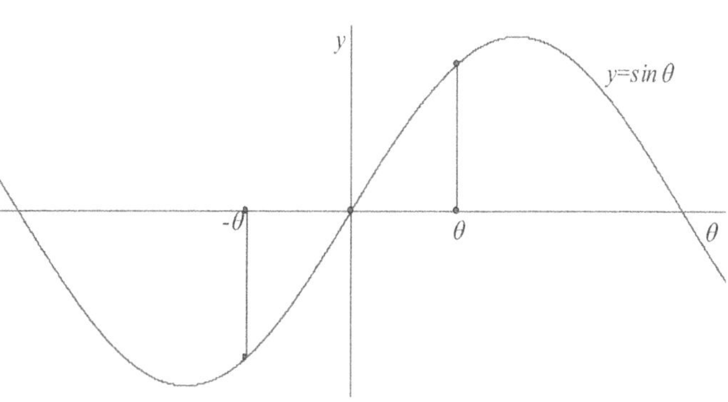

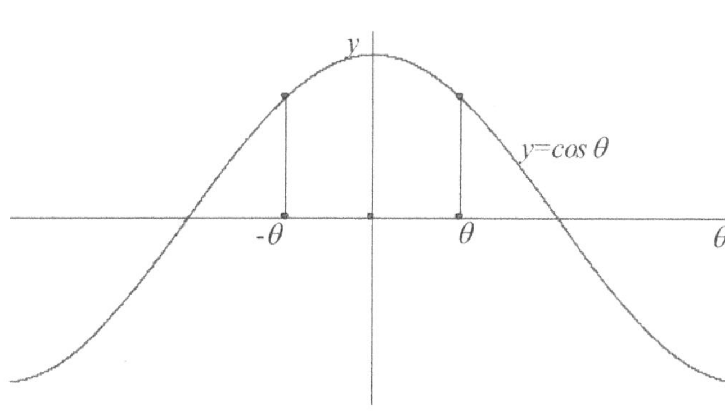

The reflectional symmetry (reflection in the y-axis) of the graph of $y = \cos\theta$ shows that for all values of θ, $\cos(-\theta) = \cos\theta$.

The rotational symmetry (half-turn about the origin) of the graph of $y = \tan\theta$ shows that for all values of θ,

$\tan(-\theta) = -\tan\theta$.

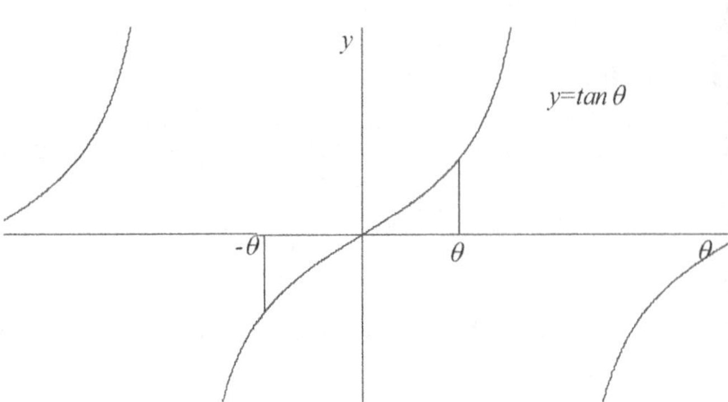

(a) $\sin\left(-\dfrac{\pi}{3}\right) = -\sin\dfrac{\pi}{3} = -\dfrac{\sqrt{3}}{2}$

(b) $\cos\left(-\dfrac{4\pi}{3}\right) = \cos\dfrac{4\pi}{3} = \cos\left(\pi + \dfrac{\pi}{3}\right) = -\cos\dfrac{\pi}{3} = -\dfrac{1}{2}$

(c) $\tan\left(-\dfrac{11\pi}{6}\right) = -\tan\left(\dfrac{11\pi}{6}\right) = \dfrac{1}{\sqrt{3}}$

Exercise 3.3

1. (a) $\dfrac{5}{8}$ (b) $x^2 + 5^2 = 8^2 \Rightarrow x = \sqrt{39}$ (c) $\cos\theta = \dfrac{\sqrt{39}}{8}$

   (d) $\cos^2\theta + \sin^2\theta = \left(\dfrac{\sqrt{39}}{8}\right)^2 + \left(\dfrac{5}{8}\right)^2 = \dfrac{39}{64} + \dfrac{25}{64} = \dfrac{39+25}{64} = 1$

2. (a) $\cos^2\theta = 1 - \sin^2\theta = 1 - \dfrac{4}{25} = \dfrac{21}{25} \Rightarrow \cos\theta = \pm\dfrac{\sqrt{21}}{5}$ but, as θ is acute $\cos\theta > 0$ so

   $\cos\theta = \dfrac{\sqrt{21}}{5}$.

   (b) As θ is obtuse $\cos\theta < 0$ so $\cos\theta = -\dfrac{\sqrt{21}}{5}$.

3. (a) $\sin^2\theta = 1 - \cos^2\theta \Rightarrow \sin^2\theta = 1 - \dfrac{4}{9} = \dfrac{5}{9} \Rightarrow \sin\theta = \pm\dfrac{\sqrt{5}}{3}$ but as $0 < \theta < \dfrac{\pi}{2}$, $\sin\theta = \dfrac{\sqrt{5}}{3}$ and

   $k = 5$.

   (b) $\sin^2\theta = 1 - \dfrac{81}{1681} = \dfrac{1600}{1681} \Rightarrow \sin\theta = \pm\dfrac{40}{41}$ but as $\pi < \theta < 2\pi$ so $\sin\theta < 0$ and $\sin\theta = -\dfrac{40}{41}$.

4. (a) $\sin^2\theta = 1 - \dfrac{25}{169} = \dfrac{144}{169} \Rightarrow \sin\theta = \pm\dfrac{12}{13}$. As $0 < \theta < \pi$, $\sin\theta > 0$, so $\sin\theta = \dfrac{12}{13}$.

   (b) $\sin 2\theta = 2\cos\theta\sin\theta = 2\left(-\dfrac{5}{13}\right)\left(\dfrac{12}{13}\right) = -\dfrac{120}{169}$

   (c) $\cos 2\theta = 2\cos^2\theta - 1 = 2\left(\dfrac{25}{169}\right) - 1 = -\dfrac{119}{169}$

5. $\cos^2\theta = 1 - \dfrac{9}{25} = \dfrac{16}{25} \Rightarrow \cos\theta = \pm\dfrac{4}{5}$, $\cos 2\theta = 2\cos^2\theta - 1 = 2\left(\dfrac{16}{25}\right) - 1 = \dfrac{7}{25}$

6. (a) $\cos^2\theta = 1 - \dfrac{49}{625} = \dfrac{576}{625} \Rightarrow \cos\theta = \pm\dfrac{24}{25}$

   (b) $\tan\theta = \dfrac{\sin\theta}{\cos\theta} = \dfrac{7/25}{\pm 24/25} = \pm\dfrac{7}{24}$

   (c) $\cos 2\theta = 1 - 2\sin^2\theta = 1 - 2\left(\dfrac{49}{625}\right) = \dfrac{527}{625}$

   (d) $\sin 2\theta = 2\cos\theta\sin\theta = 2\left(\pm\dfrac{24}{25}\right)\left(\dfrac{7}{25}\right) = \pm\dfrac{336}{625}$

7. (a) $1 + \tan\theta = 1 + \dfrac{\sin\theta}{\cos\theta} = \dfrac{\cos\theta + \sin\theta}{\cos\theta}$

   (b) $\left(\cos\theta + \sin\theta\right)^2 = \cos^2\theta + 2\cos\theta\sin\theta + \sin^2\theta = \left(\cos^2\theta + \sin^2\theta\right) + 2\cos\theta\sin\theta$
   $= 1 + \sin 2\theta$

   (c) $1 - \tan^2\theta = 1 - \dfrac{\sin^2\theta}{\cos^2\theta} = \dfrac{\cos^2\theta - \sin^2\theta}{\cos^2\theta} = \dfrac{\cos^2\theta - \sin^2\theta}{1 - \sin^2\theta} = \dfrac{\left(\cos\theta - \sin\theta\right)\left(\cos\theta + \sin\theta\right)}{\left(1 - \sin\theta\right)\left(1 + \sin\theta\right)}$

   (d) $\cos 2A = 1 - 2\sin^2 A$. If $A = 3\theta$, then $\cos 6\theta = 1 - 2\sin^2 3\theta \Rightarrow 2\sin^2 3\theta = 1 - \cos 6\theta$

   (e) $1 + \cos\theta + \cos 2\theta = 1 + \cos\theta + 2\cos^2\theta - 1 = \cos\theta + 2\cos^2\theta = \cos\theta\left(1 + 2\cos\theta\right)$

Exercise 3.4

1. (a) A stretch of factor 5, parallel to the y-axis.
   (b) A stretch of factor 3 parallel to the x-axis.
   (c) A translation of $\dfrac{\pi}{6}$ parallel to the positive x-axis.

2. (a) A stretch of factor $\dfrac{2}{5}$ parallel to the y-axis.
   (b) A translation of π parallel to the negative x-axis.
   (c) A stretch of factor 4 parallel to the x-axis.

3. (a) A stretch of factor $\dfrac{1}{2}$ parallel to the x-axis. A translation of $+1$ parallel to the y-axis.
   (b) A stretch of factor $\dfrac{1}{3}$ parallel to the x-axis. A stretch of factor 2 parallel to the y-axis.
   (c) A translation of $\dfrac{\pi}{2}$ parallel to the positive x-axis. A stretch of factor 2 parallel to y-axis.

4.   (a) 4π (b) π (c) π (d) $\dfrac{\pi}{3}$ (e) 6

   (f) $\cos 2x = 1 - 2\sin^2 x \Rightarrow \sin^2 x = \dfrac{1}{2} - \dfrac{1}{2}\cos 2x$. Therefore the period is π.

   (g) Use your calculator to draw the graph of $f(x) = \cos x + \sin x$, measure the period $= 2\pi$.

5.   (a) $y = 2\sin x$ (b) $y = 1 + \cos x$ (c) $y = \tan\left(x - \dfrac{\pi}{3}\right)$

   (d) $y = \sin\dfrac{x}{2}$ (e) $y = \tan 3x$ (f) $y = -\cos x$

6.   (a) Period is $\dfrac{2\pi}{3}$.   (b) (i) 5   (ii) -5 (c) (i) $f(0) = -\dfrac{5\sqrt{3}}{2}$ (ii) $f\left(\dfrac{\pi}{2}\right) = \dfrac{5}{2}$

7.   (a) $\dfrac{2\pi}{3} = \dfrac{2\pi}{b} \Rightarrow b = 3$ (b) $5 = a\sin(3\times 0) + c \Rightarrow c = 5$; $9 = a\sin\dfrac{3\pi}{2} + 5 \Rightarrow a = -4$

8.   (a) (i) $a = \dfrac{17-5}{2} = 6$ (ii) $d = \dfrac{17+5}{2} = 11$ (b) The period of y is π, $\dfrac{2\pi}{b} = \pi \Rightarrow b = 2$

   (c) $14 = 6\sin(2\times 0 + c) + 11 \Rightarrow 6\sin c = 3 \Rightarrow \sin c = \dfrac{1}{2} \Rightarrow c = \dfrac{\pi}{6}$

9.   (a) (i) Period is π (ii) amplitude is 4   (b) (i) $a = 4$ (ii) $b = 2$

   (c) $p = \dfrac{\pi}{2}$ (d) $q = \dfrac{\pi}{12}$

10.   (a) Period is $\dfrac{2\pi}{120\pi} = \dfrac{1}{60}$ second   (b) 60   (c) (i) 0   (ii) 4.52

11.   (a) $t(1) = -1.96$. Therefore sunset is 1h 57 minutes before 18:00 which is 16:03.

   (b) $t(46) = -1.11$. Therefore sunset is 1h 7minutes before 18:00 which is 16:53.

   (c) $t(176) = 1.99$. Therefore sunset is 2h 0minutes after 18:00 which is 20:00.

Exercise 3.5

1.   (a) $x = \dfrac{\pi}{6}, \dfrac{5\pi}{6}$ (b) $x = \dfrac{\pi}{4}, \dfrac{5\pi}{4}$ (c) $x = 0$ (d) $x = 0,\ \pi$ (e) $x = \dfrac{3\pi}{4}, \dfrac{5\pi}{4}$

   (f) $x = \dfrac{5\pi}{6}, \dfrac{11\pi}{6}$ (g) $x = \dfrac{2\pi}{3}, \dfrac{4\pi}{3}$ (h) $x = \dfrac{\pi}{4}, \dfrac{3\pi}{4}$ (i) $x = 0,\ \pi$

2.   (a) $x = \dfrac{\pi}{2},\ -\dfrac{\pi}{2}$ (b) $x = \dfrac{3\pi}{4},\ -\dfrac{\pi}{4}$ (c) $x = -\dfrac{\pi}{6},\ -\dfrac{5\pi}{6}$ (d) $x = -\dfrac{\pi}{6}, \dfrac{\pi}{6}$

(e) $x = -\dfrac{\pi}{2}$ (f) $x = \dfrac{\pi}{3},\ -\dfrac{2\pi}{3}$ (g) $x = -\dfrac{\pi}{4},\ \dfrac{\pi}{4}$ (h) $x = \dfrac{\pi}{6},\ \dfrac{5\pi}{6}$

3.

(a) $\sin\left(x + \dfrac{\pi}{3}\right) = 1 \Rightarrow x + \dfrac{\pi}{3} = \dfrac{\pi}{2} \Rightarrow x = \dfrac{\pi}{6}$

(b) $\tan\left(x - \dfrac{\pi}{2}\right) = 1 \Rightarrow x - \dfrac{\pi}{2} = \dfrac{\pi}{4} \Rightarrow x = \dfrac{3\pi}{4},\ x - \dfrac{\pi}{2} = \dfrac{5\pi}{4} \Rightarrow x = \dfrac{7\pi}{4}$

(c) $\sin x = -\dfrac{1}{2} \Rightarrow x = \dfrac{7\pi}{6},\ \dfrac{11\pi}{6}$ (d) $\cos x = \dfrac{1}{\sqrt{2}} \Rightarrow x = \dfrac{\pi}{4},\ \dfrac{7\pi}{4}$

(e) $\sin^2 x = \dfrac{1}{2} \Rightarrow \sin x = \pm\dfrac{1}{\sqrt{2}} \Rightarrow x = \dfrac{\pi}{4},\ \dfrac{3\pi}{4},\ \dfrac{5\pi}{4},\ \dfrac{7\pi}{4}$

(f) $\cos^2 x = \dfrac{1}{4} \Rightarrow \cos x = \pm\dfrac{1}{2} \Rightarrow x = \dfrac{\pi}{3},\ \dfrac{2\pi}{3},\ \dfrac{4\pi}{3},\ \dfrac{5\pi}{3}$

(g) $\tan^2 x = \dfrac{1}{3} \Rightarrow \tan x = \pm\dfrac{1}{\sqrt{3}} \Rightarrow x = \dfrac{\pi}{6},\ \dfrac{5\pi}{6},\ \dfrac{7\pi}{6},\ \dfrac{11\pi}{6}$

4.

(a) $\cos x = \dfrac{1}{\sqrt{2}} \Rightarrow x = \dfrac{\pi}{4},\ \dfrac{7\pi}{4},\ \sin x = -\dfrac{1}{\sqrt{2}} \Rightarrow x = \dfrac{5\pi}{4},\ \dfrac{7\pi}{4}$. So $\tan x = -1 \Rightarrow x = \dfrac{7\pi}{4}$

(b) $\cos 2x = \dfrac{1}{2} \Rightarrow 2\cos^2 x - 1 = \dfrac{1}{2} \Rightarrow \cos^2 x = \dfrac{3}{4} \Rightarrow \cos x = \pm\dfrac{\sqrt{3}}{2} \Rightarrow x = \pm\dfrac{\pi}{6}$

5.   (a)

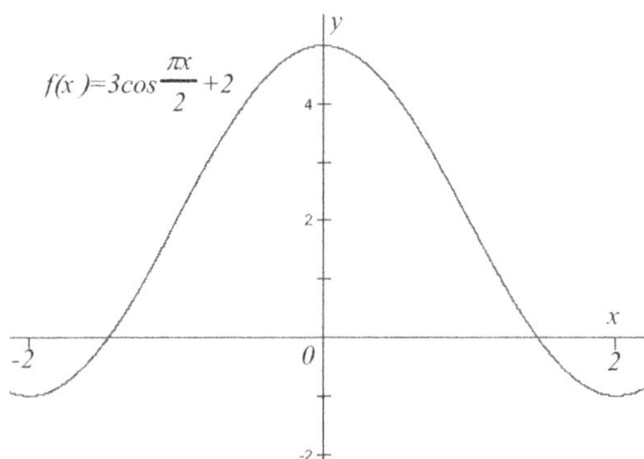

(b)   (i) The period of f is 4   (ii) the amplitude of f is 3

(c) $f(2) = 3\cos\pi + 2 = 3\times(-1) + 2 = -1$

(d) $3\cos\left(\dfrac{\pi x}{2}\right) + 2 = \dfrac{1}{2} \Rightarrow \cos\left(\dfrac{\pi x}{2}\right) = -\dfrac{1}{2} \Rightarrow \dfrac{\pi x}{2} = \pm\dfrac{2\pi}{3} \Rightarrow x = \pm\dfrac{4}{3}$

(e) Using the double angle formula $\cos\theta = 1 - 2\sin^2\left(\dfrac{\theta}{2}\right)$, where $\theta = \dfrac{\pi x}{2}$,

$$f(x) = 3\cos\theta + 2 = 3\left(1 - 2\sin^2\left(\dfrac{\theta}{2}\right)\right) + 2 = 3 - 6\sin^2\left(\dfrac{\pi x}{4}\right) + 2 \Rightarrow f(x) = 5 - 6\sin^2\left(\dfrac{\pi x}{4}\right)$$

Exercise 3.6

1.　(a)　$\sin\theta(\sin\theta+1)=0\Rightarrow\sin\theta=0,\ \sin\theta=-1\Rightarrow\theta=0,\ \pi,\ \dfrac{3\pi}{2}$

　　(b)　$2\cos^2\theta+\cos\theta-1=0\Rightarrow 2\cos^2\theta+2\cos\theta-\cos\theta-1=0$

　　　　$\Rightarrow 2\cos\theta(\cos\theta+1)-1(\cos\theta+1)=0\Rightarrow(\cos\theta+1)(2\cos\theta-1)=0$

　　　　$\Rightarrow\cos\theta=-1\ \text{and}\ \cos\theta=\dfrac{1}{2}\Rightarrow\theta=\dfrac{\pi}{3},\ \pi,\ \dfrac{5\pi}{3}$

　　(c)　$2\sin^2\theta=1\Rightarrow\sin^2\theta=\dfrac{1}{2}\Rightarrow\sin\theta=\pm\dfrac{1}{\sqrt{2}}\Rightarrow\theta=\dfrac{\pi}{4},\ \dfrac{3\pi}{4},\ \dfrac{5\pi}{4},\ \dfrac{7\pi}{4}$

2.　(a)　$6\sin^2x+3\sin x-2\sin x-1=0\Rightarrow 3\sin x(2\sin x+1)-1(2\sin x+1)=0$

　　　　$(2\sin x+1)(\sin x-1)=0\Rightarrow\sin x=-\dfrac{1}{2},\ x=-150°,-30°,$

　　　　$\sin x=\dfrac{1}{3}\Rightarrow,\ x=19.47122...°,\ 180-19.47122...°=19.5°,\ 161°$

　　(b)　$3\cos^2x+3\cos x-\cos x-1=0\Rightarrow 3\cos x(\cos x+1)-1(\cos x+1)=0$

　　　　$\Rightarrow(\cos x+1)(3\cos x-1)=0\Rightarrow\cos x=-1,\Rightarrow x=180°$

　　　　$\cos x=\dfrac{1}{3}\Rightarrow x=\pm70.52877...°=\pm70.5°$

　　(c)　$4\tan^2x-4\tan x+3\tan x-3=0\Rightarrow 4\tan x(\tan x-1)+3(\tan x-1)=0$

　　　　$\Rightarrow(\tan x-1)(4\tan x+3)=0\Rightarrow\tan x=1\Rightarrow x=-135°,\ 45°$

　　　　$\tan x=-\dfrac{3}{4},\Rightarrow x=-36.86989...°=-36.9°,\ x=180°-36.86989...°=143°$

3.　$2\sin^2x=1-\cos x\ \text{but}\ \cos^2x+\sin^2x=1\Rightarrow\sin^2x=1-\cos^2x$

　　$\Rightarrow 2(1-\cos^2x)=1-\cos x\Rightarrow 2-2\cos^2x=1-\cos x\Rightarrow 2\cos^2x-\cos x-1=0$

　　$\Rightarrow 2\cos^2x-2\cos x+\cos x-1=0\Rightarrow 2\cos x(\cos x-1)+1(\cos x-1)=0$

　　$\Rightarrow(\cos x-1)(2\cos x+1)=0\Rightarrow\cos x=1,\ \cos x=-\dfrac{1}{2}\Rightarrow x=-\dfrac{2\pi}{3},\ 0,\ \dfrac{2\pi}{3}$

4.　$\sin 2x=\cos x\Rightarrow 2\cos x\sin x=\cos x\Rightarrow 2\cos x\sin x-\cos x=0$

　　$\Rightarrow\cos x(2\sin x-1)=0\Rightarrow\cos x=0,\ \sin x=\dfrac{1}{2}.\ \sin x=\dfrac{1}{2}\Rightarrow x=\dfrac{\pi}{6},\ \text{as required, but also}$

　　$x=\dfrac{5\pi}{6}.\ \text{When}\ \cos x=0,\ x=\dfrac{\pi}{2}.$

5.　(a)　$\sin x=4\cos x\Rightarrow\tan x=4\Rightarrow x=75.96375...°=76°,\ x=180°+75.96375°=256°$

　　(b)　$3\cos x+2\sin x=0\Rightarrow\tan x=-\dfrac{3}{2}$

　　　　$\Rightarrow x=180°-56.30993...°=124°,\ x=360°-56.30993...°=304°\Rightarrow x=124°,\ 304°$

(c) $\sin 2x = 3\cos^2 x \Rightarrow 2\sin x\cos x = 3\cos^2 x \Rightarrow 2\sin x\cos x - 3\cos^2 x = 0$

$\Rightarrow \cos x(2\sin x - 3\cos x) = 0 \Rightarrow \cos x = 0$, $x = 90°$, $270°$ and $2\sin x - 3\cos x = 0$

$\Rightarrow \tan x = \dfrac{3}{2} \Rightarrow x = 56.30993...°=56.3°$, $180° + 56.30993...° = 236°$,

6. As the graphs intersect in two places, there are two solutions in the required interval. The least value of x which satisfies this equation is found using the graphing calculator. $x = 75.2754°$, correct to four decimal places.

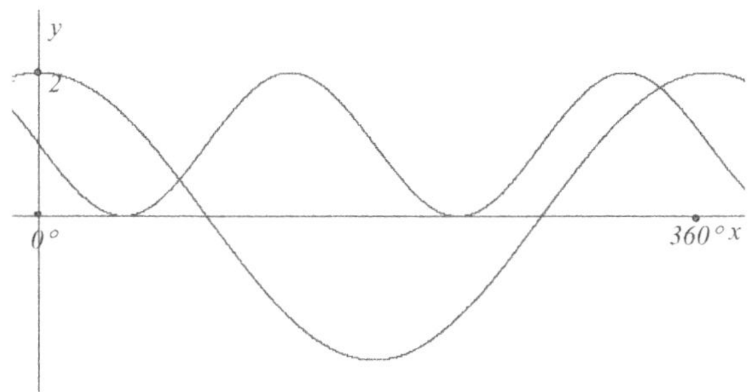

7. (a) $18:00 \Rightarrow t = 0 \Rightarrow \cos\left(\dfrac{2\pi}{365}(d+11)\right) = 0 \Rightarrow \dfrac{2\pi}{365}(d+11) = \dfrac{\pi}{2}, \dfrac{3\pi}{2}$

$\Rightarrow d+11 = \dfrac{365}{4} = 91.25$, $d+11 = \dfrac{3\times 365}{4} = 273.75 \Rightarrow d = 80.25, \ 262.75$

Now $(31 + 28) + 21.25 = 80.25$ and $(31 + 28 + 31 + 30 + 31 + 30 + 31) + 19.75$. Therefore, the dates are March 21 and September 20.

(b) $t = -1 \Rightarrow -2\cos\left(\dfrac{2\pi}{365}(d+11)\right) = -1 \Rightarrow \cos\left(\dfrac{2\pi}{365}(d+11)\right) = \dfrac{1}{2} \Rightarrow \dfrac{2\pi}{365}(d+11)$

$= \dfrac{\pi}{3}, \dfrac{5\pi}{3} \Rightarrow d+11 = 60.83, \ 304.17 \Rightarrow d = 49.83, \ 293.17$. Now $31 + 18.83 = 49.83$ and $(31 + 28 + 31 + 30 + 31 + 30 + 31 + 31 + 30) + 20.17$. Therefore, the dates are February 19 and October 20.

(c) An alternative, and perhaps quicker, approach is to use your calculator. 19:31 is equivalent to 1.51667h after 18:00 so $t = 1.51667$. This method could also have been used in parts (a) and (b)

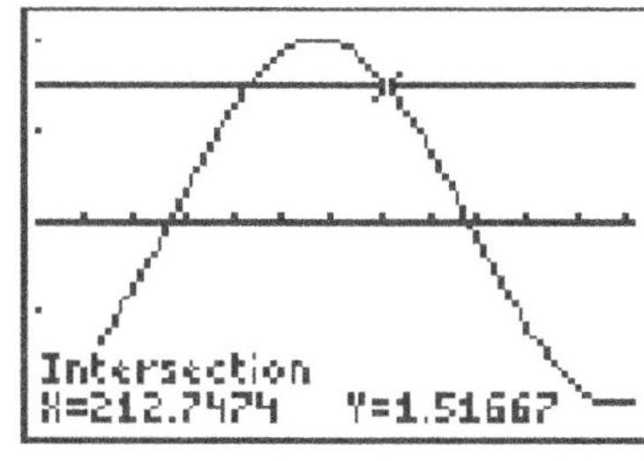

The diagram shows that $d = 212.75$ when $t = 1.51667$. The other value of d for $t = 1.51667$ can be found similarly and is $d = 130.25$. $(31 + 28 + 31 + 30) + 10.25 = 130.25$ and $(31 + 28 + 31 + 30 + 31 + 30 + 31) + 0.75 = 212.75$, and so the dates are May 10 and August 1.

8. (a) $h(0) = 16 \Rightarrow a = 16$, $a+b = 25 \Rightarrow b = 9$

(b) The period of the function is $\dfrac{25}{2}$ so that, $\dfrac{2\pi}{k} = \dfrac{25}{2} \Rightarrow k = \dfrac{4\pi}{25}$.

(c) $h(t) = 16 + 9\sin\dfrac{4\pi t}{25}$. $h(6) = 17.1$, $h(14) = 22.2$ so the depth of the water in the harbor at

06:00 is 17.1m and at 14:00 it is 22.2m.

(d) The depth of the water in the harbor is below 10m between 07.702 and 11.048, which is a time interval of 3.346h. However, there is another equal time interval later in the day so that the total time during which the depth is below 10m is 6.692h in a total time of 25 hours. Therefore, the proportion of time during which the depth of water is less than 10m is

$$\frac{6.692}{25} = 0.26768 = 26.8\%.$$

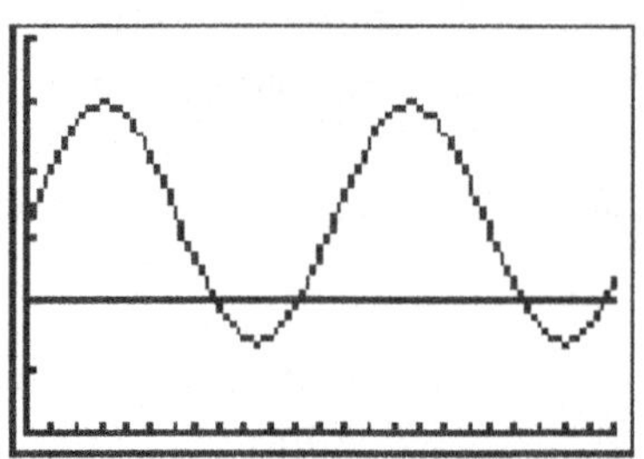

Exercise 3.7

1. (a) See diagram on right.
    (b) Curve cuts x-axis in two places so there are two solutions.
    (c) $x = 1.422083... = 1.4221$

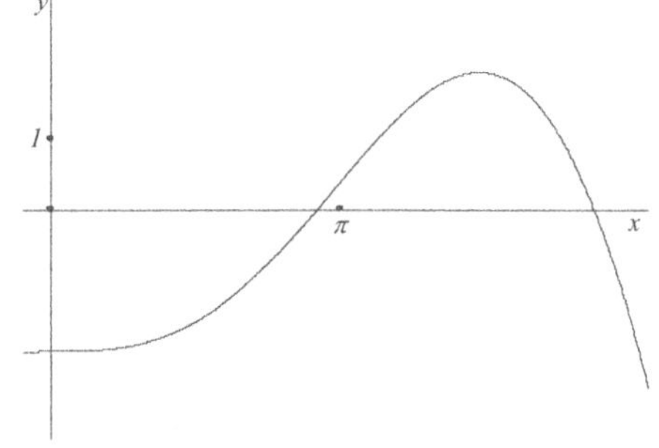

2. (a)

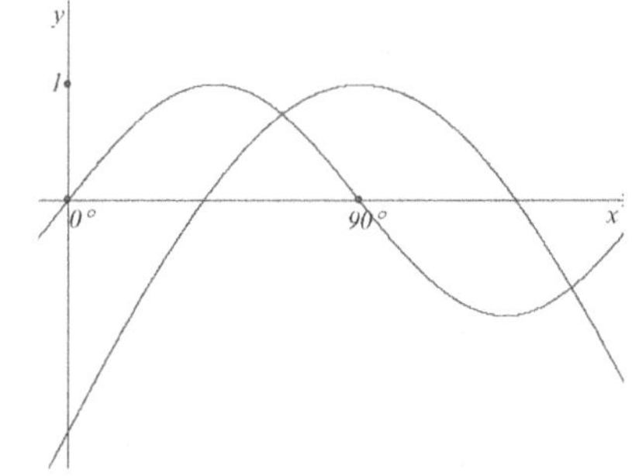

(b) $x = 66.05617...° = 66.1°,\ 155.4828...° = 155°$

3. (a)

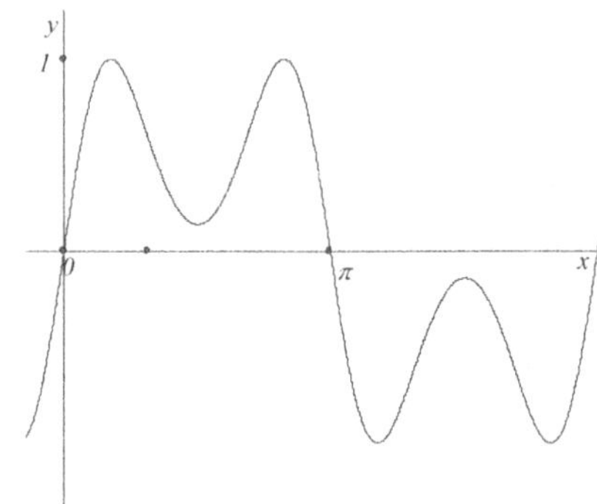

(b) (i) 2 (ii) 1 (iii) 4 (iv) 7

4. (a)

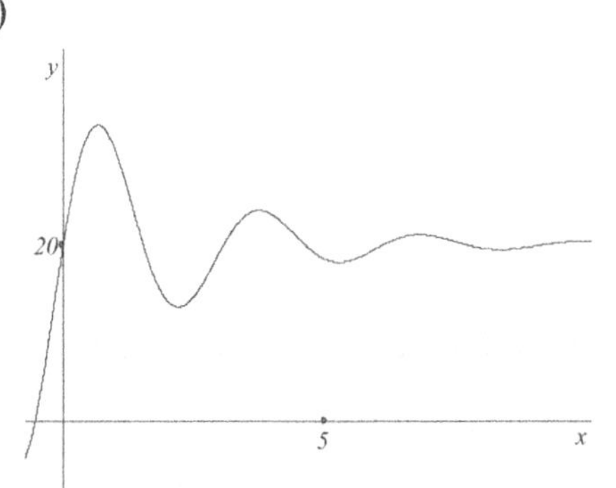

(b) Minimum clearance is 12.8cm, occurring when $t = 2.257497... = 2.26$.

(c) The graph shows an enlarged portion of the function $h(t)$ and $y = 18$ for the given window.

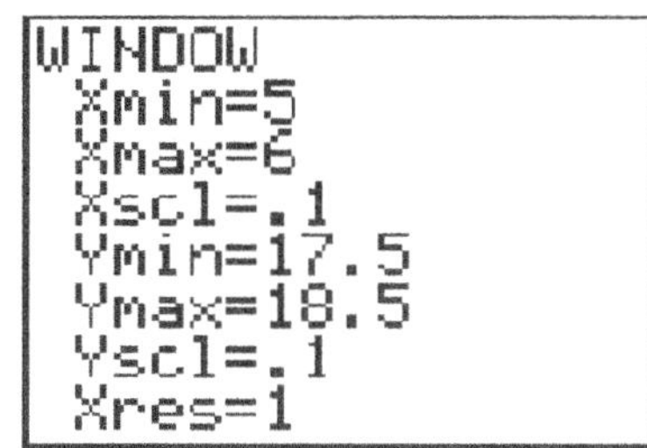

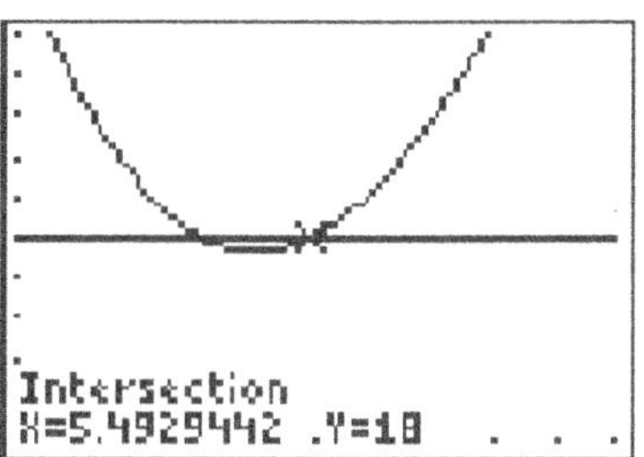

This shows that after $5.4929442... = 5.49$ seconds the vibration is less than $\pm 2\,$cm.

Exercise 3.8

1. (a) $\hat{A} = 29.9°$, $\hat{B} = 56.3°$, $\hat{C} = 93.8°$ (b) $c = 13.9$, $\hat{A} = 78.5°$, $\hat{B} = 48.5°$

    (c) $a = 21.6$, $c = 22.3$, $\hat{C} = 79°$ (d) $\hat{A} = 35.7°$, $\hat{B} = 50.1°$, $\hat{C} = 94.1°$

    (e) $a = 6.00$, $b = 3.96$, $\hat{C} = 118°$ (f) $a = 2.36$, $\hat{B} = 47.6°$, $\hat{C} = 17.4°$

2. (a)

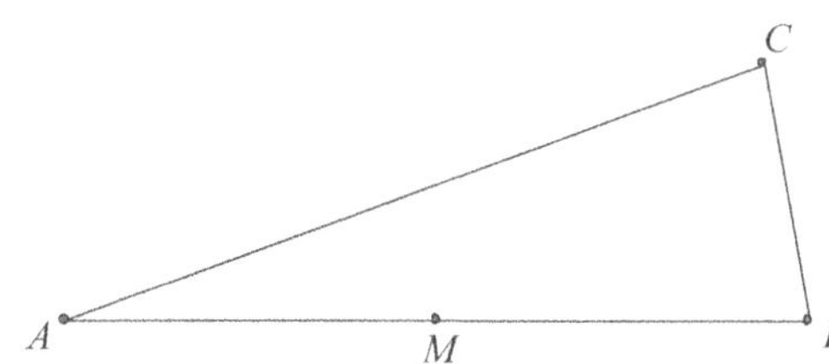

(b) $BC = 2BN$, $N\hat{A}B = 10°$. Using sine rule $\sin 10° = \dfrac{NB}{15} \Rightarrow NB = 15\sin 10°$

    $\Rightarrow BC = 30\sin 10° = 5.21$cm

(c) Consider ΔAMC, $AM = 7.5$cm, $AC = 15$cm, $C\hat{A}M = 20°$

    $\Rightarrow CM^2 = 7.5^2 + 15^2 - 2 \times 7.5 \times 15 \cos 20° \Rightarrow CM = 8.355786... = 8.36$cm

3. $AB = AN + NB$. $\tan A\hat{C}N = \dfrac{AN}{5} \Rightarrow AN = 5\tan A\hat{C}N$, $\tan B\hat{C}N = \dfrac{NB}{5} \Rightarrow AN = 5\tan B\hat{C}N$,

    but $A\hat{C}N = 60°$, $N\hat{C}B = 40° \Rightarrow AB = 5\tan 60° + 5\tan 40° = 12.85575... = 12.9\,$m.

4. (a)

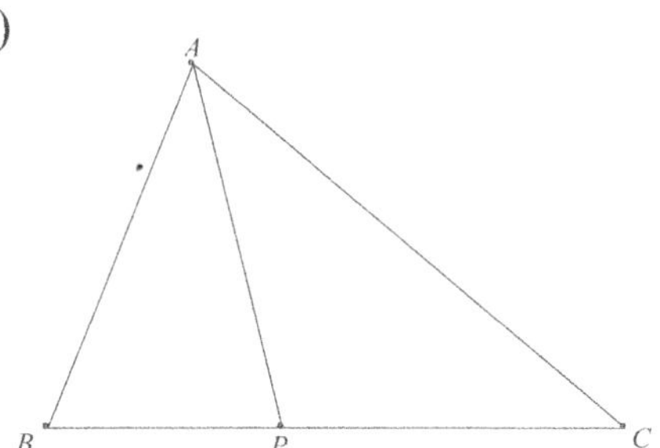

(b) Using cosine rule, $AC^2 = 7.6^2 + 11.2^2 - 2 \times 7.6 \times 11.2 \cos 68° \Rightarrow AC = 10.92826... = 10.9\,$cm.

(c) $\cos A = \left(7.6^2 + 119.427 - 11.2^2\right) / \left(2 \times 7.6 \times 10.928\right) \Rightarrow \hat{A} = 71.84899...°$

Therefore $B\hat{A}P = 35.92449...° \Rightarrow A\hat{P}B = 76.07551...° \Rightarrow \dfrac{AP}{\sin 68°} = \dfrac{7.6}{\sin A\hat{P}B}$

$$\Rightarrow AP = \dfrac{7.6\sin 68°}{\sin A\hat{P}B} = 7.259939... \Rightarrow AP = 7.26\text{cm}.$$

(d) $A\hat{C}P = 180° - \left(\hat{A} + \hat{B}\right) = 40.15101...°.$

In $\triangle APC$ $\dfrac{PC}{\sin B\hat{A}P} = \dfrac{AP}{\sin A\hat{C}P} \Rightarrow PC = 6.605942... = 6.61\text{cm}$

5. (a)

(b) Using cosine rule, $\cos \hat{A} = \dfrac{\left(30^2 + 33^2 - 38^2\right)}{2 \times 30 \times 33} \Rightarrow \hat{A} = 74.02293...°$ but $\cos A = \dfrac{AQ}{30}$

$\Rightarrow AQ = 30\cos \hat{A} = 8.257575...$ and $\cos A = \dfrac{AP}{33} \Rightarrow AP = 33\cos \hat{A} = 9.083333...$

Using cosine rule again,
$PQ^2 = AP^2 + AQ^2 - 2 \times AP \times AQ\cos \hat{A} \Rightarrow PQ = 10.45959... = 10.5$

6. Proof of the cosine rule:

$$a^2 = BN^2 + h^2, \quad b^2 = \left(c + BN\right)^2 + h^2 \Rightarrow b^2 - a^2 = \left(\left(c + BN\right)^2 + h^2\right) - \left(BN^2 + h^2\right)$$

$\Rightarrow b^2 - a^2 = c^2 + 2cBN$ but $\cos\left(180° - B\right) = \dfrac{BN}{a} \Rightarrow BN = a\cos\left(180° - B\right)$

$\Rightarrow b^2 - a^2 = c^2 + 2ca\cos\left(180° - B\right) \Rightarrow b^2 = c^2 + a^2 + 2ca\cos\left(180° - B\right)$ but as

$\cos\left(180° - B\right) = -\cos B$, so $b^2 = c^2 + a^2 - 2ca\cos B$.

Proof of the sine rule:

$\sin A = \dfrac{h}{b}, \quad \sin\left(180° - B\right) = \dfrac{h}{a} \Rightarrow h = b\sin A$ and $h = a\sin\left(180° - B\right)$

$\Rightarrow a\sin\left(180° - B\right) = b\sin A$ but $\sin\left(180° - B\right) = \sin B \Rightarrow a\sin B = b\sin A \Rightarrow \dfrac{a}{\sin A} = \dfrac{b}{\sin B}.$

Exercise 3.9

1.   (a) 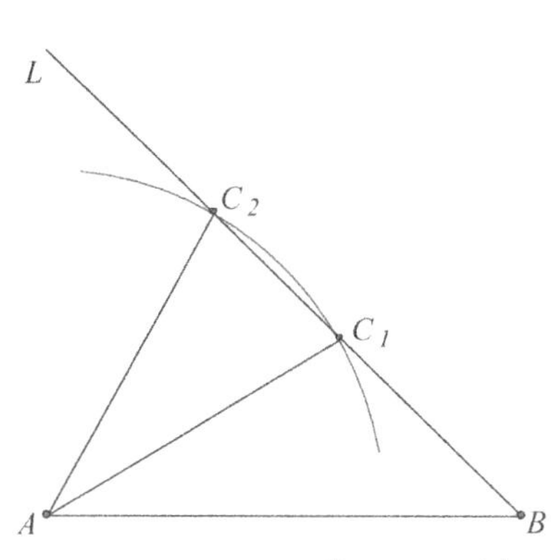   (b) $BC_1 = 5.5\text{cm}, BC_2 = 10.3\text{cm}$.

  (c) Using sine rule, $\dfrac{8}{\sin 44°} = \dfrac{11}{\sin C}$

$$\Rightarrow \sin C = \dfrac{11}{8}\sin 44° \Rightarrow C = 72.77616...°, \ 107.2238...°$$

$$\Rightarrow A = 63.22384...°, \ 28.77616...°$$

$$\Rightarrow a = \dfrac{8\sin C_2 \hat{A}B}{\sin 44°} = 10.28158..., \text{ or } a = \dfrac{8\sin C_1 \hat{A}B}{\sin 44°} = 5.543893...$$

Therefore, $BC_1 = 5.54\text{cm}, \ BC_2 = 10.3\text{cm}$.

2.   (a)

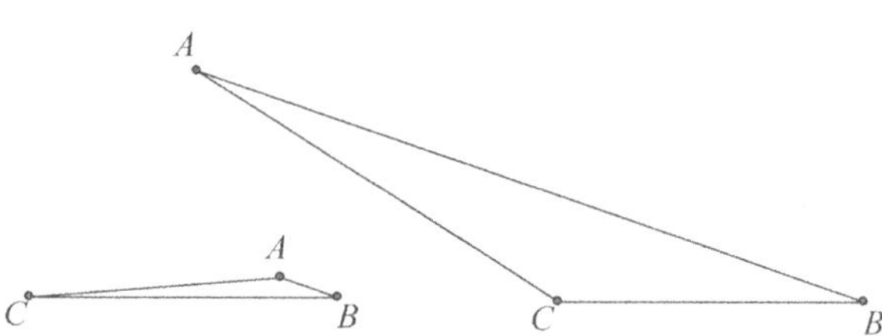

  (b) $27^2 = c^2 + 33^2 - 2\times 33\cos 19°\, c \Rightarrow c^2 - 62.404c + 360 = 0 \Rightarrow c = 6.43, \ 56.0$

3.   Using sine rule: $\dfrac{15}{\sin 11°} = \dfrac{51}{\sin R} \Rightarrow \sin R = \dfrac{51}{15}\sin 11° \Rightarrow R = 40.44746...°, \ 139.5525...°$.

Therefore, the possible values of $\hat{R}$ are $40.4°$ and $140°$.

4.   The figure shows the two possible positions of vertex B.

$$\dfrac{13.9}{\sin 33°} = \dfrac{16.6}{\sin B} \Rightarrow \sin B = \dfrac{16.6\sin 33°}{13.9}$$

$$\Rightarrow B = 40.57419...°, \ 139.4258...°$$

For the large triangle:

$$\hat{C} = 180° - \left(33° + 40.57419...°\right) = 106.42581...°.$$

$$c^2 = 16.6^2 + 13.9^2 - 2\times 16.6 \times 13.9\cos \hat{C}$$

$\Rightarrow c = 24.47987...$. Therefore the large triangle ABC has

$a = 13.9, \ b = 16.6, \ c = 24.5, \ \hat{A} = 33°, \ \hat{B} = 40.6°, \ \hat{C} = 106°$.

For the small triangle: $\hat{C} = 180^\circ - \left(33^\circ + 139.4258...^\circ\right) = 7.57419...^\circ$.

Then $\quad c^2 = 16.6^2 + 13.9^2 - 2 \times 16.6 \times 13.9 \cos \hat{C} \Rightarrow c = 3.363986...$.

So, the small triangle ABC has $a = 13.9$, $b = 16.6$, $c = 3.36$, $\hat{A} = 33^\circ$, $\hat{B} = 139^\circ$, $\hat{C} = 7.57^\circ$.

Exercise 3.10

1.   (a) 216   (b) 1.48   (c) 4850

2. $\cos \hat{A} = \dfrac{4.03^2 + 3.86^2 - 2.18^2}{2 \times 4.03 \times 3.86} \Rightarrow \hat{A} = 31.98621...^\circ$

       Area $= \dfrac{1}{2} \times 3.86 \times 4.03 \times \sin 31.986^\circ = 4.120071... = 4.12\text{cm}^2$

3. $\dfrac{15}{\sin \hat{C}} = \dfrac{9}{\sin 30^\circ} \Rightarrow \sin \hat{C} = \dfrac{15 \sin 30^\circ}{9} \Rightarrow \hat{C} = 56.44269...^\circ,\ 123.5573...^\circ$.

       Therefore $\hat{B} = 93.55731...^\circ,\ 26.44269...^\circ$.

       Therefore, the area of the smaller triangle is $\dfrac{1}{2} \times 15 \times 9 \times \sin \hat{B} = 30.05792...$ and the area of the

       larger triangle is $\dfrac{1}{2} \times 15 \times 9 \times \sin \hat{B} = 67.36994... = 67.4$.

4. $\hat{R} = 36^\circ$, using sine rule $\dfrac{8}{\sin 36^\circ} = \dfrac{p}{\sin 75^\circ} \Rightarrow p = 13.14664...$

       Therefore the area is $\dfrac{1}{2} \times 8 \times p \times \sin 69^\circ = 49.09378... = 49.1\text{m}^2$.

5. $10\sqrt{3} = \dfrac{1}{2} \times 5 \times 8 \times \sin A \Rightarrow \sin A = \dfrac{\sqrt{3}}{2} \Rightarrow A = 60^\circ,\ 120^\circ$. Therefore there are two possible

       triangles which have the given data. For the triangle with $\hat{A} = 60^\circ$, using cosine rule,
       $BC^2 = 5^2 + 8^2 - 2 \times 5 \times 8 \times \cos 60^\circ \Rightarrow BC = 7$. For the triangle with $\hat{A} = 120^\circ$,
       $BC^2 = 5^2 + 8^2 - 2 \times 5 \times 8 \times \cos 120^\circ \Rightarrow BC = \sqrt{129}$. Therefore, the possible values of BC are 7,
       $\sqrt{129}$.

6.   (a) Area $\Delta A_1 B_1 C_1 = \dfrac{1}{2} \times 1 \times 1 \times \sin 60^\circ = \dfrac{\sqrt{3}}{4}$

       (b) Area $\Delta A_2 B_2 C_2$ is equilateral and $A_2 B_2 = B_2 C_2 = C_2 A_2 = \dfrac{1}{2}$ so area

            $\Delta A_2 B_2 C_2 = \dfrac{1}{2} \times \left(\dfrac{1}{2} \times \dfrac{1}{2} \right) \sin 60^\circ = \dfrac{\sqrt{3}}{16}$.

       (c) Area $\Delta A_3 B_3 C_3 = \dfrac{1}{2} \times \left(\dfrac{1}{4} \times \dfrac{1}{4} \right) \sin 60^\circ = \dfrac{\sqrt{3}}{64}$

(d) The areas form a geometric sequence in which $a = \dfrac{\sqrt{3}}{4}$, $r = \dfrac{1}{4}$. Therefore the area of

$$\Delta A_{12}B_{12}C_{12} = u_{12} = ar^{11} = \frac{\sqrt{3}}{4}\left(\frac{1}{4}\right)^{11} = \frac{\sqrt{3}}{4^{12}} = \frac{\sqrt{3}}{2^{24}} \Rightarrow n = 24 \,.$$

7. $XY^2 = 16^2 + 20^2 - 2 \times 16 \times 20 \cos 125^\circ \Rightarrow XY = 31.98576...$
$YZ^2 = 20^2 + 9^2 = 481 \Rightarrow YZ = 21.93171...,\quad ZX^2 = 16^2 + 9^2 = 337 \Rightarrow ZX = 18.35755...$
Therefore, $XY = 32.0$cm, $YZ = 21.9$cm, $ZX = 18.4$cm .

The area of $\Delta XYZ = \dfrac{1}{2} \times ZY \times ZX \times \cos X\hat{Z}Y$.

$\cos X\hat{Z}Y = \dfrac{481 + 337 - XY^2}{2 \times YZ \times ZX} \Rightarrow X\hat{Z}Y = 104.7556...^\circ$. Therefore area of

$\Delta XYZ = \dfrac{1}{2} \times YZ \times ZX \times \sin X\hat{Z}Y = 194.6673... = 195\text{cm}^2$, correct to three significant figures.

Solutions to Unit 3 Review Exercise

Part 1

1. (a) $\sin 130^\circ = \sin\left(180^\circ - 50^\circ\right) = \sin 50^\circ = p$

(b) $\cos 160^\circ = \cos\left(180^\circ - 20^\circ\right) = -\cos 20^\circ = -q$

(c) $\cos 40^\circ = \cos\left(2 \times 20^\circ\right) = 2\cos^2 20^\circ - 1 = 2q^2 - 1$

(d) $p = \sin 50^\circ = \cos\left(90^\circ - 50^\circ\right) = \cos 40^\circ = 2q^2 - 1 = p$. So $p = 2q^2 - 1$.

2. (a) $\cos 210^\circ = \cos\left(180^\circ + 30^\circ\right) = -\cos 30^\circ = -\dfrac{\sqrt{3}}{2}$

(b) $\cos 300^\circ = \cos\left(360^\circ - 60^\circ\right) = \cos 60^\circ = \dfrac{1}{2}$

(c) $\cos 135^\circ = \cos\left(180^\circ - 45^\circ\right) = -\cos 45^\circ = -\dfrac{1}{\sqrt{2}}$

(d) $\sin 150^\circ = \sin\left(180^\circ - 30^\circ\right) = \sin 30^\circ = \dfrac{1}{2}$

3. (a) $\sin^2 10^\circ + \cos^2 10^\circ = 1 \Rightarrow \sin^2 10^\circ = 1 - \cos^2 10^\circ = 1 - a^2 \Rightarrow \sin 10^\circ = \pm\sqrt{1 - a^2}$, but
$\sin 10^\circ > 0$ so $\sin 10^\circ = \sqrt{1 - a^2}$

(b) $\sin 170^\circ = \sin\left(180^\circ - 10^\circ\right) = \sin 10^\circ = \sqrt{1 - a^2}$

(c) $\sin 20^\circ = \sin 2 \times 10^\circ = 2\cos 10^\circ \sin 10^\circ = 2a\sqrt{1 - a^2}$
but since $\sin 20^\circ = b$, $b = 2a\sqrt{1 - a^2} \Rightarrow b^2 = 4a^2\left(1 - a^2\right) = 4a^2 - 4a^4$
so that $4a^4 = 4a^2 - b^2$.

4.　(a) $\cos^2 65° + \sin^2 65° = 1 \Rightarrow \cos^2 65° = 1 - \sin^2 65° = 1 - u^2 \Rightarrow \cos 65° = \pm\sqrt{1-u^2}$, but

$\cos 65° > 0$ so $\cos 65° = \sqrt{1-u^2}$.

(b) $v = \cos 130° = 1 - 2\sin^2 65° = 1 - 2u^2 \Rightarrow v = 1 - 2u^2$

(c) $\sin 130° = 2\sin 65° \cos 65° = 2u\sqrt{1-u^2}$ and therefore $\tan 130° = \dfrac{\sin 130°}{\cos 130°} = \dfrac{2u\sqrt{1-u^2}}{1-2u^2}$.

5.　(a) Period of h is $\dfrac{2\pi}{3}$

(b)

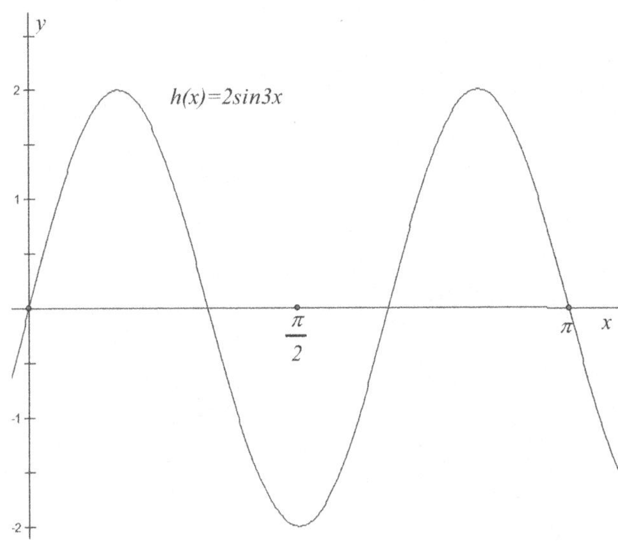

6.　(a) $AB = \sqrt{5^2 - 1^2} = \sqrt{24} = \sqrt{4 \times 6} = \sqrt{4}\sqrt{6} = 2\sqrt{6}$

(b) $\cos\theta = \dfrac{AB}{AC} = \dfrac{2\sqrt{6}}{5}$

(c) $\sin 2\theta = 2\cos\theta\sin\theta = 2 \times \dfrac{2\sqrt{6}}{5} \times \dfrac{1}{5} = \dfrac{4\sqrt{6}}{25}$.

7.　(a) $8x^2 - 6x + 1 = 0 \Rightarrow (4x-1)(2x-1) = 0 \Rightarrow x = \dfrac{1}{2}, \ x = \dfrac{1}{4}$.

(b) $8\cos^4\theta - 6\cos^2\theta + 1 = 0 \Rightarrow \cos^2\theta = \dfrac{1}{4}, \ \cos^2\theta = \dfrac{1}{2}$. So, $\cos\theta = \pm\dfrac{1}{2}, \ \cos\theta = \pm\dfrac{1}{\sqrt{2}}$.

and therefore, in the interval, $0 \le \theta < \pi$, $\theta = \dfrac{\pi}{4}, \ \dfrac{\pi}{3}, \ \dfrac{2\pi}{3}, \ \dfrac{3\pi}{4}$.

8.　(a) (i) Area of sector OAB $= \dfrac{1}{2} \times 6^2 \times \theta = 18\theta$

(ii) Arc length $= 6\theta$

(b) $18\theta = 10 \Rightarrow \theta = \dfrac{5}{9}$ so $\theta = \dfrac{5}{9}$ radians or $\theta = \dfrac{5}{9} \times \dfrac{180°}{\pi} = \left(\dfrac{100}{\pi}\right)°$.

9.   (a) Area of sector $= \dfrac{1}{2}r^2\theta$ and the area of $\triangle OBA = \dfrac{1}{2}r^2\sin\theta$ so the area of the shaded

region $= \dfrac{1}{2}r^2\theta - \dfrac{1}{2}r^2\sin\theta = \dfrac{1}{2}r^2(\theta - \sin\theta)$

(b) The diameter of $C_1 = 2$ so $2\left(r + r\cos\dfrac{1}{2}\theta\right) = 2 \Rightarrow r\left(1 + \cos\dfrac{1}{2}\theta\right) = 1$ but $\dfrac{1}{2}\theta = \dfrac{\pi}{3}$ and

$\cos\dfrac{\pi}{3} = \dfrac{1}{2} \Rightarrow r\left(1 + \dfrac{1}{2}\right) = 1 \Rightarrow \dfrac{3}{2}r = 1 \Rightarrow r = \dfrac{2}{3}$

(c) The area of shaded region $= \dfrac{1}{2}\left(\dfrac{2}{3}\right)^2\left(\dfrac{2\pi}{3} - \sin\dfrac{2\pi}{3}\right) = \dfrac{2}{9}\left(\dfrac{2\pi}{3} - \dfrac{\sqrt{3}}{2}\right)$ and the area of the

overlap between C_1 and $C_2 = 2 \times \dfrac{2}{9}\left(\dfrac{2\pi}{3} - \dfrac{\sqrt{3}}{2}\right) = \dfrac{4}{9}\left(\dfrac{2\pi}{3} - \dfrac{\sqrt{3}}{2}\right)$.

10.   (a) $3 = a + b\sin 0 \Rightarrow a = 3$ (b) $b = 3 - 1 = 2$ (c) $\dfrac{2\pi}{k} = \dfrac{\pi}{3} \Rightarrow k = 6$

(d)

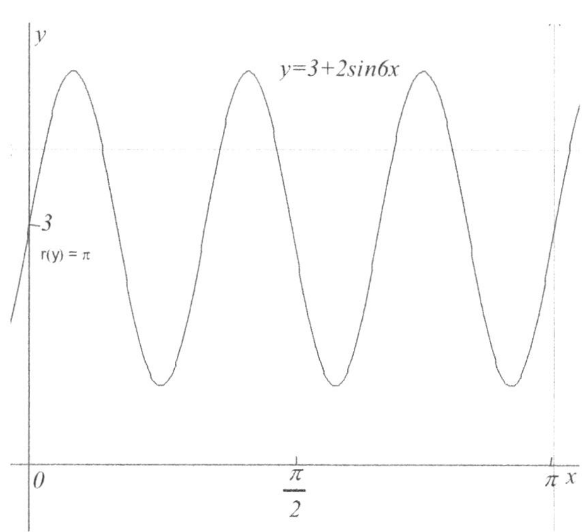

(e) The graph in part (d) also shows the line representing $y = 4$ and indicates that it intersects
the graph of $y = 3 + 2\sin 6x$ in 6 places in the interval $0 \le x < \pi$ so that there are 6
solutions.

11.   (a) Area of sector is $\dfrac{1}{2}r^2\theta$

(b) Perimeter of sector is $2r + r\theta$

(c) (i) $\dfrac{1}{2}r^2\theta = r(2 + \theta) \Rightarrow r\theta = 4 + 2\theta \Rightarrow \theta(r - 2) = 4 \Rightarrow \theta = \dfrac{4}{r - 2}$.

(ii) $16 = r\theta + 2r \Rightarrow 16 = \dfrac{4r}{r - 2} + 2r \Rightarrow 16(r - 2) = 4r - 2r(r - 2)$.

Therefore, $16r - 32 = 4r + 2r^2 - 4r \Rightarrow r^2 - 8r + 16 = 0 \Rightarrow (r - 4)^2 = 0 \Rightarrow r = 4$

Part 2

12.   (a) $PS^2 = OS^2 - OP^2 = 5^2 - 4^2 = 9 \Rightarrow PS = 3$ cm

      (b) $\cos P\hat{O}S = \dfrac{4}{5} \Rightarrow P\hat{O}S = 0.6435011\ldots = 0.644$

      (c) The area of $\triangle OSP = \dfrac{1}{2} \times 3 \times 4 = 6$ cm$^2 \Rightarrow$ area $OPSQ = 2 \times 6 = 12$ cm^2

      (d) The area of the sector $OPQ = \dfrac{1}{2} \times 4^2 \times P\hat{O}S \times 2 = 10.29601\ldots = 10.3$ cm^2

      (e) The area of the shaded region is area $OPSQ -$ area of sector OPQ

         $= 12 - 10.29601\ldots = 1.70399\ldots = 1.70$ cm^2

13.   (a)

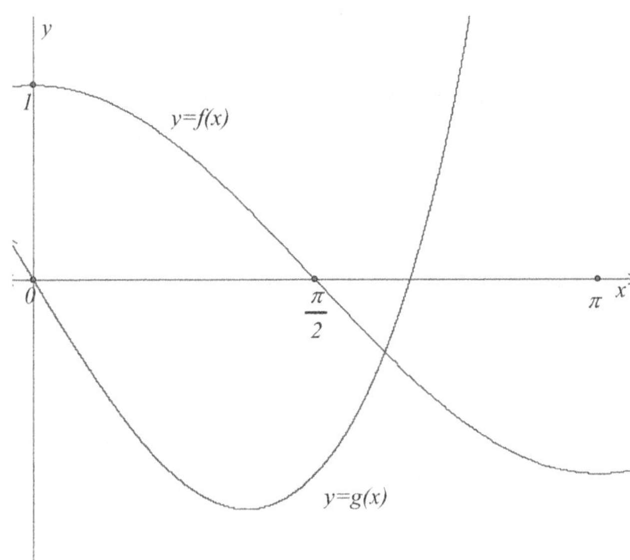

      (b) $x = 1.953913\ldots = 1.95$

14.   (a) $A\hat{B}D = 180° - 35° = 145°$, so $A\hat{D}B = 180° - \left(18° + 145°\right) = 17°$

      (b) $\dfrac{BD}{\sin 18°} = \dfrac{10}{\sin 17°} \Rightarrow BD = \dfrac{10 \sin 18°}{\sin 17°} = 10.56931\ldots = 10.6$

      (c) Area of $\triangle BCD = \dfrac{1}{2} BC \times BD \sin 35° = 30.31153\ldots = 30.3$

15.   (a) $5c^2 - 2c - 3 = 5c^2 - 5c + 3c - 3 = 5c(c-1) + 3(c-1) = (c-1)(5c+3)$

      (b) $5\cos^2 \theta - 2\cos \theta - 3 = (\cos \theta - 1)(5\cos \theta + 3) = 0 \Rightarrow \cos \theta = 1,\ \cos \theta = -\dfrac{3}{5}$.

        Therefore, $\theta = 0°$, $\theta = 126.8698\ldots° = 127°$, $\theta = 306.8698\ldots° = 307°$.

16. (a) $6^2 = 8^2 + x^2 - 2 \times 8 \times x \cos 30°$. Therefore, $36 = 64 + x^2 - 2 \times 8 \times \dfrac{\sqrt{3}}{2} x$ and

$0 = 64 + x^2 - 8\sqrt{3}x - 36 \Rightarrow x^2 - 8\sqrt{3}x + 28 = 0$.

(b) $x = \dfrac{8\sqrt{3} \pm \sqrt{192 - 4 \times 1 \times 28}}{2} = \dfrac{8\sqrt{3} \pm \sqrt{80}}{2} = \dfrac{8\sqrt{3} \pm 4\sqrt{5}}{2} = 4\sqrt{3} \pm 2\sqrt{5}$

$= 11.40033..., \ 2.456067...$ but since $B\hat{A}D$ is obtuse we know that $x = 11.4$.

If $x = 2.46$ then $B\hat{A}D$ would be acute.

(c) Using cosine rule, $\cos D\hat{A}B = \dfrac{6^2 + 8^2 - x^2}{96} = \dfrac{100 - 11.40033^2}{96} \Rightarrow DAB = 108.1895...°$

So $D\hat{A}B = 108°$.

(d) $A\hat{D}B = 180° - \left(30° + 108.1895...°\right) = 41.8105...°$ and since $B\hat{D}C = 138.1895...°$ and

$D\hat{C}B = 180° - \left(138.1895...° + 15°\right) = 26.8105...°$, then $\dfrac{DC}{\sin 15°} = \dfrac{11.4}{\sin D\hat{C}B}$

$\Rightarrow DC = \dfrac{11.4 \sin 15°}{\sin D\hat{C}B} = 6.541614... = 6.54$

(e) Area of $\Delta DBC = \dfrac{1}{2} 11.4 \times 6.54 \sin B\hat{D}C = 24.8$

17. (a) $f(\theta) = \sin\theta + 2\cos 2\theta + 1$ and, since $\cos 2\theta = 1 - 2\sin^2\theta$,

$f(\theta) = \sin\theta + 2\left(1 - 2\sin^2\theta\right) + 1 = \sin\theta + 2 - 4\sin^2\theta + 1 = 3 + \sin\theta - 4\sin^2\theta$.

(b) $g(x) = 3 + x - 4x^2 = (1 - x)(3 + 4x) = 0 \Rightarrow x = 1, \ x = -\dfrac{3}{4}$

(c) Let $x = \sin\theta$, then, from part (b) $\sin\theta = 1$ and $\sin\theta = -\dfrac{3}{4}$ so that $\theta = 90°$ and

$\theta = -48.590...°$, so, correct to the nearest degree, $\theta = -49°$

18. (a) $a = 4$

(b) The period is 12 hours so that $\dfrac{2\pi}{k\pi} = 12 \Rightarrow k = \dfrac{1}{6}$

(c) $5.5 = 4 + b\sin\left(\dfrac{\pi 2}{6}\right) \Rightarrow b\sin\left(\dfrac{\pi}{3}\right) = 1.5 \Rightarrow b\dfrac{\sqrt{3}}{2} = \dfrac{3}{2} \Rightarrow b = \sqrt{3}$

(d) $17{:}00$ is when $t = 5$ so that $h = 4 + \sqrt{3}\sin\left(\dfrac{5\pi}{6}\right) = 4.866025... = 4.87$ and the tide has height

of 4.87 m.

(e) High tide occurs when $\dfrac{\pi t}{6} = \dfrac{\pi}{2} \Rightarrow t = 3$ or $\dfrac{\pi t}{6} = \dfrac{5\pi}{2} \Rightarrow t = 15$ so that high tide occurs

at 03:00 and 15:00.

19.   (a)

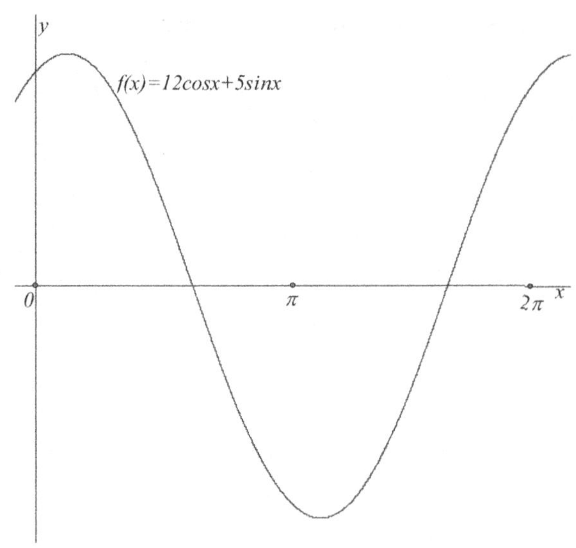

(b) Max value of f is 13 and occurs when $x = 0.3947914...$, and min point is -13 and occurs when $x = 3.536383...$. So max point is $(0.395, 13)$ and min point is $(3.54, -13)$

(c) (i) $a = 13$ (ii) $b = \dfrac{\pi}{2} - 0.3947914... = 1.176004... = 1.18$

(d) $13\sin(x+b) = 0 \Rightarrow x + b = 0, \pi, 2\pi$

    $\Rightarrow x = -1.176004..., \ x = 1.965588... = 1.97, \ x = 5.107181... = 5.11$

    But since $x = -1.18$ is outside the required interval the solutions are $x = 1.97, \ 5.11$

(e) $12\cos x + 5\sin x = 0 \Rightarrow \tan x = -\dfrac{12}{5} \Rightarrow x = 1.965587... = 1.97, \ x = 5.107180... = 5.10$ which

    are equal to the solutions found in (d). We should expect this since

    $12\cos x + 5\sin x = 13\sin(x+b)$ as was shown in part (c).

20.   (a) $f(-x) = 1 - 4\big(\sin(-x)\big)^2 = 1 - 4(-\sin x)^2 = 1 - 4\sin^2 x = f(x)$. So if $f(-x) = f(x)$ the

    graph is symmetrical about the y-axis.

  (b) $f(x) = 1 - 4\sin^2 x$ but $\cos 2x = 1 - 2\sin^2 x \Rightarrow 2\sin^2 x = 1 - \cos 2x$

    $\Rightarrow f(x) = 1 - 4\sin^2 x = 1 - 2(1 - \cos 2x) = -1 + 2\cos 2x \Rightarrow a = -1, \ b = 2$

  (c)

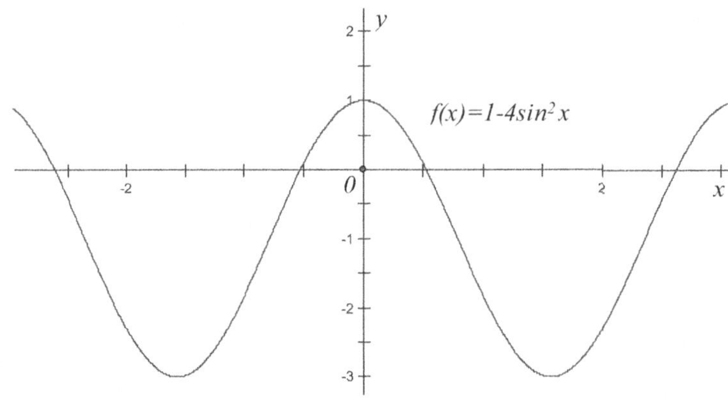

344

(d) 4 solutions

(e) The solutions are $\dfrac{\pi}{6}$, $-\dfrac{\pi}{6}$, $\dfrac{5\pi}{6}$, $-\dfrac{5\pi}{6}$

Solutions to Unit 4 Exercises

Exercise 4.1

1.   (a)   (b)   (c)

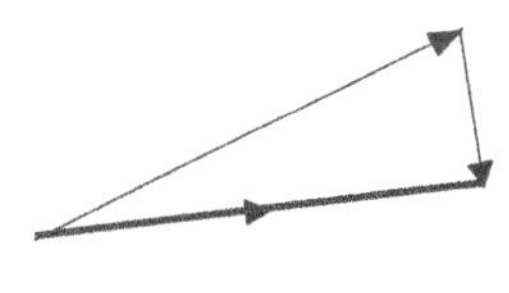 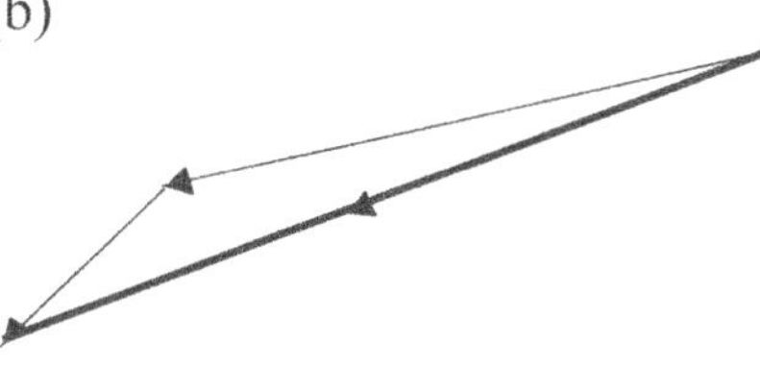 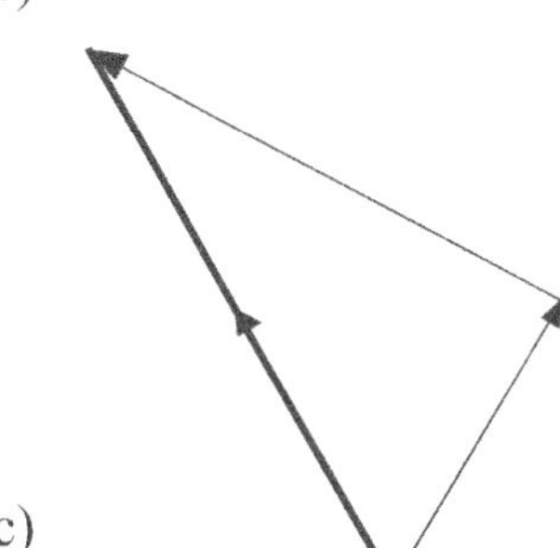

2.   (a)   (b)   (c)

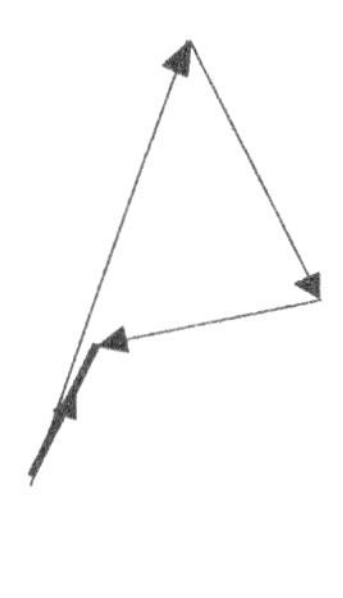  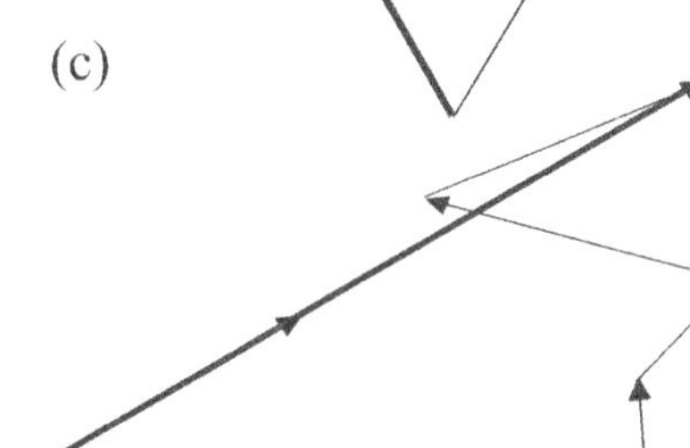

3. $|\vec{v}| = \sqrt{36^2 + 10^2} = 37.36308...$, $\tan\theta = 5/18 \Rightarrow \theta = 15.52411...$

So the velocity of the harpoon is $37.4\,\text{ms}^{-1}$, at an angle west of north of $15.5°$.

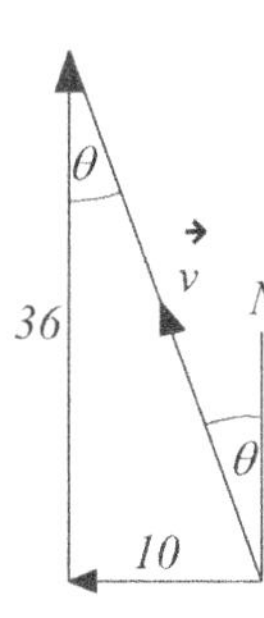

4.   (a)   (b) $P\hat{R}Q = 18° + \left(180° - 100°\right) = 98°$.

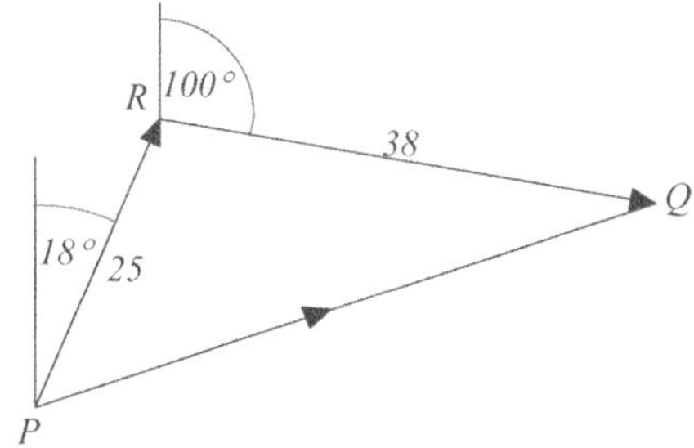

Using cosine rule
$$PQ^2 = 25^2 + 38^2 - 2 \times 25 \times 38 \times \cos 98°$$
$$\Rightarrow PQ = 48.30557....$$

Using sine rule
$$\frac{38}{\sin R\hat{P}Q} = \frac{48.30557...}{\sin 98°} \Rightarrow \sin R\hat{P}Q = \frac{38\sin 98°}{PQ}$$
and as $R\hat{P}Q < 90°$, $R\hat{P}Q = 51.16938... = 51.17°$.
Therefore $\overrightarrow{PQ}$ has magnitude 48.3km and
direction $18° + 51.2° = 69.2°$ east of north.

5. (a)

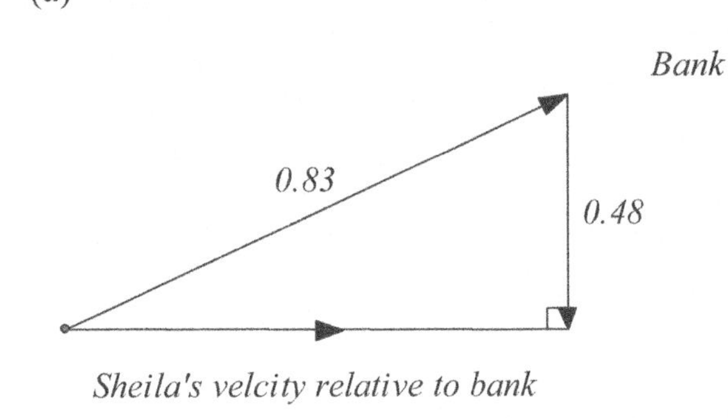

(b) $|\vec{v}| = \sqrt{0.83^2 - 0.48^2} = 0.6771262....$

Therefore, Sheila's speed relative to the bank is 0.677 ms^{-1}.

(c) As the width of the river is 35 m and Sheila is crossing the river at 0.677 ms^{-1}, she will take $\dfrac{35}{0.6771262...} = 51.68903...$ seconds, or approximately 52 seconds, to cross.

Exercise 4.2

1. (a) $\overrightarrow{PQ}$ (b) $\overrightarrow{OQ}$ (c) $\overrightarrow{OP} - \overrightarrow{OQ} = \overrightarrow{QP}$

   (d) Let $\_ = x$. Then $x + \overrightarrow{OQ} = \overrightarrow{OR} - \overrightarrow{PR} \Rightarrow x + \overrightarrow{OQ} = \overrightarrow{OR} + \overrightarrow{RP} = \overrightarrow{OP} \Rightarrow x = \overrightarrow{OP} - \overrightarrow{OQ} = \overrightarrow{QP}$

2. (a) $\overrightarrow{AD}$ (b) $\overrightarrow{OB} + \overrightarrow{BC} = \overrightarrow{OC}$ (c) $\overrightarrow{AB} + \overrightarrow{BA} = \vec{0}$

3. (a) $\overrightarrow{AC} = \vec{b}$ (b) $\vec{a}$ (c) $\overrightarrow{OC} = \vec{a} + \vec{b}$ (d) $\overrightarrow{AB} = \vec{b} - \vec{a}$

4. (a) $\overrightarrow{AB} = \vec{b} - \vec{a}$ (b) $\overrightarrow{BA} = \vec{a} - \vec{b}$ (c) $\overrightarrow{BC} = \overrightarrow{AB} = \vec{b} - \vec{a}$

   (d) $\overrightarrow{AC} = 2\overrightarrow{AB} = 2\vec{b} - 2\vec{a}$ (e) $\overrightarrow{OC} = \vec{a} + \left(2\vec{b} - 2\vec{a}\right) = 2\vec{b} - \vec{a}$

5. (a) $\overrightarrow{AC} = \vec{u} + \vec{v}$ (b) $\overrightarrow{CD} = -\vec{u} + \vec{v}$ (c) $\overrightarrow{AD} = \vec{u} + \vec{v} + \overrightarrow{CD} = 2\vec{v}$

   (d) $\overrightarrow{BF} = -\vec{u} + \overrightarrow{AF} = -\vec{u} + \overrightarrow{CD} = \vec{v} - 2\vec{u}$

6. (a)

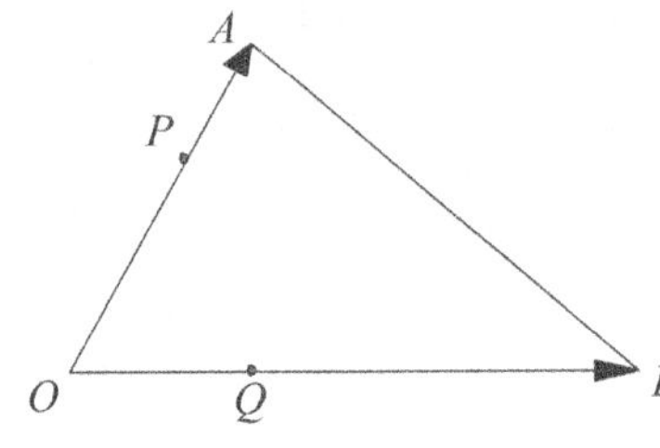

(b) (i) $\overrightarrow{BA} = \vec{a} - \vec{b}$

     (ii) $\overrightarrow{PB} = \overrightarrow{PO} + \overrightarrow{OB} = -\dfrac{2}{3}\vec{a} + \vec{b} = \vec{b} - \dfrac{2}{3}\vec{a}$

     (iii) $\overrightarrow{AQ} = \overrightarrow{AO} + \overrightarrow{OQ} = -\vec{a} + \dfrac{1}{3}\vec{b}$

     (iv) $\overrightarrow{PQ} = \overrightarrow{PO} + \overrightarrow{OQ} = -\dfrac{2}{3}\vec{a} + \dfrac{1}{3}\vec{b}$

7. (a) $\overrightarrow{AB} = \overrightarrow{AO} + \overrightarrow{OB} = \overrightarrow{OB} - \overrightarrow{OA} = \vec{b} - \vec{a}$

   (b) $\vec{a} + \dfrac{1}{2}\vec{c} = \dfrac{4}{5}\vec{b} \Rightarrow \vec{c} = \dfrac{8}{5}\vec{b} - 2\vec{a}$. Therefore, $\overrightarrow{BC} = \overrightarrow{BO} + \overrightarrow{OC} = \overrightarrow{OC} - \overrightarrow{OB} = \vec{c} - \vec{b}$

     $= \left(\dfrac{8}{5}\vec{b} - 2\vec{a}\right) - \vec{b} = \dfrac{3}{5}\vec{b} - 2\vec{a}$

(c) $\overrightarrow{CA} = \overrightarrow{CO} + \overrightarrow{OA} = -\vec{c} + \vec{a} = \vec{a} - \left(\dfrac{8}{5}\vec{b} - 2\vec{a}\right) = 3\vec{a} - \dfrac{8}{5}\vec{b}$

(d) $\overrightarrow{OA} + \overrightarrow{AB} + \overrightarrow{BC} + \overrightarrow{CO} = \vec{a} + \left(\vec{b} - \vec{a}\right) + \left(\dfrac{3}{5}\vec{b} - 2\vec{a}\right) + -\vec{c} = \vec{a} + \vec{b} - \vec{a} + \dfrac{3}{5}\vec{b} - 2\vec{a} - \left(\dfrac{8}{5}\vec{b} - 2\vec{a}\right) = \vec{0}$

8.   (a) $\overrightarrow{AB} = \vec{b} - \vec{a}$ (b) $\overrightarrow{OP} = \overrightarrow{OA} + \overrightarrow{AP} = \vec{a} + \dfrac{1}{2}\overrightarrow{AB} = \vec{a} + \dfrac{1}{2}\left(\vec{b} - \vec{a}\right) = \dfrac{1}{2}\left(\vec{a} + \vec{b}\right)$

       (c) $\overrightarrow{AQ} = \overrightarrow{AO} + \overrightarrow{OQ} = -\vec{a} + \dfrac{7}{8}\vec{b}$ (d) $\overrightarrow{QR} = \overrightarrow{QO} + \overrightarrow{OR} = -\dfrac{7}{8}\vec{b} + \dfrac{1}{2}\vec{a}$

Exercise 4.3

1.

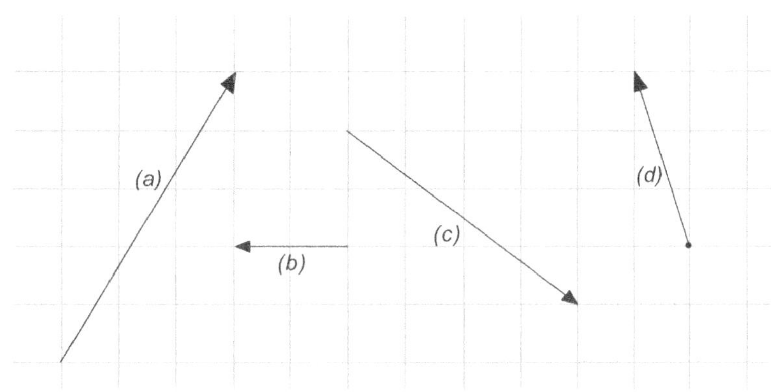

2.

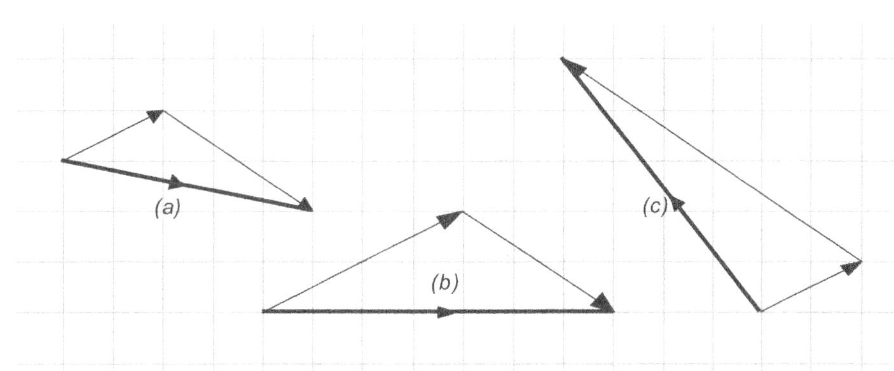

3.   (a) (i) $\vec{u} + \vec{v} = \begin{pmatrix} 4 \\ 3 \end{pmatrix} + \begin{pmatrix} -2 \\ 7 \end{pmatrix} = \begin{pmatrix} 2 \\ 10 \end{pmatrix}$

       (ii) $2\vec{u} + 3\vec{v} = 2\begin{pmatrix} 4 \\ 3 \end{pmatrix} + 3\begin{pmatrix} -2 \\ 7 \end{pmatrix} = \begin{pmatrix} 8 \\ 6 \end{pmatrix} + \begin{pmatrix} -6 \\ 21 \end{pmatrix} = \begin{pmatrix} 2 \\ 27 \end{pmatrix}$

       (iii) $\dfrac{1}{2}\left(\vec{u} - \vec{v}\right) = \dfrac{1}{2}\left(\begin{pmatrix} 4 \\ 3 \end{pmatrix} - \begin{pmatrix} -2 \\ 7 \end{pmatrix}\right) = \dfrac{1}{2}\begin{pmatrix} 6 \\ -4 \end{pmatrix} = \begin{pmatrix} 3 \\ -2 \end{pmatrix}$

    (b) (i) $2\vec{u} + \vec{v} = 2\left(5\vec{i} - 2\vec{j} + \vec{k}\right) + \left(3\vec{i} + \vec{j} + \vec{k}\right) = 13\vec{i} - 3\vec{j} + 3\vec{k}$

       (ii) $-\vec{u} + 2\vec{v} = -5\vec{i} + 2\vec{j} - \vec{k} + 2\left(3\vec{i} + \vec{j} + \vec{k}\right) = \vec{i} + 4\vec{j} + \vec{k}$

(iii) $\dfrac{1}{3}\vec{u} - \dfrac{1}{2}\vec{v} = \dfrac{1}{3}\left(5\vec{i} - 2\vec{j} + \vec{k}\right) - \dfrac{1}{2}\left(3\vec{i} + \vec{j} + \vec{k}\right) = \dfrac{1}{6}\vec{i} - \dfrac{7}{6}\vec{j} - \dfrac{1}{6}\vec{k}$

4. $\left|\vec{a}\right| = 3\left|\vec{b}\right|$ so, since $\vec{a}$ is parallel to $\vec{b}$, $\vec{a} = 3\vec{b}$ or $\vec{a} = -3\vec{b}$. If $\vec{a} = 3\vec{b}$ then $\vec{b} = \dfrac{1}{3}\vec{a}$ and

$$\vec{b} = \dfrac{1}{3}\begin{pmatrix} -2 \\ 0 \\ 3 \end{pmatrix} = \begin{pmatrix} -\tfrac{2}{3} \\ 0 \\ 1 \end{pmatrix}. \text{ If } \vec{a} = -3\vec{b} \text{ then } b = -\dfrac{1}{3}\begin{pmatrix} -2 \\ 0 \\ -3 \end{pmatrix} = \begin{pmatrix} \tfrac{2}{3} \\ 0 \\ -1 \end{pmatrix}.$$

5.   In each case, let $\vec{v}$ be the vector and $\vec{u}$ the corresponding unit vector.

(a) $\left|\vec{v}\right| = \sqrt{8^2 + 15^2} = 17 \Rightarrow \vec{u} = \dfrac{1}{17}\begin{pmatrix} 8 \\ 15 \end{pmatrix} = \begin{pmatrix} \tfrac{8}{17} \\ \tfrac{15}{17} \end{pmatrix}$

(b) $\left|\vec{v}\right| = \sqrt{(-7)^2 + 3^2} = 58 \Rightarrow \vec{u} = \dfrac{1}{\sqrt{58}}\begin{pmatrix} -7 \\ 3 \end{pmatrix} = \begin{pmatrix} -0.9191450... \\ 0.3939192... \end{pmatrix} = \begin{pmatrix} -0.919 \\ 0.394 \end{pmatrix}$

(c) $\vec{u} = \begin{pmatrix} 1 \\ 0 \end{pmatrix}$

(d) $\left|\vec{v}\right| = \sqrt{6^2 + (-5)^2} = \sqrt{61} \Rightarrow \vec{u} = \dfrac{6}{\sqrt{61}}\vec{i} - \dfrac{5}{\sqrt{61}}\vec{j} = 0.7682212...\vec{i} - 0.6401843...\vec{j}$

$\qquad\qquad = 0.768\vec{i} - 0.640\vec{j}$

(e) $\left|\vec{v}\right| = \sqrt{(-0.43)^2 + 0.37^2} = 0.5672741... \Rightarrow \vec{u} = -\dfrac{0.43}{\left|\vec{v}\right|} + \dfrac{0.37}{\left|\vec{v}\right|}$

$\qquad \Rightarrow \vec{u} = -0.7580109\vec{i} + 0.6522420...\vec{j} = -0.758\vec{i} + 0.652\vec{j}$

6.   (a) (i) $\vec{a} + \vec{b} = \begin{pmatrix} 9 \\ 6 \end{pmatrix} \Rightarrow \left|\vec{a} + \vec{b}\right| = \sqrt{9^2 + 6^2} = 10.81665... = 10.8$

          (ii) $\left|\vec{a}\right| = \sqrt{85}, \left|\vec{b}\right| = \sqrt{58} \Rightarrow \left|\vec{a}\right| + \left|\vec{b}\right| = \sqrt{85} + \sqrt{58} = 16.835318... = 16.8$

(b) Therefore as $10.8 < 16.8$, $\left|\vec{a} + \vec{b}\right| < \left|\vec{a}\right| + \left|\vec{b}\right|$

(c) Consider the arbitrary vectors $\vec{a}$ and $\vec{b}$. The diagram
shows the sides of a vector triangle in which the
lengths of the sides are $\left|\vec{a}\right|$, $\left|\vec{b}\right|$ and $\left|\vec{a} + \vec{b}\right|$. It is
clear from the diagram that, in order to form a
triangle, $\left|\vec{a} + \vec{b}\right| < \left|\vec{a}\right| + \left|\vec{b}\right|$. The equality occurs when

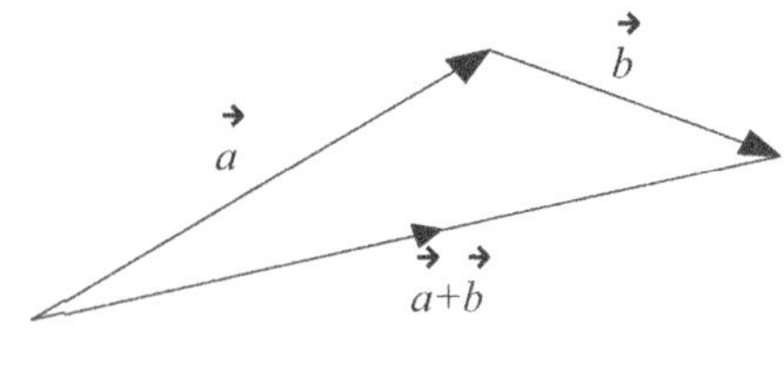

$\vec{a}$ and $\vec{b}$ are parallel. Therefore, in general, $\left|\vec{a} + \vec{b}\right| \le \left|\vec{a}\right| + \left|\vec{b}\right|$.

7. $2ab + 1 = 7 \Rightarrow ab = 3$; $-3a + 4 = 2 \Rightarrow a = \dfrac{2}{3} \Rightarrow \dfrac{2}{3}b = 3 \Rightarrow b = \dfrac{9}{2}$

8. $m\left(5\vec{i}+\vec{j}\right)+n\left(3\vec{i}+3\vec{j}\right)=12\left(-\vec{i}-2\vec{j}\right)$. Therefore, $5m+3n=-12$, $m+3n=-24$.

Subtracting we get $4m=12\Rightarrow m=3$. Then $3+3n=-24\Rightarrow 1+n=-8\Rightarrow n=-9$.

9.   (a) $\overrightarrow{OA}=\begin{pmatrix}1\\4\\2\end{pmatrix}$, $\overrightarrow{OB}=\begin{pmatrix}-3\\1\\6\end{pmatrix}$, $\overrightarrow{OC}=\begin{pmatrix}5\\5\\-6\end{pmatrix}$

(b) $\left|\overrightarrow{OA}\right|=\sqrt{1^2+4^2+2^2}=\sqrt{21}$; $\left|\overrightarrow{OB}\right|=\sqrt{(-3)^2+1^2+6^2}=\sqrt{46}$; $\left|\overrightarrow{OC}\right|=\sqrt{86}$

10.   (a) $\overrightarrow{AB}=\overrightarrow{OB}-\overrightarrow{OA}=\begin{pmatrix}-3\\6\end{pmatrix}-\begin{pmatrix}4\\5\end{pmatrix}=\begin{pmatrix}-7\\1\end{pmatrix}$, $\overrightarrow{BC}=\begin{pmatrix}10\\-8\end{pmatrix}$, $\overrightarrow{CA}=\begin{pmatrix}-3\\7\end{pmatrix}$

$\overrightarrow{AB}+\overrightarrow{BC}+\overrightarrow{CA}=\begin{pmatrix}-7\\1\end{pmatrix}+\begin{pmatrix}10\\-8\end{pmatrix}+\begin{pmatrix}-3\\7\end{pmatrix}=\begin{pmatrix}0\\0\end{pmatrix}=\vec{0}$

(b) $\left|\overrightarrow{OA}\right|=\sqrt{41}$, $\left|\overrightarrow{OB}\right|=\sqrt{45}$; $\overrightarrow{AB}+2\overrightarrow{BC}=\begin{pmatrix}-7\\1\end{pmatrix}+2\begin{pmatrix}10\\-8\end{pmatrix}=\begin{pmatrix}13\\-15\end{pmatrix}$

$\Rightarrow\left|\overrightarrow{AB}+2\overrightarrow{BC}\right|=\sqrt{13^2+(-15)^2}=\sqrt{394}=19.84943...=19.8$

Exercise 4.4

1.   In each case, assume that the vectors are $\vec{a}$ and $\vec{b}$ and the angle between them is θ.

(a) $\vec{a}\bullet\vec{b}=1\times2+5\times2=12$, $|\vec{a}|=\sqrt{6}$, $|\vec{b}|=\sqrt{8}$, $\theta=\arccos\left(\dfrac{12}{\sqrt{26}\sqrt{8}}\right)=33.69006...=33.7°$

(b) $\vec{a}\bullet\vec{b}=-12-2=-14$, $|\vec{a}|=\sqrt{10}$, $|\vec{b}|=\sqrt{20}$, $\theta=\arccos\left(\dfrac{-14}{\sqrt{10}\sqrt{20}}\right)=171.8698...=172°$

(c) $\vec{a}\bullet\vec{b}=5-6=-1$, $|\vec{a}|=\sqrt{34}$, $|\vec{b}|=\sqrt{5}$, $\theta=\arccos\left(\dfrac{-1}{\sqrt{34}\sqrt{5}}\right)=94.39870...=94.4°$

(d) $\vec{a}\bullet\vec{b}=7\times0+4\times3=12$, $|\vec{a}|=\sqrt{65}$, $|\vec{b}|=3$, $\theta=\arccos\left(\dfrac{12}{3\sqrt{65}}\right)=60.25511...=60.3°$

2.   In each case, assume that the vectors are $\vec{a}$ and $\vec{b}$ and the angle between them is θ.

(a) $\vec{a}\bullet\vec{b}=1\times2+1\times3+(-1)\times2=3$; $|\vec{a}|=\sqrt{1^2+1^2+(-1)^2}=\sqrt{3}$, $|\vec{b}|=\sqrt{2^2+3^2+2^2}=\sqrt{17}$

$\Rightarrow\theta=\arccos\left(\dfrac{3}{\sqrt{3}\sqrt{17}}\right)=65.16010...=65.2°$

(b) $\vec{a}\bullet\vec{b}=6\times(-2)+(-2)\times(-5)+1\times4=2$, $|\vec{a}|=\sqrt{6^2+(-2)^2+1^2}=\sqrt{41}$,

$|\vec{b}|=\sqrt{(-2)^2+(-5)^2+4^2}=\sqrt{45}\Rightarrow\theta=\arccos\left(\dfrac{2}{\sqrt{41}\sqrt{45}}\right)=87.33122...=87.3°$

(c) $\vec{a}\cdot\vec{b}=(-4)\times5+3\times1+(-3)\times(-6)=1,\ |\vec{a}|=\sqrt{(-4)^2+3^2+(-3)^2}=\sqrt{34},$

$|\vec{b}|=\sqrt{5^2+1^2+(-6)^2}=\sqrt{62}\Rightarrow\theta=\arccos\left(\dfrac{1}{\sqrt{34}\sqrt{62}}\right)=88.75197\ldots=88.8°$

(d) $\vec{a}\cdot\vec{b}=4\times3+5\times4+6\times6=68,\ |\vec{a}|=\sqrt{4^2+5^2+6^2}=\sqrt{77},$

$|\vec{b}|=\sqrt{3^2+4^2+6^2}=\sqrt{61}\Rightarrow\theta=\arccos\left(\dfrac{68}{\sqrt{77}\sqrt{61}}\right)=7.161519\ldots=7.16°$

3. (a) $\overrightarrow{AB}=\begin{pmatrix}0\\2\end{pmatrix}-\begin{pmatrix}1\\2\end{pmatrix}=\begin{pmatrix}-1\\0\end{pmatrix},\ \overrightarrow{CD}=\begin{pmatrix}3\\5\end{pmatrix}-\begin{pmatrix}-2\\4\end{pmatrix}=\begin{pmatrix}5\\1\end{pmatrix}\ \overrightarrow{AB}\cdot\overrightarrow{CD}=-5,\ |\overrightarrow{AB}|=1,\ |\overrightarrow{CD}|=\sqrt{26}$

$\Rightarrow\theta=\arccos\left(\dfrac{-5}{1\sqrt{26}}\right)=168.6900\ldots=169°$

(b) $\overrightarrow{AC}=\begin{pmatrix}-2\\4\end{pmatrix}-\begin{pmatrix}1\\2\end{pmatrix}=\begin{pmatrix}-3\\2\end{pmatrix};\ \overrightarrow{CD}=\begin{pmatrix}5\\1\end{pmatrix}\ \overrightarrow{AC}\cdot\overrightarrow{CD}=-13,\ |\overrightarrow{AC}|=\sqrt{13},\ |\overrightarrow{CD}|=\sqrt{26}$

$\Rightarrow\theta=\arccos\left(\dfrac{-13}{\sqrt{13}\sqrt{26}}\right)=135°$

(c) $\overrightarrow{AD}=\begin{pmatrix}3\\5\end{pmatrix}-\begin{pmatrix}1\\2\end{pmatrix}=\begin{pmatrix}2\\3\end{pmatrix};\ \overrightarrow{BC}=\begin{pmatrix}-2\\4\end{pmatrix}-\begin{pmatrix}0\\2\end{pmatrix}=\begin{pmatrix}-2\\2\end{pmatrix}\ \overrightarrow{AD}\cdot\overrightarrow{BC}=2,\ |\overrightarrow{AD}|=\sqrt{13},\ |\overrightarrow{BC}|=\sqrt{8}$

$\Rightarrow\theta=\arccos\left(\dfrac{2}{\sqrt{8}\sqrt{13}}\right)=78.69006\ldots=78.7°$

4. (a) $\overrightarrow{AB}=\begin{pmatrix}2\\1\\5\end{pmatrix}-\begin{pmatrix}0\\1\\3\end{pmatrix}=\begin{pmatrix}2\\0\\2\end{pmatrix};\ \overrightarrow{DC}=\begin{pmatrix}-3\\2\\0\end{pmatrix}-\begin{pmatrix}1\\4\\1\end{pmatrix}=\begin{pmatrix}-4\\-2\\-1\end{pmatrix}\ \overrightarrow{AB}\cdot\overrightarrow{DC}=-10,\ |\overrightarrow{AB}|=\sqrt{8},\ |\overrightarrow{DC}|=\sqrt{21}$

$\Rightarrow\theta=\arccos\left(\dfrac{-10}{\sqrt{8}\sqrt{21}}\right)=140.4902\ldots=140°$

(b) $\overrightarrow{CB}=\begin{pmatrix}2\\1\\5\end{pmatrix}-\begin{pmatrix}-3\\2\\0\end{pmatrix}=\begin{pmatrix}5\\-1\\5\end{pmatrix};\ \overrightarrow{AD}=\begin{pmatrix}1\\4\\1\end{pmatrix}-\begin{pmatrix}0\\1\\3\end{pmatrix}=\begin{pmatrix}1\\3\\-2\end{pmatrix}\ \overrightarrow{CB}\cdot\overrightarrow{AD}=-8,\ |\overrightarrow{CB}|=\sqrt{51},\ |\overrightarrow{AD}|=\sqrt{14}$

$\Rightarrow\theta=\arccos\left(\dfrac{-8}{\sqrt{51}\sqrt{14}}\right)=107.4211\ldots=107°$

(c) $\overrightarrow{BC}=\begin{pmatrix}-5\\1\\-5\end{pmatrix};\ \overrightarrow{BD}=\begin{pmatrix}1\\4\\1\end{pmatrix}-\begin{pmatrix}2\\1\\5\end{pmatrix}=\begin{pmatrix}-1\\3\\-4\end{pmatrix}\ \overrightarrow{BC}\cdot\overrightarrow{BD}=28,\ |\overrightarrow{BC}|=\sqrt{51},\ |\overrightarrow{BD}|=\sqrt{26}$

$\Rightarrow\theta=\arccos\left(\dfrac{28}{\sqrt{51}\sqrt{26}}\right)=39.74218\ldots=39.7°$

(d) $\overrightarrow{DC} = \begin{pmatrix} -4 \\ -2 \\ -1 \end{pmatrix}$; $\overrightarrow{DA} = \begin{pmatrix} -1 \\ -3 \\ 2 \end{pmatrix}$; $\overrightarrow{DC} \cdot \overrightarrow{DA} = -8$; $|\overrightarrow{DC}| = \sqrt{21}$ $|\overrightarrow{DA}| = \sqrt{14}$

$$\Rightarrow \theta = \arccos\left(\frac{8}{\sqrt{21}\sqrt{14}}\right) = 62.18815... = 62.2°$$

5. (a) $\vec{a} \cdot \vec{b} = 0 \Rightarrow 10 \times -8 + 5k = 0 \Rightarrow 5k - 80 = 0 \Rightarrow k = 16$

(b) $\vec{u} \cdot \vec{v} = 0 \Rightarrow 3 \times (-5) + 1 \times (-9) + 8c = 0 \Rightarrow 8c = 24 \Rightarrow c = 3$

6. $\vec{a} \cdot \vec{c} = 3 + 2m$; $\vec{a} \cdot \vec{b} = -5$. Therefore, $\begin{pmatrix} -10 \\ n \end{pmatrix} = (3 + 2m)\begin{pmatrix} 1 \\ -4 \end{pmatrix} + 5\begin{pmatrix} 1 \\ m \end{pmatrix} \Rightarrow -10 = 3 + 2m + 5$

$\Rightarrow 2m = -18 \Rightarrow m = -9$. $n = -4(3 + 2m) + 5m \Rightarrow n = -12 - 8m + 5m = -12 - 3m = -12 + 27 = 15$
So, $m = -9$, $n = 15$.

7. (a) $\begin{pmatrix} \frac{3}{\sqrt{34}} \\ \frac{-5}{\sqrt{34}} \end{pmatrix}$ or $\begin{pmatrix} \frac{-3}{\sqrt{34}} \\ \frac{5}{\sqrt{34}} \end{pmatrix}$ (b) $\begin{pmatrix} \frac{11}{\sqrt{185}} \\ \frac{8}{\sqrt{185}} \end{pmatrix}$ or $\begin{pmatrix} \frac{-11}{\sqrt{185}} \\ \frac{-8}{\sqrt{185}} \end{pmatrix}$ (c) $\begin{pmatrix} -0.909 \\ 0.418 \end{pmatrix}$ or $\begin{pmatrix} 0.909 \\ -0.418 \end{pmatrix}$

(d) $\begin{pmatrix} 0 \\ 1 \end{pmatrix}$ or $\begin{pmatrix} 0 \\ -1 \end{pmatrix}$ (e) $-\frac{3}{5}\vec{i} + \frac{4}{5}\vec{j}$ or $\frac{3}{5}\vec{i} - \frac{4}{5}\vec{j}$ (f) $-i$ or i

8. (a)1, (c)2 (b)1, (a)2 (c)1, (b)2 (d)1, (e)2 (e)1, (d)2

9. $\begin{pmatrix} 1 \\ 1 \\ -3 \end{pmatrix} \cdot \begin{pmatrix} 1 \\ -10 \\ -3 \end{pmatrix} = 1 - 10 + 9 = 0$; $\begin{pmatrix} 1 \\ 1 \\ -3 \end{pmatrix} \cdot \begin{pmatrix} 3 \\ 0 \\ 1 \end{pmatrix} = 3 - 3 = 0$; $\begin{pmatrix} 1 \\ -10 \\ -3 \end{pmatrix} \cdot \begin{pmatrix} 3 \\ 0 \\ 1 \end{pmatrix} = 3 - 3 = 0$

10. (a) $\vec{u} \cdot \vec{v} = 21 - 20 - c = 0 \Rightarrow c = 1$ (b) $\vec{u} \cdot \vec{v} = c - 6 + 8 = 0 \Rightarrow c = -2$
(c) $\vec{u} \cdot \vec{v} = 8 + 5c - 54 = 0 \Rightarrow 5c = 46 \Rightarrow c = 9.2$ (d) $\vec{u} \cdot \vec{v} = 2c - 12 + 2c = 0 \Rightarrow 4c = 12 \Rightarrow c = 3$

11. (a) $\vec{u} \cdot \vec{v} = 3 + 12 + 27 = 42$, $|\vec{u}| = \sqrt{1^2 + 2^2 + (-3)^2} = \sqrt{14}$; $|\vec{v}| = \sqrt{3^2 + 6^2 + (-9)^2} = \sqrt{126}$
$|\vec{u}||\vec{v}| = \sqrt{14}\sqrt{126} = \sqrt{2}\sqrt{7}\sqrt{2}\sqrt{7}\sqrt{9} = 2 \times 7 \times 3 = 42$. Therefore $\vec{u} \cdot \vec{v} = |\vec{u}||\vec{v}|$.
(b) $\vec{w} = -\vec{i} - 2\vec{j} + 3\vec{k}$

Exercise 4.5

1. (a) $x = 1 + 2t$, $y = 2 - t \Rightarrow t = 2 - y \Rightarrow x = 1 + 2(2 - y) \Rightarrow x + 2y - 5 = 0$

(b) $x = -5 + 4t$, $y = -2 + 9t \Rightarrow 9x - 4y + 37 = 0$

(c) $x = 13 + 8(t - 1)$, $y = 5 - 3(t - 1) \Rightarrow 3x + 8y - 79 = 0$

(d) $\quad x=3+5t,\ y=-4+t \Rightarrow x-5y-23=0$

(e) $\quad x=-7+3t,\ y=2 \Rightarrow y=2$

2.   In each case, $\vec{v}_1$ and $\vec{v}_2$ are vectors parallel to the corresponding lines and θ is the angle between the lines.

(a) $\vec{v}_1 = \begin{pmatrix} -1 \\ 1 \end{pmatrix},\ \vec{v}_2 = \begin{pmatrix} 1 \\ 1 \end{pmatrix} \Rightarrow \vec{v}_1 \bullet \vec{v}_2 = 0 \Rightarrow \theta = 90°$

(b) $\vec{v}_1 = \begin{pmatrix} -3 \\ 2 \end{pmatrix},\ \vec{v}_2 = \begin{pmatrix} 1 \\ 4 \end{pmatrix} \Rightarrow \vec{v}_1 \bullet \vec{v}_2 = 5 \Rightarrow \cos\theta = \dfrac{5}{\sqrt{17}\sqrt{13}} \Rightarrow \theta = 70.34617... = 70.3°$

(c) $\vec{v}_1 = \begin{pmatrix} 6 \\ 5 \end{pmatrix},\ \vec{v}_2 = \begin{pmatrix} 4 \\ -3 \end{pmatrix} \Rightarrow \vec{v}_1 \bullet \vec{v}_2 = 9 \Rightarrow \cos\theta = \dfrac{9}{5\sqrt{61}} \Rightarrow \theta = 76.67546... = 76.7°$

(d) $\vec{v}_1 = \begin{pmatrix} 1 \\ 4 \end{pmatrix},\ \vec{v}_2 = \begin{pmatrix} -3 \\ 2 \end{pmatrix} \Rightarrow \vec{v}_1 \bullet \vec{v}_2 = 5 \Rightarrow \cos\theta = \dfrac{5}{\sqrt{17}\sqrt{13}} \Rightarrow \theta = 70.34617... = 70.3°$

3.   (a) $\overrightarrow{AB} = \overrightarrow{OC} = \begin{pmatrix} 6 \\ -3 \end{pmatrix},\ \overrightarrow{OB} = \overrightarrow{OA} + \overrightarrow{AB} = \begin{pmatrix} -2 \\ 5 \end{pmatrix} + \begin{pmatrix} 6 \\ -3 \end{pmatrix} = \begin{pmatrix} 4 \\ 2 \end{pmatrix}$

(b) $\overrightarrow{AC} = \begin{pmatrix} 6 \\ -3 \end{pmatrix} - \begin{pmatrix} -2 \\ 5 \end{pmatrix} = \begin{pmatrix} 8 \\ -8 \end{pmatrix} = 8\begin{pmatrix} 1 \\ -1 \end{pmatrix}$. Therefore, $k = -1$.

(c) (i) $x = -2+1 = -1$, $y = 5-1 = 4$, so, after 1 second, P has coordinates $(-1,\ 4)$.

     (ii) $t = 3$

(d) $\vec{r} = \begin{pmatrix} 4 \\ 2 \end{pmatrix} + s\begin{pmatrix} 1 \\ 1 \end{pmatrix}$

(e) Let the point of intersection of L_1 and L_2 be N. At N, $-2+t = 4+s$ and $5-t = 2+s$. By subtraction, $-7+2t = 2 \Rightarrow t = 4.5$. Then, at N $\ x = -2+4.5 = 2.5$ and $y = 5-4.5 = 0.5$. Therefore, the coordinates of N are $(2.5,\ 0.5)$.

Min dist is $\left|\overrightarrow{BN}\right| = \sqrt{(4-2.5)^2 + (2-0.5)^2} = 2.121322... = 2.12$ and it occurs when $t = 4.5$.

4.   (a) $\vec{r} = (3\vec{i} + 7\vec{j}) + t(6\vec{i} - 6\vec{j})$ (b) When $x = 1$, $\ 1 = 3+6t \Rightarrow 6t = -2 \Rightarrow t = -\frac{1}{3}$, then

$$y = 7-6t = 7-6\left(-\tfrac{1}{3}\right) = 9 \ \text{ so } P \text{ is } (1,\ 9).$$

(c) $\vec{r} = (\vec{i} + 9\vec{j}) + s(\vec{i} + \vec{j})$ (d) At the x-axis, $y = 0 \Rightarrow 9+s = 0 \Rightarrow s = -9$. Then

$$x\vec{i} + y\vec{j} = \vec{i} + s\vec{i} = -8\vec{i} \ \Rightarrow x = -8,\ y = 0 \text{ and } Q \text{ is}$$

$(-8,\ 0)$.

5.   (a) Antelope's speed is $\sqrt{4^2 + 3^2} = 5\text{ms}^{-1}$.

(b) Antelope starts when $t = 0$ and therefore, starting point is $(-9,\ 11)$.

(c) $\vec{r} = \begin{pmatrix} -9 \\ 11 \end{pmatrix} + 4\begin{pmatrix} 4 \\ 3 \end{pmatrix} = \begin{pmatrix} 7 \\ 23 \end{pmatrix}$. Therefore, its position after 4 seconds is $(7,\ 23)$.

(d) After 11 seconds, the antelope's position vector is $\begin{pmatrix} x \\ y \end{pmatrix} = \begin{pmatrix} -9 \\ 11 \end{pmatrix} + 11 \begin{pmatrix} 4 \\ 3 \end{pmatrix} = \begin{pmatrix} 35 \\ 44 \end{pmatrix}$.

So, the coordinates of the point where the cheetah catches the antelope are $(35,\ 44)$.

(e) Vector equation of the path of the cheetah is $\begin{pmatrix} x \\ y \end{pmatrix} = \begin{pmatrix} 0 \\ 25 \end{pmatrix} + t \begin{pmatrix} u \\ v \end{pmatrix}$, where $\begin{pmatrix} u \\ v \end{pmatrix}$ is the

cheetah's velocity. Therefore, if the cheetah is to catch the antelope after 11 seconds

$11u = 35$ and $-25 + 11v = 44 \Rightarrow u = \dfrac{35}{11}$, $v = \dfrac{69}{11}$. Therefore, the speed of the cheetah is

$\sqrt{\left(\tfrac{35}{11}\right)^2 + \left(\tfrac{69}{11}\right)^2} = 7.033567\ldots = 7.03 \text{ms}^{-1}$.

(f) Animals do not move in straight lines at constant speed.

6. (a) $\overrightarrow{BC} = \overrightarrow{BA} + \overrightarrow{AC} = \overrightarrow{AC} - \overrightarrow{AB} = \begin{pmatrix} 5 \\ 5 \end{pmatrix} - \begin{pmatrix} 4 \\ 2 \end{pmatrix} = \begin{pmatrix} 1 \\ 3 \end{pmatrix}$

(b) $\overrightarrow{OB} = \overrightarrow{OA} + \overrightarrow{AB} = \begin{pmatrix} -3 \\ -1 \end{pmatrix} + \begin{pmatrix} 4 \\ 2 \end{pmatrix} = \begin{pmatrix} 1 \\ 1 \end{pmatrix}$

(c) An equation for L is $\vec{r} = \begin{pmatrix} 1 \\ 1 \end{pmatrix} + t \begin{pmatrix} 1 \\ 3 \end{pmatrix}$

(d) (i) $\overrightarrow{AD} = \overrightarrow{OD} - \overrightarrow{OA} = \begin{pmatrix} h \\ k \end{pmatrix} - \begin{pmatrix} -3 \\ -1 \end{pmatrix} = \begin{pmatrix} h+3 \\ k+1 \end{pmatrix}$

   (ii) $\overrightarrow{AD} \bullet \overrightarrow{BC} = 0 \Rightarrow \begin{pmatrix} h+3 \\ k+1 \end{pmatrix} \bullet \begin{pmatrix} 1 \\ 3 \end{pmatrix} = 0 \Rightarrow (h+3)1 + (k+1)3 = 0 \Rightarrow h+3+3k+3 = 0$

   $\Rightarrow h + 3k + 6 = 0$

   (iii) but, from (c) $h = 1+t$ and $k = 1+3t$, therefore,

   $1+t+3+9t+6 = 0 \Rightarrow 10t+10 = 0 \Rightarrow t = -1$, so $h = 1+(-1) = 0$ and

   $k = 1+3(-1) = -2$

(e) $\overrightarrow{OD} = \begin{pmatrix} 0 \\ -2 \end{pmatrix}$ so $\overrightarrow{AD} = \overrightarrow{OD} - \overrightarrow{OA} = \begin{pmatrix} 0 \\ -2 \end{pmatrix} - \begin{pmatrix} -3 \\ -1 \end{pmatrix} = \begin{pmatrix} 3 \\ -1 \end{pmatrix} \Rightarrow |\overrightarrow{AD}| = \sqrt{3^2 + (-1)^2} = \sqrt{10}$

Exercise 4.6

1. Assume, in each case, that $\vec{b_1}$ and $\vec{b_2}$ are vectors parallel to the two lines and that θ is the angle
between the two vectors.

(a) $\vec{b_1} = \vec{i} - 2\vec{j} + 5\vec{k}$, $\vec{b_2} = 3\vec{i} + 3\vec{j} + 2\vec{k}$, $\vec{b_1} \bullet \vec{b_2} = 7$, $|\vec{b_1}| = \sqrt{30}$, $|\vec{b_2}| = \sqrt{22}$

$\Rightarrow \cos\theta = \dfrac{7}{\sqrt{30}\sqrt{22}} \Rightarrow \theta = 74.18842\ldots = 74.2°$

(b) $\vec{b_1} = 6\vec{i} + \vec{j} - 3\vec{k}$, $\vec{b_2} = 5\vec{i} + 4\vec{j} + 3\vec{k}$, $\vec{b_1} \bullet \vec{b_2} = 25$, $|\vec{b_1}| = \sqrt{46}$, $|\vec{b_2}| = \sqrt{50}$

$\Rightarrow \cos\theta = \dfrac{25}{\sqrt{46}\sqrt{50}} \Rightarrow \theta = 58.58144\ldots = 58.6°$

(c) $\vec{b}_1 = -\vec{i} - 8\vec{j} + 2\vec{k}$, $\vec{b}_2 = -3\vec{i} + 8\vec{j} - 7\vec{k}$, $\vec{b}_1 \cdot \vec{b}_2 = -75$, $\left|\vec{b}_1\right| = \sqrt{69}$, $\left|\vec{b}_2\right| = \sqrt{122}$

$\Rightarrow \cos\theta = \dfrac{-75}{\sqrt{69}\sqrt{122}} \Rightarrow \theta = 144.8295\ldots = 144.8°$. Therefore the acute angle between the

lines is $180° - 144.8° = 35.2°$.

2.   (a) $\vec{b}_1 = \begin{pmatrix} 3 \\ 5 \\ 1 \end{pmatrix}$, $\vec{b}_2 = \begin{pmatrix} 2 \\ -1 \\ 1 \end{pmatrix} \Rightarrow \vec{b}_1 \cdot \vec{b}_2 = 2$, $\left|\vec{b}_1\right| = \sqrt{35}$, $\left|\vec{b}_2\right| = \sqrt{6} \Rightarrow \theta = 82.1°$

   (b) $\vec{b}_1 = \begin{pmatrix} -3 \\ -10 \\ 5 \end{pmatrix}$, $\vec{b}_2 = \begin{pmatrix} 0 \\ 1 \\ 0 \end{pmatrix}$, $\Rightarrow \vec{b}_1 \cdot \vec{b}_2 = -10$, $\left|\vec{b}_1\right| = \sqrt{134}$, $\left|\vec{b}_2\right| = 1 \Rightarrow \theta = 149.8°$

Therefore the acute angle between the lines is $180° - 149.8° = 30.2°$

   (c) $\vec{b}_1 = \begin{pmatrix} -3 \\ 4 \\ -2 \end{pmatrix}$, $\vec{b}_2 = \begin{pmatrix} 0.7 \\ 1.2 \\ -0.3 \end{pmatrix}$, $\Rightarrow \vec{b}_1 \cdot \vec{b}_2 = 3.3$, $\left|\vec{b}_1\right| = \sqrt{29}$, $\left|\vec{b}_2\right| = \sqrt{2.02} \Rightarrow \theta = 64.5°$

3.   (i) Vector equation of line is $\vec{r} = \left(3\vec{i} - \vec{j} + \vec{k}\right) + t\left(2\vec{i} + 3\vec{j} - 2\vec{k}\right)$. When $x = 7$, $3 + 2t = 7$

$\Rightarrow t = 2$. Therefore, $y = -1 + 3t = 5$ and $z = 1 - 2t = -3$, so that L passes through the point

$(7, 5, -3)$.

(ii) Vector parallel to l is $\begin{pmatrix} 2 \\ 1 \\ 4 \end{pmatrix} - \begin{pmatrix} 1 \\ -3 \\ -2 \end{pmatrix} = \begin{pmatrix} 1 \\ 4 \\ 6 \end{pmatrix}$. Therefore the vector equation of l is

$\begin{pmatrix} x \\ y \\ z \end{pmatrix} = \begin{pmatrix} 1 \\ -3 \\ -2 \end{pmatrix} + c\begin{pmatrix} 1 \\ 4 \\ 6 \end{pmatrix}$ (there are many other possibilities). When $x = 0$, $1 + c = 0 \Rightarrow c = -1$.

So $y = -3 + 4(-1) = -7$ and $z = -2 + 6(-1) = -8$ and the point $(0, -7, -8)$ lies on l.

4.   (a) $M = \left(\dfrac{1+5}{2}, \dfrac{3-1}{2}, \dfrac{-4+0}{2}\right) = (3, 1, -2)$

   (b) $\vec{r} = \begin{pmatrix} 6 \\ -3 \\ 5 \end{pmatrix} + t\begin{pmatrix} 3 \\ -4 \\ 7 \end{pmatrix}$

   (c) $\overrightarrow{AB} = \begin{pmatrix} 5-1 \\ -1-3 \\ 0--4 \end{pmatrix} = \begin{pmatrix} 4 \\ -4 \\ 4 \end{pmatrix} \Rightarrow \left|\overrightarrow{AB}\right| = 4\sqrt{3}$, $\overrightarrow{CB} = \begin{pmatrix} 5-6 \\ -1+3 \\ 0-5 \end{pmatrix} = \begin{pmatrix} -1 \\ 2 \\ -5 \end{pmatrix} \Rightarrow \left|\overrightarrow{CB}\right| = \sqrt{30}$

(d) $\cos A\hat{B}C = \dfrac{\overrightarrow{AB}\bullet\overrightarrow{CB}}{\left|\overrightarrow{AB}\right|\left|\overrightarrow{CB}\right|} = \dfrac{4\times-1+-4\times 2+4\times-5}{4\sqrt{3}\sqrt{30}} = \dfrac{-32}{4\sqrt{90}} = -\dfrac{8}{\sqrt{90}} \Rightarrow A\hat{B}C = 147.4874... \approx 147°$

(e) Area of $\triangle ABC = \dfrac{1}{2}\left|\overrightarrow{AB}\right|\left|\overrightarrow{CB}\right|\sin A\hat{B}C = 10.19806... = 10.2$

5. (a) $\overrightarrow{OP} = 350\vec{i} - 217\vec{j} + 785\vec{k}$

(b) $|\vec{v}| = \sqrt{(-9)^2 + 7^2 + (-4)^2} = 12.08304....$ Therefore the speed of the submarine is $12.1\ \text{ms}^{-1}$.

(c) $\vec{r} = \left(350\vec{i} - 217\vec{j} + 785\vec{k}\right) + t\left(-9\vec{i} + 7\vec{j} - 4\vec{k}\right)$

(d) At the surface, $-217 + 7t = 0 \Rightarrow t = 31$. Then $x = 350 + 31\times(-9) = 71$ and

$z = 785 + 31\times(-4) = 661$. Therefore, the submarine surfaces at a point whose

coordinates are $(71,\ 0,\ 661)$ and the distance from the fishing boat is

$\sqrt{(71-10)^2 + (661-600)^2} = 86.26702... = 86.3\ \text{meters}$.

6. (a) Skier's speed is $\sqrt{0.711^2 + (-0.283)^2 + 1.782^2} = 1.939364... = 1.94\text{ms}^{-1}$.

(b) $\vec{r}(10) = \begin{pmatrix} 5.83 \\ 2.11 \\ 3.74 \end{pmatrix} + 10\begin{pmatrix} 0.711 \\ -0.283 \\ 1.782 \end{pmatrix} = \begin{pmatrix} 12.94 \\ -0.72 \\ 21.56 \end{pmatrix}$. So, after 10 seconds, the skier's position has

coordinates $(12.94, -0.72,\ 21.56)$.

(c) $5.83 + 0.711t = 43.51 \Rightarrow t = 52.99578...$, or, $2.11 - 0.283t = -12.89 \Rightarrow t = 53.00353...$
or $3.74 + 1.782t = 98.19 \Rightarrow t = 53.00224....$ Therefore, the time taken is 53.0 seconds.

7. (a) $\overrightarrow{QP} = \overrightarrow{QO} + \overrightarrow{OP} = \overrightarrow{OP} - \overrightarrow{OQ} = \left(21\vec{i} - 8\vec{j} + 8\vec{k}\right) - \left(4\vec{i} + 15\vec{k}\right) = 17\vec{i} - 8\vec{j} - 7\vec{k}$

(b) $\left|\overrightarrow{QP}\right| = \sqrt{17^2 + (-8)^2 + (-7)^2} = 20.04993...$

$\overrightarrow{QP} = \dfrac{1}{\left|\overrightarrow{QP}\right|}\left(17\vec{i} - 8\vec{j} - 7\vec{k}\right) = 0.8478829...\vec{i} - 0.3990037...\vec{j} - 0.3491284...\vec{k}$

$= 0.848\vec{i} - 0.399\vec{j} - 0.349\vec{k}$

(c) The time to point P is 20.05 seconds. The time from P to the surface is 8 seconds.
Therefore, he needs to hold his breath for 28 seconds.

Exercise 4.7

1. (a) $x = 1 - 3p = 7 + q \Rightarrow 3p + q = -6 \ \ldots\ldots\ldots\ (1)$

$y = 1 + 2p = 5 - q \Rightarrow 2p + q = 4 \ \ldots\ldots\ldots\ (2)$

Solving (1) and (2) gives $p = -10$, $q = 24$. $z = -4 + p = -4 - 10 = -14$, and

$z = 10 - q = 10 - 24 = -14$. Because the values of p and q found in (1) and (2) give a

consistent value for z, the lines intersect. The point of intersection is $(31, -19, -14)$.

(b) $x = p = -3 + 2q \Rightarrow p - 2q = -3$ ...... (1)

$y = 2 + p = -1$ .... (2)

Solving (1) and (2) gives $p = -3$ and $q = 0$, $z = 1 + p = 1 - 3 = -2$ and
$z = 2 + 3q = 2 + 3 \times 0 = 2 \neq -2$. Therefore the values of p and q found in (1) and (2) give
inconsistent values of z, so the lines do not intersect.

(c) $x = 7 + 2p = 2 + 2q \Rightarrow 2p - 2q = -5$ ...... (1)

$y = 2 = 2 + 2q \Rightarrow q = 0$ .... (2)

Solving (1) and (2) gives $p = -\dfrac{5}{2}$, $q = 0$. $z = 1$, and $z = 1 + q = 1 + 0 = 1$. Because the

values of p and q found in (1) and (2) give a consistent value for z, the lines intersect.

The point of intersection is $(2,\ 2,\ 1)$.

(d) $x = 3p = 6 + 3q \Rightarrow p - q = 2$ .... (1)

$y = -5p = -11$ .... (2)

Solving (1) and (2) gives, $p = \dfrac{11}{5}$, $q = \dfrac{1}{5}$. $z = 8p = \dfrac{88}{5}$ and $z = -q = -\dfrac{1}{5} \neq \dfrac{88}{5}$. The values

of p and q found in (1) and (2) give inconsistent values of z, so the lines do not intersect.

2.　　(a) Equating x components: $1 + pc = -2 + 5q$ ...... (1)

Equating y components: $-1 + p = q$ .... (2)

Equating z components: $0 + p = 1 + 3q$ .... (3)

Solving (2) and (3) $\Rightarrow p - q = 1$, $p - 3q = 1 \Rightarrow q = 0$, $p = 1$. Then, in (1), $1 + c = -2 \Rightarrow c = -3$.

(b) $p = 3 + 2q$, $-p = c - q$, $-1 + p = -3 - 3q$. Therefore, $p = 1$, $q = -1$ and $c = q - p = -2$.

(c) $1 + 2p = 2 + 6q$, $1 + p = cq$, $1 + p = 5 + q$. Therefore $p = \dfrac{23}{4}$, $q = \dfrac{7}{4}$ and $c = \dfrac{27}{7}$.

3.　　Let the first equation represent l_1 and the second l_2, in each pair.

(a) $(1,\ 3)$ lies on l_1 so equating the x components of l_1 and l_2 gives $1 = 3 - t \Rightarrow t = 2$ and
equating the y components gives $3 = 3 + 2t \Rightarrow t = 0$. Because the values of t are
inconsistent, $(1,\ 3)$ does not lie on l_2 and so the lines are distinct.

(b) $(2,\ 0,\ 5)$ lies on l_1. Equating components: $2 = -7 + 3t \Rightarrow t = 3$, $0 = -6 + 2t \Rightarrow t = 3$, $5 = 5$.
Because all three equations give consistent results, $(2,\ 0,\ 5)$ lies on l_2 so that l_1 and l_2
are coincident.

(c) $(2, -5, -1)$ lies on l_1. Equating components: $2 = -1 + 3t \Rightarrow t = 1$, $-5 = 1 - 6t \Rightarrow t = 1$,
$-1 = -13 + 12t \Rightarrow t = 1$. Because all three equations give consistent results, $(2, -5, -1)$ lies
on l_2 so that l_1 and l_2 are coincident.

(d) $(0,\ 4)$ lies on l_1. Equating components: $0 = -1 + 3t \Rightarrow t = \dfrac{1}{3}$, $4 = -11 + t \Rightarrow t = 15 \neq \dfrac{1}{3}$

Because the values of t are inconsistent, $(0,\ 4)$ does not lie on l_2 and so the lines are
distinct.

(e) $(3,\ 10, -1)$ lies on l_1. Equating components: $3 = -5 + 3t \Rightarrow t = \dfrac{8}{3}$, $10 = 10$,

$-1 = 17 - t \Rightarrow t = 18 \neq \dfrac{8}{3}$. Because the values of t are inconsistent, $(3,\ 10, -1)$ does not lie

on l_2 and so the lines are distinct.

4   (a) $\overrightarrow{PQ} = \overrightarrow{OQ} - \overrightarrow{OP} = \begin{pmatrix} -2 \\ 8 \\ -4 \end{pmatrix} - \begin{pmatrix} 1 \\ 5 \\ -2 \end{pmatrix} = \begin{pmatrix} -3 \\ 3 \\ -2 \end{pmatrix}$ (b) $\vec{r} = \begin{pmatrix} 1 \\ 5 \\ -2 \end{pmatrix} + s \begin{pmatrix} -3 \\ 3 \\ -2 \end{pmatrix}$

    (c) (i) At R $x = 1 - 3s = 1 + t$ .... (1)

                $y = 5 + 3s = 5 + kt$ .... (2)

                $z = -2 - 2s = 3 + t$ .... (3)

        subtracting (1) from (3) $-3 + s = 2 \Rightarrow s = 5$ and then $t = -3s = -15$

        and, in (2) $5 + 3 \times 5 = 5 - 15k \Rightarrow k = -1$

    (ii) At R $x = 1 - 3 \times 5 = -14$, $y = 5 + 3 \times 5 = 20$, $z = -2 - 2 \times 5 = -12$ so R is $(-14,\ 20, -12)$

5.   (a) $\vec{r} = \begin{pmatrix} 9.2 \\ 0 \\ 17 \end{pmatrix} + t \begin{pmatrix} 2 \\ 15 \\ 5 \end{pmatrix}$

    (b) Let t_1 be the time when the aircraft reaches Q and t_2 be the time when the missile

        reaches Q. Equating x and y components: $7 + 6t_1 = 9.2 + 2t_2$ ......(1)

                                $6 = 15t_2$ .... (2)

        Solving (1) and (2) gives $6t_1 - 2t_2 = 2.2$, $t_2 = 0.4 \Rightarrow t_1 = 0.5$

        Now the z component of the aircraft's path is $15 + 8t_1 = 19$ and the z component of the

        missile's path is $17 + 5t_2 = 19$, and, because these values are consistent, the paths intersect.

        The point of intersection Q is $(10,\ 6,\ 19)$.

    (c) At Q, $t_1 = 0.5$ and $t_2 = 0.4$ so that the aircraft reaches Q after 30 seconds. However, the

        missile reaches Q after 24 seconds, so that the missile goes in front of the aircraft.

Solutions to Unit 4 Review Exercise

Part 1

1. $\left(\vec{a} + t\vec{b} \right) \cdot \vec{c} = \begin{pmatrix} 1 + 4t \\ 1 + t \\ 2t \end{pmatrix} \cdot \begin{pmatrix} -3 \\ 2 \\ 6 \end{pmatrix} = 0 \Rightarrow -3 - 12t + 2 + 2t + 12t = 0 \Rightarrow t = \dfrac{1}{2}$

2. (a) $\overrightarrow{AB} = \begin{pmatrix} 2 \\ -2 \end{pmatrix} - \begin{pmatrix} 1 \\ 3 \end{pmatrix} = \begin{pmatrix} 1 \\ -5 \end{pmatrix}$,

(b) $\begin{pmatrix} 5 \\ 1 \end{pmatrix} \cdot \begin{pmatrix} 1 \\ -5 \end{pmatrix} = 5 \times 1 + 1 \times (-5) = 0$ so $\begin{pmatrix} 5 \\ 1 \end{pmatrix}$ is perpendicular to $\overrightarrow{AB}$.

(c) The direction of L is $\begin{pmatrix} 5 \\ 1 \end{pmatrix}$ so the vector equation of L is $\vec{r} = \begin{pmatrix} 1 \\ 3 \end{pmatrix} + t \begin{pmatrix} 5 \\ 1 \end{pmatrix}$.

3. (a) $\vec{a} \cdot \vec{b} = \left(6\vec{i} - 8\vec{j} + 2\vec{k}\right) \cdot \left(\vec{i} + 4\vec{j} + p\vec{k}\right) = 6 - 32 + 2p = 0 \Rightarrow p = 13$

(b) $\vec{a} + q\vec{b} = \begin{pmatrix} 6 \\ -8 \\ 2 \end{pmatrix} + q \begin{pmatrix} 1 \\ 4 \\ p \end{pmatrix} = \begin{pmatrix} 6+q \\ -8+4q \\ 2+qp \end{pmatrix}$. If q is parallel to the x-axis then $\vec{a} = k\vec{i}$ for some $k \in \mathbb{R}$

Therefore
$$6 + q = k \ \ldots\ldots\ldots\ldots\ (1)$$
$$-8 + 4q = 0 \ \ldots\ldots\ldots\ldots\ (2)$$
$$2 + pq = 0 \ \ldots\ldots\ldots\ldots\ (3)$$
From (2) $q = 2$, from (3) $2 + 2p = 0 \Rightarrow p = -1$ therefore $p = -1$, $q = 2$.

4. (a) Assume that L_1 and L_2 intersect at P. Then at P:
$$x = 1 + 2s = 2 + 4t \ \ldots\ldots\ldots.(1)$$
$$y = 1 = -2 \ \ldots\ldots\ldots\ (2)$$
$$z = -1 + 3s = 7 + ct \ \ldots\ldots\ (3)$$
Equation (2) is obviously is inconsistent so that the assumption that L_1 and L_2 intersect is wrong and so L_1 and L_2 do not intersect.

(b) If L_1 and L_2 are parallel, $c = 6$.

5. (a) $\vec{r} = \begin{pmatrix} 4 \\ 0 \\ 5 \end{pmatrix} + t \begin{pmatrix} -3 \\ 1 \\ 1 \end{pmatrix}$

(b) $\begin{pmatrix} 2 \\ -2 \\ c \end{pmatrix} = \begin{pmatrix} 4 \\ 0 \\ 5 \end{pmatrix} + t \begin{pmatrix} -3 \\ 1 \\ 1 \end{pmatrix} \Rightarrow -2 = t$ and $c = 5 + t \Rightarrow c = 5 - 2 = 3$.

(c) At C, $13 = 4 - 3t \Rightarrow t = -3$. $2 = 5 + t \Rightarrow t = -3$. Since, in each case, $t = -3$, the results are consistent and C lies on L.

6. A; (iii) perpendicular to L B: (ii) coincident with L C: (iv) intersects L
D: (i) distinct but parallel to L E: (iv) none of these.

7. (a) At C: $x = 4 + 4s = 7 + t \ \ldots\ldots\ldots\ldots\ (1)$
$$y = 4 + 3s = 0 + 7t \ \ldots\ldots\ldots\ (2)$$

Solving (1) and (2) we obtain $s = 1$, $t = 1$ so coordinates of C are $(8, 7)$

(b) L_3 is perpendicular to L_2 and $\begin{pmatrix} 7 \\ -1 \end{pmatrix}$ is a vector parallel to $[AB]$ so L_3 is $\vec{r} = \begin{pmatrix} 0 \\ 1 \end{pmatrix} + p \begin{pmatrix} 7 \\ -1 \end{pmatrix}$.

(c) At B: $\qquad 0 + 7p = 7 + t \ \dots\dots\dots (3)$

$\qquad\qquad\quad 1 - p = 7t \ \dots\dots\dots\dots (4)$

Solving (3) and (4) we obtain $t = 7p - 7 \Rightarrow 1 - p = 7(7 - p) \Rightarrow 1 - p = 49p - 49 \Rightarrow p = 1$

and so the coordinates of B are $(7,\ 0)$.

(d) $\overrightarrow{OD} = \overrightarrow{OC} + \overrightarrow{CD} = \overrightarrow{OC} + \overrightarrow{BA}$, but $\overrightarrow{BA} = \begin{pmatrix} -7 \\ 1 \end{pmatrix}$ so $\overrightarrow{OD} = \begin{pmatrix} 8 \\ 7 \end{pmatrix} + \begin{pmatrix} -7 \\ 1 \end{pmatrix} = \begin{pmatrix} 1 \\ 8 \end{pmatrix}$ and the coordinates

of D are $(1,\ 8)$.

8. (a) $\left(2\vec{i} + \vec{j} + \vec{k}\right) \bullet \left(\vec{i} - \vec{j} - \vec{k}\right) = 2 \times 1 + 1 \times (-1) + 1 \times (-1) = 2 - 1 - 1 = 0$ so L_1 and L_2 are

perpendicular.

(b) Assume that L_1 and L_2 intersect at P. Then at P:

$\qquad\qquad 1 + 2s = 2 + t \ \dots\dots\dots\dots\dots (1)$

$\qquad\qquad 0 + s = -1 - t \ \dots\dots\dots\dots (2)$

$\qquad\qquad -2 + s = 0 - t \ \dots\dots\dots\dots (3)$

$(2) - (3)$ gives $2 = -1$ which is inconsistent so our assumption is wrong and L_1 does not

intersect L_2.

9. (a) $|\vec{a}| = \sqrt{1^2 + 2^2 + n^2} = \sqrt{5 + n^2} = 3 \Rightarrow 5 + n^2 = 9 \Rightarrow n^2 = 4 \Rightarrow n = \pm 2$

(b) $\vec{p} \bullet \vec{r} = 0 \Rightarrow 1 \times 16 + 3m + n = 0 \ \dots\dots\dots\dots (1)$

$\qquad \vec{q} \bullet \vec{r} = 0 \Rightarrow 2 \times 16 + m - 5n = 0 \ \dots\dots\dots\dots (2)$

Solving (1) and (2) we obtain $m = -7,\ n = 5$

10. (a) $\left(\vec{a} + \vec{b}\right) \bullet \vec{c} = \left(4\vec{i} - 2\vec{j} + 4\vec{k}\right) \bullet \left(-2\vec{i} + \vec{j} - 2\vec{k}\right) = -8 - 2 - 8 = -18$

(b) $\left|\vec{a} + \vec{b}\right| = \sqrt{4^2 + (-2)^2 + 4^2} = 6$, $|\vec{c}| = \sqrt{(-2)^2 + 1^2 + (-2)^2} = 3$ so $\left|\vec{a} + \vec{b}\right||\vec{c}| = 6 \times 3 = 18$

(c) Since $\left|\vec{a} + \vec{b}\right||\vec{c}| = \left(\vec{a} + \vec{b}\right) \bullet \vec{c}$ $\left(\vec{a} + \vec{b}\right) \bullet \vec{c} = \left|\vec{a} + \vec{b}\right||\vec{c}| \cos\theta \Rightarrow \cos\theta = -1 \Rightarrow \theta = 180° $ and $\vec{a} + \vec{b}$ is

parallel to $\vec{c}$.

Part 2

11. (a) $\overrightarrow{AB} = \begin{pmatrix} 2 \\ -1 \\ 4 \end{pmatrix} - \begin{pmatrix} 1 \\ 3 \\ 5 \end{pmatrix} = \begin{pmatrix} 1 \\ -4 \\ -1 \end{pmatrix}$ $\qquad$ (b) $\left|\overrightarrow{AB}\right| = \sqrt{1^2 + (-4)^2 + (-1)^2} = \sqrt{18} = \sqrt{9}\sqrt{2} = 3\sqrt{2}$

(c) $\overrightarrow{AC} = \overrightarrow{OC} - \overrightarrow{OA} \Rightarrow \overrightarrow{OC} = \overrightarrow{OA} + \overrightarrow{AC} = \left(-\vec{i} + 2\vec{j} - 3\vec{k}\right) + \left(\vec{i} + 3\vec{j} + 5\vec{k}\right) = 5\vec{j} + 2\vec{k}$

(d) $\cos B\hat{A}C = \dfrac{\overrightarrow{AB} \bullet \overrightarrow{AC}}{\left|\overrightarrow{AB}\right|\left|\overrightarrow{AC}\right|} = \dfrac{1 \times -1 + (-4) \times 2 + (-1) \times (-3)}{3\sqrt{2}\sqrt{14}} = -\dfrac{1}{\sqrt{7}} \Rightarrow B\hat{A}C = 112.2076...° = 112°$

12.　(a) $\overrightarrow{AB} = \begin{pmatrix} 3 \\ 2 \\ -5 \end{pmatrix} - \begin{pmatrix} 1 \\ 2 \\ -4 \end{pmatrix} = \begin{pmatrix} 2 \\ 0 \\ -1 \end{pmatrix}$

(b) $\vec{r} = \begin{pmatrix} 1 \\ 2 \\ -4 \end{pmatrix} + s \begin{pmatrix} 2 \\ 0 \\ -1 \end{pmatrix}$

(c) Both L_1 and L_2 are parallel to $\begin{pmatrix} 2 \\ 0 \\ -1 \end{pmatrix}$ so they are parallel to each other.

(d) Assume that C lies on L_2, then $-2 = -8 - 2t \Rightarrow 2t = -6 \Rightarrow t = -3$
$$3 = 3$$
$$-10 = -7 + t \Rightarrow t = -3$$
Since these results are consistent C lies on L_2.

(e) $\overrightarrow{AC} = \overrightarrow{OC} - \overrightarrow{OA} = \begin{pmatrix} -2 \\ 3 \\ -10 \end{pmatrix} - \begin{pmatrix} 1 \\ 2 \\ -4 \end{pmatrix} = \begin{pmatrix} -3 \\ 1 \\ -6 \end{pmatrix}$

(f) $\begin{pmatrix} -3 \\ 1 \\ -6 \end{pmatrix} \cdot \begin{pmatrix} 2 \\ 0 \\ -1 \end{pmatrix} = -6 + 0 + 6 = 0$ and since L_1 and L_2 are both parallel to $\begin{pmatrix} 2 \\ 0 \\ -1 \end{pmatrix}$, L_1 and L_2 are

both perpendicular to $\overrightarrow{AC}$.

(g) The distance between L_1 and L_2 is $\left|\overrightarrow{AC}\right| = \sqrt{(-3)^2 + 1^2 + (-6)^2} = \sqrt{46} = 6.782329... = 6.78$

13.　(a) $\overrightarrow{BC} = \overrightarrow{OC} - \overrightarrow{OB} = \begin{pmatrix} 3 \\ 3 \\ 1 \end{pmatrix} - \begin{pmatrix} 1 \\ 8 \\ 2 \end{pmatrix} = \begin{pmatrix} 2 \\ -5 \\ -1 \end{pmatrix}$

(b) (i) $\overrightarrow{OA} = -\overrightarrow{BC} = \begin{pmatrix} -2 \\ 5 \\ 1 \end{pmatrix}$　(ii) $\cos A\hat{O}B = \dfrac{\overrightarrow{OA} \cdot \overrightarrow{OB}}{\left|\overrightarrow{OA}\right|\left|\overrightarrow{OB}\right|} = \dfrac{1 \times (-2) + 8 \times 5 + 2 \times 1}{\sqrt{1^2 + 8^2 + 2^2}\sqrt{(-2)^2 + 5^2 + 1^2}} = \dfrac{40}{\sqrt{69}\sqrt{30}}$

$$\Rightarrow A\hat{O}B = 28.45712...^\circ = 28.5^\circ$$

(c) $\overrightarrow{BD} = 2\overrightarrow{BC} = 2\begin{pmatrix} 2 \\ -5 \\ -1 \end{pmatrix} = \begin{pmatrix} 4 \\ -10 \\ -2 \end{pmatrix}$, so $\overrightarrow{AD} = \overrightarrow{AB} + \overrightarrow{BD} = \overrightarrow{OC} + \overrightarrow{BD} = \begin{pmatrix} 3 \\ 3 \\ 1 \end{pmatrix} + \begin{pmatrix} 4 \\ -10 \\ -2 \end{pmatrix} = \begin{pmatrix} 7 \\ -7 \\ -1 \end{pmatrix}$

14. (a) $\overrightarrow{OA} = \begin{pmatrix} 1 \\ 5 \end{pmatrix}$, $\overrightarrow{OC} = \begin{pmatrix} 8 \\ 2 \end{pmatrix}$ (b) $\overrightarrow{OA} + \overrightarrow{OC} = \overrightarrow{OA} + \overrightarrow{AB} = \overrightarrow{OB} = \begin{pmatrix} 1 \\ 5 \end{pmatrix} + \begin{pmatrix} 8 \\ 2 \end{pmatrix} = \begin{pmatrix} 9 \\ 7 \end{pmatrix}$

(c) $\cos A\hat{M}B = \dfrac{\overrightarrow{AM} \cdot \overrightarrow{BM}}{|\overrightarrow{AM}||\overrightarrow{BM}|} = \dfrac{\overrightarrow{AC} \cdot \overrightarrow{OB}}{|\overrightarrow{AC}||\overrightarrow{OB}|} = \dfrac{7 \times 9 + (-3) \times 7}{\sqrt{58}\sqrt{130}} = \dfrac{42}{\sqrt{58}\sqrt{130}}$, noting that $\overrightarrow{AC} = \begin{pmatrix} 7 \\ -3 \end{pmatrix}$.

$\Rightarrow A\hat{M}B = 118.9264... = 119°$

(d)

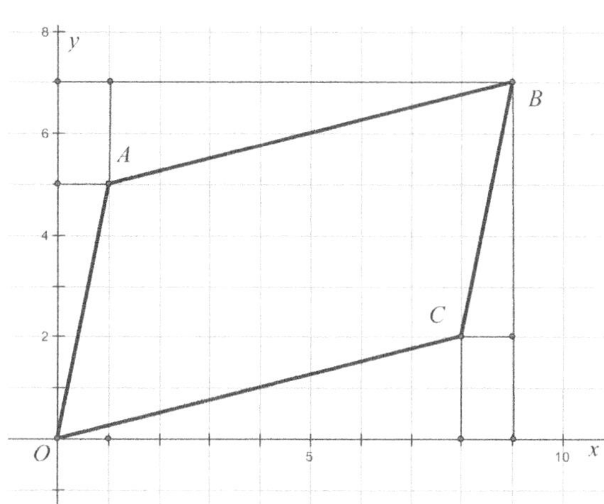

The diagram suggests an easy way to find the area of the parallelogram $OABC$.
Consider a rectangle of $7 \times 9 = 63$ and subtract from it two rectangles of $1 \times 2 = 2$,

Two right triangles of $\dfrac{1}{2} \times 2 \times 8 = 8$ and two right triangles of $\dfrac{1}{2} \times 1 \times 5 = \dfrac{5}{2}$. So the area of

the parallelogram is $63 - 2\left(2 + 8 + \frac{5}{2}\right) = 63 - (4 + 16 + 5) = 63 - 25 = 38$.

15. (a) At P, equating x-components: $-2 - 4p = 6 = 2 - s$
$$\Rightarrow 4p = -8 \Rightarrow p = -2 \text{ and } s = -4$$
equating y-components: $3 + ap = -4 + q = 3 + s$
$$\Rightarrow 3 - 2a = -4 + q = -1$$
$$\Rightarrow 2a = 4 \Rightarrow a = 2 \text{ and } q = 3$$
equating z-components: $4 + p = b + 2q = c + 3s$
$$\Rightarrow 2 = b + 6 = c - 12$$
$$\Rightarrow b = -4 \text{ and } c = 14$$
Therefore $a = 2, \ b = -4, \ c = 14$

(b) The coordinates of P are $(6, -1, \ 2)$

(c) The vectors which represent the directions of L_1 and L_2 are $\left(-4\vec{i} + 2\vec{j} + \vec{k}\right)$ and $\vec{j} + 2\vec{k}$, so

the acute angle between L_1 and L_2 is θ, where $\cos\theta = \dfrac{(-4) \times 0 + 2 \times 1 + 1 \times 2}{\sqrt{21}\sqrt{5}} = \dfrac{4}{\sqrt{105}}$,

$\Rightarrow \theta = 67.02309... = 67.0°$

16. (a) Speed of balloon is $\sqrt{200^2 + 200^2 + 100^2} = 100\sqrt{2^2 + 2^2 + 1^2} = 300$ m/min

(b) When $z = 300$, $0 + 100t = 300 \Rightarrow t = 3$

(c) Balloon reaches height of 500m after 5 minutes so then its position is $(1150, 1250, 500)$. It then goes down and reaches the ground when $500 + (t-5)(-20) = 0 \Rightarrow t = 30$. So, after a further 25 minutes the balloon reaches the ground and the time the balloon was in the air is 30 minutes.

(d) Its position when $t = 30$ is $(1150 - 250, 1250 - 500, 500 - 500) = (900, 750, 0)$ so distance of balloon A is $\sqrt{(900 - 150)^2 + (750 - 250)^2} = 901.3878... = 901$m

17. (a) $\overrightarrow{OP} = t\begin{pmatrix} 4 \\ 3 \end{pmatrix}$ so at $t = 1$, $\overrightarrow{OP} = \begin{pmatrix} 4 \\ 3 \end{pmatrix}$,

at $t = 2$, $\overrightarrow{OP} = \begin{pmatrix} 8 \\ 6 \end{pmatrix}$ and at $t = 5$, $\overrightarrow{OP} = \begin{pmatrix} 20 \\ 15 \end{pmatrix}$

At $t = 1$, $\overrightarrow{AP} = \begin{pmatrix} -6 \\ -10 \end{pmatrix}$, at $t = 2$, $\overrightarrow{AP} = \begin{pmatrix} -2 \\ -7 \end{pmatrix}$

and at $t = 5$ $\overrightarrow{AP} = \begin{pmatrix} 10 \\ 2 \end{pmatrix}$.

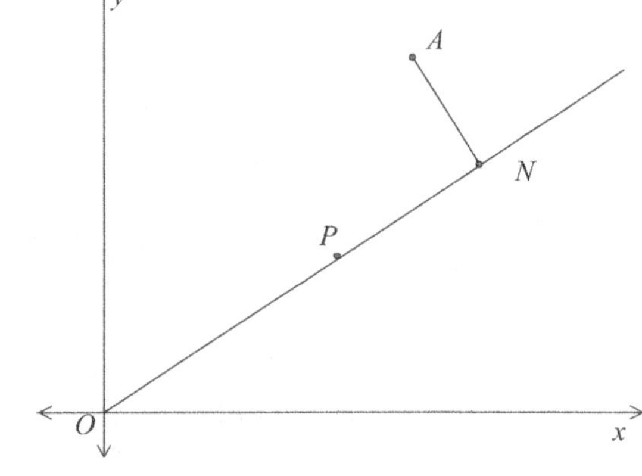

(b) $\overrightarrow{AP} = \overrightarrow{OP} - \overrightarrow{OA} = t\begin{pmatrix} 4 \\ 3 \end{pmatrix} - \begin{pmatrix} 10 \\ 13 \end{pmatrix} = \begin{pmatrix} 4t - 10 \\ 3t - 13 \end{pmatrix}$

(c) $|\overrightarrow{AP}| = 15 \Rightarrow \sqrt{(-10 + 4t)^2 + (-13 + 3t)^2} = 15 \Rightarrow 25t^2 - 158t + 44 = 0$

$\Rightarrow t = 0.2919693..., \ 6.028030....$ Therefore the ship is 15km from the rock for the first time when $t = 0.2919693...$ which is $60 \times .2919693... = 17.51815...$ 18 minutes, to the nearest minute, after leaving the harbor. So the first time when the ship is 15 km from rock is 09:18.

(d) $\overrightarrow{AP} \cdot \begin{pmatrix} 4 \\ 3 \end{pmatrix} = \begin{pmatrix} 4t - 10 \\ 3t - 13 \end{pmatrix} \cdot \begin{pmatrix} 4 \\ 3 \end{pmatrix} = 25t - 79$

(e) The ship is nearest to A when $\overrightarrow{AP} \cdot \begin{pmatrix} 4 \\ 3 \end{pmatrix} = 0 \Rightarrow 25t - 79 = 0 \Rightarrow t = \dfrac{79}{25} = 3.16$ so the time will be 12:10

Solutions to Unit 5 Exercises

Exercise 5.1

In each case, the graph of the function is drawn and the tangents found at regular intervals. The gradients of these tangents are then entered into lists and the suggested regression carried out.

1.

x	0	1	2	3	4	5	6
m	0	3	12	27	48	75	108

$f'(x) = 3x^2$

2.

x	-3	-2	-1	0	1	2
m	-13	-7	-1	5	11	17

$f'(x) = 6x + 5$

3.

x	-2	-1	0	1	2
m	-32	-4	0	4	32

$f'(x) = 4x^3$

4

x	1	2	3	4	5
m	0.5	0.3535	0.2887	0.25	0.2236

$f'(x) = 0.5x^{-0.5}$

5. $\quad f'(x) = 3ax^2 + 2bx + c$

Exercise 5.2

1. $\quad f'(x) = \lim_{h \to 0}\left(\dfrac{f(x+h) - f(x)}{h}\right) \Rightarrow f'(x) = \lim_{h \to 0}\left(\dfrac{(x+h)^2 + 3(x+h) - (x^2 + 3x)}{h}\right)$

$= \lim_{h \to 0}\left(\dfrac{x^2 + 2xh + h^2 + 3x + 3h - x^2 - 3x}{h}\right) = \lim_{h \to 0}\left(\dfrac{2xh + h^2 + 3h}{h}\right)$

$\Rightarrow f'(x) = \lim_{h \to 0}\left(\dfrac{h(2x + h + 3)}{h}\right) = \lim_{h \to 0}(2x + 3 + h) \Rightarrow f'(x) = 2x + 3$

2. $$f'(x) = \lim_{h \to 0}\left(\frac{f(x+h)-f(x)}{h}\right) = \lim_{h \to 0}\left(\frac{(3-2(x+h))-(3-2x)}{h}\right)$$

$$\Rightarrow f'(x) = \lim_{h \to 0}\left(\frac{-2h}{h}\right) = -2$$

3. $$f'(x) = \lim_{h \to 0}\left(\frac{\left((x+h)^3 + 4(x+h)^2\right) - \left(x^3 + 4x^2\right)}{h}\right)$$

$$\Rightarrow f'(x) = \lim_{h \to 0}\left(\frac{\left(x^3 + 3x^2h + 3xh^2 + h^3 + 4(x^2 + 2xh + h^2)\right) - \left(x^3 + 4x^2\right)}{h}\right)$$

$$\Rightarrow f'(x) = \lim_{h \to 0}\left(\frac{3x^2h + 3xh^2 + h^3 + 8xh + 4h^2}{h}\right) = \lim_{h \to 0}\left(\frac{h\left(3x^2 + 3xh + h^2 + 8x + 4h\right)}{h}\right)$$

$$\Rightarrow f'(x) = \lim_{h \to 0}\left(3x^2 + 3xh + h^2 + 8x + 4h\right) = 3x^2 + 8x$$

Exercise 5.3

1. (a) $f'(x) = 7x^6$ (b) $f'(x) = 45x^4$ (c) $f'(x) = 1 - 9x^2$ (d) $f'(x) = 6x$

    (e) $f'(x) = 8x^3 + 6x - 6$ (f) $f'(x) = \dfrac{5}{3}x^4 - \dfrac{6}{5}x^2 - \dfrac{1}{2}$ (g) $f'(x) = x^2$

    (h) $f'(x) = 1 + \dfrac{3}{x^2}$ (i) $f'(x) = 5x^4 + \dfrac{5}{x^6}$ (j) $f'(x) = (2k+1)x^{2k}$

2. (a) $g'(x) = \dfrac{3}{2\sqrt{x}}$ (b) $g'(x) = 4x^3 - \dfrac{1}{x^2}$

    (c) $g'(x) = -\dfrac{3}{x^4} + \dfrac{2}{x^2}$ (d) $g'(x) = \dfrac{1}{2\sqrt{x}} - \dfrac{1}{2\sqrt{x^3}}$

3. (a) $f'(x) = 4x^3 \Rightarrow f'(2) = 32$ (b) $f'(x) = 4x^3 \Rightarrow f'(2) = 32$

    (c) $f'(x) = 2x - 9 \Rightarrow f'(2) = -5$ (d) $f'(x) = -\dfrac{2}{3}x^{-3} \Rightarrow f'(2) = -\dfrac{1}{12}$

    (e) $f'(x) = 2x^{-\frac{1}{2}} \Rightarrow f'(2) = \sqrt{2}$

Exercise 5.4

1. (a) $\dfrac{dy}{dx} = 4x^3$ (b) $\dfrac{dy}{dx} = 24x^5$ (c) $\dfrac{dy}{dx} = -21x^2$ (d) $\dfrac{dy}{dx} = 48x^3 - 21x^2 - 6x + 5$

    (e) $y = 4x^{-3} + 3x^{-2} - \dfrac{2}{3}x^{-1} \Rightarrow \dfrac{dy}{dx} = -12x^{-4} - 6x^{-3} + \dfrac{2}{3}x^{-2} = -\dfrac{12}{x^4} - \dfrac{6}{x^3} + \dfrac{2}{3x^2}$

(f) $y = 3x^{\frac{1}{2}} + x^{\frac{3}{2}} + x^{\frac{5}{2}} \Rightarrow \dfrac{dy}{dx} = \dfrac{3}{2}x^{-\frac{1}{2}} + \dfrac{3}{2}x^{\frac{1}{2}} + \dfrac{5}{2}x^{\frac{3}{2}} = \dfrac{3}{2\sqrt{x}} + \dfrac{3\sqrt{x}}{2} + \dfrac{5\sqrt{x^3}}{2}$

2. (a) $y = x^4 + 3x^3 \Rightarrow \dfrac{dy}{dx} = 4x^3 + 9x^2$ (b) $y = 2x^3 - 4x^2 + 3x - 6 \Rightarrow \dfrac{dy}{dx} = 6x^2 - 8x + 3$

(c) $y = x^5 + 2x^3 + x \Rightarrow \dfrac{dy}{dx} = 5x^4 + 6x^2 + 1$ (d) $y = 1 + 3x^{-1} \Rightarrow \dfrac{dy}{dx} = -3x^{-2} = -\dfrac{3}{x^2}$

3. (a) $f(x) = 1 - 2x - 3x^2 \Rightarrow f'(x) = -2 - 6x$

(b) $f(x) = x^4 + x^3 + x^2 + x \Rightarrow f'(x) = 4x^3 + 3x^2 + 2x + 1$

(c) $f(x) = 24 + 3x - 8x^3 - x^4 \Rightarrow f'(x) = 3 - 24x^2 - 4x^3$

(d) $f(x) = x^{\frac{5}{2}} + x^{\frac{3}{2}} + x^{-\frac{1}{2}} \Rightarrow f'(x) = \dfrac{5}{2}x^{\frac{3}{2}} + \dfrac{3}{2}x^{\frac{1}{2}} - \dfrac{1}{2}x^{-\frac{3}{2}} = \dfrac{5\sqrt{x^3}}{2} + \dfrac{3\sqrt{x}}{2} - \dfrac{1}{2\sqrt{x^3}}$

(e) $f(x) = x^2 + 1 + \dfrac{1}{4}x^{-2} \Rightarrow f'(x) = 2x - \dfrac{1}{2}x^{-3} = 2x - \dfrac{1}{2x^3}$

4. (a) $\dfrac{dy}{dx} = 2x$, $x = 3$, so gradient is 6. (b) $f'(x) = -2x^{-3} - 12x^{-4} \Rightarrow f'(1) = -14$

(c) $g'(x) = \dfrac{3}{2}x^{-\frac{1}{2}} + x^{-2} \Rightarrow g'\left(\dfrac{1}{4}\right) = 3 + 16 = 19$ (d) $\dfrac{dy}{dx} = 3x^2 + 5$, $x = -2$, so gradient is 17.

5. (a) $\dfrac{dy}{dx} = 3x^2 = 48 \Rightarrow x = \pm 4$, so coordinates are $(-4,\ -64)$ and $(4, 64)$.

(b) $f'(x) = 2x + 5 = 8 \Rightarrow x = \dfrac{3}{2}$, so coordinates are $\left(\dfrac{3}{2}, \dfrac{63}{4}\right)$.

(c) $g'(x) = -4x^{-2} = -1 \Rightarrow x^2 = 4 \Rightarrow x = \pm 2$, so coordinates are $(-2,\ -1)$ and $(2, 3)$

6. $\dfrac{dy}{dx} = 2ax + b$. At $x = -2$, the gradient is -3, so $-4a + b = -3$. In addition, $-5 = 4a - 2b$.

Therefore, $a = \dfrac{11}{4}$, $b = 8$.

7. $f(x) = 2x^2 - 7x + 3 \Rightarrow f'(x) = 4x - 7 \Rightarrow f'(5) = 13$. Therefore, the tangent is

$y - 18 = 13(x - 5) \Rightarrow y = 13x - 47$ and $m = 13$, $c = -47$.

8. $\dfrac{dy}{dx} = 2x = 1$ at the tangent point. Therefore, at the tangent point, $x = \dfrac{1}{2}$ and $y = \dfrac{9}{2}$.

Then, $\dfrac{9}{2} = \left(\dfrac{1}{2}\right)^2 + k \Rightarrow k = \dfrac{17}{4}$.

Exercise 5.5

1.   (i)   (a) $f(a)<0$ (b) $f'(a)>0$ (c) $f''(a)>0$

     (ii)   (a) $f(a)>0$ (b) $f'(a)>0$ (c) $f''(a)=0$

     (iii) (a) $f(a)>0$ (b) $f'(a)=0$ (c) $f''(a)<0$

     (iv) (a) $f(a)<0$ (b) $f'(a)=0$ (c) $f''(a)=0$

2.   (a)   (b)

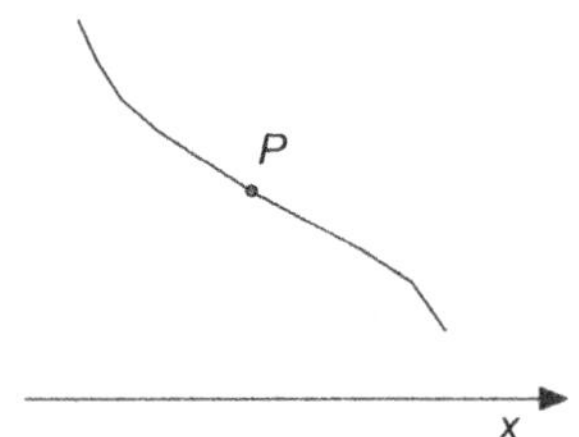   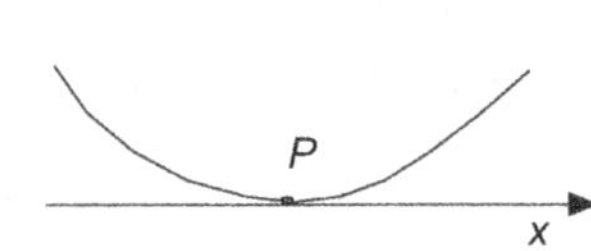

   (c)   (d)

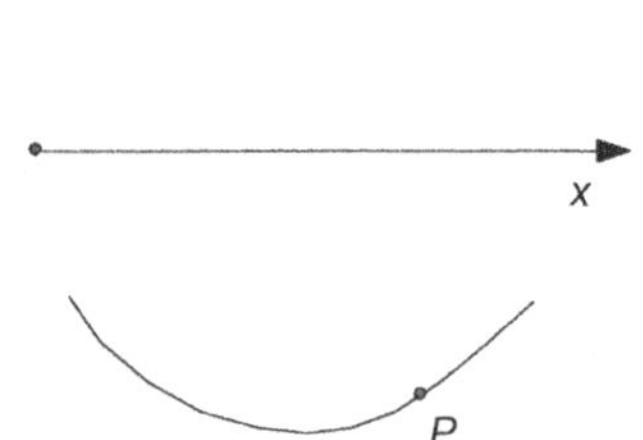   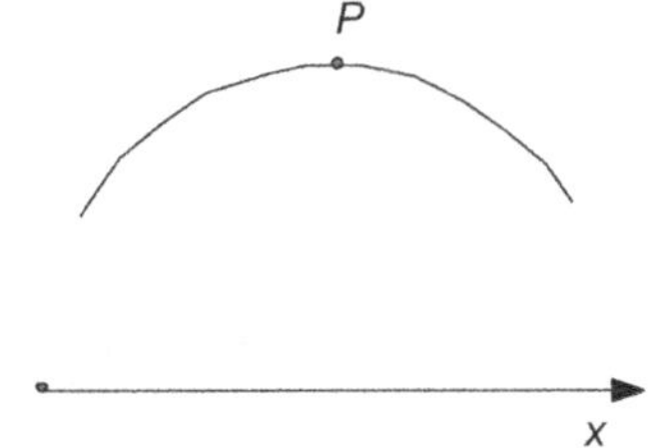

3.   (a), (iv)   (b), (i)   (c), (ii)   (d), (iii)

4.   (a)   (b)

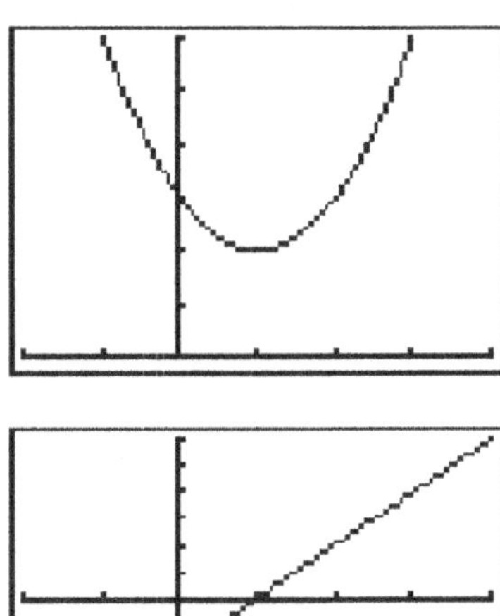   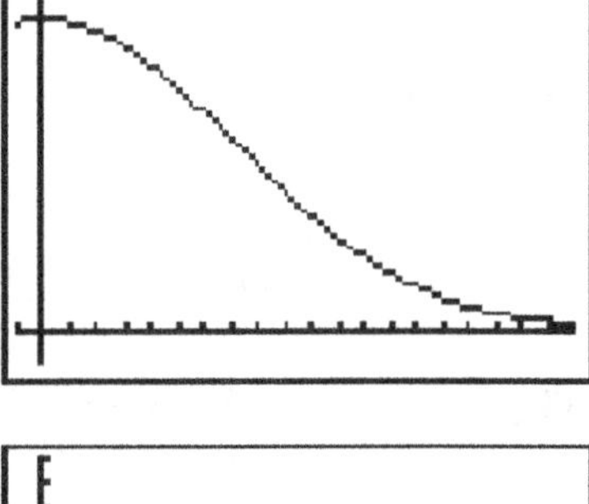

(c)

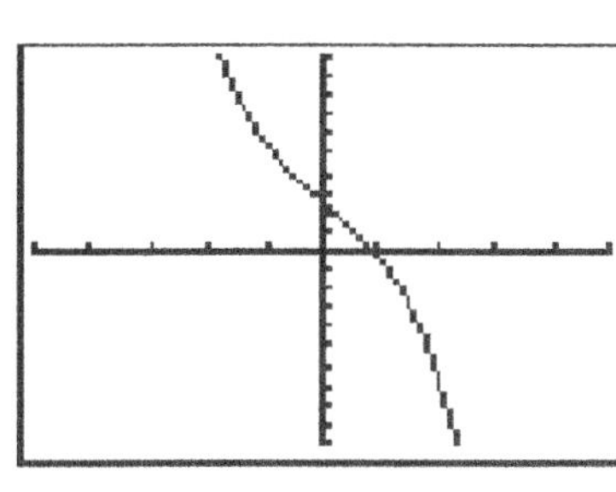

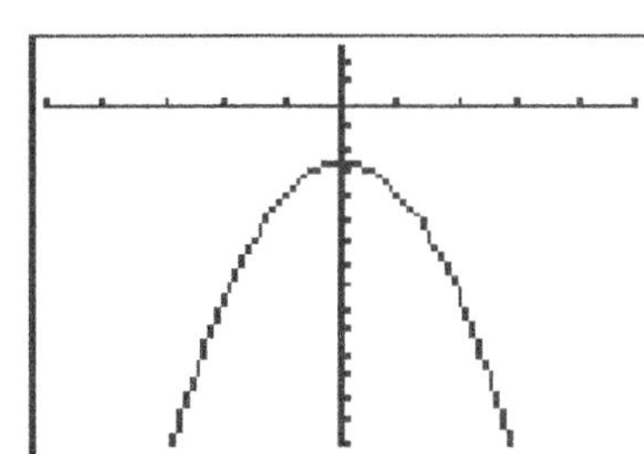

(d)

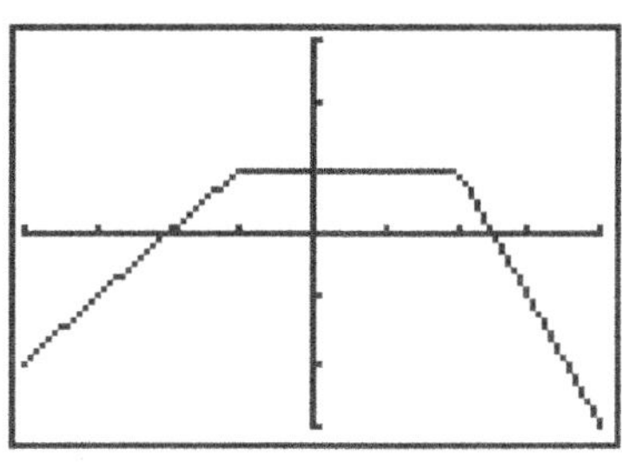

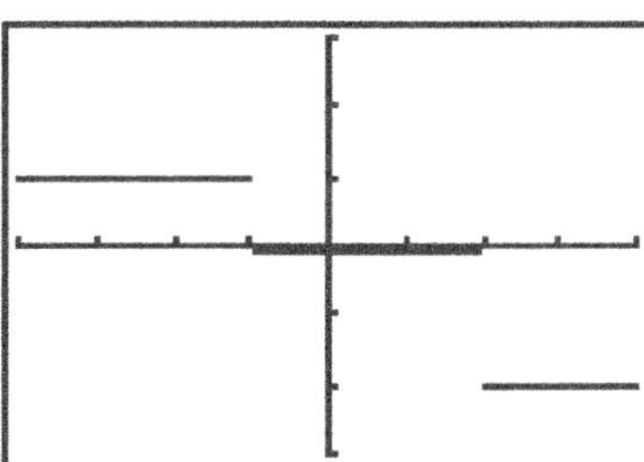

Exercise 5.6

1. (a) $f'(x) = -2 - 2x = 0 \Rightarrow x = -1$. So $f(-1) = 7$ is a maximum value

x	-2	-1	0
$f'(x)$	2	0	-2
Tangent	↗	→	↘

(b) $f'(x) = 2x + 7 = 0 \Rightarrow x = -\dfrac{7}{2}$. So $f(-3.5) = -\dfrac{65}{4}$ is a minimum value.

x	-4	-3.5	-3
$f'(x)$	-1	0	1
Tangent	↘	→	↗

(c) $f'(x) = 3x^2 - 10x - 8 = 0 \Rightarrow (3x + 2)(x - 4) = 0 \Rightarrow x = -\dfrac{2}{3}, \ x = 4$

x	-1	-2/3	0
$f'(x)$	5	0	-8
Tangent	↗	→	↘

x	3	4	5
$f'(x)$	-11	0	17
Tangent	↘	→	↗

So $f'\left(-\dfrac{2}{3}\right) = \dfrac{373}{27}$ is a local maximum, and $f(4) = -37$ is a local minimum.

(d) $f'(x) = 6x^2 + 2x - 4 = 0 \Rightarrow 3x^2 + x - 2 = 0 \Rightarrow (x + 1)(3x - 2) = 0 \Rightarrow x = -1, \ x = \dfrac{2}{3}$

x	-2	-1	0
$f'(x)$	16	0	-4
Tangent	↗	→	↘

x	0	0.667	1
$f'(x)$	-4	0	4
Tangent	↘	→	↗

So $f(-1) = -2$ is a local maximum, and $f\left(\dfrac{2}{3}\right) = -\dfrac{179}{27}$ is a local minimum.

(e) $f'(x) = 4x^3 - 256 = 0 \Rightarrow x^3 = 64 \Rightarrow x = 4$. So $f(4) = -753$
   is a minimum.

x	3	4	5
$f'(x)$	-148	0	244
Tangent	↘	→	↗

(f) $f'(x) = \dfrac{1}{2\sqrt{x}} - 2 = 0 \Rightarrow 4\sqrt{x} = 1 \Rightarrow x = \dfrac{1}{16}$. So $f\left(\dfrac{1}{16}\right) = \dfrac{1}{8}$
   is a maximum.

x	1/36	1/16	1
$f'(x)$	1	0	-1.5
Tangent	↗	→	↘

2.   (a)   (i) $\dfrac{dy}{dx} = 2x + 6$

       (ii) $2x + 6 = 0 \Rightarrow x = -3$

       (iii) $(-3,\ -1)$, $f''(x) = 2 > 0$ so $(-3,\ -1)$ has a minimum value.

   (b)   (i) $\dfrac{dy}{dx} = 3x^2 - 12$

       (ii) $3x^2 - 12 = 0 \Rightarrow x = \pm 2$

       (iii) $(-2,\ 19)$, $(2,\ -13)$, $f''(x) = 6x \Rightarrow f''(-2) = -12 < 0$ and $f''(2) = 12 > 0$. So
   $(-2,\ 19)$ is a local maximum and $(2,\ -13)$ is a local minimum.

   (c)   (i) $\dfrac{dy}{dx} = 5x^4 - 80$

       (ii) $5x^4 - 80 = 0 \Rightarrow x = \pm 2$

       (iii) $(-2,\ 128)$, $(2,\ -128)$, $f''(x) = 20x^3 \Rightarrow f''(-2) = -160 < 0$ and $f''(2) = 160 > 0$.
   So $(-2,\ 128)$ is a local maximum and $(2,\ -128)$ is a local minimum.

3.   (a)   (i) $f'(x) = 3x^2 + 9 \Rightarrow f''(x) = 6x$

       (ii) At points of inflection $f''(x) = 0$ so that $6x = 0 \Rightarrow x = 0$. Then $f(0) = -7$ and the
   coordinates of the point of inflection are $(0,\ -7)$.

   (b)   (i) $f'(x) = 2x + x^{-2} \Rightarrow f''(x) = 2 - 2x^{-3} = 2\left(1 - \dfrac{1}{x^3}\right)$.

       (ii) At points of inflection $f''(x) = 0$ so that $1 - \dfrac{1}{x^3} = 0 \Rightarrow x^3 = 1 \Rightarrow x = 1$.

       Then $f(1) = 0$ and the coordinates of the point of inflection are $(1,\ 0)$.

   (c)   (i) $f'(x) = 4x^3 + 9x^2 + 6x - 1 \Rightarrow f''(x) = 12x^2 + 18x + 6$

       (ii) At points of inflection $f''(x) = 0$ so that $2x^2 + 3x + 1 = 0 \Rightarrow (2x+1)(x+1) = 0$
   $\Rightarrow x = -1,\ x = -\dfrac{1}{2}$. Then $f(-1) = 7$ and $f\left(-\dfrac{1}{2}\right) = \dfrac{95}{16}$ and the coordinates of
   the two points of inflection are $(-1,\ 7)$ and $\left(-0.5,\ \tfrac{95}{16}\right)$.

Exercise 5.7

1. (a) $y = \dfrac{8-x}{3}$
    (b) $W = \dfrac{8x}{3} - \dfrac{x^2}{3}$
    (c) $\dfrac{dW}{dx} = \dfrac{8}{3} - \dfrac{2x}{3}$
    (d) $\dfrac{8}{3} - \dfrac{2x}{3} = 0 \Rightarrow x = 4$

    (e) Therefore W has a maximum value of $\dfrac{16}{3}$ when $x = 4$.

x	3	4	5
$\dfrac{dW}{dx}$	$\dfrac{2}{3}$	0	$-\dfrac{2}{3}$
Tangent	↗	→	↘

2. (a) $P(x) = (400 - 2x)x - \left(10000 - 400x + 8x^2\right)$
    $$= 800x - 10000 - 10x^2$$
    (b) $P(x) = 800 - 20x = 0 \Rightarrow x = 40$. Therefore, as $400 - 2 \times 40 = 320$, the profit is maximized
    when each digital camera is sold for \$320.
    (c) $P''(x) = -20 < 0 \Rightarrow P$ has a maximum value when $x = 40$.

3. (a) Area $A = y^2$, but $NP = \sqrt{y^2 - x^2}$ and as $x + \sqrt{y^2 - x^2} = 8$, $\sqrt{y^2 - x^2} = 8 - x$
    $$\Rightarrow y^2 - x^2 = (8-x)^2 = 64 - 16x + x^2 \Rightarrow A = y^2 = 64 - 16x + 2x^2.$$
    (b) $A'(x) = -16 + 4x$, $A'(x) = 0 \Rightarrow -16 + 4x = 0 \Rightarrow x = 4$. So A is
    minimized when $x = 4$.

x	3	4	5
$A'(x)$	-4	0	4
Tangent	↘	→	↗

4. (a) Each side of the square is x. There are 4 sides, so the perimeter of the square cross-section
    is $4x$. Hence the total perimeter is $4x + l$. Therefore, $4x + l = 6$.
    (b) $V = lx^2 = x^2(6 - 4x) = 6x^2 - 4x^3$
    (c) $V'(x) = 12x - 12x^2$, $V'(x) = 0 \Rightarrow 12x - 12x^2 = 0 \Rightarrow 12x(1-x) = 0 \Rightarrow x = 0,\ 1$
    (d) $V''(x) = 12 - 24x \Rightarrow V''(1) = 12 - 24 < 0$. When $x = 1$, the volume has a maximum value.
    This value is $V(1) = 2$ and the maximum volume is 2m^2.

5. (a) $2x^2h = 36 \Rightarrow h = \dfrac{18}{x^2}$

    (b) Surface area, $S = 2xh + 4xh + 2x^2 = 6xh + 2x^2 = 6x\left(\dfrac{18}{x^2}\right) + 2x^2 = \dfrac{108}{x} + 2x^2$

    $S'(x) = 4x - \dfrac{108}{x^2} = 0 \Rightarrow 4x^3 = 108 \Rightarrow x = 27$. Then $h = 2$ and dimensions of the
    box that minimize the surface area are, in centimeters, $3\text{ cm} \times 6\text{ cm} \times 2\text{ cm}$.
    (c) $S''(x) = 4 + \dfrac{216}{x^3} \Rightarrow S''(3) = 12 > 0$, and so the dimensions give a minimum rather
    than a maximum surface area.

6. (a) $\pi x^2 h = 2000\pi \Rightarrow h = \dfrac{2000}{x^2}$

    (b) $A = 2\pi x^2 + 2\pi xh = 2\pi x^2 + 2\pi x\left(\dfrac{2000}{x^2}\right) = 2\pi x^2 + \dfrac{4000\pi}{x}$

(c) $A'(x) = 4\pi x - \dfrac{4000\pi}{x^2} = 0 \Rightarrow x^3 = \dfrac{4000\pi}{4\pi} = 1000 \Rightarrow x = 10$,

$A''(x) = 4\pi + \dfrac{8000\pi}{x^3} \Rightarrow A''(10) = 12\pi > 0$, so that when $x = 10$, A is a minimum of

$A(10) = 600\pi = 1884.955... = 1880 \text{ cm}^2$.

7.   (a) Area, $A = PQ \times PS$, but if $OP = x$ and $PS = y$,

PQ $+ 2x = 1$ and $PS = x - x^2$.

$\Rightarrow A = (1 - 2x)(x - x^2) = 2x^3 - 3x^2 + x$

(b) $A'(x) = 6x^2 - 6x + 1$ and $A'(x) = 0 \Rightarrow 6x^2 - 6x + 1 = 0$.

$\Rightarrow \left(x - \dfrac{1}{2}\right)^2 = \dfrac{1}{12} \Rightarrow x - \dfrac{1}{2} = \pm\dfrac{1}{2\sqrt{3}}$

$\Rightarrow x = 0.2113248...,\ x = 0.7886751....\ \ A''(x) = 12x - 6$

$\Rightarrow A''(0.211) = -3.46 < 0$, so that when $x = 0.211$, the

area is a maximum. This area is $A(.211) = 0.09622504....$ When $x = 0.789$,

$PQ = 1 - 2x = 1 - 2 \times 0.789 < 0$, so this is not a practical possibility.

Exercise 5.8

1.   (a) $f'(x) = 9(3x + 1)^2$ (b) $f'(x) = -16x(1 - 4x^2)$

(c) $f(x) = (x^2 + 5)^{\frac{1}{2}} \Rightarrow f'(x) = \dfrac{1}{2}(x^2 + 5)^{-\frac{1}{2}} \times 2x = \dfrac{x}{\sqrt{x^2 + 5}}$

(d) $g'(x) = -e^{-x}$ (e) $g'(x) = 12e^{4x}$ (f) $g'(x) = -2xe^{1-x^2}$ (g) $h'(x) = 3\cos 3x$

(h) $h'(x) = -2\sin\left(2x - \dfrac{\pi}{6}\right)$ (i) $h'(x) = 2\sin x \cos x$ (j) $f'(x) = \dfrac{1}{1 + 5x} \times 5 = \dfrac{5}{1 + 5x}$

(k) $f'(x) = \dfrac{4}{x^2 + 3} \times 2x = \dfrac{8x}{x^2 + 3}$ (l) $f'(x) = \dfrac{1}{1 + e^x} \times e^x = \dfrac{e^x}{1 + e^x}$

2.   (a) $\dfrac{dy}{dx} = 4(2x - 5)$ (b) $\dfrac{dy}{dx} = 6x(x^2 - 4)^2$ (c) $\dfrac{dy}{dx} = -20(3 - 5x)^3$ (d) $\dfrac{dy}{dx} = 2xe^{x^2}$

(e) $\dfrac{dy}{dx} = -2e^{1-2x}$ (f) $\dfrac{dy}{dx} = -3\cos^2 x \sin x$ (g) $\dfrac{dy}{dx} = 12\cos 4x$ (h) $\dfrac{dy}{dx} = 6x\cos 3x^2$

(i) $y = \ln\dfrac{1}{x} = \ln x^{-1} = -\ln x \Rightarrow \dfrac{dy}{dx} = -\dfrac{1}{x}$ (j) $\dfrac{dy}{dx} = \dfrac{3x^2}{(x^3 - 3)}$ (k) $\dfrac{dy}{dx} = \dfrac{4}{4x} = \dfrac{1}{x}$

(l) $\dfrac{dy}{dx} = -2(\cos x)^{-2}(-\sin x) = \dfrac{2\sin x}{\cos^2 x}$

370

3. (a) $f'(x) = -28(1-7x)^3$ (b) $f'(x) = -20e^{-4x}$ (c) $\dfrac{dy}{dx} = \dfrac{1+\cos x}{x+\sin x}$ (d) $\dfrac{dy}{dx} = \dfrac{-2x}{\left(x^2-3\right)^2}$

(e) $g'(x) = 2\sin x(2-\cos x)$ (f) $f'(x) = e^{-x}\sin\left(e^{-x}\right)$ (g) $f'(x) = -6\cos\left(\dfrac{\pi}{2}-6x\right)$

(h) $\dfrac{dy}{dx} = 3e^{x}\left(1+e^{x}\right)^2$ (i) $\dfrac{dy}{dx} = \dfrac{\cos x}{2\sqrt{\sin x+1}}$ (j) $f'(x) = \sin x e^{-\cos x}$ (k) $\dfrac{dy}{dx} = \dfrac{-2\sin 2x}{\cos 2x}$

(l) $g'(x) = 6\sin 3x\cos 3x$

Exercise 5.9

1. (a) $f'(x) = x^2\cos x + 2x\sin x$

(b) $f'(x) = x^3 2(x+1) + 3x^2(x+1)^2 = x^2(x+1)(2x+3(x+1)) = x^2(x+1)(5x+3)$

(c) $f'(x) = e^x(-\sin x) + e^x\cos x = e^x(\cos x - \sin x)$

(d) $f'(x) = x\dfrac{1}{x} + 1\ln x = 1 + \ln x$

(e) $f'(x) = 4xe^x + 4e^x = 4e^x(x+1)$

(f) $f'(x) = 1(\cos x + \sin x) + x(-\sin x + \cos x) = (1-x)\sin x + (1+x)\cos x$

2. (a) $f'(x) = \dfrac{2x}{\left(x^2+1\right)^2}$ (b) $f'(x) = \dfrac{\ln x - 1}{\left(\ln x\right)^2}$ (c) $f'(x) = \dfrac{x\cos x - \sin x}{x^2}$

(d) $f'(x) = \dfrac{-\ln x}{x^2}$ (e) $f'(x) = \dfrac{1}{\cos^2 x}$ (f) $f'(x) = \dfrac{-1-2\sin x}{\left(2+\sin x\right)^2}$

3. (a) $\dfrac{dy}{dx} = \dfrac{(2x+1)1 - 2x}{(2x+1)^2} = \dfrac{1}{(2x+1)^2}$

(b) $\dfrac{dy}{dx} = x\left(-2(2x+1)^{-2}\right) + 1(2x+1)^{-1} = \dfrac{-2x+(2x+1)}{(2x+1)^2} = \dfrac{1}{(2x+1)^2}$

4. (a) The denominator is a constant, so differentiate directly to obtain $\dfrac{dy}{dx} = \dfrac{1}{3}(1+\cos x)$.

(b)   (i) $\dfrac{dy}{dx} = \dfrac{(x\sin x)0 - (1+\cos x)3}{(x+\sin x)^2} = \dfrac{-3(1+\cos x)}{(x+\sin x)^2}$

(ii) $\dfrac{dy}{dx} = 3\left(-1(x+\sin x)^{-2}\right)(1+\cos x) = \dfrac{-3(1+\cos x)}{(x+\sin x)^2}$

5. (a) $f'(x) = 2x\cos 2x - x^2 2\sin 2x = 2x(\cos 2x - x\sin 2x)$ (b) $g'(x) = \dfrac{2-2x^2}{\left(x^2+1\right)^2}$

(c) $\dfrac{dy}{dx} = \dfrac{1-\ln x}{x^2}$ (d) $\dfrac{dy}{dx} = x3(1-4x)^2(-4) + (1-4x)^3 = (1-4x)^2(1-16x)$

(e) $g'(x) = x + 2$ (f) $f'(x) = -e^{-x}(x+2)$ (g) $\dfrac{dy}{dx} = \dfrac{1}{(2x+3)^2}$

(h) $g(x) = \ln(x^2-1) - \ln(x^2+1) \Rightarrow g'(x) = \dfrac{2x}{x^2-1} - \dfrac{2x}{x^2+1} = \dfrac{4x}{x^4-1}$ (i) $\dfrac{dy}{dx} = \dfrac{e^{2x}(5-6x)}{(1-3x)^2}$

(j) $f'(x) = e^{-x}(2\cos 2x - \sin 2x)$ (k) $\dfrac{dy}{dx} = \dfrac{6}{\cos^2 3x}$

(l) $g'(x) = 2\cos 2x \cos x - \sin 2x \sin x$

6.   (a) $\dfrac{dy}{dx} = \dfrac{(1+e^x)e^x - e^x \times e^x}{(1+e^x)^2} = \dfrac{e^x + e^{2x} - e^{2x}}{(1+e^x)^2} = \dfrac{e^x}{(1+e^x)^2}$

(b) $f'(x) = \dfrac{(\sin x + \cos x)(\cos x + \sin x) - (\sin x - \cos x)(\cos x - \sin x)}{(\sin x + \cos x)^2}$

$= \dfrac{1 + 2\sin x \cos x + 1 - 2\sin x \cos x}{\sin^2 x + 2\sin x \cos x + \cos^2 x} = \dfrac{2}{1+\sin 2x}$

(c) $g(x) = \ln x - \ln(2x+1) \Rightarrow g'(x) = \dfrac{1}{x} - \dfrac{2}{2x+1} = \dfrac{2x+1-2x}{x(2x+1)} = \dfrac{1}{x(2x+1)}$

Exercise 5.10

1.   (a) $f'(x) = -2 < 0$ for all $x \in \mathbb{R}$, so $f(x)$ is a decreasing function.

(b) $f'(x) = -\dfrac{3}{x^4}$ but $x^4 > 0$ for all x so $-\dfrac{3}{x^4} < 0$ for all x, so $f(x)$ is a decreasing function.

(c) $f'(x) = \dfrac{1}{x} > 0$ for $x > 0$, so $f(x)$ is an increasing function.

(d) $f'(x) = -e^{-x}$. As $e^{-x} > 0$ for all $x \in \mathbb{R}$, $-e^x < 0$ for all x, so $f(x)$ is a decreasing function.

(e) $f'(x) = 3x^2 + 1$, but $x^2 \geq 0 \Rightarrow 3x^2 + 1 > 0$, so $f(x)$ is an increasing function.

2.   (a) $\dfrac{dy}{dx} = e^x(1-x)$. For $x > 1$, $1-x < 0$, $e^{-x} > 0 \Rightarrow \dfrac{dy}{dx} < 0$. Therefore, $y = xe^{-x}$ is a decreasing function for $x > 1$.

(b) $\dfrac{dy}{dx} = k - \cos x$. For $y = kx - \sin x$ to be an increasing function, $k - \cos x > 0$ for all x. But $-1 \leq \cos x \leq 1$, therefore, if $k > 1$, $k - \cos x > 0$ for all x. Therefore, $k > 1$.

3.   (a) $12x - y - 16 = 0$ (b) $3x - y - 2 = 0$ (c) $2x - y + 4 = 0$
(d) $4x + y + 32 = 0$ (e) $x - 4y + 4 = 0$

4.   (a) $x + 12y - 98 = 0$ (b) $x + 3y - 4 = 0$ (c) $x + 2y + 2 = 0$
(d) $x - 4y - 196 = 0$ (e) $4x + y - 18 = 0$

5. $y = x$

6. $\dfrac{dy}{dx} = 2x + 2$. At $x = -2$ the gradient of the tangent is -2, so the gradient of the tangent is $\dfrac{1}{2}$.

When $x = -2$, $y = 5$, the equation of the normal is $y - 5 = \dfrac{1}{2}(x + 2) \Rightarrow 2y - 10 = x + 2$

$\Rightarrow x - 12y + 12 = 0$.

7. $\dfrac{dy}{dx} = x(1 + 2\ln x)$, so when $x = e$, the gradient is $e(1 + 2) = 3e$ and $y = e^2$. Therefore, the

equation of the tangent is $y - e^2 = 3e(x - e) \Rightarrow y = 3ex - 2e^2$.

8. $f'(x) = xe^{-x}(2 - x) \Rightarrow f'(1) = \dfrac{1}{e}$. $f(1) = \dfrac{1}{e}$. Therefore, the equation of the tangent is

$y - \dfrac{1}{e} = \dfrac{1}{e}(x - 1) \Rightarrow y = \dfrac{x}{e}$.

9. $f'(x) = \dfrac{e^{2x}(2x - 1)}{x^2} \Rightarrow f'(1) = e^2$. $f(x) = e^2$, so the equation of the tangent is

$y - e^2 = e^2(x - 1) \Rightarrow y = e^2 x$.

10. $f'(x) = 1 + \ln x \Rightarrow f'(e) = 2$, $f(e) = e$. The gradient of the normal is $-\dfrac{1}{2}$. Therefore, the

equation of the normal is $y - e = -\dfrac{1}{2}(x - e) \Rightarrow 2y - 2e = -x + e \Rightarrow x + 2y - 3e = 0$.

Exercise 5.11

1.   (a) $f'(x) = \dfrac{2(4 - x^2)}{(x^2 + 4)^2}$ $f'(x) = 0 \Rightarrow 2(4 - x^2) = 0 \Rightarrow x^2 = 4 \Rightarrow x = \pm 2$

(b)   (c) Minimum value is $f(-2) = -\dfrac{1}{2}$.

x	1	2	3
$f'(x)$	6/25	0	-10/169
Tangent	↗	→	↘

2.   (a) $f'(x) = e^{-x}(1 - x)$

(b) $f'(x) = 0 \Rightarrow e^{-x}(1 - x) = 0 \Rightarrow x = 1$. Therefore, a maximum or minimum occurs when $x = 1$.

(c) $f''(x) = e^{-x}(x - 2) \Rightarrow f''(1) = -e^{-1} < 0$, so $f(1)$ is a maximum.

(d) Maximum value is $f(1) = \dfrac{1}{e}$.

3.   (a)

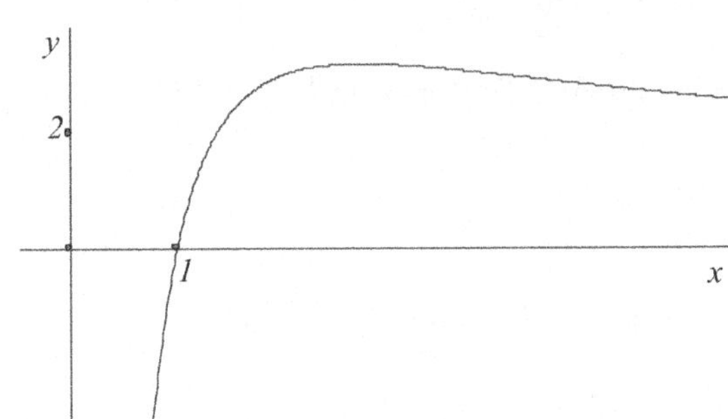

(b) $\dfrac{dy}{dx} = \dfrac{1 - \ln x}{x^2}$, $\dfrac{d^2 y}{dx^2} = \dfrac{x^2(-1/x) - (1 - \ln x)2x}{x^4} = \dfrac{-x - 2x + 2x \ln x}{x^4} = \dfrac{2 \ln x - 3}{x^3}$

(c) $\dfrac{dy}{dx} = 0 \Rightarrow 1 - \ln x = 0 \Rightarrow \ln x = 1 \Rightarrow x = e$. When $x = e$, $\dfrac{d^2 y}{dx^2} = -\dfrac{1}{e^3} < 0$ so y has a maximum value.

(d) $y = \dfrac{1}{e}$

4.   (a) $f'(x) = e^{-x} \cos x - e^{-x} \sin x$

(b) $f'(x) = 0 \Rightarrow e^{-x}(\cos x - \sin x) = 0 \Rightarrow \tan x = 1 \Rightarrow x = \dfrac{\pi}{4}$ in given domain.

(c) $f''(x) = e^{-x}(-\sin x - \cos x) - e^{-x}(\cos x - \sin x) = -2e^{-x} \cos x$ so that

$f''\left(\dfrac{\pi}{4}\right) = -2e^{-\frac{\pi}{4}}\left(\cos \dfrac{\pi}{4}\right) = -\sqrt{2}e^{-\frac{\pi}{4}} < 0$ so that f is a maximum value at $x = \dfrac{\pi}{4}$.

5.   (a)

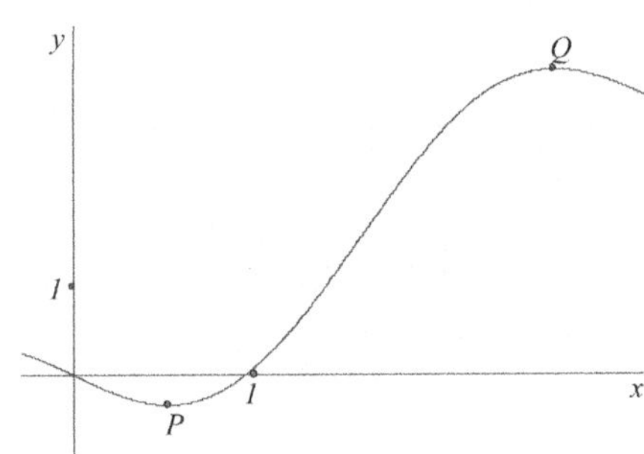

(b) $g'(x) = 1 - 2\cos 2x$

$g'(x) = 0 \Rightarrow 1 - 2\cos 2x = 0 \Rightarrow \cos 2x = \dfrac{1}{2} \Rightarrow 2x = \dfrac{\pi}{3}, \dfrac{5\pi}{3} \Rightarrow x = \dfrac{\pi}{6}, \dfrac{5\pi}{6}$

(c) $g''(x) = 4\sin 2x \Rightarrow g''\left(\dfrac{\pi}{6}\right) = 4\sin \dfrac{\pi}{3} = 2\sqrt{3} > 0$. So value of function at P is

$f\left(\dfrac{\pi}{6}\right) = \dfrac{\pi}{6} - \sin \dfrac{\pi}{3} = \dfrac{\pi}{6} - \dfrac{\sqrt{3}}{2} = \dfrac{\pi - 3\sqrt{3}}{6}$ and the coordinates of P are $\left(\dfrac{\pi}{6}, \dfrac{\pi - 3\sqrt{3}}{6}\right)$.

(d) Q is $\left(\dfrac{5\pi}{6}, \dfrac{5\pi + 3\sqrt{3}}{6}\right)$, and is a maximum value, because $f''\left(\dfrac{5\pi}{6}\right) = 4\sin\dfrac{5\pi}{3} = -2\sqrt{3} < 0$.

6. (a) $\pi r + 2x = 5$

   (b) $A = \pi r^2 + x^2$, $r = \dfrac{5 - 2x}{\pi} \Rightarrow A = \pi\left(\dfrac{5 - 2x}{\pi}\right)^2 + x^2 = x^2 + \dfrac{(5 - 2x)^2}{\pi}$

   (c) $A'(x) = 2x - \dfrac{4}{\pi}(5 - 2x) = 0 \Rightarrow 2\pi x - 20 + 8x = 0 \Rightarrow x = \dfrac{10}{\pi + 4} \approx 1.400247... = 1.40$.

   $A''(x) = 2 + \dfrac{8}{\pi} > 0$, so $A(1.40)$ has a minimum value.

7. (a) $L = \sqrt{(x - 0)^2 + (9.5 - x^2)^2} = \sqrt{x^2 + 9.5^2 - 19x^2 + x^4} = \sqrt{x^4 - 18x^2 + 90.25}$

   (b) $\dfrac{dL}{dx} = \dfrac{2x(x^2 - 9)}{\sqrt{x^4 - 18x^2 + 90.25}}$

   (c) $\dfrac{dL}{dx} = 0 \Rightarrow 2x(x^2 - 9) = 0 \Rightarrow x = 0,\ x = 3$ and $x = -3$

   (d) Points closest to P are $(3, 9)$, $(-3,\ 9)$.

   (e) O is a local maximum. The origin, O is further from P than all other points close to O.

8. (a) $P(x) = 0.85 f(x) - (8x + 30) = 425\left(1 - e^{-0.2x}\right) - 8x - 30$

   (b) $P'(x) = 85 e^{-0.2x} - 8$, $P'(x) = 0 \Rightarrow 85 e^{0.2x} = 8 \Rightarrow x = 11.81604...$, but as $x \in \mathbb{Z}^+$, $x = 12$.

   Therefore, 12 employees gives maximum profit of $P(12) = 260.44$.

   (c)

9. (a) $P(x) = 8x - \left(3x + \dfrac{x(x - 400)^2}{80000}\right) = 5x - \dfrac{x(x - 400)^2}{80000}$

   (b) $P(300) = 5 \times 300 - \dfrac{300(-100)^2}{80000} = 1500 - \dfrac{300}{8} = 1500 - 37.5 = 1462.50$

   (c) $\dfrac{dP}{dx} = 5 - \dfrac{1}{80000}\left(x2(x - 400) + (x - 400)^2\right) = 5 - \dfrac{(x - 400)(3x - 400)}{80000}$

   $\dfrac{dP}{dx} = 0 \Rightarrow 5 - \dfrac{(x - 400)(3x - 400)}{80000} = 0 \Rightarrow (x - 400)(3x - 400) = 400\ 000$

   $\Rightarrow 3x^2 - 1600x - 240000 = 0 \Rightarrow x = -122.0634...,\ 655.3967....$ But $x > 0$, so that the maximum profit is when 655 liters of wine is produced and profit is $ P(655) = \$2742.61$.

(d)

Exercise 5.12

1.   (a) $y(0) = 5$

      (b) $v(t) = y'(t) = 12 - 2t$.

      (c) $v(t) = 0 \Rightarrow 12 - 2t = 0 \Rightarrow t = 6$ and $y(6) = 41\,\text{m}$.

      (d) $12 - 2t = 10 \Rightarrow t = 1$

2.   (a) $x(0) = 0$

      (b) $v(t) = 3t^2 - 10t + 3 \Rightarrow v(0) = 3$.

      (c) $3t^2 - 10t + 3 = 0 \Rightarrow (3t - 1)(t - 3) = 0 \Rightarrow t = \dfrac{1}{3}, 3$. $x\left(\dfrac{1}{3}\right) = \dfrac{13}{27}$, $x(3) = -9$ so the distance

         travelled is $\dfrac{13}{27} - (-9) = 9.481481... = 9.48$ meters.

3.   (a) $v(t) = \dfrac{1}{2} \Rightarrow \sin\dfrac{\pi t}{3} = \dfrac{1}{2} \Rightarrow \dfrac{\pi t}{3} = \dfrac{\pi}{6} \Rightarrow t = \dfrac{1}{2}$

      (b) $a(t) = \dfrac{\pi}{3}\cos\dfrac{\pi t}{3}$, for $t \geq 0$.

      (c) $a(1) = \dfrac{\pi}{3}\cos\dfrac{\pi}{3} = \dfrac{\pi}{6}$.

4.   (a) $s(5.2) = 1.4 - 5(5.2)^2 = -133.8$. So the estimated depth of the mine shaft is 133.8 m or
         about 134m.

      (b) $v(t) = -10t$ so $v(5.2) = -52$ and the speed of the stone is $52\,\text{ms}^{-1}$.

5.   (a) $v(t) = x'(t) = \dfrac{4t}{1 + 2t^2}$ (b) $v'(t) = \dfrac{(1 + 2t^2)4 - 4t(4t)}{(1 + 2t^2)^2} = \dfrac{4 + 8t^2 - 16t^2}{(1 + 2t^2)^2} = \dfrac{4(1 - 2t^2)}{(1 + 2t^2)^2}$.

      (c) $a(t) = 0 \Rightarrow 1 - 2t^2 = 0 \Rightarrow t^2 = \dfrac{1}{2} \Rightarrow t = \pm\dfrac{1}{\sqrt{2}}$ but $t \geq 0$ so $t = \dfrac{1}{\sqrt{2}}$

         $\Rightarrow v\left(\dfrac{1}{\sqrt{2}}\right) = \dfrac{4/\sqrt{2}}{1 + 2\times\frac{1}{2}} = \dfrac{4/\sqrt{2}}{2} = \dfrac{4}{2\sqrt{2}} = \dfrac{2}{\sqrt{2}} = \sqrt{2}\,\text{ms}^{-1}$

(d) Average speed $= \dfrac{\text{distance}}{\text{time}} = \dfrac{\ln 19 - \ln 3}{3 - 1} = 0.9229133\ldots = 0.923$ ms^{-1}.

Solutions to Unit 6 Exercises

Exercise 6.1

For questions $1 - 4$ the frequencies will depend on the random numbers which are generated. The column for tally marks has been omitted.

1.

Score	1	2	3	4	5	6
Frequency	16	11	6	6	6	15

2.

Score	2	3	4	5	6	7	8	9	10	11	12
Frequency	0	1	5	8	12	13	7	6	4	4	0

3.

Number of Tosses	1	2	3	4	5	6	7	8
Frequency	11	10	7	0	1	0	0	1

4.

Number	0-9	10-19	20-29	30-39	40-49
Frequency	7	10	7	8	9
Number	50-59	60-69	70-79	80-89	90-99
Frequency	6	11	5	9	8

5.

Class Interval	Tally Marks	Frequency
300 – 349	111	3
350 – 399	11	2
400 – 449	11111	5
450 – 499	11111 111	8
500 – 549	11111 111	8
550 – 599	11111 1111	9
600 – 649	11111 11	7
650 – 699	11	2
700 – 749	11111	5
750 – 799		0
800 – 850	1	1

6.

Time	0.5-1	1-1.5	1.5-2	2-2.5	2.5-3	3-3.5
Frequency	3	10	19	5	6	3
Time	3.5-4	4-4.5	4.5-5	5-5.5	5.5-6	
Frequency	0	1	2	0	1	

Exercise 6.2

1.   (a)

Number	170-174	175-179	180-184	185-189	190-194
Frequency	1	6	11	10	5

   (b)

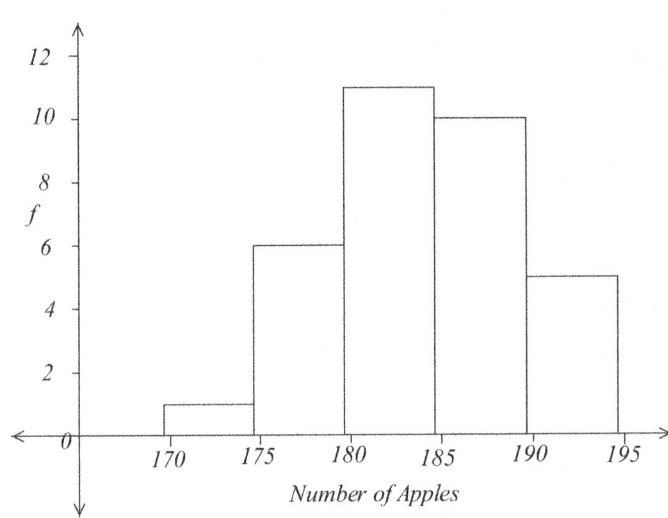

2.   The frequencies and hence the shape of the frequency histogram will depend on the random numbers generated.

   (a)

Class Interval	0-9	10-19	20-29	30-39	40-49
Frequency	8	5	11	19	9
Class Interval	50-59	60-69	70-79	80-89	90-99
Frequency	9	16	6	14	3

(b)

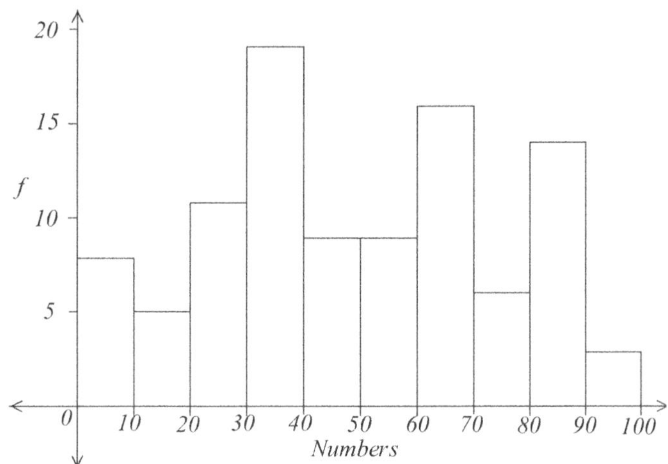

(c) As the random numbers have an equal probability of being any integer between 0 and 99, one might expect the height of the rectangles of the frequency histogram to be roughly equal. This frequency histogram does not conform to that expectation.   If many more random numbers were generated, then the heights of the rectangles are likely to be more equal.

3.   (a)

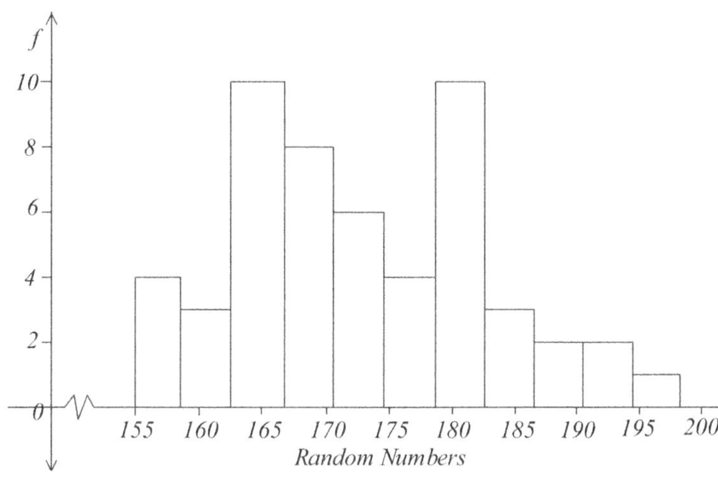

(b)

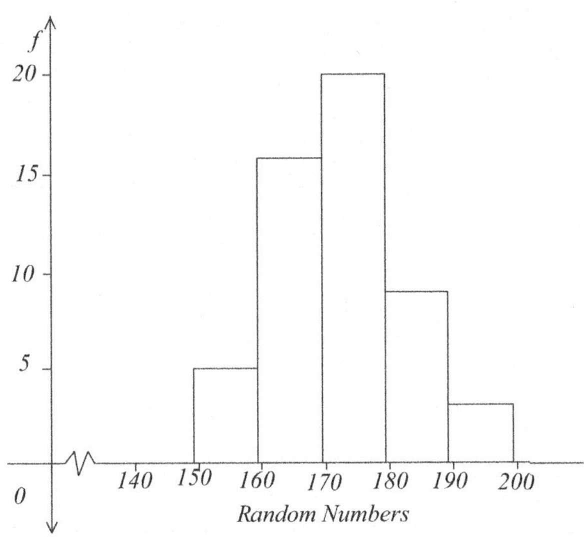

Exercise 6.3

1.

Grade	Cumulative Frequency
1	2
2	7
3	14
4	25
5	39
6	45
7	47

2.   (a)

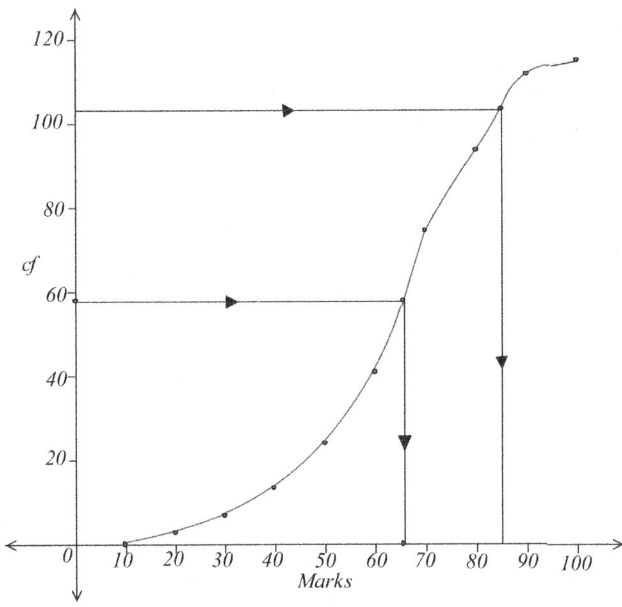

(b) The construction lines for part (b) are shown on the cumulative frequency curve for part (a).
    (i) $0.90 \times 115 = 103.5$. Therefore the 90th percentile is 85.
    (ii) $0.5 \times 115 = 57.5$. Therefore the 50th percentile is 66.

3.   (a)   (b)

Upper Class Boundary	Cum Freq
0	0
9.95	2
19.95	13
29.95	30
39.95	37
49.95	48
59.95	55
69.95	61
79.95	70
89.95	74
99.95	82
109.95	87

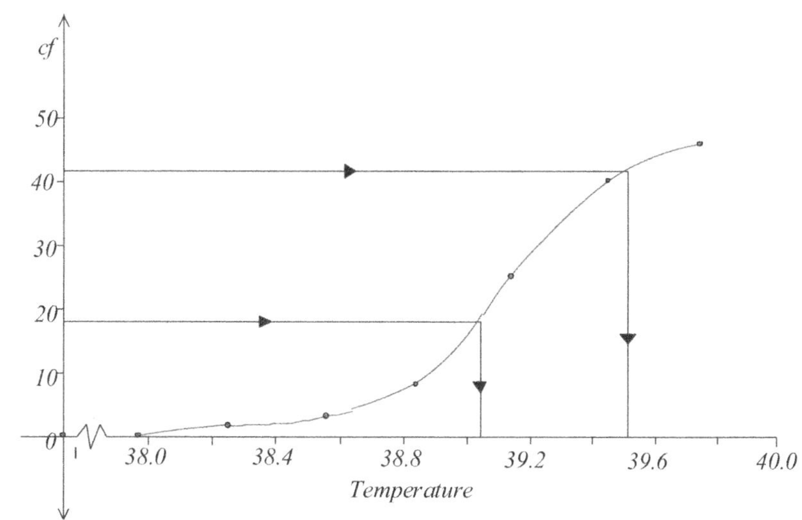

4.   The grouping of the data will depend on the class intervals chosen. This will also affect the cumulative frequency table and curve.

(a)   (b)

Class Interval	Frequency
$37.95 - 38.25$	2
$38.25 - 38.55$	1
$38.55 - 38.85$	6
$38.85 - 39.15$	16
$39.15 - 39.45$	15
$39.45 - 39.75$	6

Upper Class Boundary	Cumulative Frequency
≤ 37.95	0
≤ 38.25	2
≤ 38.55	3
≤ 38.85	9
≤ 39.15	25
≤ 39.45	40
≤ 39.75	46

(c)

381

(d) The construction lines are shown on the cumulative frequency curve drawn for part (c).

    (i) $0.9 \times 46 = 41.4$. Therefore, 90th percentile is $39.52°C$.

    (ii) $1 - 0.6 = 0.4$, $0.4 \times 46 = 18.4$. Therefore, the temperature is $39.04°C$.

These values will vary slightly depending on the way the curve is drawn.

5.   (a)

Upper Class Limit	Cumulative Frequency
0	0
≤ 3.0	74
≤ 3.5	129
≤ 4.0	168
≤ 4.5	179
≤ 5.0	186
≤ 6.0	189
≤ 7.0	190

(b)

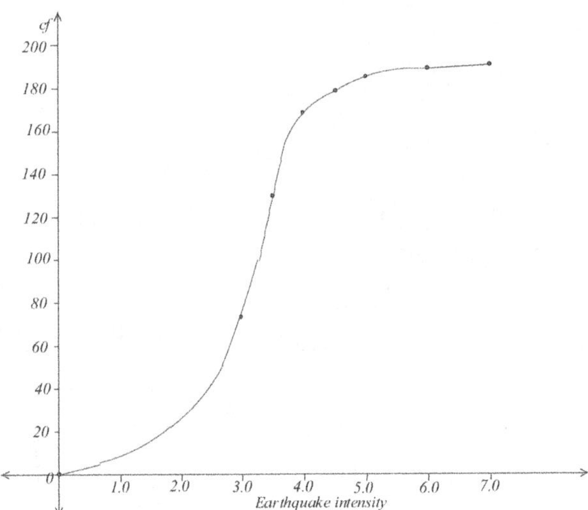

(c) (i) $0.95 \times 190 = 180.5$. Therefore, the 95th percentile is 4.6.

    (ii) $0.6 \times 190 = 114$. Therefore, the 60th percentile is 3.3.

    (iii) $0.9 \times 190 = 171$. Therefore, the 90th percentile is 4.1.

(d) $0.25 \times 190 = 47.5$. Therefore, the lower quartile is 2.7.

    $0.75 \times 190 = 142.5$. Therefore, the upper quartile is 3.7.

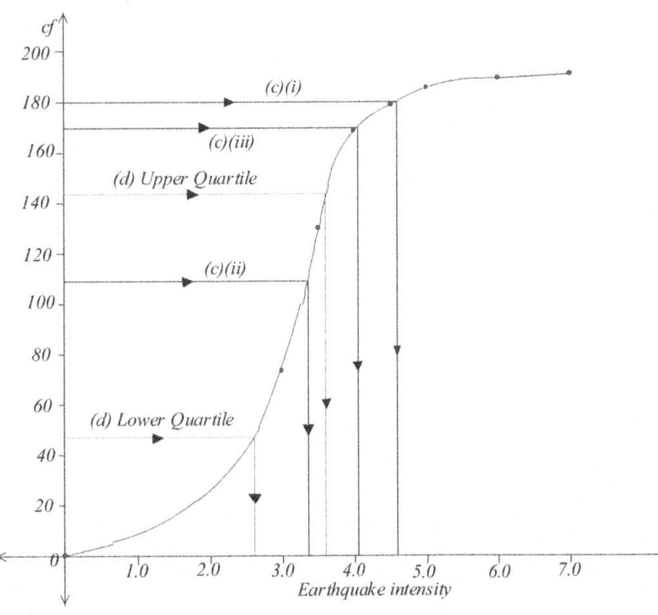

Exercise 6.4

1. (a) $17.02857... = 17.0$ (b) 0.50925 (c) $1.151333... \times 10^8 = 1.15 \times 10^{-8}$

2.

x_i	f_i	$f_i x_i$
0	8	0
1	5	5
2	4	8
3	4	12
4	0	0
5	1	5
6	0	0
7	1	7

$$\bar{x} = \frac{\sum f_i x_i}{\sum f_i} = \frac{37}{23} = 1.608695... = 1.61$$

3. $$\bar{x} = \frac{\sum f_i x_i}{\sum f_i} = \frac{362}{45} = 8.04$$

Mid-Interval Value, x_i	Frequency, f_i	$f_i x_i$
4	3	12
6	12	72
8	15	120
10	11	110
12	4	48

4. (a) $\bar{x} = 3.63398... = 3.63$ (b) $\bar{x} = 46.12903... = 46.1$ (c) $\bar{x} = 0.3970588... = 0.397$

5. $1.966666... = 1.97 \text{mg}$

6. (a) $\bar{x} = 4.7608$
    (b) This is most easily done by sorting the data in ascending order in your graphing calculator.
    (c) $\bar{x} = 4.76$
    (d) The mean calculated in part (c) assumes that the data in each class interval are evenly distributed across the interval. In practice, this is rarely true. For example, the three data points in the class interval $0 - 1$ are 0.036, 0.298 and 0.902, which are spread quite well across the interval but not very evenly. However, it can be seen from the small difference between the two means that the assumption is generally valid.

Class Interval	Mid-Interval Value	Freq
$0 - 1$	0.5	3
$1 - 2$	1.5	5
$2 - 3$	2.5	6
$3 - 4$	3.5	10
$4 - 5$	4.5	4
$5 - 6$	5.5	5
$6 - 7$	6.5	6
$7 - 8$	7.5	3
$8 - 9$	8.5	4
$9 - 10$	9.5	4

Exercise 6.5

1. (a) 42 (b) 1.935 (c) 589

2. (a) 6 (b) 3

3.   (a)

Upper Boundary Limit	Cumulative Frequency
0	0
20	28
30	87
40	124
50	148
60	159
70	172

(b)

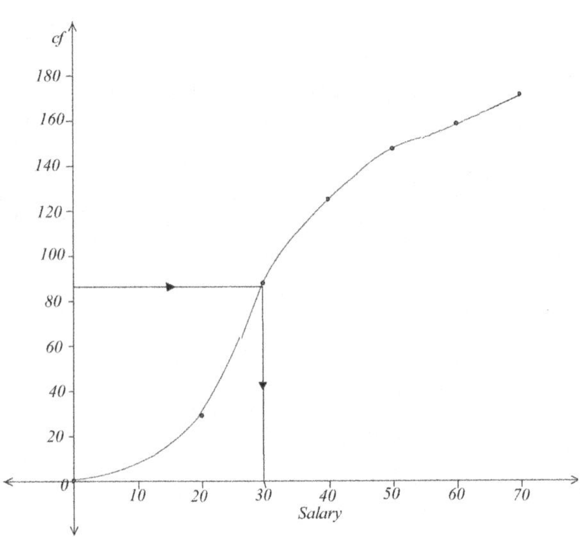

(c) Median salary is $30000
(d) Modal class of salaries is $20000 to $30000

4.

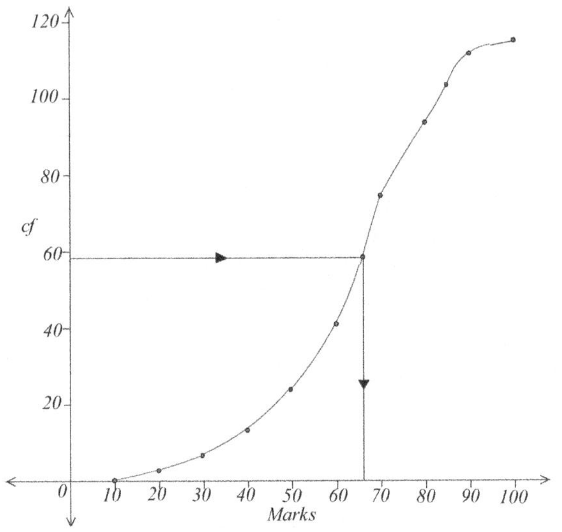

Median mark is 66

384

5.   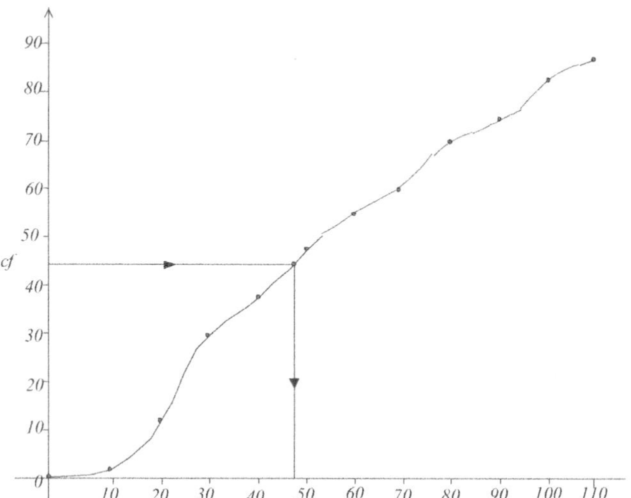

Median is 47ha.

Exercise 6.6

1.   (a) The median is 385, the lower quartile is 380 and the upper
          quartile is 396.5. Therefore, the inter-quartile range is 16.5.

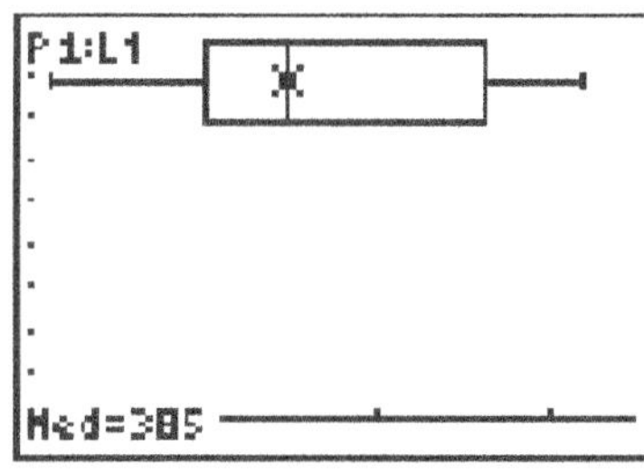

      (b) 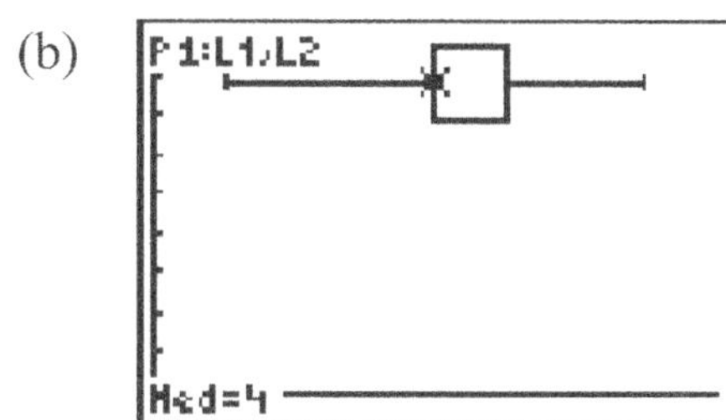   The median is 4, the lower
          quartile is 4 and the upper
          quartile is 5. Therefore, the
          inter-quartile range is 1.

2.   (a)

Upper Boundary Limit	Cumulative Frequency
$-40°$	0
$-35°$	5
$-30°$	16
$-25°$	42
$-20°$	79
$-15°$	97
$-10°$	100

(b) The lower quartile is $-27.5°C$ and the
upper quartile is $-20.5°C$. Therefore,
the interquartile range is $7°C$.

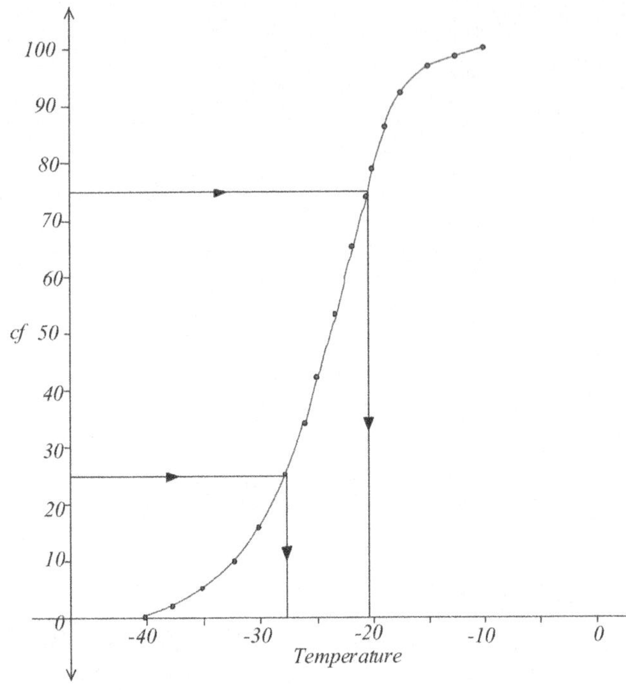

3. (a)

x_i	$x_i - \bar{x}$	$(x_i - \bar{x})^2$
8.5	0.5	0.25
7.8	-0.2	0.04
8.1	0.1	0.01
7.9	-0.1	0.01
7.7	-0.3	0.09
8.0	0	0

$$\bar{x} = \frac{8.5 + 7.8 + 8.1 + 7.9 + 7.7 + 8.0}{6} = 8,$$

$$\sum (x_i - \bar{x})^2 = 0.4$$

Therefore, $s = \sqrt{\dfrac{1}{6} \times 0.4} = 0.2581988... = 0.258$.

(b) $\bar{x} = \dfrac{123 + 134 + 109 + 115 + 117 + 124 + 130 + 122 + 125 + 111}{10}$

$= 121$

$$\sum (x_i - \bar{x})^2 = 576.$$

Therefore $s = \sqrt{\dfrac{1}{10} \times 576} = 7.589466... = 7.59$

x_i	$x_i - \bar{x}$	$(x_i - \bar{x})^2$
123	2	4
134	13	169
109	-12	144
115	-6	36
117	-4	16
124	3	9
130	9	81
122	1	1
125	4	16
111	10	100

4.

x_i	f_i	$f_i x_i$	$f_i(x_i - \bar{x})^2$
10	2	20	800
20	11	220	1100
30	15	450	0
40	9	360	900
50	3	150	1200

$\bar{x} = \dfrac{1200}{40} = 30$, $\sum f_i(x_i - \bar{x})^2 = 4000$. Therefore, $s = \sqrt{\dfrac{1}{40} \times 4000} = 10$.

5.	(a) $s = 0.07615317... = 0.0762$

(b)

Class Interval	Frequency
0.8 – 0.9	3
0.9 – 1.0	6
1.0 – 1.1	2
1.1 – 1.2	1

Standard deviation is $0.08620067... = 0.0862$

(c) In grouping the data, it is assumed that all data points in a class interval are located at the center of that interval. For example, all three data points in the interval $0.8 - 0.9$ are assumed to be at 0.85, whereas, in fact, they are 0.869, 0.876, 0.887. Therefore, the answer to part (b) differs slightly from the answer of part (a).

6.	(a) (i) $s = 2.291287... = 2.29$ (ii) $s = 0.05969309... = 0.0597$

(b) (i) $s = 2.29$ (ii) $s = \dfrac{1}{10}(0.0597) = 0.00597$

7.	(a) The standard deviation of the English exam marks is $16.46483... = 16.5$
The standard deviation of the Mathematics exam marks is $10.39761... = 10.4$

(b) By looking at the frequency table, you can see that there are more low marks and more high marks in the English exam than in the Math exam. Therefore, the mark distribution of the English exam is spread more widely than the Math exam, and so you would expect the standard deviation of the English exam marks to be higher than the standard deviation of the Math exam marks.

8.	$s_A = 9.500888... = 9.50$, $s_B = 4.716961... = 4.72$. By looking at the frequency table, you can see that, in Pluvaria, there are more years of low rainfall and more years of high rainfall than in Torrentia. Therefore, you would expect that the standard deviation of rainfall in Pluvaria to be higher than in Torrentia, where the rainfall is less widely distributed.

Exercise 6.7

1   (a) (i) and (iii)   (b) (i) and (iii)

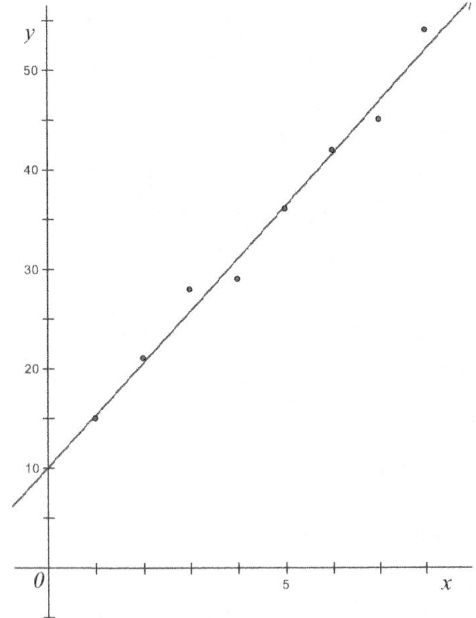   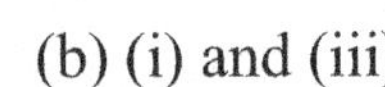

(ii) $y = 5.26x + 10.1$ (ii) $y = 87.8 - 0.384x$

2.   (a)

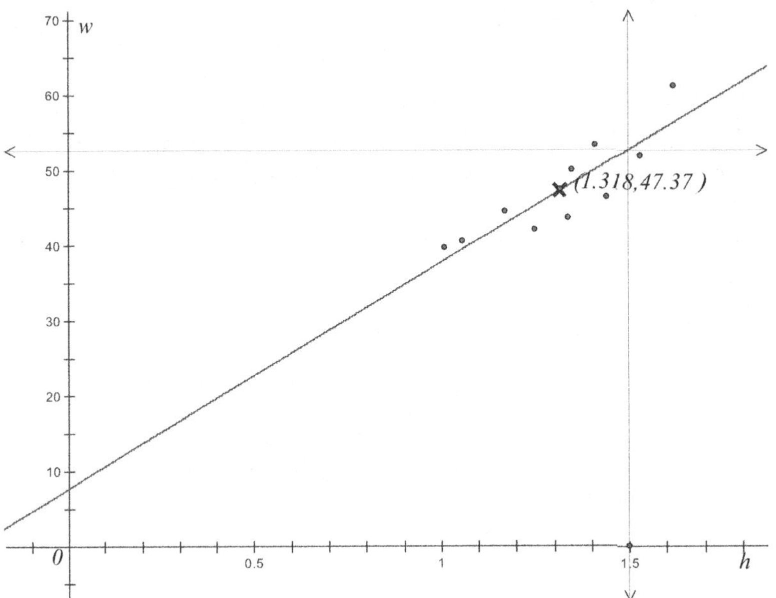

(b) $w = 30.1h + 7.70$ (c) $30.1 \times 1.5 + 7.70 = 52.85$ so child's weight is estimated at 52.9kg.

3.   (a) (i), (ii) $r = 0.9982783... = 0.998$

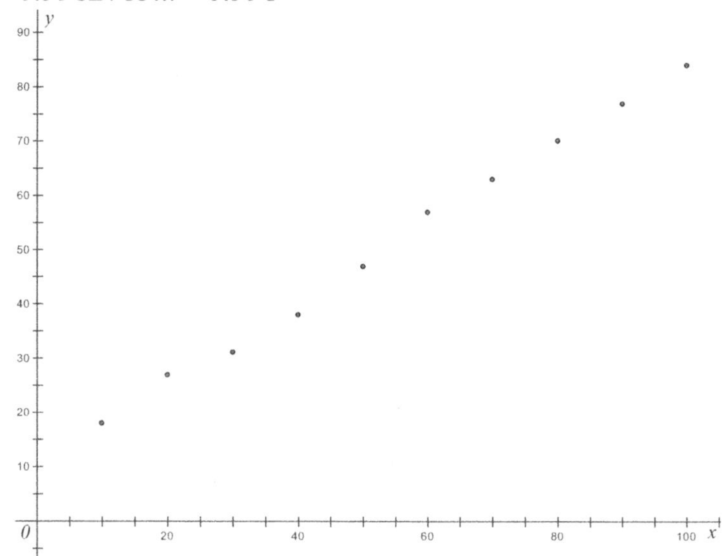

(b) (i), (ii) $r = -0.9035091... = -0.904$

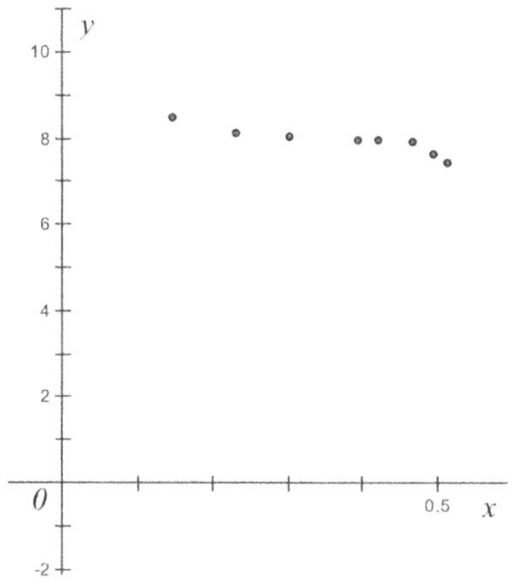

(c) (i), (ii) $r = -0.9977960... = -0.998$

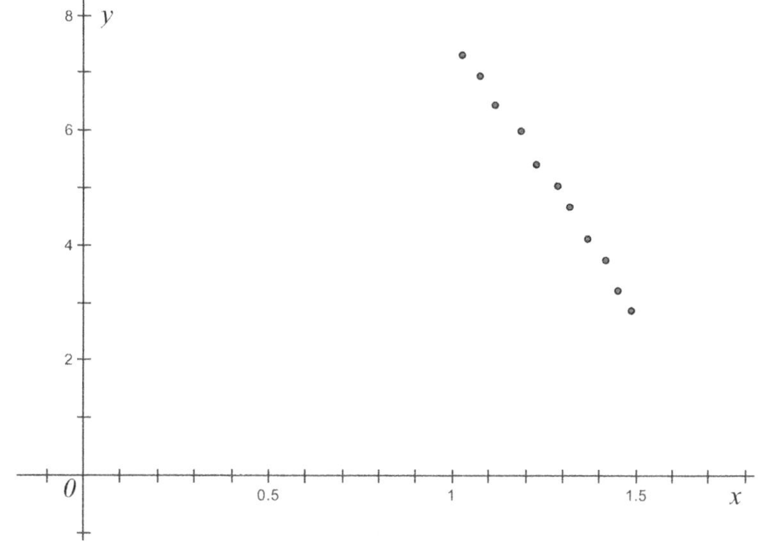

4.   (a)   (b)

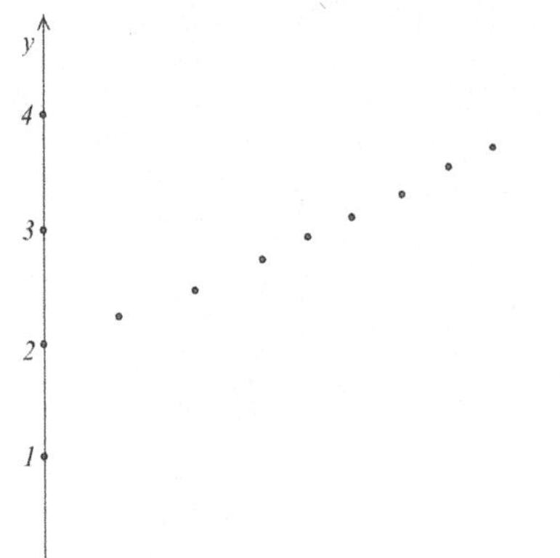

(c)

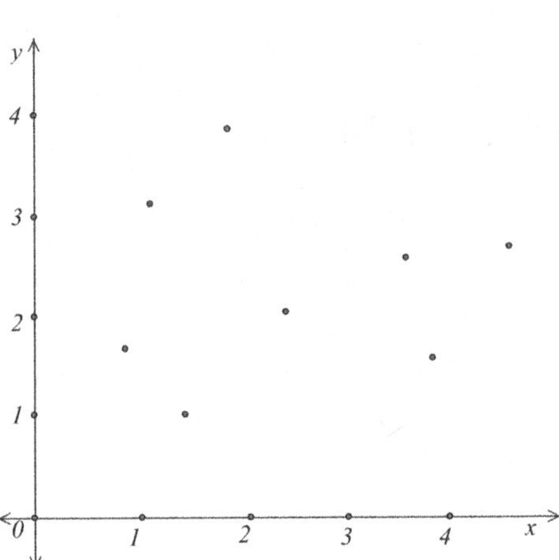

5.   (a) (i)

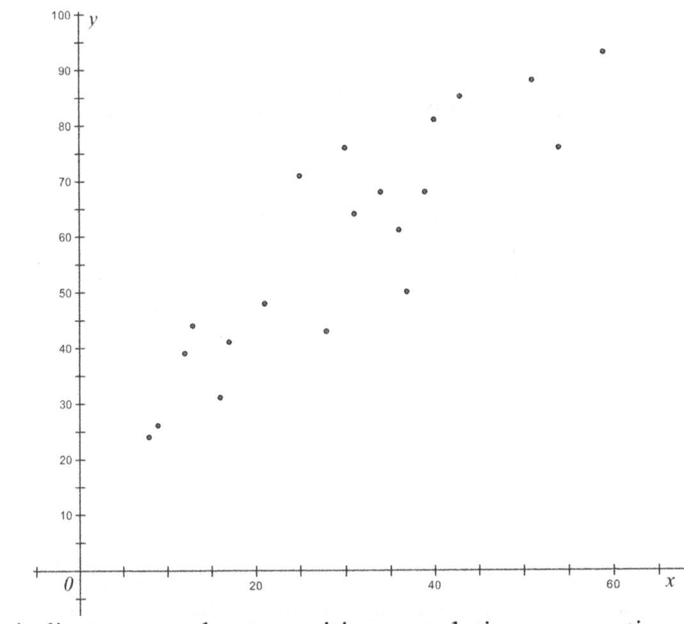

(ii) The data indicates a moderate positive correlation, suggesting that the more goals that a team scores the more games it wins.

(iii) $r = 0.8882123... = 0.888$

(b) $a = 1.250986...$, $b = 21.13276...$ so the regression line of y on x is $y = 1.25x + 21.1$

(c) Team U wins 30 games so the estimated number of goals scored is 58.6 or 59 goals.

Solutions to Unit 7 Exercises

Exercise 7.1

1. (a) $x^3 + c$ (b) $\dfrac{1}{5}x^5 + c$ (c) $-x^{-1} + c$ (d) $\dfrac{1}{3}x^3 + \dfrac{1}{2}x^2 + x + c$

   (e) $2x - \dfrac{5}{2}x^2 + c$ (f) $x^3 + x^2 + c$ (g) $\dfrac{3}{4}x^{\frac{4}{3}} + c$ (h) $2\sqrt{x} + c$

   (i) $\dfrac{1}{2}x^4 - x^3 + c$ (j) $\dfrac{2}{3}\sqrt{x^3} + 2\sqrt{x} + c$

2. (a) $-3\cos x + c$ (b) $\dfrac{1}{5}\sin x + c$ (c) $\sin x - \cos x + x + c$ (d) $4e^x + c$

   (e) $e^x - \sin x + c$ (f) $2\sin x + \cos x + e^x + c$

3. (a) $2\ln x + c$ (b) $3x + 2\ln x + c$ (c) $\dfrac{1}{4}x^4 + \dfrac{3}{2}x^2 - \ln x + c$ (d) $\dfrac{1}{4}\ln x + c$

   (e) $-\dfrac{1}{3x} + \dfrac{4}{3}\ln x - \dfrac{1}{3}x + c$ (f) $\dfrac{1}{2}\ln x + 2x + c$

4. (a) $\displaystyle\int \sin\dfrac{1}{2}x \cos\dfrac{1}{2}x\,dx = \int \dfrac{1}{2}\sin x\,dx = -\dfrac{1}{2}\cos x + c$

   (b) $\displaystyle\int \sin^2\dfrac{1}{2}x\,dx = \dfrac{1}{2}\int (1 - \cos x)\,dx = \dfrac{1}{2}(x - \sin x) + c = \dfrac{1}{2}x - \dfrac{1}{2}\sin x + c$

   (c) $\displaystyle\int \cos^2\dfrac{1}{2}x\,dx = \dfrac{1}{2}\int (1 + \cos x)\,dx = \dfrac{1}{2}(x + \sin x) + c = \dfrac{1}{2}x + \dfrac{1}{2}\sin x + c$

Exercise 7.2

1. $\dfrac{1}{8}(2x+1)^4 + c$ 2. $-\dfrac{1}{3}(1-x)^3 + c$ 3. $-\dfrac{1}{2}\cos 2x + c$ 4. $\dfrac{1}{4}\sin 2x + c$

5. $\dfrac{1}{3}\ln(3x-2) + c$ 6. $e^{x+2} + c$ 7. $-\dfrac{1}{5}e^{3-5x} + c$ 8. $-20e^{-0.05x} + c$

9. $-4\ln(10-x) + c$ 10. $\dfrac{1}{3}\sin\left(3x - \dfrac{\pi}{4}\right) + c$ 11. $\dfrac{1}{2}\cos\left(\dfrac{\pi}{3} - 2x\right) + c$

12. $\dfrac{1}{6}\sqrt{(4x+3)^3} + c$ 13. $3\ln(x+2) + c$ 14. $-\dfrac{1}{20}(3-4x)^5 + c$ 15. $2\sqrt{2x+1} + c$

16. $\dfrac{1}{3}e^{3x} + c$ 17. $\dfrac{1}{10}(5x-2)^2 + c$ 18. $5\sin\dfrac{1}{5}x + \dfrac{5}{2}\cos\dfrac{2}{5}x + c$

19. $\dfrac{3}{\pi}\sin\dfrac{\pi x}{3} + c$ 20. $-\dfrac{180}{\pi}\cos x^\circ + c$

Exercise 7.3

1.   (a) $u = x^2 + 3 \Rightarrow \dfrac{du}{dx} = 2x \Rightarrow \dfrac{dx}{du} = \dfrac{1}{2x} \Rightarrow \int x(x^2 + 3)\,dx = \int xu\,\dfrac{dx}{du}\,du = \int xu\,\dfrac{1}{2x}\,du = \dfrac{1}{2}\int u\,du$

$$\Rightarrow \dfrac{1}{2}\int u\,du = \dfrac{1}{4}u^2 + c = \dfrac{1}{4}(x^2 + 3)^2 + c$$

   (b) $u = x^2 + 2 \Rightarrow \dfrac{du}{dx} = 2x \Rightarrow \dfrac{dx}{du} = \dfrac{1}{2x} \Rightarrow \int \dfrac{x}{x^2+2}\,dx = \int \dfrac{x}{u}\,\dfrac{dx}{du}\,du = \int \dfrac{x}{u} \times \dfrac{1}{2x}\,du = \dfrac{1}{2}\int \dfrac{1}{u}\,du$

$$\Rightarrow \dfrac{1}{2}\int \dfrac{1}{u}\,du = \dfrac{1}{2}\ln u + c = \dfrac{1}{2}\ln(x^2 + 2) + c$$

   (c) $u = 3 - e^x \Rightarrow \dfrac{du}{dx} = -e^x \Rightarrow \dfrac{dx}{du} = -\dfrac{1}{e^x} \Rightarrow \int e^x u^4\,\dfrac{dx}{du}\,du = \int e^x u^4 \left(-\dfrac{1}{e^x}\right)du = -\int u^4\,du = -\dfrac{1}{5}u^5 + c$

$$= -\dfrac{1}{5}(3 - e^x)^5 + c$$

   (d) $u = 1 + \sin x \Rightarrow \dfrac{dy}{dx} = \cos x \Rightarrow \dfrac{dx}{du} = \dfrac{1}{\cos x} \Rightarrow \int \cos x(1 + \sin x)^2\,dx = \int (\cos x)\,u^2\,\dfrac{dx}{du}\,du$

$$= \int u^2 \cos x\,\dfrac{1}{\cos x}\,du = \int u^2\,du = \dfrac{1}{3}u^3 + c = \dfrac{1}{3}(1 + \sin x)^3 + c$$

   (e) $u = \ln x \Rightarrow \dfrac{du}{dx} = \dfrac{1}{x} \Rightarrow \dfrac{dx}{du} = x \Rightarrow \int \dfrac{1}{x}\ln x\,dx = \int \dfrac{1}{x}\,u\,\dfrac{dx}{du}\,du = \int \dfrac{1}{x}\,ux\,du = \int u\,du = \dfrac{1}{2}u^2 + c$

$$= \dfrac{1}{2}(\ln x)^2 + c$$

   (f) $u = \cos x \Rightarrow \dfrac{du}{dx} = -\sin x \Rightarrow \dfrac{dx}{du} = \dfrac{-1}{\sin x} \Rightarrow \int \sin x\,e^{\cos x}\,dx = \int \sin x\,e^u\,\dfrac{dx}{du}\,du$

$$= \int \sin x\,e^u \left(\dfrac{-1}{\sin x}\right)du = \int -e^u\,du = -e^u + c = -e^{\cos x} + c$$

2.   (a) $u = x^3 + 1 \Rightarrow \dfrac{du}{dx} = 3x^2 \Rightarrow \dfrac{dx}{du} = \dfrac{1}{3x^2} \Rightarrow \int \dfrac{1}{3}x^2(x^3 + 1)^2\,dx = \dfrac{1}{3}\int x^2 u^2\,\dfrac{dx}{du}\,du$

$$= \dfrac{1}{3}\int x^2 u^2 \left(\dfrac{1}{3x^2}\right)du = \dfrac{1}{9}\int u^2\,du = \dfrac{1}{27}u^3 + c = \dfrac{1}{27}(x^3 + 1)^3 + c$$

   (b) $u = x^3 + 1 \Rightarrow \dfrac{du}{dx} = 3x^2 \Rightarrow \dfrac{dx}{du} = \dfrac{1}{3x^2} \Rightarrow \int \dfrac{1}{3}\dfrac{x^2}{(x^3 + 1)}\,dx = \dfrac{1}{3}\int \dfrac{x^2}{u}\,\dfrac{dx}{du}\,du = \dfrac{1}{3}\int \dfrac{x^2}{u}\,\dfrac{1}{3x^2}\,du$

$$= \dfrac{1}{3}\int \dfrac{1}{u}\,du = \dfrac{1}{3}\ln u + c = \dfrac{1}{3}\ln(x^3 + 1) + c$$

   (c) $u = x^2 + x + 1 \Rightarrow \dfrac{du}{dx} = 2x + 1 \Rightarrow \dfrac{dx}{du} = \dfrac{1}{2x+1} \Rightarrow \int (2x+1)\sqrt{x^2 + x + 1}\,dx = \int (2x+1)u^{\frac{1}{2}}\,\dfrac{dx}{du}\,du$

$$\int (2x+1)u^{\frac{1}{2}} \left(\dfrac{1}{2x+1}\right)du = \int u^{\frac{1}{2}}\,du = \dfrac{2}{3}u^{\frac{3}{2}} + c = \dfrac{2}{3}\sqrt{(x^2 + x + 1)^3} + c$$

   (d) $u = e^{2x} + 1 \Rightarrow \dfrac{du}{dx} = 2e^{2x} \Rightarrow \dfrac{dx}{du} = \dfrac{1}{2e^{2x}} \Rightarrow \int \dfrac{2e^{2x}}{e^{2x} + 1}\,dx = \int \dfrac{2e^{2x}}{u}\,\dfrac{dx}{du}\,du = \int \dfrac{2e^{2x}}{u}\,\dfrac{1}{2e^{2x}}\,du$

$$\int \frac{1}{u} du = \ln u + c = \ln\left(e^{2x}+1\right)+c$$

3. $\quad u = \cos x \Rightarrow \dfrac{du}{dx} = -\sin x \Rightarrow \dfrac{dx}{du} = \dfrac{-1}{\sin x} \Rightarrow \int \tan x\, dx = \int \dfrac{\sin x}{\cos x} dx = \int \dfrac{\sin x}{u} \dfrac{dx}{du} du$

$\quad = \int \dfrac{\sin x}{u}\left(\dfrac{-1}{\sin x}\right) dx = -\int \dfrac{1}{u} du = -\ln u + c = -\ln \cos x + c$

Exercise 7.4

1. $e^{x^2} + c$

2. $\dfrac{1}{3}\left(\ln x\right)^3 + c$

3. $\ln\left(x^3 + 1\right) + c$

4. $\ln \sin x + c$

5. $-\dfrac{1}{2}\cos x^2 + c$

6. $-\dfrac{1}{3}\cos^3 x + c$

7. $\dfrac{1}{2}\left(x^2 + 4\right)^2 + c$

8. $\dfrac{1}{4}\times\dfrac{2}{3}\left(2x^2 + 3\right)^{\frac{3}{2}} + c = \dfrac{1}{6}\left(2x^2 + 3\right)^{\frac{3}{2}} + c$

9. $\ln\left(3 - \cos x\right) + c$

10. $2\ln\left(e^x + 3\right) + c$

11. $\dfrac{1}{3}\ln\left(x^3 - 1\right) + c$

12. $\dfrac{1}{6}\left(4 + 3\sin 2x\right) + c$

13. $\int \sin^3 x\, dx = \int \sin x\left(1 - \cos^2 x\right) dx = \int \sin x\, dx + \int \left(-\sin x\right)\cos^2 x\, dx = -\cos x + \dfrac{1}{3}\cos^3 x + c$

Exercise 7.5

1.   (a) $y = 2x^2 + x + c,\ 3 = 0 + 0 + c \Rightarrow c = 3 \Rightarrow y = 2x^2 + x + 3$

    (b) $y = 3x - \dfrac{2}{3}x^3 + c,\ 15 = 9 - 18 + c \Rightarrow c = 24 \Rightarrow y = 24 + 3x - \dfrac{2}{3}x^3$

    (c) $y = -\cos x + c,\ 1 = -\cos\dfrac{\pi}{3} + c \Rightarrow c = 1 + \cos\dfrac{\pi}{3} = \dfrac{3}{2} \Rightarrow y = \dfrac{3}{2} - \cos x$

    (d) $y = -2e^{-2x} + c,\ 1 = -2 + c \Rightarrow c = 3 \Rightarrow y = 3 - 2e^{-2x}$

    (e) $y = \left(2x + 1\right)^{\frac{1}{2}} + c,\ 4 = 3 + c \Rightarrow c = 1 \Rightarrow y = \sqrt{2x + 1} + 1$

2. $\quad y = x^4 - 3x^2 + 5x + c,\ 8 = 81 - 27 + 15 + c \Rightarrow c = -61 \Rightarrow y = x^4 - 3x^2 + 5x - 61$

3. $\quad x = t^3 + \dfrac{2}{t} + c,\ 5 = 1 + 2 + c \Rightarrow c = 2 \Rightarrow x = t^3 + \dfrac{2}{t} + 2$

4. $\quad y = e^x + c,\ 2 = 1 + c \Rightarrow c = 1 \Rightarrow y = e^x + 1$

5. $\quad y = 4\ln x + c,\ 5 = 4\ln e^2 + c \Rightarrow 5 = 8 + c \Rightarrow c = -3 \Rightarrow y = 4\ln x - 3$

Exercise 7.6

1. (a) $a=10 \Rightarrow v(t)=-10t+c, \ v(0)=16 \Rightarrow v(t)=16-10t$

    (b) $v(2)=16-20=-4$. Therefore the velocity is -4ms^{-1}.

    (c) $s(t)=16t-5t^2+c, \ s(0)=2 \Rightarrow 2=0+0+c \Rightarrow c=2$. The maximum velocity occurs when
    $$v(t)=0 \Rightarrow t=1.6 \Rightarrow s(1.6)=16\times1.6-5(1.6)^2+2=14.8.$$

2. (a) $a(5)=5$

    (b) $v(t)=20t-\dfrac{3}{2}t^2+c, \ 3=0+0+c \Rightarrow c=3 \Rightarrow v(t)=20t-\dfrac{3}{2}t^2+3$

    (c) $v(5)=65.5$

    (d) $20t-\dfrac{3}{2}t^2+3=0 \Rightarrow t=-0.1483494..., \ 13.481682....$ But, $t>0$, so $t=13.5$.

3. (a) $v(10)=306.8263...$, so the velocity of the rocket is 307ms^{-1}

    (b) $s(t)=\dfrac{1}{2}t^2+4e^{0.5t}+c, \ s(0)=0+4+c=0 \Rightarrow c=-4 \Rightarrow s(t)=\dfrac{1}{2}t^2+4e^{0.5t}-4$

    (c) $s(10)=639.6526...$, therefore the altitude is 640m.

    (d) For a further 50s the rocket travels at 306.8ms^{-1}. Therefore after 1 minute, its height is
    $639.7+306.8\times50=15976.7$ or approximately 16 000m.

4. (a) (i) $v(t)=-4\cos t+c, \ 8=-4+c \Rightarrow c=12$ and $v(t)=12-4\cos t$.

    (ii) Maximum velocity is 16ms^{-1} occurring when $\cos t=-1 \Rightarrow t=\pi \approx 3.14s$.

    (iii) $s(t)=12t-4\sin t+c, \ c=0 \Rightarrow s(t)=12t-4\sin t$

    (b)

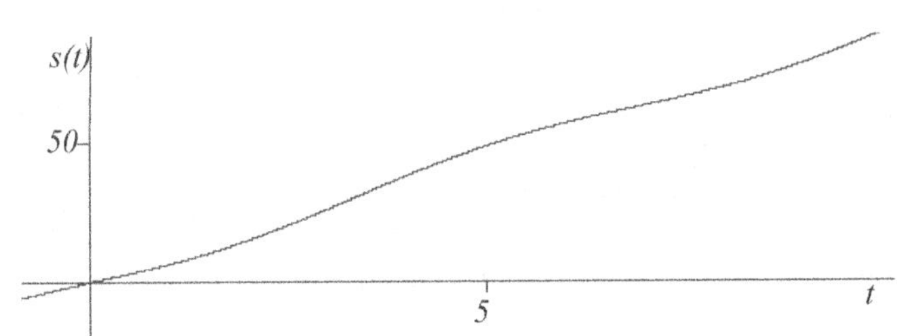

5. $v(0)=10 \Rightarrow 10=c, \ v(3)=8 \Rightarrow 8=9a+3b+c, \ v(6)=16 \Rightarrow 16=36a+6b+c$. Therefore,

    $9a+3b=-2, \ 36a+6b=6 \Rightarrow 6a+b=1$. $\ b=1-6a \Rightarrow 9a+3(1-6a)=-2 \Rightarrow -9a+3=-2$

    $\Rightarrow a=\dfrac{5}{9}, \ b=1-6a=-\dfrac{7}{3}$. Therefore, $v(t)=\dfrac{5}{9}t^2-\dfrac{7}{3}t+10 \Rightarrow s(t)=\dfrac{5}{27}t^3-\dfrac{7}{6}t^2+10t+c$.

    But, $s(0)=0 \Rightarrow c=0$ so $\Rightarrow s(t)=\dfrac{5}{27}t^3-\dfrac{7}{6}t^2+10t$ and $s(10)=168.5185...$. So, after 10

    seconds, the particle is 168.5 m from O.

6.   (a) $s(t) = -12.5e^{-0.4t} + c, \; 0 = -12.5 + c \Rightarrow c = 12.5 \; \Rightarrow s(t) = 12.5\left(1 - e^{-0.4t}\right)$ and

        $s(8) = 12.5\left(1 - e^{-3.2}\right) = 11.99047... = 11.99$

   (b)   12.5

   (c) $a(t) = -2e^{-0.4t} \Rightarrow a(1) = -1.340640... = 1.34$

Exercise 7.7

1.   (a)   3.75   (b)   2   (c) $\dfrac{1}{6}$ (d) $\dfrac{8}{3}$ (e)   2.5   (f)   14.25

2.   (a) $\displaystyle\int_{-1}^{1} e^x \, dx = \left[e^x\right]_{-1}^{1} = e - \dfrac{1}{e} = 2.35$ (b) $\displaystyle\int_{0}^{2} \dfrac{1}{2x+3}\, dx = \left[\dfrac{1}{2}\ln(2x+3)\right]_{0}^{2} = \dfrac{1}{2}\ln\dfrac{7}{3} = 0.424$

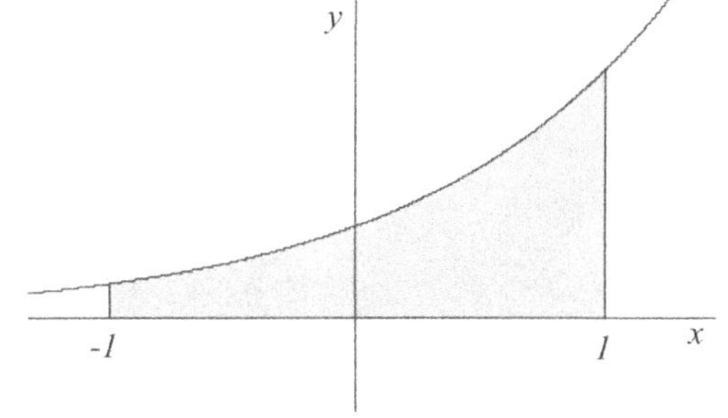

   (c) $\displaystyle\int_{0}^{\frac{\pi}{3}} \cos x \, dx = \left[\sin x\right]_{0}^{\frac{\pi}{3}} = \dfrac{\sqrt{3}}{2} = 0.866$

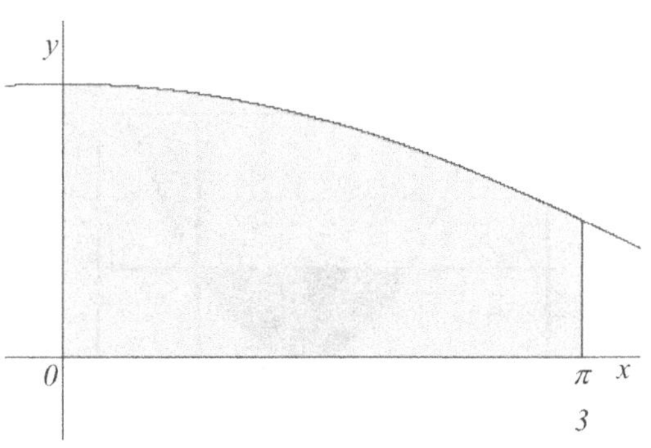

3.   (a) $\displaystyle\int_{0}^{4} e^{-\frac{1}{2}t} \, dt = \left[-2e^{-\frac{1}{2}t}\right]_{0}^{4} = -2e^{-2} - (-2) = 2 - \dfrac{2}{e^2} = 2\left(1 - \dfrac{1}{e^2}\right)$

   (b) $\displaystyle\int_{1}^{5} \dfrac{4}{(3x+1)}\, dx = \left[\dfrac{4}{3}\ln(3x+1)\right]_{1}^{5} = \dfrac{4}{3}\left(\ln 16 - \ln 4\right) = \dfrac{4}{3}\ln 4$

   (c) $\displaystyle\int_{0}^{\frac{\pi}{2}} (1 + \sin 2\theta)\, d\theta = \left[\theta - \dfrac{1}{2}\cos 2\theta\right]_{0}^{\frac{\pi}{2}} = \left(\dfrac{\pi}{2} - \dfrac{1}{2}\cos\pi\right) - \left(0 - \dfrac{1}{2}\cos 0\right) = 1 + \dfrac{\pi}{2}$

4. The large sketches are not drawn, but the shape of the functions can be seen from the graphing calculator screen displays shown below.

(a) 1.76   (b) 2.40   (c) 19.7

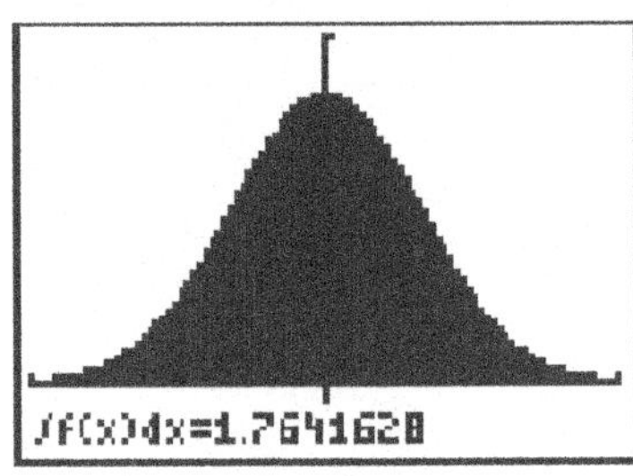

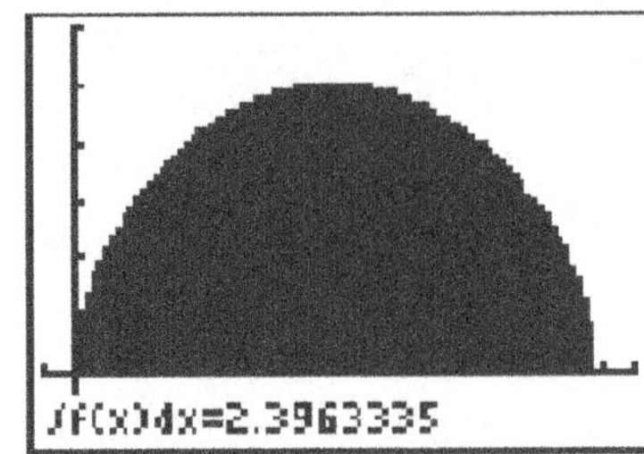

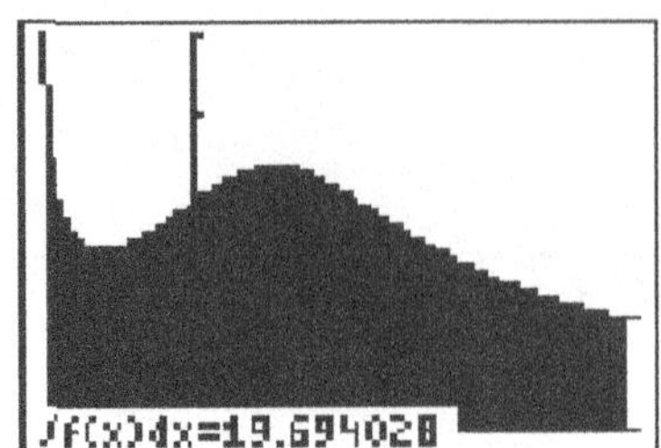

5. (a)

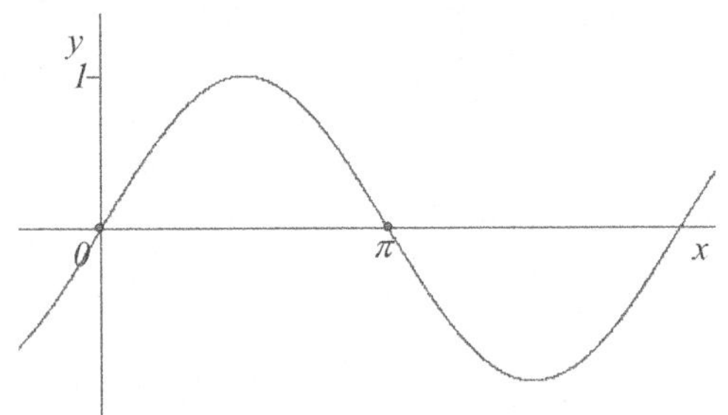

(b)   (i) $\left[-\cos x\right]_0^\pi = 2$ (ii) $\left[-\cos x\right]_\pi^{2\pi} = -2$ (iii) $\left[-\cos x\right]_0^{2\pi} = 0$

(c) Area below the x-axis is measured negatively so that the net area between the curve and the x-axis between 0 and 2π is $2 + (-2) = 0$.

6. (a)   (b)

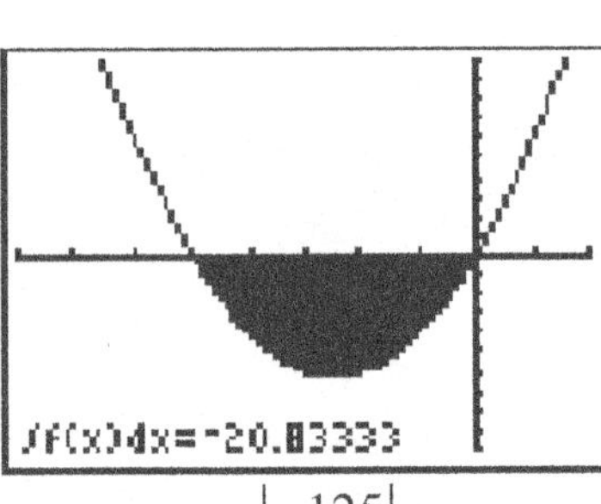

$$\text{Area} = \left|-\frac{125}{6}\right| = 20.8$$

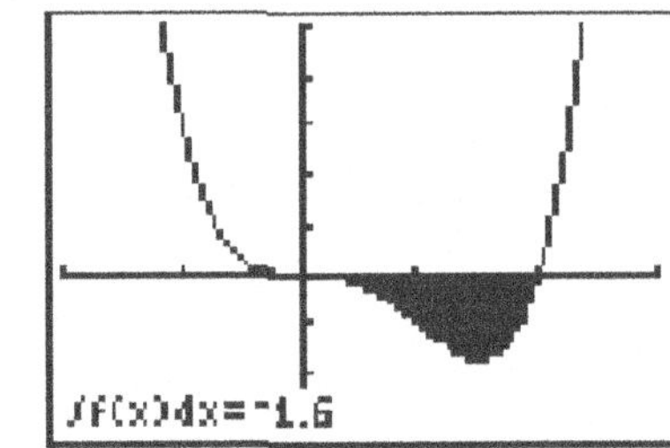

$$\text{Area} = \left|-1.6\right| = 1.6$$

(c)

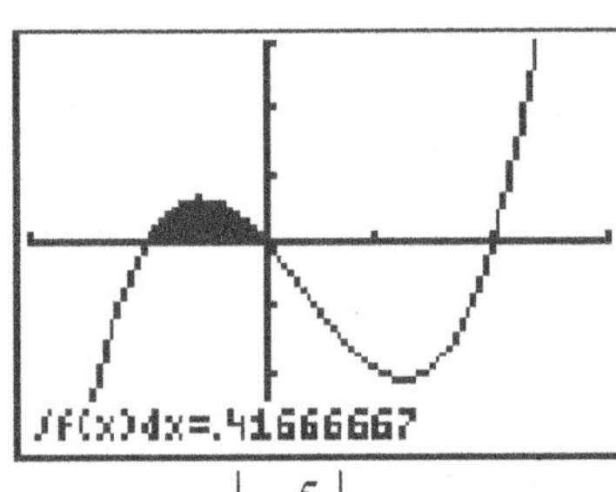

$$\text{Area} = \left|-\frac{5}{12}\right| \approx 0.417$$

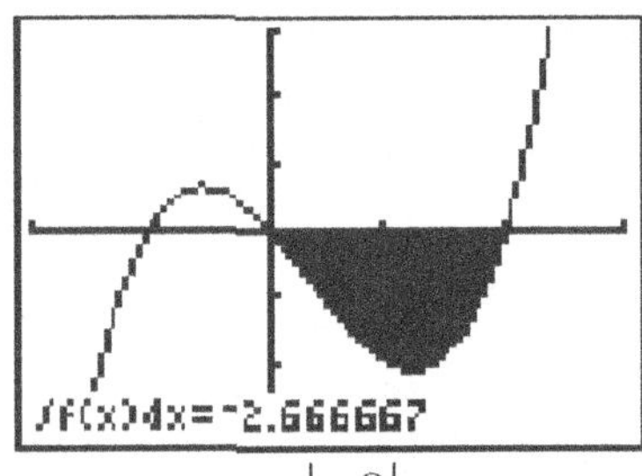

$$\text{Area} = \left|-\frac{8}{3}\right| \approx 2.67$$

Therefore total area enclosed is $\dfrac{37}{12} = 3.08$

(d)

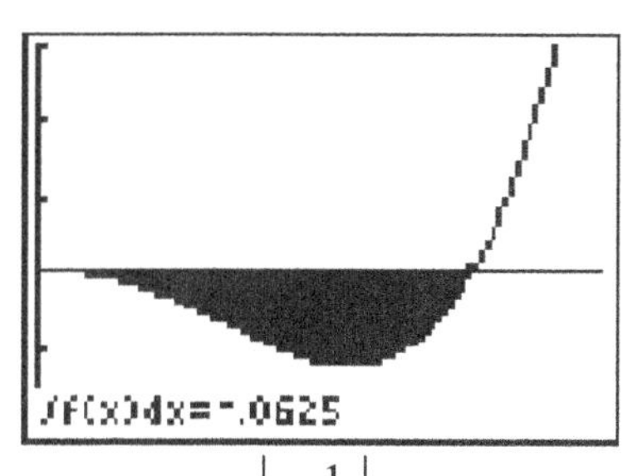

$$\text{Area} = \left| -\frac{1}{16} \right| = 0.0625$$

(e)

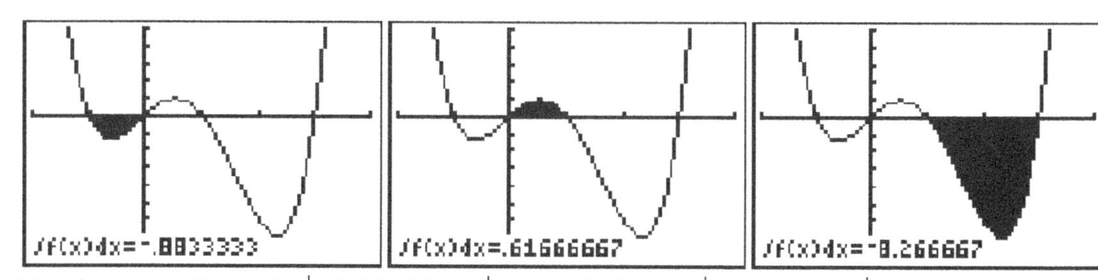

Total area enclosed is $\left| -0.8833333... \right| + 0.6166666... + \left| -8.266666... \right| = 9.766665... = 9.77$

(f) The curve intersects the *x*-axis in the following values of *x*:: -0.561070, 0.59924, 1.34805

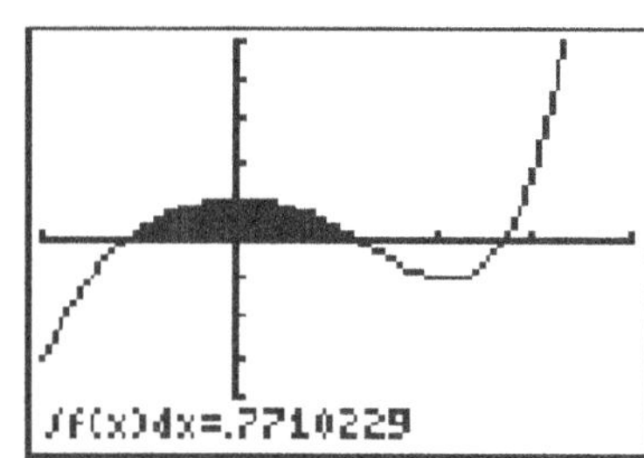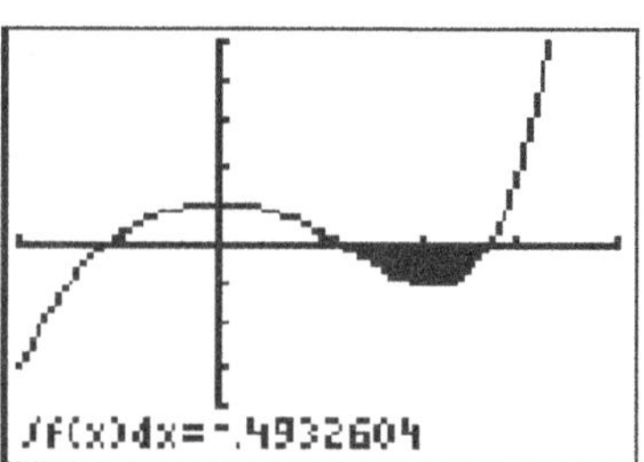

Therefore the total area enclosed is $0.7710229... + \left| -0.4932604... \right| = 1.264283... = 1.26$

Exercise 7.8

1.   (a) 162π (b) 111π (c) 18.6π (d) $\dfrac{\pi \ln 9}{4}$

2.   It is advisable to use your graphing calculator.   (a) 9.87   (b) 182   (c) 1.36   (d) 197

3. $V = \pi \int_0^3 \left(3x - x^2\right)^2 dx = \pi \int_0^3 \left(9x^2 - 6x^3 + x^4\right) dx = \pi \left[3x^3 - \frac{3}{2}x^4 + \frac{1}{5}x^5 \right]_0^3 = \pi \left(81 - 121.5 + \frac{243}{5} \right)$

$\qquad = \dfrac{40.5}{5}\pi = 8.1\pi$

4. $V = \pi \int_0^a x\, dx = \pi \left[\frac{1}{2}x^2 \right]_0^a = 18\pi \Rightarrow \frac{1}{2}a^2 - 0 = 18 \Rightarrow a^2 = 36 \Rightarrow a = \pm 6$, but $a > 0$ so $a = 6$

5. $\quad V = \int_0^1 \pi\left(x^2 - x^4\right)dx = \pi\left[\dfrac{1}{3}x^3 - \dfrac{1}{5}x^5\right]_0^1 = \pi\left(\dfrac{1}{3} - \dfrac{1}{5}\right) = \dfrac{2}{15}\pi$

6. $\quad V = \pi\int_0^a (x+1)^2 dx = 100\pi \Rightarrow \dfrac{\pi}{3}\Big[(x+1)^3\Big]_0^a = 100\pi$

$\Rightarrow (a+1)^3 - 1^3 = 300 \Rightarrow (a+1) = \sqrt[3]{301} \Rightarrow a = 5.701759\ldots = 5.70$

7. $\quad$ (a) $\dfrac{d}{dx}\left(\dfrac{1}{4}x^2 - \dfrac{1}{4}x\sin 2x - \dfrac{1}{8}\cos 2x\right) = \dfrac{1}{2}x - \dfrac{1}{4}\left(x \times 2\cos 2x + 1 \times \sin 2x\right) - \dfrac{1}{8}\left(-2\sin 2x\right)$

$= \dfrac{1}{2}x - \dfrac{1}{2}x\cos 2x - \dfrac{1}{4}\sin 2x + \dfrac{1}{4}\sin 2x = \dfrac{1}{2}x - \dfrac{1}{2}x\cos 2x = \dfrac{1}{2}x\left(1 - \cos 2x\right)$

$= \dfrac{1}{2}x\left(1 - \left(1 - 2\sin^2 x\right)\right) = x\sin^2 x$

$\quad$ (b) $V = \pi\int_0^\pi x\sin^2 x\,dx = \pi\left[\dfrac{1}{4}x^2 - \dfrac{1}{4}x\sin 2x - \dfrac{1}{8}\cos 2x\right]_0^\pi$

$= \pi\left\{\left(\dfrac{1}{4}\pi^2 - \dfrac{1}{4}\pi\sin 2\pi - \dfrac{1}{8}\cos 2\pi\right) - \left(0 - 0 - \dfrac{1}{8}\cos 0\right)\right\} = \pi\left\{\dfrac{1}{4}\pi^2 - \dfrac{1}{8} + \dfrac{1}{8}\right\} = \dfrac{1}{4}\pi^3$

Solutions to Units 5 & 7 Review Exercise

Part 1

1. $\quad \dfrac{dy}{dx} = 3x^2 + 2x - 4 \Rightarrow y = x^3 + x^2 - 4x + c$, but $y(1) = 5 \Rightarrow 5 = 1 + 1 - 4 + c \Rightarrow c = 7$ and

$y = x^3 + x^2 - 4x + 7$

2. $\quad$ (a) $2\sin\left(a - \dfrac{\pi}{3}\right) = 0 \Rightarrow a - \dfrac{\pi}{3} = 0 \Rightarrow a = \dfrac{\pi}{3}$

$\quad$ (b) $\int_0^a 2\sin\left(x - \dfrac{\pi}{3}\right)dx = \left[-2\cos\left(x - \dfrac{\pi}{3}\right)\right]_0^a = \left(-2\cos\left(\dfrac{\pi}{3} - \dfrac{\pi}{3}\right)\right) - \left(-2\cos\left(0 - \dfrac{\pi}{3}\right)\right)$

$= -2\cos 0 - \left(-2\cos\left(-\dfrac{\pi}{3}\right)\right) = -2 + 2 \times \dfrac{1}{2} = -2 + 1 = -1$

$\quad$ (c) Therefore, the area of the shaded region is $|-1| = 1$

3.   (a) $x = 0,\ y = 2$ (b) $2 - \dfrac{1}{x} = 0 \Rightarrow \dfrac{1}{x} = 2 \Rightarrow x = \dfrac{1}{2}$

(c)

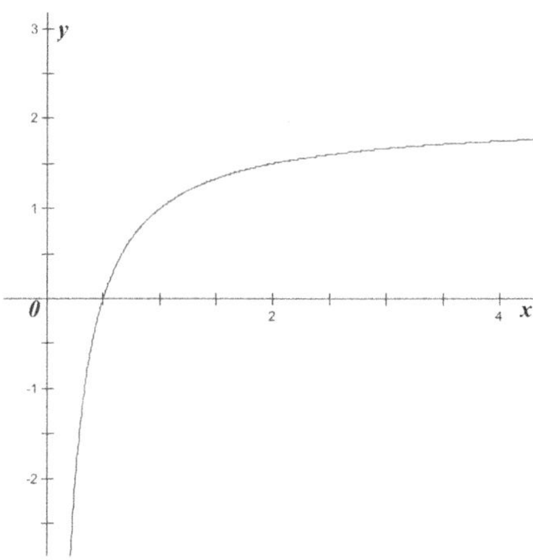

(d) (i) $V = \pi \displaystyle\int_1^2 y^2\,dx = \pi \int_1^2 \left(2 - \dfrac{1}{x} \right)^2 dx$

(ii) $V = \pi \displaystyle\int_1^2 \left(4 - \dfrac{4}{x} + \dfrac{1}{x^2} \right) dx = \pi \left[4x - 4\ln x - \dfrac{1}{x} \right]_1^2$

$= \pi \left\{ \left(8 - 4\ln 2 - \dfrac{1}{2} \right) - (4 - 1) \right\} = \pi \left((7.5 - 3) - \ln 2^4 \right) = \pi \left(\dfrac{9}{2} - \ln 16 \right)$

4.   (a) $s(0) = 9$, therefore initial displacement is 9 m.

(b) $v(t) = \dfrac{ds}{dt} = 8 - 2t \Rightarrow v(6) = 8 - 12 = -4$, therefore the velocity is $4\ \text{ms}^{-1}$ south.

(c) The maximum distance occurs when $v = 0 \Rightarrow s'(t) = 0 \Rightarrow 8 - 2t = 0 \Rightarrow t = 4$ so that

the maximum distance is $s(4) = 9 + 32 - 16 = 25$ m.

5.   (a) $\dfrac{dy}{dx} = 2ax + b$ (b) $c = 3$ (c) At B $9 = 9a + 3b + 3 \Rightarrow 9a + 3b = 6 \Rightarrow 3a + b = 2$.

(d) $6a + b = 0 \Rightarrow -3a = 2 \Rightarrow a = -\dfrac{2}{3}$, therefore, $b = -6\left(-\dfrac{2}{3} \right) = 4$

(e) (i) $\dfrac{dy}{dx} = 2ax + b = -\dfrac{4x}{3} + 4$ so, at A, the gradient is $0 + 4 = 4$

(ii) the equation of the tangent is $y = 4x + 3$

6.   (a) $\displaystyle\int_{-1}^2 f(x)\,dx + \int_{-1}^2 1\,dx = 11 + [x]_{-1}^2 = 11 + (2 - (-1)) = 14$

(b) $5\displaystyle\int_{-1}^2 f(x)\,dx = 55$

(c) $\displaystyle\int_{-1}^2 f'(x)\,dx = f(2) - f(-1) = 4 - 1 = 3$

7. $\quad s = -\dfrac{2}{3}\cos 3t + t + c$, $s(0) = 4 \Rightarrow 4 = -\dfrac{2}{3} + 0 + c \Rightarrow c = \dfrac{12}{3} + \dfrac{2}{3} = \dfrac{14}{3}$

$\quad \Rightarrow s = \dfrac{14}{3} + t - \dfrac{2}{3}\cos 3t$

8. $\quad$ (a) (i) $f'(1) = 0$ $\quad$ (ii) $f'(-1) = 1$

$\quad$ (b)

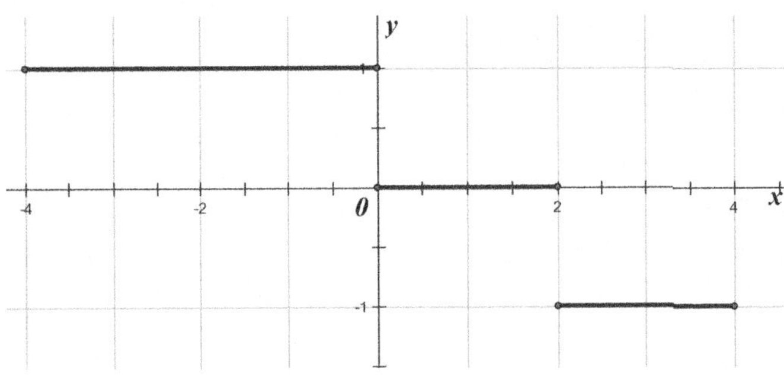

$\quad$ (c)

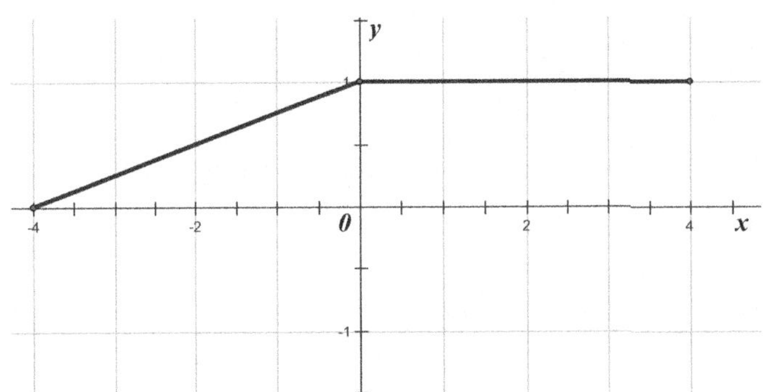

9. $\quad$ (a) $-\dfrac{1}{2} \times \dfrac{1}{(2x+1)} + c = -\dfrac{1}{2(2x+1)} + c$

$\quad$ (b) $\displaystyle\int_1^2 \dfrac{1}{(2x+1)^2}\, dx = \left[-\dfrac{1}{2(2x+1)} \right]_1^2 = \left(-\dfrac{1}{10} \right) - \left(-\dfrac{1}{6} \right) = \dfrac{5-3}{30} = \dfrac{1}{15} \Rightarrow k = 15$

10.

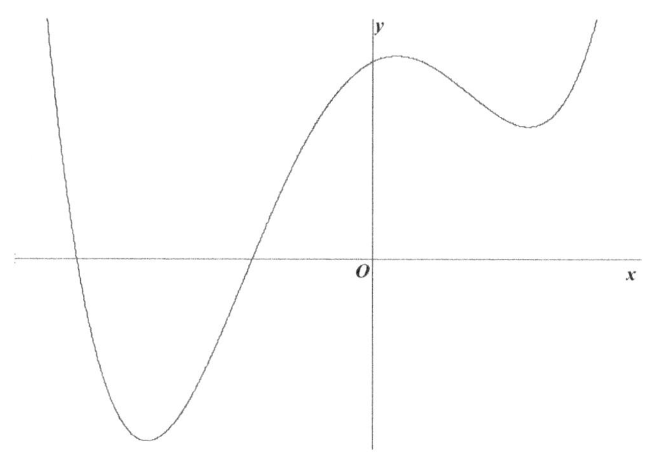

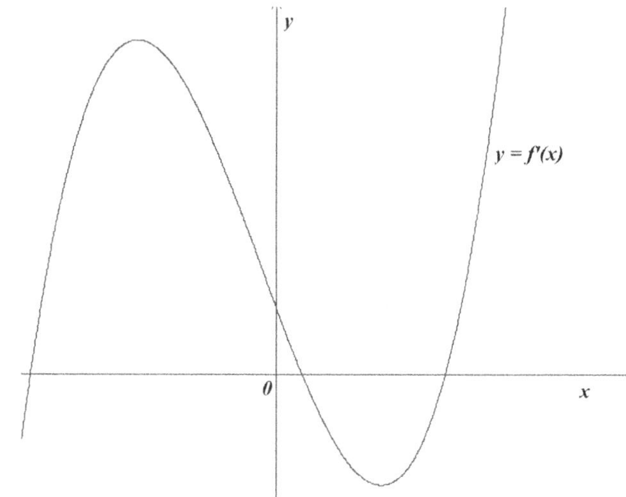

11. $\quad V = \pi \int_1^k \dfrac{4}{(x+1)} dx = \Big[4\pi \ln(x+1) \Big]_1^k = 4\pi \ln\left(\dfrac{k+1}{2} \right) = 4\pi \ln 3 \Rightarrow \dfrac{k+1}{2} = 3 \Rightarrow k = 5$

12.   (a) Area of rectangle $ABCD = AB \times AD = 36 \Rightarrow AD = \dfrac{36}{x}$

      (b) Perimeter, $y = 2AB + 2AD = 2x + \dfrac{72}{x}$

      (c) $\dfrac{dy}{dx} = 2 - \dfrac{72}{x^2}$

      (d) When the perimeter is a maximum or a minimum

$\qquad \dfrac{dy}{dx} = 0 \Rightarrow 2 - \dfrac{72}{x^2} = 0 \Rightarrow x^2 = 36 \Rightarrow x = \pm 6$ but since $x > 0$, $x = 6$. For a minimum

$\qquad$ perimeter we require $\dfrac{d^2 y}{dx^2} > 0$ at $x = 6$. $\dfrac{d^2 y}{dx^2} = \dfrac{144}{x^3} = \dfrac{144}{6^3} = \dfrac{2}{3} > 0$ so when $x = 6$ the

$\qquad$ perimeter is a minimum.

Part 2

13.   (a) $\dfrac{dv}{dt}=2e^{-t}+4 \Rightarrow v=-2e^{-t}+4t+c_1$. If $v(0)=3$ then $3=-2+0+c_1 \Rightarrow c_1=5$ and

$$v=-2e^{-t}+4t+5$$

(b) Assume that y is the displacement from origin, O then $\dfrac{dy}{dt}=-2e^{-t}+4t+5$

$\Rightarrow s=2e^{-t}+2t^2+5t+c_2$. If $y(0)=0$ then $0=2+0+0+c_2$

$\Rightarrow c_2=-2$ and $y=2e^{-t}+2t^2+5t-2$

(c) $y(2)=2e^{-2}+8+10-2=16.27067...=16.3$, therefore, distance from O is 16.3 m.

14.   (a) (i) $f'(x)=0$ (ii) $f'(x)=x^2\left(-e^{-x}\right)+2xe^{-x}=e^{-x}\left(2x-x^2\right)$

(iii) At B, $e^{-x}\left(2x-x^2\right)=0 \Rightarrow 2x-x^2=0 \Rightarrow x(2-x)=0 \Rightarrow x=0,\ x=2$.

(b) $f''(x)=e^{-x}(2-2x)+\left(-e^{-x}\right)\left(2x-x^2\right)=e^{-x}\left(2-4x+x^2\right)$

(c) At A and $C \Rightarrow e^{-x}\left(2-4x+x^2\right)=0 \Rightarrow 2-4x+x^2=0$.

$\Rightarrow x=3,414213...,\ x=0.5857864....$ so x-coordinate of A is 0.586, x-coordinate of C is 3.41

15.   (a) $f'(x)=\dfrac{2x^2}{x^2+5}+\ln\left(x^2+5\right)$

(b)

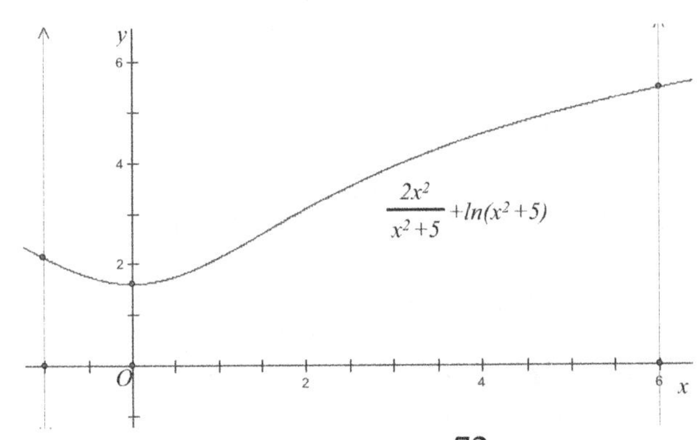

$f'(0)=\ln 5=1.609437...=1.61$ and $f'(6)=\dfrac{72}{41}+\ln 41=5.469669...=5.47$ so range of

$y=f'(x)$ is $1.61\le f'\le 5.47$.

16.   (a) Since graph of f passes through $(1,\ 5)$, $\dfrac{15}{a+b}=5 \Rightarrow 15=5a+5b$. In addition, since

there is an asymptote at $x=-2$, $-2a+b=0$. So, $b=2a \Rightarrow 15=5a+10a=15a$ and
$a=1$. Therefore, $b=2$

(b) (i) $f(x)=15(x+2)^{-1} \Rightarrow f'(x)=-15(x+2)^{-2}$

(ii) Whatever the value of x, $(x+2)^2>0$ so that $-\dfrac{15}{(x+2)^2}<0$, therefore the gradient of the graph of f is always negative and f is a decreasing function.

(c) $V = \pi \displaystyle\int_0^4 \left(\dfrac{15}{x+2}\right)^2 dx = 75\pi$

17.   (a) A is $\left(\dfrac{\pi}{2},\ 0\right)$, D is $\left(\dfrac{3\pi}{2},\ 0\right)$

    (b) $f'(x) = -e^{-x}\cos x + e^{-x}(-\sin x) = -e^{-x}(\cos x + \sin x)$.

    At B, $f'(x) = 0 \Rightarrow -e^{-x}(\cos x + \sin x) = 0 \Rightarrow \cos x + \sin x = 0 \Rightarrow 1 + \tan x = 0$

    $\Rightarrow \tan x = -1$. Now B is between A and D so that the x-coordinate of B is between

    $\dfrac{\pi}{2}$ and $\dfrac{3\pi}{2}$. Therefore, $\Rightarrow x = \pi - \dfrac{\pi}{4} = \dfrac{3\pi}{4}$

    (c) $f''(x) = -e^{-x}(-\sin x + \cos x) + e^{x}(\cos x + \sin x)$

        $= e^{-x}(\cos x + \sin x - \cos x + \sin x) = 2e^{-x}\sin x$

    At C, $f''(x) = 0 \Rightarrow 2e^{-x}\sin x = 0 \Rightarrow \sin x = 0 \Rightarrow x = 0,\ \pi,\ 2\pi$ but x-coordinate of C is

    between $\dfrac{\pi}{2}$ and $\dfrac{3\pi}{2}$ so that $x = \pi$ and coordinates of C are $\left(\pi,\ -e^{-\pi}\right)$

    (d) Area is $\left|-0.1084314...\right| = 0.108$

18.   (a) $y = 0$

    (b) $f'(x) = \dfrac{\left(1+e^{x}\right)\times 1 - xe^{x}}{\left(1+e^{x}\right)^{2}} = \dfrac{1 + e^{x} - xe^{x}}{\left(1+e^{x}\right)^{2}} = \dfrac{1 + e^{x}(1-x)}{\left(1+e^{x}\right)^{2}}$

    (c) At P, $f'(x) = 0 \Rightarrow 1 + e^{x}(1-x) = 0 \Rightarrow x = 1.278464... = 1.28$, $f(1.278464...) = 0.2784645...$

    Therefore the coordinates of P are $(1.28,\ 0.278)$

    (d) At Q, $f''(x) = 0$, therefore the graph of $y = f'(x)$ has a minimum at Q.

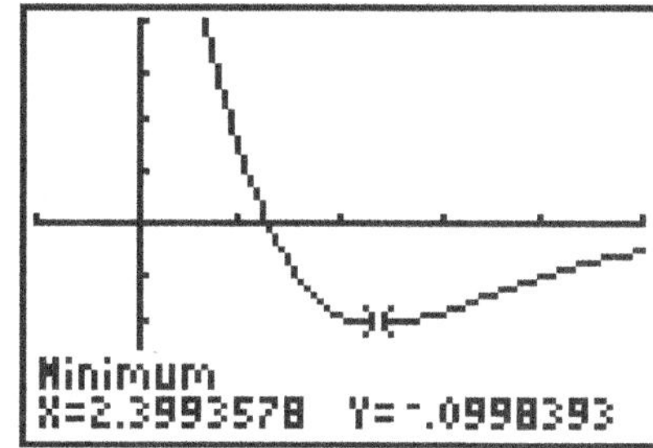

    This minimum point has x-coordinate $x = 2.399355... = 2.40$

    (e) $f'(0) = \dfrac{1}{2}$, therefore the gradient of the normal at O is -2. The equation of the

    normal to $y = f(x)$ is therefore, $\dfrac{y-0}{x-0} = -2 \Rightarrow y = -2x$.

(f) (i)

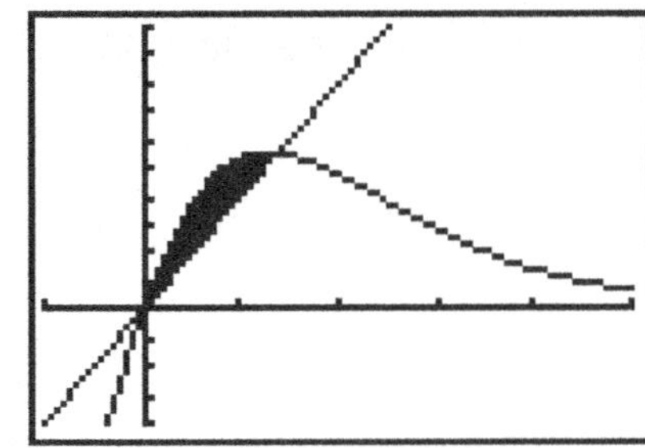

The screen display shows the relevant region which is shaded. The area of this

region is $\displaystyle\int_0^\alpha\left(\frac{x}{1+e^x}-\frac{x}{5}\right)dx$, where $\alpha=1.386294\ldots$

(ii) This area is $0.085042\ldots=0.0850$

19.   The radius of the bowl is 1 unit so its volume is $\dfrac{2}{3}\pi r^3=\dfrac{2}{3}\pi$ and the volume of the flour being

a half of this is $\dfrac{1}{3}\pi$. If $OD=h$, then $\displaystyle\pi\int_0^h y^2 dx=\pi\int_0^h\left(1-x^2\right)dx\approx\dfrac{1}{3}\pi$

$$\Rightarrow\int_0^h\left(1-x^2\right)dx=\frac{1}{3}\Rightarrow\left[x-\frac{1}{3}x^3\right]_0^h=\frac{1}{3}\Rightarrow h-\frac{1}{3}h^3=\frac{1}{3}\Rightarrow 3h-h^3=1\Rightarrow h=0.3472963\ldots$$

Therefore the distance $DB=1-0.347=0.653$, correct to 3 significant figures.

Solutions to Unit 8 Exercises

Exercise 8.1

1.   (a) $P(A)=\dfrac{1}{6}$, $P(B)=\dfrac{1}{6}$ (i) not exhaustive   (ii) mutually exclusive

    (b) $P(A)=\dfrac{1}{6}$, $P(B)=\dfrac{1}{2}$ (i) not exhaustive   (ii) not mutually exclusive

    (c) $P(A)=\dfrac{1}{2}$, $P(B)=\dfrac{1}{2}$ (i) exhaustive   (ii) mutually exclusive

    (d) $P(A)=\dfrac{2}{3}$, $P(B)=\dfrac{1}{6}$ (i) not exhaustive   (ii) mutually exclusive

    (e) $P(A)=\dfrac{1}{2}$, $P(B)=\dfrac{2}{3}$ (i) exhaustive   (ii) not mutually exclusive

2.

	1	2	3	4	5	6
1	2	3	4	5	6	7
2	3	4	5	6	7	8
3	4	5	6	7	8	9
4	5	6	7	8	9	10
5	6	7	8	9	10	11
6	7	8	9	10	11	12

	R	R		G	B
R	(RR)	(RR)	(RG)	(RG)	(RB)
R	(RR)	(RR)	(RG)	(RG)	(RB)
G	(GR)	(GR)	(GG)	(GG)	(GB)
G	(GR)	(GR)	(GG)	(GG)	(GB)
B	(BR)	(BR)	(BG)	(BG)	(BB)

(a) $P(A)=\dfrac{5}{18}$, $P(B)=\dfrac{1}{9}$ (i) not exhaustive   (ii) not mutually exclusive

(b) $P(A)=\dfrac{11}{36}$, $P(B)=\dfrac{1}{36}$ (i) not exhaustive   (ii) mutually exclusive

(c) $P(A)=\dfrac{1}{6}$, $P(B)=\dfrac{5}{6}$ (i) exhaustive   (ii) mutually exclusive

(d) $P(A)=\dfrac{1}{36}$, $P(B)=\dfrac{1}{36}$ (i) not exhaustive   (ii) not mutually exclusive

(e) $P(A)=1$, $P(B)=\dfrac{1}{6}$ (i) exhaustive   (ii) not mutually exclusive

3.   (a) $P(W)=1-(0.2+0.5)=0.3$

    (b) $P(A\cup W)=P(A)+P(W)$ because A and W

      are mutually exclusive. Therefore, the probability is $0.2+0.3=0.5$.

4.   (a) $\dfrac{2}{5}$ (b) $\dfrac{3}{10}$ (c) $\dfrac{2}{5}$ (d) $\dfrac{3}{5}$

5.   (a) 0.19   (b) 0.26

6.   (a) $\dfrac{4}{25}$ (b) $\dfrac{9}{25}$

7.   (a) $\dfrac{3}{10}$ (b) $\dfrac{1}{10}$ (c) $\dfrac{2}{5}$

8.   (a) $\dfrac{1}{10}$ (b) $\dfrac{3}{10}$

Exercise 8.2

1.   (a) $\dfrac{5}{8}$ (b) $\dfrac{3}{8}$ (c) $\dfrac{3}{4}$ (d) $\dfrac{1}{4}$

2.   (a) $\dfrac{1}{2}$ (b) $\dfrac{1}{4}$ (c) $\dfrac{5}{8}$ (d) $\dfrac{5}{8}$

3. (a) 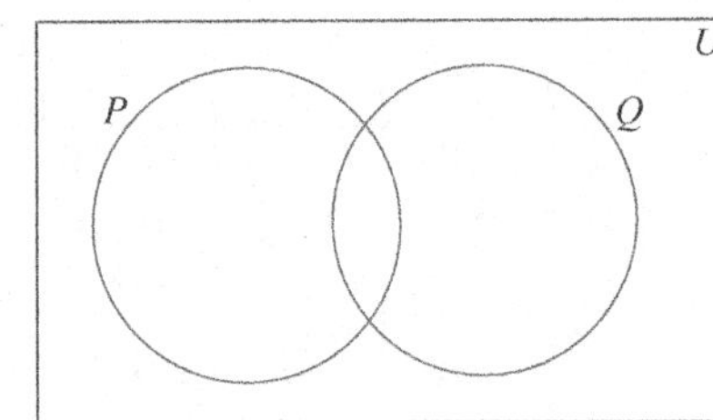

(b) $n(P \cup Q) = 50 - 18 = 32$

$n(P \cup Q) = n(P) + n(Q) - n(P \cap Q).$

$\Rightarrow 32 = 20 + 20 - n(P \cap Q)$

$\Rightarrow n(P \cap Q) = 8$

(c) $P(P \cap Q) = \dfrac{4}{25}$

4. (a) 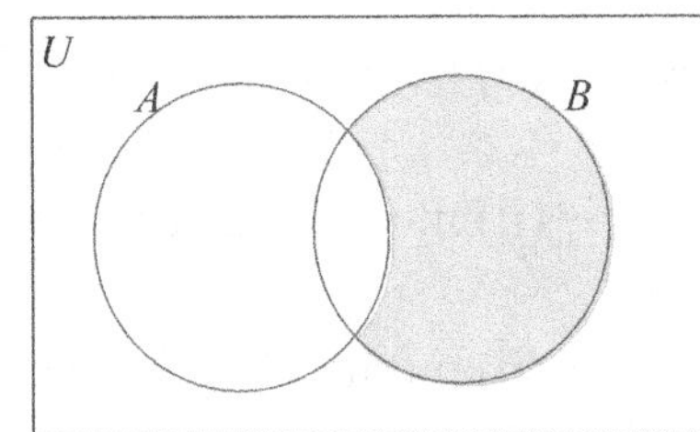

(b) $n(A' \cap B) = n(B) - n(A \cap B)$

(c) (i) 10 (ii) $\dfrac{1}{8}$

(d) A and B are not mutually exclusive because $n(A \cap B) \neq 0$. (Alternatively, because $P(A \cap B) \neq 0$.)

5. (a) $P(A \cup B) = P(A) + P(B) - P(A \cap B) \Rightarrow 0.95 = 0.62 + 0.51 - P(A \cap B)$

$\Rightarrow P(A \cap B) = 0.18$

(b) $P(A \cup B') = 1 - P(B) + P(A \cap B) = 1 - 0.51 + 0.18 = 0.67$

6. (a) $P(A \cup B) = P(A) + P(B) - P(A \cap B) = \dfrac{1}{3} + \dfrac{3}{4} - \dfrac{1}{5} = \dfrac{53}{60}$

(b) $P(A' \cup B') = 1 - P(A \cap B) = 1 - \dfrac{1}{5} = \dfrac{4}{5}$

7. (a) $\dfrac{3}{4} = \dfrac{1}{2} + P(B) - \dfrac{1}{6} \Rightarrow P(B) = \dfrac{3}{4} + \dfrac{1}{6} - \dfrac{1}{2} = \dfrac{5}{12}$

(b) $P(A' \cap B) = P(B) - P(A \cap B) = \dfrac{5}{12} - \dfrac{1}{6} = \dfrac{1}{4}$

8. $P(A' \cup B') = 1 - P(A \cap B) = 0.8 \Rightarrow P(A \cap B) = 0.2 \Rightarrow P(A \cup B) = 0.5 + 0.4 - 0.2 = 0.7$

9. $P(A \cap B') + P(A' \cap B) + P(A \cap B) = P(A \cup B) \Rightarrow 0.23 + 0.42 + P(A \cap B) = 0.93$

$\Rightarrow p(A \cap B) = 0.28$

10. The probability that a randomly chosen student studies math not physics is $\dfrac{80}{200} = \dfrac{2}{5}$

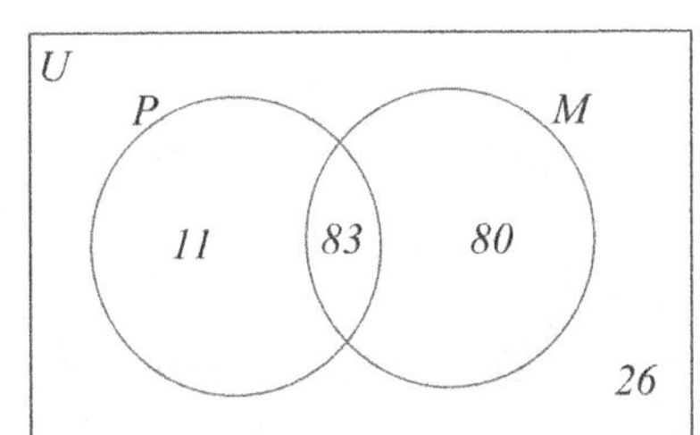

11. Let L be the event "fails due to bad lights" and let B be the event "fails due to bad brakes". Then $P(L) = 0.16$, $P(B) = 0.28$, $P(L \cap B) = 0.08$, $P(L \cup B) = 0.16 + 0.28 - 0.08 = 0.36$. Therefore the probability that a randomly selected car fails is 0.36.

Exercise 8.3

1. (a) $\dfrac{39}{125}$ (b) $\dfrac{31}{57}$ (c) $\dfrac{1}{3}$

2. (a) 0.52   (b) 0.33   (c) 0.68

3. (a) $\dfrac{11}{72}$ (b) $\dfrac{7}{18}$

4. (a) $P(A \mid B) = \dfrac{P(A \cap B)}{P(B)} = \dfrac{1}{4}$

   (b) $P(A \cap B') = P(A) - P(A \cap B) \Rightarrow P(A \mid B') = \dfrac{P(A \cap B')}{P(B')} = \dfrac{P(A) - P(A \cap B)}{P(B')} = \dfrac{5}{8}$

5. (a) $P(A \mid B) = \dfrac{P(A \cap B)}{P(B)} = \dfrac{3}{4}$ (b) $P(B \mid A) = \dfrac{P(B \cap A)}{P(A)} = \dfrac{P(A \cap B)}{P(A)} = \dfrac{5}{8}$

6. (a) $P(P \mid Q) = \dfrac{P(P \cap Q)}{P(Q)} \Rightarrow \dfrac{3}{7} = \dfrac{P(P \cap Q)}{\frac{2}{3}} \Rightarrow P(P \cap Q) = \dfrac{2}{7}$

   (b) $P(Q \mid P) = \dfrac{P(P \cap Q)}{P(P)} \Rightarrow P(P) = \dfrac{P(P \cap Q)}{P(Q \mid P)} = \dfrac{5}{14}$

7. (a) $P(A \cap B) = \dfrac{3}{8} \times \dfrac{2}{3} = \dfrac{1}{4}$ (b) $P(A' \cap B') = \dfrac{5}{8} \times \dfrac{3}{5} = \dfrac{3}{8}$

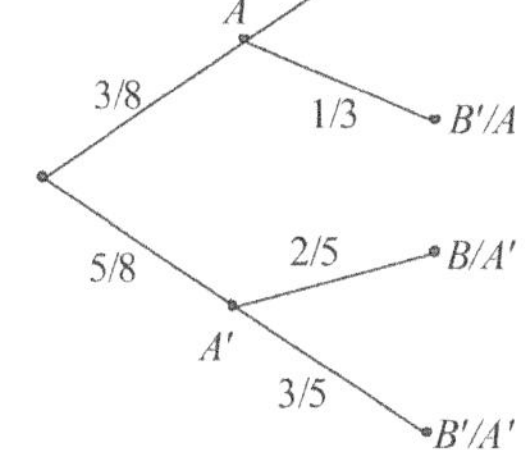

8. (a) $P(B \mid A) = \dfrac{P(A \cap B)}{P(A)} = \dfrac{0.36}{0.48} = 0.75$

   (b) $P(B' \mid A') = \dfrac{P(A' \cap B')}{P(A')} = \dfrac{0.26}{0.52} = 0.5$

   (c) $P(B) = P(A) \times P(B \mid A) + P(A') \times P(B \mid A') = 0.62$

   (d) $P(A \mid B) = \dfrac{P(A \cap B)}{P(B)} = \dfrac{0.36}{0.62} = 0.581$

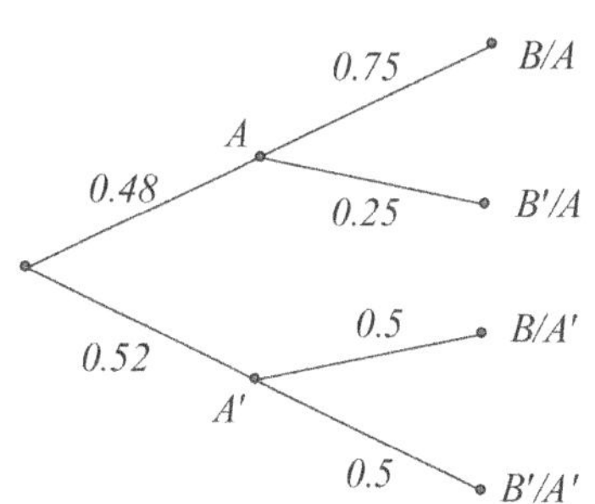

9. $P(B\,|\,A)=\dfrac{P(A\cap B)}{P(A)}=\dfrac{2}{5},$

$P(A\,|\,B)=\dfrac{P(A\cap B)}{P(B)}\Rightarrow P(A\cap B)=\dfrac{1}{3}\times\dfrac{4}{5}=\dfrac{4}{15}\ \Rightarrow P(A)=\dfrac{4/15}{2/5}=\dfrac{2}{3}$

10.   (a)   (b)

$$P(D)=P(D/V)\times P(V)+P(D/V')\times P(V')$$
$$=0.155$$

(c) $P(V\cap D)=P(V)\times P(D\,|\,V)=0.015$

Therefore, $P(V\,|\,D)=\dfrac{P(V\cap D)}{P(D)}=0.0.968$

11.   (a) $P(A)=0.7,\ \ P(D\,|\,A')=0.05,\ \ P(D\,|\,A)=0.01$

(b) $P(D)=P(D\,|\,A)\times P(A)+P(D\,|\,A')\times P(A')=0.022$ and $P(A'\cap D)=\dfrac{0.015}{0.022}=\dfrac{15}{22}$

12.   (a) $P(A)=\dfrac{1}{100},\ P(B\,|\,A)=1,\ P(B\,|\,A')=\dfrac{1}{1024}$

(b) $P(B)=P(B\,|\,A)\times P(A)+P(B\,|\,A')\times P(A')=\dfrac{1123}{102400}$

$\Rightarrow P(A\,|\,B)=\dfrac{P(A\cap B)}{P(B)}=\dfrac{0.01\times1\times102400}{1123}=0.9118432...=0.912$

Exercise 8.4

1.   (a) $P(A\cap B)=\dfrac{1}{36}$ (b) $P(A\cup B)=\dfrac{23}{72}$

2.   (a) $P(B)=\dfrac{11}{16}$ (b) $P(A\cap B)=\dfrac{33}{80}$

3. $P(A\cap B)=0.3$, $P(A)\,p(B)=0.3$. Therefore A and B are independent events.

4. $P(B)=0.6$, $P(A)P(B)=0.45\times0.6=0.27=P(A\cap B)$. So A and B are independent events.

5.   (a) Let $P(Q)=x$, (a) $0.8=x+3x-3x^2\Rightarrow 3x^2-4x+0.8=0\Rightarrow x=0.2450,\ 1.0883$ but $|x|\le1$
        so that $P(Q)=0.245$.

(b) $P(P\cup Q)=1\Rightarrow 1=4x-3x^2\Rightarrow(3x-1)(x-1)=0\Rightarrow P(Q)=\dfrac{1}{3}$. (If $x=1$, then $P(P)=3$.)

6. (a) $\dfrac{1}{8}$ (b) $\dfrac{3}{8}$ (c) $\dfrac{5}{8}$ (d) $\dfrac{3}{8}$

7. (a) $\dfrac{8}{81}$ (b) $\dfrac{16}{81}$ (c) $\dfrac{1}{9}$ (d) $\dfrac{29}{81}$

8. Letters in common are M, I and S. Then the probability that Ms are chosen is $\frac{1}{11}\times\frac{1}{8}=\frac{1}{88}$. The probability that Is are chosen is $\frac{4}{11}\times\frac{2}{8}=\frac{1}{11}$. The probability that Ss are chosen is $\frac{4}{11}\times\frac{2}{8}=\frac{1}{11}$. Therefore, the probability that the letters chosen are the same is $\frac{1}{88}+\frac{1}{11}+\frac{1}{11}=\frac{17}{88}=0.1931818...$, and so the probability that the same letters are chosen is 19.3%.

9. (a) $P(TTTH)=\dfrac{1}{16}$

   (b) $P(n\text{ tails})=\left(\dfrac{1}{2}\right)^{n}\le0.1 \Rightarrow \log\left(\dfrac{1}{2}\right)^{n}\le-1 \Rightarrow n\ge3.321928...$. Therefore, $n=4$.

10. The probability of not scoring on n throws is $\left(\dfrac{1}{2}\right)^{n}$. This probability must be less than 0.01.

   Therefore, $\left(\dfrac{1}{2}\right)^{n}\le0.01 \Rightarrow \log\left(\dfrac{1}{2}\right)^{n}\le-2 \Rightarrow n\ge\dfrac{-2}{\log0.5}=6.643856...$ and the player would need to take 7 free throws.

11. (a) $\left(1-\dfrac{1}{6}\right)^{n}\le0.05 \Rightarrow n\log\left(\dfrac{5}{6}\right)\le\log0.05 \Rightarrow n\ge16.43107...$.

   Therefore, the die should be thrown 17 times.

   (b) If die is to be thrown at least 10 times, we require 9 non-sixes of which the probability is $\left(\dfrac{5}{6}\right)^{9}=0.1938066...$. Therefore, the probability is 0.194.

12. Let C be the event: agreement on chemical weapons. Let N be the event: agreement on nuclear weapons. Then $P(C)=0.5$, $P(N)=0.7$ and $P(C'\cap N')=0.2$. Therefore, as

   $$P(C'\cap N')=1-P(C\cup N),\ P(C\cup N)=0.8 \Rightarrow P(C\cap N)=0.5+0.7-0.8=0.4.$$

   So the probability of agreement on both chemical and nuclear weapons is 0.4.

   As $P(C)\times P(N)=0.35\ne0.4$, the two issues are not treated independently.

Solutions to Unit 9 Exercises

Exercise 9.1

1.   (a) $c = 0.1$ (b) $c = 0.05$ (c) $c = 0.5$

2.   (a)

	1	2	3	4	5	6
1	0	1	2	3	4	5
2	1	0	1	2	3	4
3	2	1	0	1	2	3
4	3	2	1	0	1	2
5	4	3	2	1	0	1
6	5	4	3	2	1	0

(b)

X	0	1	2	3	4	5
$P(X = x)$	1/6	5/18	2/9	1/6	1/9	1/18

3.   (a)

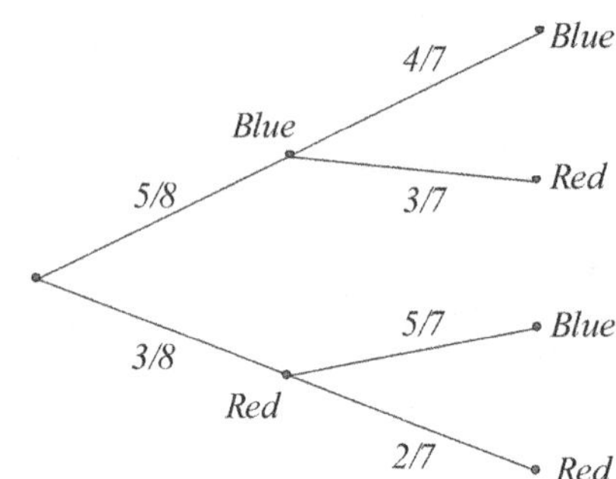

(b)

X	0	1	2
$P(X = x)$	$\frac{3}{28}$	$\frac{15}{28}$	$\frac{5}{14}$

4.   (a)

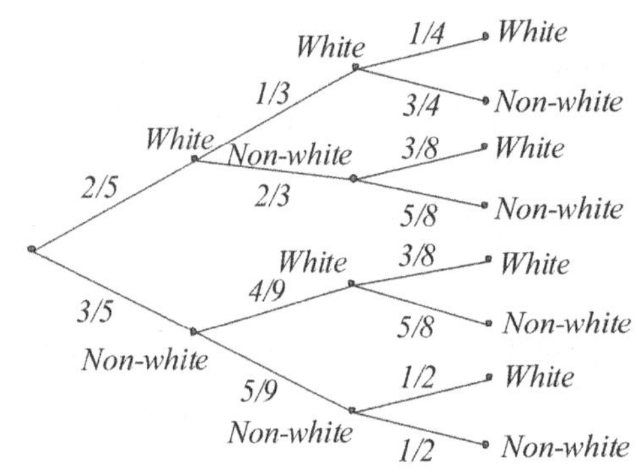

(b)

X	0	1	2	3
$P(X = x)$	$\frac{1}{6}$	$\frac{1}{2}$	$\frac{3}{10}$	$\frac{1}{30}$

(c) $E(X) = 1.2$

5.   (a) The probability of 'TTH' is $\left(\dfrac{1}{2}\right)^3 = \dfrac{1}{8}$.

     (b) 3 points can be scored either by 'HTT' or by 'THH'.

     Therefore, $P(X = 3) = \left(\dfrac{1}{2}\right)^3 + \left(\dfrac{1}{2}\right)^3 = \dfrac{1}{8} + \dfrac{1}{8} = \dfrac{1}{4}$.

(c)

X	0	1	2	3	4	5	6
$P(X=x)$	$\frac{1}{8}$	$\frac{1}{8}$	$\frac{1}{8}$	$\frac{1}{4}$	$\frac{1}{8}$	$\frac{1}{8}$	$\frac{1}{8}$

(d) $E(X)=3$

6.   Let X be the random variable representing the score. Then $P(X=5)=\frac{1}{2}\times\frac{1}{2}=\frac{1}{4}$,

$P(X=3)=\frac{1}{2}\times\frac{1}{2}=\frac{1}{4}$ and $P(X=1)=\frac{1}{2}\times\frac{1}{2}=\frac{1}{4}$. So, $E(X)=5\times\frac{1}{4}+3\times\frac{1}{4}+1\times\frac{1}{2}=\frac{5}{2}=2.5$.

7.   (a) \$2, \$10, \$50 and \$1000 are prizes but it costs \$1 to play. Therefore, if the player loses $X=-1$, if the player wins, $X=1,\ 9,\ 49,\ 999$, depending on the extent of his winning.

     (b)

x	-1	1	9	49	999
$P(X=x)$	0.8445	0.125	0.027	0.003375	0.000125

     (c) $E(X)=-0.18625$ so that the owner makes \$186.25.

8.   (a) Let X be the event "the lavatory is occupied"

     (b) $E(X)=1.2$

X	0	1	2	3
$P(X=x)$	0.216	0.432	0.288	0.064

9.   (a) The 13 possible PIN numbers are: 106, 610, 160, 601, 133, 313, 331, 124, 142, 214, 241, 412, 421. Let X be the event "the correct PIN is guessed", then $P(X=1)=\frac{1}{13}$.

$P(X=2)=\frac{12}{13}\times\frac{1}{12}=\frac{1}{13}$ $P(X=3)=\frac{12}{13}\times\frac{11}{12}\times\frac{1}{11}=\frac{1}{13}$ and so on. Therefore,

$P(X=x)=\frac{1}{13}$ for all values of x.

     (b) $E(X)=\frac{1}{13}(1+2+3+\ \ldots\ +13)=7$

Exercise 9.2

1.   (a) $\frac{1}{4}$ (b) $\frac{5}{16}$

2.   (a) $\frac{1}{64}$ (b) $\frac{5}{16}$ (c) $\frac{3}{32}$

3.   (a) 0.372   (b) 0.260   (c) 0.135

4.   (a) 0.478   (b) 0.124   (c) 0.0230

Exercise 9.3

1.   (a) 0.00412   (b) 0.165

2.   0.313

3.   0.121

4.   (a) 0.277   (b) 0.231   (c) 0.127

5.   (a) 0.665   (b) 0.619   (c) 0.597   (d) 0.584

6.   Let X be the event 'a person who has booked a ticket fails to take up the booking'. $n = 50$, $p = \frac{1}{16}$, $P(X \le 1) = 0.1719435\ldots$. Therefore, the probability that there will not be enough seats is 0.172.

7.   (a) 0.133   (b) 0.311

8.   (a) 0.123   (b) 0.00856   (c) 0.991   (d) 0.945   (e) 0.900

9.   (a) $P(X \ge 1) = 1 - P(X = 0) = 1 - (0.992)^7 = 0.05467377\ldots \approx 5\%$
    (b) (i) 0.799   (ii) 0.185   (iii) 0.0160

Exercise 9.4

1.   (a) $E(X) = 25$ (b) $E(X) = 10$

2.   (i) (a) 1   (b) 0.402
    (ii) (a) 10   (b) 0.137

3.   (a) Let X be the number of defective components. Then $X \sim B(120,\ 0.06)$

$$E(X) = np = 120 \times 0.06 = 7.2$$

    (b) Let Y be the number of families that are willing to host a refugee.
        Then $Y \sim B(25\ 000,\ 0.35)$ and $E(Y) = 25000 \times 0.35 = 8750$

4.   (a) $np = 15$, $p = 0.3 \Rightarrow n = \dfrac{15}{0.3} = 50$ (b) $8p = 5 \Rightarrow p = \dfrac{5}{8}$

5. $E(X) = E(Y) \Rightarrow 10p = 30p^2 \Rightarrow 3p^2 - p = 0 \Rightarrow p(3p - 1) = 0 \Rightarrow p = 0,\ 3p - 1 = 0$

    so the non-zero value of p is $\dfrac{1}{3}$.

6.   (a) $\binom{3}{2}(1 - p)p^2 = 0.2 \Rightarrow 3p^2 - 3p^3 = 0.2 \Rightarrow 15p^3 - 15p^2 + 1 = 0$

    (b) The solutions are -0.233, 0.311, 0.921.

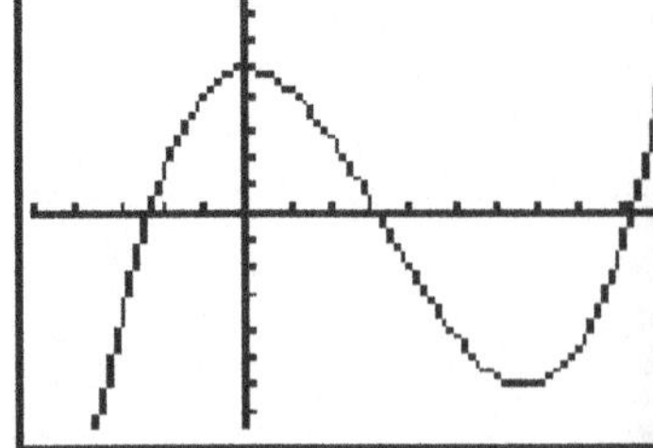

(c)   As $0 \le p \le 1$, the only valid solutions are 0.311, 0.921.

7.   (a) $n = 3$. This should be evident from the observation that
$X \in \{0, 1, 2, 3\}$.

(b) $E(X) = 1 \times \dfrac{48}{125} + 2 \times \dfrac{12}{125} + 3 \times \dfrac{1}{125} = \dfrac{48 + 24 + 3}{125} = \dfrac{3}{5} = 0.6$

(c) $E(X) = np \Rightarrow 0.6 = 3p \Rightarrow p = \dfrac{0.6}{3} = 0.2 = \dfrac{1}{5}$.

8.   (a) $\bar{x} = \dfrac{1}{n} \displaystyle\sum_{i=1}^{N} f_i x_i = \dfrac{0 \times 16 + 1 \times 33 + 2 \times 31 + 3 \times 18 + 4 \times 14 + 5 \times 7 + 6 \times 1}{120} = 2.05$

(b) Let X be a random variable representing the number of red flowers in a packet of six.
$E(X) = np \Rightarrow 2.05 = 6p \Rightarrow p = 0.3416666... = 0.342$

(c) $P(X = 0) = (1 - 0.3416666)^6 = 0.08140954... = 0.814$ so number of samples with 0 red

flowers is $0.0814 \times 120 \approx 10$. $P(X = 1) = \dbinom{6}{1} \times 0.658^5 \times 0.342 = 0.2535029... = 0.253$ and

number of samples with 1 red flower is $0.253 \times 120 \approx 30$. The other frequencies are
calculated in a similar way and the frequency distribution is

Number of red flowers, x_i	0	1	2	3	4	5	6
Number of samples with x_i red flowers	10	30	39	27	11	2	0

Exercise 9.5

1.   (a) 0.989   (b) 0.637   (c) 0.923   (d) 0.189   (e) 0.00336   (f) 0.318

(g) 0.298   (h) 0.965   (i) 0.0195   (j) 0.819   (k) 0.576   (l) 0.0789

2   (a) $P(|Z| < 1) = P(-1 < Z < 1) = 0.683$

(b) $P(|Z| < 2.5) = P(-2.5 < Z < 2.5) = 0.988$

(c) $P(|Z| > 3) = 1 - P(|Z| < 3) = 1 - P(-3 < Z < 3) = 0.002699934... = 0.00270$

(d) $P(|Z - 1| < 0.1) = P(-0.1 < Z - 1 < 0.1) = P(0.9 < Z < 1.1) = 0.0484$

(e) $P(|Z - 0.5| > 1.5) = 1 - P(|Z - 0.5| < 1.5) = 1 - P(-1 < Z < 2) = 0.181$

Exercise 9.6

1.   (i)   (a) 0.955   (b) 0.212   (c) 0.661
(ii)   (a) 0.949   (b) 0.975   (c) 0.0441
(iii) (a) 0.309   (b) 0.579   (c) 0.0797
(iv) (a) 0.0881   (b) 0.192   (c) 0.853

2.	(i)	(a) 0.383		(b) 0.819
	(ii)	(a) 0.683		(b) 0.997

3.	(a) 3.34%		(b) 18.7%	(c) 49.9%

4.	Let X be the arrival time of the bus relative to 08:04.

(a) $P(|X|<2)=P(-2<X<2)=0.6826826\ldots$. Therefore the probability that the bus arrives within 2 minutes of 08:04 is 0.683.

(b) $P(X>1)=0.3085375\ldots$ so the bus arrives after 08:05 on 30.9% of the time.

(c) $P(X<-4)=0.02275006\ldots$ so that, in 90 days the man can expect to miss the bus $90\times0.02275006=2.047505\ldots$ or approximately twice.

5.	The variance of the sample mean is $\dfrac{\sigma^2}{n}=\dfrac{9}{10}=0.9$ so the standard deviation of the sample mean is $\sqrt{0.9}=0.9486832\ldots$. $P(\bar{X}<43)=0.01750741\ldots$ so that the probability that the mean of the sample is less than 43g is 0.0175.

6.	Let X be the weight of a randomly chosen orange.

(a) $P(X>126)=0.1150697\ldots$ and $P(X<60)=0.07493374\ldots$

Therefore, approximately 11.5% of the oranges are grade A, approximately 7.49% are grade C and the remainder, 81.0% are grade B.

Let the receipts be $\$Y$, then

(b) $E(Y)=80\,000(0.07493\times0.15+0.8100\times0.25+0.11507\times0.30)=19860.84$

So the expected receipts are $\$19\,860$.

Exercise 9.7

1.	(a) 2.58	(b) −0.786	(c) −0.407	(d) 1.20	(e) 0.674	(f) 1.64

2.	(a) 18.1	(b) 8.02	(c) 15.7	(d) 11.2	(e) 0.956

3.	$P(X<4.04)=0.772$ but since $P(Z<z_1)=0.772 \Rightarrow z_1=0.7454495\ldots$ and $Z=\dfrac{X-\mu}{\sigma}$, we can write $0.7454495=\dfrac{4.04-\mu}{0.56}$. Therefore $4.04-\mu=0.7454495\times0.56=0.4174517\ldots$ and $\mu=3.622548\ldots=3.62$.

4.	Let T be the random variable representing the time taken to complete the survey. Then $P(T<8)=0.50 \Rightarrow \mu=8$. Also $P(T<10)=0.85$, but $P(Z<z_1)=0.85 \Rightarrow z_1=1.036433\ldots$

$$\Rightarrow 1.036433=\frac{10-8}{\sigma} \Rightarrow \sigma=\frac{2}{1.036433}=1.929695\ldots=1.93$$

5.   (a)

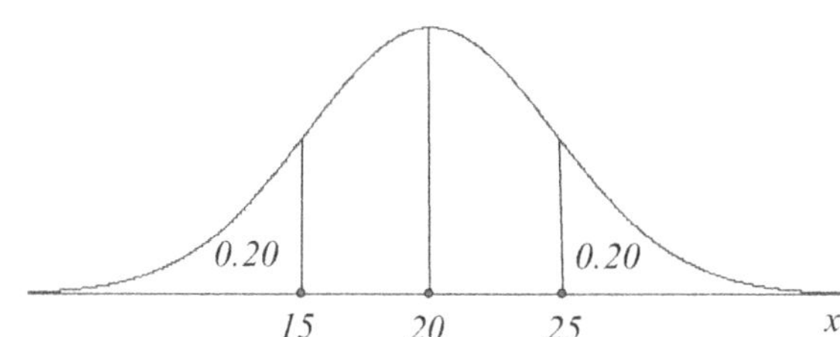

(b) $P(|X-20|>5)=0.4 \Rightarrow P(X<25)=0.8$, but $P(Z<z_1)=0.8 \Rightarrow z_1=0.8416212...$

$\Rightarrow 0.8416212 = \dfrac{25-20}{\sigma} \Rightarrow \sigma = \dfrac{5}{0.8416212} = 5.940914... = 5.94$

6. $P(X>53)=0.08 \Rightarrow P(X<53)=0.92$ but $P(Z<z_1)=0.92 \Rightarrow z_1=1.405071...$

$P(X<41)=0.15$ but $P(Z<z_2)=0.15 \Rightarrow z_2=-1.036433....$ Now $Z=\dfrac{X-\mu}{\sigma}$ so that

$$1.405071\sigma = 53-\mu \ \dots\dots\dots(1)$$
$$-1.036433\sigma = 41-\mu \ \dots\dots\dots(2)$$

Therefore $2.441504\sigma = 12 \Rightarrow \sigma = 4.915003...$ and $\mu = 41+1.036433\times4.915003 = 46.09407....$
So the mean is 46.1 and the standard deviation is 4.92.

7.   Let X be the examination mark of a randomly selected student. Then $P(X<45)=0.16$, but

$P(Z<z_1)=0.16 \Rightarrow z_1=-0.9944578...$ and since $Z=\dfrac{X-\mu}{\sigma}$, we can write

$-0.9944758\sigma = 45-\mu \Rightarrow -0.9944578\times14 = 45-\mu \Rightarrow \mu = 45+0.9944578\times14 = 58.92240....$
Therefore the mean mark is 58.9%.

8.   The proportion of years in which rainfall exceeded 30 cm is $\dfrac{5}{134} = 0.3731343....$

The proportion of years in which rainfall was less than 10 cm is $\dfrac{12}{134} = 0.08955223...$

Let X be the rainfall in a randomly given year.
Then $P(X>30)=0.03731 \Rightarrow P(X<30)=1-0.03731=0.9626865....$ In addition,

$P(X<10)=0.08955223....$ But $P(Z<z_1)=0.9626865 \Rightarrow z_1=1.782750...$ and

$P(Z<z_2)=0.08955223 \Rightarrow z_2=-1.343517...$ therefore, since $Z=\dfrac{X-\mu}{\sigma}$

$$1.782750\sigma = 30-\mu \ \dots\dots\dots\dots(1)$$
$$-1.343517\sigma = 10-\mu \ \dots\dots\dots\dots(2)$$

and solving simultaneously $3.126267\sigma = 20 \Rightarrow \sigma = 6.397406....$
Then $\mu = 10+1.343517\times6.397406 = 18.59502...$ and mean November rainfall is 18.6 cm.

9.   Let k be the minimum weight of a grade A tomato and let X be the weight of a randomly
selected tomato. Then $P(X>k)=0.25 \Rightarrow P(X<k)=0.75$.

Therefore, since $P(Z<z_1)=0.75 \Rightarrow z_1=0.6744897...$ and $Z=\dfrac{X-\mu}{\sigma}$ we can write

$$0.6744897 = \frac{k-85}{15} \Rightarrow k = 85 + 15 \times 0.6744897 = 95.11734... \text{ and the minimum weight of a}$$

grade A tomato is 95.1g.

10.   Let X be the diameter of a randomly chosen washer.

For the original machine $P(X > 3.758) = 0.009815306...$ so for the upgraded machine we also

require that $P(X > 3.758) = 0.009815306$, therefore $P(X < 3.758) = 0.9901846...$ but

$$P(Z < z_1) = 0.9901846 \Rightarrow z_1 = 2.333330... \text{ and as } Z = \frac{X - \mu}{\sigma}, \text{ we have } 2.333330 = \frac{3.758 - \mu}{0.002}.$$

Therefore $\mu = 3.758 - 2.333330 \times 0.002 = 3.753333...$.

So, on the upgraded machine, the mean should be set at 3.753cm.

Solutions to Unit 6, 8 & 9 Review Exercise

Part 1

1.   (a) $P(A \cup B) = P(A) + P(B) - P(A \cap B) \Rightarrow \frac{5}{6} = \frac{2}{3} + P(B) - \frac{1}{4} \Rightarrow P(B) = \frac{5}{6} + \frac{1}{4} - \frac{2}{3} = \frac{5}{12}$

      (b) $P(A) \times P(B) = \frac{5}{8}$, $P(A \cap B) = \frac{1}{4}$, $P(A \cap B) \neq P(A) \times P(B)$ and A and B are not independent.

      (c) $P(B \mid A) = \dfrac{P(A \cap B)}{P(A)} = \dfrac{3}{8}$

2.   (a)

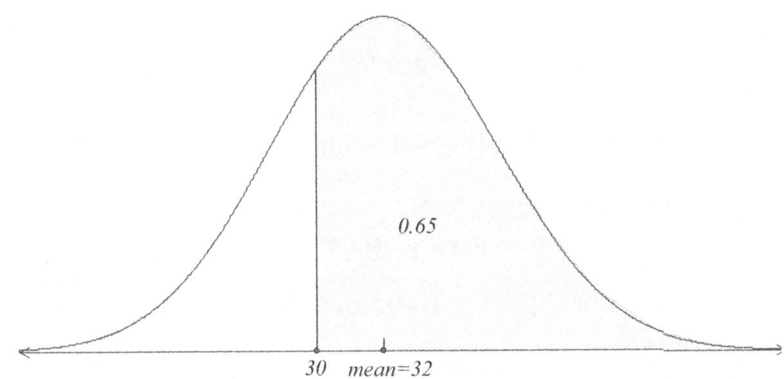

      (b) $P(X < 34) = P(X > 30)$, by symmetry so $P(X < 34) = 0.65$

      (c) $P(|X - 32| < 2) = P(30 < X < 34) = 2(0.65 - 0.50) = 0.30$

3.   (a) (i) $\sum xy = 200$ (ii) $\sum x^2 = 55$ (iii) $\sum x = 15$ (iv) $\bar{x} = 3$ (v) $\sum y = 50$ (vi) $\bar{y} = 10$

      (b) Using the equations from page 238:
$$200 = 55a + 15b \ldots\ldots\ldots (1)$$
$$10 = 3a + b \ldots\ldots\ldots\ldots (2)$$
      Solving simultaneously $a = 5$, $b = -5$ so the regression line is $y = 5x - 5$

(c) $y(1)=0$, $y(2)=5$, $y(3)=10$, $y(4)=15$ so the regression line passes through all five data Points and therefore the correlation coefficient is 1.

4. (a) $k^2+\dfrac{3}{5}k+\dfrac{6}{10}=1\Rightarrow 10k^2+6k+6=10\Rightarrow 5k^2+3k-2=0$

Therefore, $(5k-2)(k+1)=0\Rightarrow 5k-2=0$, $k+1=0\Rightarrow k=\dfrac{2}{5}$, $k=-1$, but since k is a

probability $k\neq -1$ so that $k=\dfrac{2}{5}$.

(b) $E(X)=0k^2+1\times\dfrac{3}{10}+2\times\dfrac{3}{5}\times\dfrac{2}{5}+3\times\dfrac{1}{10}+4\times\dfrac{1}{5}=\dfrac{3}{10}+\dfrac{12}{25}+\dfrac{3}{10}+\dfrac{4}{5}=\dfrac{94}{50}=1.88$

5. (a) $P(X=2)=\dfrac{12}{50}=\dfrac{6}{25}$

(b) $P(X\leq 1)=\dfrac{16}{50}+\dfrac{15}{50}=\dfrac{31}{50}$

(c) $E(X)=0\times\dfrac{16}{50}+1\times\dfrac{15}{50}+2\times\dfrac{12}{50}+3\times\dfrac{7}{50}=\dfrac{0+15+24+21}{50}=\dfrac{60}{50}=1.2$

6. Let X be the number of blue balls selected, then $X\sim B\left(3,\ \dfrac{2}{5}\right)$

(a) $P(X=3)=\left(\dfrac{2}{5}\right)^3=\dfrac{8}{125}$
(b) $P(X<2)=P(X\leq 1)=\left(\dfrac{3}{5}\right)^3+3\left(\dfrac{3}{5}\right)^2\left(\dfrac{2}{5}\right)=\dfrac{81}{125}$

7. (a) (i) $P(A')=0.29$

(ii) Maximum value of $P(B\cap C)=0.23$ (iii) Minimum value of $P(B\cap C)=0.07$

(b) $P(B\cap C')=P(B)-P(B\cap C)=0.55-0.18=0.37$

(c) $P(C'|A)=\dfrac{P(A\cap C')}{P(A)}=\dfrac{0.71-0.23}{0.71}=\dfrac{48}{71}$

8. (a) $\dfrac{2}{8}=\dfrac{1}{4}$ (b) $\dfrac{6}{8}\times\dfrac{2}{7}=\dfrac{3}{14}$

(c) Probability that Maria selects a green ball on her first go is $\dfrac{6}{8}\times\dfrac{2}{7}=\dfrac{3}{14}$

Probability that Maria selects a green ball on her second go is $\dfrac{6}{8}\times\dfrac{5}{7}\times\dfrac{4}{6}\times\dfrac{2}{5}=\dfrac{1}{7}$

Probability that Maria selects a green ball on her second go is $\dfrac{6}{8}\times\dfrac{5}{7}\times\dfrac{4}{6}\times\dfrac{3}{5}\times\dfrac{2}{4}\times\dfrac{2}{3}=\dfrac{1}{14}$

Therefore the probability that Maria wins the game is $\dfrac{3}{14}+\dfrac{1}{7}+\dfrac{1}{14}=\dfrac{3}{7}$

9.   (a)

x	0	1	2
$P(X=x)$	$\dfrac{3}{9}$	$\dfrac{4}{9}$	$\dfrac{2}{9}$

(b) (i) $E(Y)=3\times\dfrac{1}{3}+9\times\dfrac{2}{9}=1+2=3$ (ii) $E(Z)=9\times\dfrac{4}{9}=4$

(c) Let amount won by Dora be W and amount won by Betty be U. Then $E(W)=\dfrac{4}{9}W$ and

$$E(U)=\dfrac{3}{9}\times 6+\dfrac{2}{9}W=2+\dfrac{2}{9}W\text{, but if }E(W)=E(U)\text{ then}$$

$$\dfrac{4}{9}W=2+\dfrac{2}{9}W \Rightarrow \dfrac{2}{9}W=2 \Rightarrow W=9 \text{ and Dora wins \$9.}$$

Part 2

10.   (a) $P(X>10)=0.006209679......=0.00621$

(b) $P(3<X<6)=0.5328072...=0.533$

(c) $P(X<a)=0.75 \Rightarrow a=6.348979...=6.35$

(d)   (i)

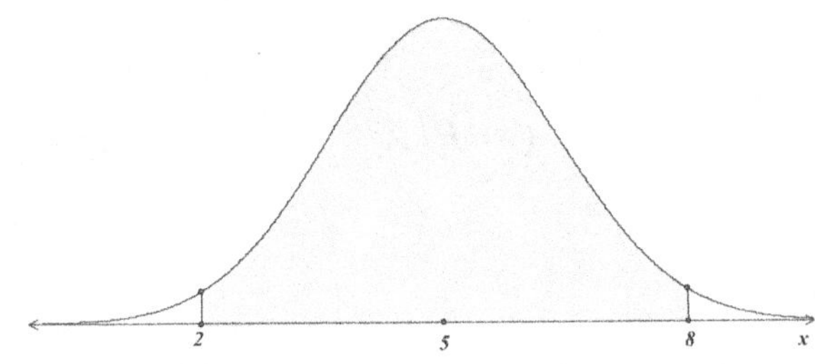

(ii) $P(|X-5|<3)=0.8663855...=0.866$

11.   (a)

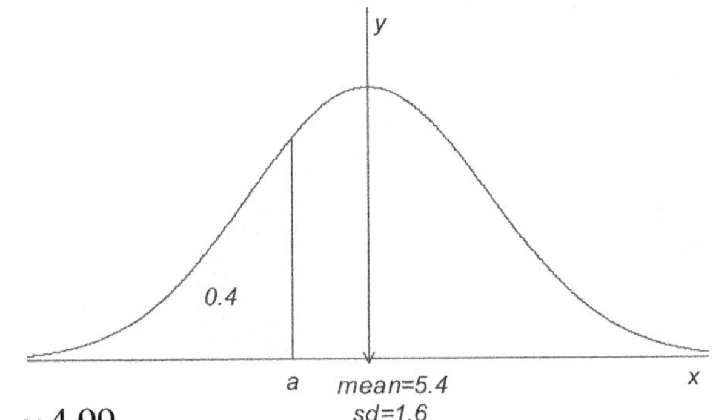

(b) $a=4.994644...\simeq 4.99$

(c) $P(Z<z_1)=0.05 \Rightarrow z_1=-1.644853...$ and

$$z_1=\dfrac{x_1-\mu}{\sigma} \Rightarrow -1.644853=\dfrac{1-\mu}{1.6} \Rightarrow \mu=1+1.644853\times 1.6=3.631765...=3.63$$

12.   (a) $E(X) = np = 12 \times \dfrac{1}{6} = 2$

    (b)   (i) $P(X = 2) = \dbinom{12}{2}\left(\dfrac{1}{6}\right)^2\left(\dfrac{5}{6}\right)^{10} = 0.2960935... = 0.296$

        (ii) $P(X > 3) = 1 - P(X \le 3) = 0.1251780... = 0.125$

        (iii) $P(X = 1) = 0.2691759... < P(X = 2)$, $P(X = 3) = 0.1973957... < P(X = 2)$ so the most likely number of 'sixes' is 2.

13.

Competitor	A	B	C	D	E	F	G	H
Judge A	9	5	7	7	8	4	5	9
Judge B	8	5	8	7	9	6	6	7

The correlation coefficient is $0.7431605... = 0.743$

14.   Let T be the random variable representing the patient's temperature. Then $T \sim N(37.3,\ 0.4^2)$.

  (a) $P(T > 38) = 0.1907869... = 0.191$

  (b) $P(T < t) = 0.70 \Rightarrow t = 37.71952... = 37.7$

  (c)

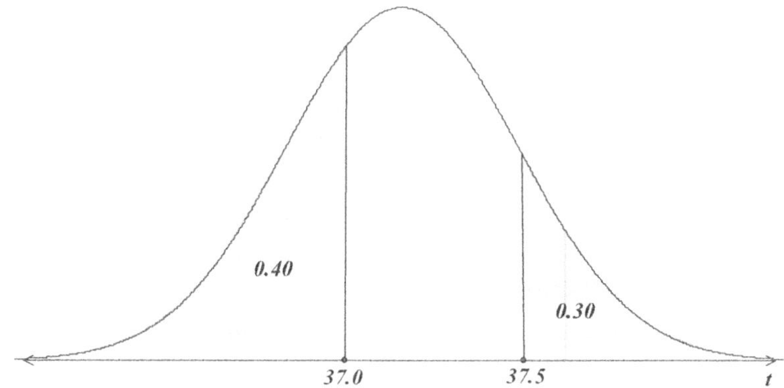

$P(Z < z_1) = 1 - 0.30 = 0.70 \Rightarrow z_1 = 0.5244005... \Rightarrow 0.5244005\sigma = 37.5 - \mu$

$P(Z < z_2) = 0.40 \Rightarrow z_2 = -0.2533471... \Rightarrow -0.2533471\sigma = 37 - \mu$

$\Rightarrow 0.777476\sigma = 0.5 \Rightarrow \sigma = 0.6428820... = 0.643$, then

$\mu = 37.5 - 0.5244005 \times 0.6428820 = 37.16287... = 37.2$

Therefore mean temperature is $37.2°C$ with standard deviation of $0.643°C$.

15.   (a) (i) median time is 106 minutes

(ii) $Q_1 = 100$, $Q_2 = 110$ so the interquartile range is $110 - 100 = 10$ minutes

(b)

Time, t (min)	Frequency, f
<90	0
$90 \le t < 95$	7
$95 \le t < 100$	8
$100 \le t < 105$	12
$105 \le t < 110$	18
$110 \le t < 115$	9
$115 \le t < 120$	6

(c) the mean time of the race $= \dfrac{92.5 \times 7 + 97.5 \times 8 + 102.5 \times 12 + 107.5 \times 18 + 112.5 \times 9 + 117.5 \times 6}{60}$

$$= 105.1666... = 105 \text{ minutes}$$

16.   (a)

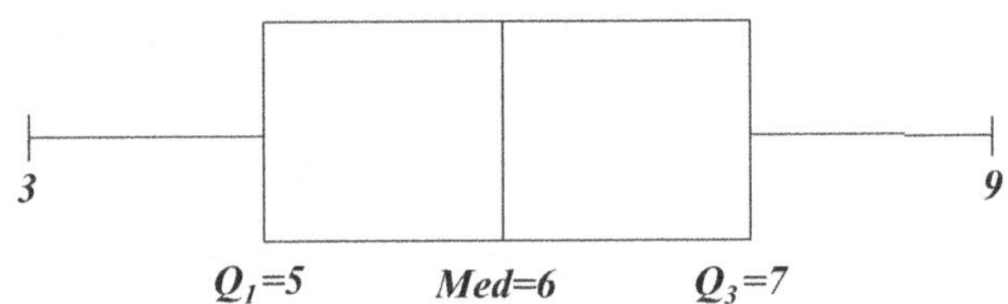

(b) mean $= 6.090909... = 6.09$, standard deviation $= 1.781447... = 1.78$

17.   (a) $x = 46,\ 47,\ 48,\ 49$

(b) y is the median and for a distinct mode we need $x = y$. In addition, $x + y + z + 205 = 350$ so that $2x + z = 145$. Now use trial and error to find that $x = y = 48$, $z = 49$.

18.   (a) $E(X) = np = 40 \times 0.03 = 1.2$

(b) $P(X = 5) = \dbinom{40}{5} 0.03^5 0.97^{35} = 0.005506151... = 0.00551$

(c) $P(X > 3) = 1 - P(X \le 3) = 0.03139680... = 0.0314$

(d) $P(2 \le X \le 6) = P(X \le 6) - P(X \le 1) = 0.3382875... = 0.338$

19.   (a) Modal value of x is 4

(b) (i) $\bar{x} = 3.45$ (ii) $s = 0.9733961... = 0.973$

20.   (a) Probability that 2 numbers are generated is $0.9 \times 0.1 = 0.09$

(b)

x	1	2	3	4
$P(X = x)$	0.1	0.09	0.081	0.729

(c) $E(X) = 1 \times 0.1 + 2 \times 0.09 + 3 \times 0.081 + 4 \times 0.729 = 3.439$

INDEX

absolute value, 125
acceleration, 187
   constant, 6
ambiguous case, 107
amplitude, 92
anti-differentiation, 221
area of a
   parallelogram, 109
   rectangle, 109, 130
   sector, 80
   triangle, 109
   under a curve, 229
arithmetic
   mean, 199
   sequence, 2
   series, 4
associative, 121
asymptote
   horizontal, 37
   vertical, 36
average, 155, 199
   velocity, 160
base, 18, 21
   vector, 124
binomial
   coefficient, 24
   distribution, 263
   expansion, 24
   expression, 23
   probability, 282
   random variable, 268
   theorem, 23
boundary condition, 226
box-and-whisker plot, 206
census, 190
chain rule, 176
circle
   arc of, 80
   sector, 80
   unit, 80, 83, 85, 86
class interval, 191
coefficient, 23 *(see binomial)*
column vector, *(see vector)*
common difference, 2
common ratio, 8
commutative, 121

complement, 245
completing the square, 51
concave down, 164
concave up, 164
cone, 233
contingency table, 250
converge, 10
correlation, 210, 216
   Pearson's product-moment, 216
cosine, 83
cosine rule, 104
covariance, 216
cumulative frequency, 195
   table, 195
   curve, 196
cylinder, 233
data
   continuous, 191
   discrete, 190
   grouped, 191
   point, 198
   raw, 190
   ungrouped, 199, 202, 206
degree, 81
derivative, 157
   second, 168
derivative of a
   graphically defined function, 166
   logarithmic function, 174
   trigonometric functions, 175
differentiation, 155, 157
   of composite functions, 175
   of polynomial functions, 159
discriminant, 57
displacement, 117, 187
distribution
   binomial, 263
   discrete probability, 258
   normal, 271
   probability, 258
diverge, 10
domain, 32
dot product, 128
element, 1, 31
event, 258
   exhaustive, 245

given, 250
    independent, 254
    mutually exclusive, 244
    non-exclusive, 247
expectation, 260, 269
exponent, 16, 17, 19
factoring a quadratic, 53
frequency histogram, 2193
function
    composite, 60, 223
    decreasing, 181
    exponential, 66
    gradient, 156, 157
    graphically defined, 166
    identity, 61
    increasing, 181
    inverse, 62, 66, 278
    periodic, 84, 85
    quadratic, 49, 50
    reciprocal, 36
    zero of a, 38, 101
geometric
    sequence, 8
    series, 10
gradient, 155, 156
    function, 157
histogram. *(see frequency histogram)*
identity, 89
image, 32
inflection
    point of, 165
integral, 222
integration, 221
    constant of, 221
    definite, 230
    indefinite, 222
    by substitution, 224
intercept, 50
interest
    compound, 12
    compounding, 14
    simple, 6
inter-quartile range, 205
inverse, 62, 63
    of the standardized normaldistribution, 278
kinematics, 187, 227
limit, 158, 176
line

of best fit, 211
    coincident, 146
    intersecting, 146
    non-intersecting, 146
    parallel, 146
    vector equation of, 134
logarithm, 16, 18, 21
mapping, 30
maximum
    local, 164
    point, 50, 164,
    value, 50, 169
mean
    arithmetic, 199
    population, 199
    sample, 199
median, 202
mid-interval value, 199
minimum
    local, 164
    point, 50, 164
    value, 50, 169
modal class, 204
mode, 204
normal
    distribution, 271
    distribution curve, 272
    inverse standardized distribution, 278
    standardized distribution, 273
obtuse angle, 107
optimization problem, 172, 184
outcome, 242, 244
parabola 49
    axis of symmetry of, 50
    vertex of, 50
Pascal's triangle, 24, 264
Pearson's product-moment correlation, 216
percentile, 196
period, 84
periodic function, 84
population
    growth, 14
    mean, 199
probability, 242
    conditional, 250
    distribution, 258
    tree diagram, 251
product rule, 178

Pythagoras' theorem, 86, 89, 104, 117, 126
quadrant, 85
quadratic equation, 54, 55, 98
quartile, 196
    lower, 196
    upper, 196
quotient rule, 179
radian, 80
random, 243
random number, 192
random variable, 259
    binomial, 268
    continuous variable, 271
    disctere variable, 259
    normal, 272
range, 32, 205
rate of change, 156, 163, 187
reflection, 40, 45, 66
regression, 210
    function, 220
    least squares, 211
    line, 211
relative frequency, 243, 244
sample, 190
    mean, 199
    space, 242
scalar, 122
    product, 128, 131
scatter diagram, 211
sequence
    arithmetic, 2
    geometric, 8
series
    arithmetic, 4
    geometric, 8
    infinite, 1
    infinite geometric, 10
sigma, 1
simulation programs, 287
simultaneous equation, 21, 54
sine, 83
sine rule, 104, 105

speed, 6, 117
sphere, 233
standard deviation, 207, 216
standardized normal variable, 273, 276
statistics, 190
stretch, 40, 42, 44
    of factor, 42
tally mark, 191
tangent, 84, 155, 181
term, 1
tetrahedron, 111
transformation, 40, 91
translation, 40, 91, 187
trial, 242, 258
triangle
    area of, 109
    isosceles, 103
    right, 103
    scalene, 104
    solution of, 102
turning point, 169
upper class boundary, 195
variance, 216, 273
vector
    addition, 118, 120
    base, 124
    column, 41, 124
    component of, 124
    equation of a line, 134
    magnitude of, 125
    parallel, 131
    perpendicular, 131
    position, 127
    unit, 126
    zero, 122
velocity
    constant, 137, 143
    mean, 155
    instantaneous, 160
volume of revolution, 233
wave, 84

Made in the USA
Monee, IL
07 July 2026

56550060R00240